Wendy A. Warr (Ed.)

Chemical Structures 2

The International Language of Chemistry

Proceedings of The Second International Conference, Leeuwenhorst Congress Center, Noordwijkerhout, The Netherlands, 3rd June to 7th June 1990

With 289 Figures and 42 Tables

Springer-Verlag
Berlin Heidelberg NewYork
London Paris Tokyo
Hong Kong Barcelona Budapest

Dr. Wendy A. Warr
6, Berwick Court
Holmes Chapel
Cheshire CW4 7HZ
United Kingdom

ISBN-13:978-3-642-78029-5 e-ISBN-13:978-3-642-78027-1
DOI: 10.1007/978-3-642-78027-1

Softcover reprint of the hardcover 1st edition 1993

Typesetting: Camera ready by editor

51/3020-543210 - Printed on acid-free paper

PREFACE

This book constitutes the Proceedings of the second conference in the series 'Chemical Structures: The International Language of Chemistry' which was held at Leeuwenhorst Congress Centre, Noordwijkerhout, in the Netherlands, between June 3 and June 7, 1990. The conference was jointly sponsored by the Chemical Structure Association; the American Chemical Society Division of Chemical Information; the Royal Netherlands Chemical Society; and the Chemical Information Groups of the Royal Society of Chemistry and the German Chemical Society.

The purpose of the conference was to bring together experts and an international professional audience to discuss and to further basic and applied research and development in the processing, storage, retrieval, and use of chemical structures; to focus international attention on the importance of chemical information and the vital research being carried out in chemical information science; and to foster co-operation among major chemical information organisations throughout the world. Subjects covered included structure-property correlations, spectral database systems, chemical nomenclature, generic structures, stereochemistry, substructure search systems, connection table formats, ring perception, information integration, three-dimensional substructure searching, similarity searching, and systems for handling chemical reaction information.

All the papers were peer-reviewed or given by invited speakers. Many internationally recognised teams in the field of chemical structure handling are represented in the chapters of this book.

The keynote address was given by Ernest Eliel of the University of North Carolina, Chapel Hill. He dealt with the history of chirality and the present status of describing it: an appropriate opening topic since so much of the conference was concerned with chemistry in three dimensions.

The second chapter in the book concerns a two-dimensional chart systematising and summarising the structure of matter. This chart was presented by Sandor Barcza of Sandoz in the poster session at the conference.

The next two chapters, from a paper by Hidetsugu Abe and co-workers at the Toyohashi University of Technology, and a poster by Heinrich Nöth and Erich Striedl of the Ludwig Maximilians University in Munich, describe systems for handling NMR spectra and chemical structures on a microcomputer.

The next three chapters concern chemical nomenclature. One from Hull University deals with integration of structure-based systems and nomenclature. Janusz Wisniewski, in a popularly acclaimed presentation, demonstrated Beilstein's AUTONOM program at the conference and his chapter describes this microcomputer program for generating IUPAC names from structural input. Val Metanomski and Joy Merritt provided a poster on Chemical Abstracts Service conventions for macromolecules.

Two chapters on generic structures follow, one from the team at Sheffield University and one from IDC, representing the papers given by Valerie Gillet and Günter Stiegler in Noordwijkerhout.

The chapters on stereochemistry are ones by Nancy Porter, concerning a chiral structure database for saccharides (the subject of her research before she joined Maxwell Online) and two by various authors from Chemical Abstracts Service, representing presentations at Noordwijkerhout by Paul Blower, about the addition of stereochemistry to the CAS Registry File and stereochemical substructure searching of that file.

Four chapters on other operational substructure search systems follow. Jean Marcali and co-workers describe the Du Pont global technical information system; Steve Welford and Clemens Jochum report on chemical structure registration for Beilstein Online; Peter Rusch also covers Beilstein Online in a paper about substructure searching on Dialog; and the DARC Toolkit which was described by Pascal Huguet is summarised in a chapter written by Bill Town with some suggestions from myself and Pierre Buffet of Questel.

John Barnard's poster on the Standard Molecular Data (SMD) File is mainly of historic interest because SMD has since progressed fast under his leadership as Technical Secretary. The Weiningers' poster also involved a connection table format. The chapter describes Daylight Chemical Information Systems' GEMINI language and interpreter.

Geoff Downs' elegant explanation of his ring perception algorithm was one of the technical highlights of the conference and his balloon visual aid will long be remembered.

The chapter by Doug Hounshell and co-workers at Molecular Design Limited concerns the representation and searching of 'difficult' structures such as polymers, non-stoichiometric mixtures, and formulations.

Three chapters on information integration follow, from talks given by Dennis Smith of Molecular Design Limited and Tom Hagadone of Upjohn; and from a poster presented by Marc Peeters and Danny Verbinnen of Janssen.

The section on three-dimensional substructure searching consists of seven chapters authored by Yvonne Martin of Abbott Laboratories; and Charles Eyermann and William Ripka of Du Pont; and by the teams of Peter Willett at Sheffield University (one of these chapters having been a poster by Andrew Poirette); Keith Davies at Chemical Design, and Frank Allen at the Cambridge Crystallographic Data Centre.

The last of these teams has also done research on 3-D similarity searching and Eleanor Mitchell presented a paper on their integrated approach to 2-D and 3-D similarity searching. Further chapters on similarity searching are derived from posters by Catherine Pepperrell of Sheffield University (in conjunction with Peter Willett), Guenter Grethe of Molecular Design Limited (in conjunction with Doug Hounshell), and Peter Walsh of the Health and Safety Executive (in conjunction with Peter Willett and co-workers). Other authors in this section are Takashi Okada of Kwansei Kakuin University, who uses a discrimantion net, and David Bawden (at that time at Pfizer Central Research) who discusses dissimilarity. Johnny Gasteiger and Wolf Ihlenfeldt of the Technical University of Munich deal with similarity criteria for reactions as well as molecules.

The final chapters cover other aspects of chemical reaction handling. Josef Brandt of the Technical University of Munich describes a tool for coding cyclisation reactions; Dan Meyer of ART Inc, and his co-workers, have studied chemical reaction retrieval using citation-based relationships; Glen Hopkinson and others at ORAC write about reaction sequence searching; and Brad Christie and Tom Moock of Molecular Design Limited discuss multistep reaction schemes in REACCS.

Only two papers given at the conference have failed to appear in these proceedings: Steve Boyer's description of IBM's Optical Character Recognition of Chemical Graphics, delayed for publication for patent reasons, and Todd Wipke's lengthy paper on computer invention of molecular structures, which I hesitated to write up from a tape-recording. Apart from these two unfortunate omissions, the book represents an excellent historic record of the Proceedings of the conference and forms a state-of-the-art picture of chemical structure handling in 1990.

I am deeply indebted to the organising committee, without whose work this book would not have been possible. The committee members were Charles Citroen of CID-TNO in the Netherlands, David Johnson of Exxon Research and Engineering, U.S.A., Reiner Luckenbach of the Beilstein Institute in Germany, Peter Rhodes, then of the Royal Society of Chemistry in England, Maarten de Hoog of the Royal Netherlands Chemical Society and myself. The interest and support of the five sponsoring organisations is also gratefully acknowledged.

The book was typeset by Hope Services Ltd. of Clifton Hampden, Oxfordshire. Their helpfulness and efficiency has been much appreciated.

I am also extremely grateful to many people at ICI Pharmaceuticals: to my overworked secretary Mary Burgess; to Joanne Chilton and others for word processing; to Martyn Wilkins and Frank Loftus for help with technicalities of hardware and software; and to John Barber, Derek Bowler, Graham Cousins and Liz Walton for literature searching and bibliography checking.

Janet Ash deserves a special mention for her efforts in preparing the index.

I myself must take most of the blame for the unfortunate late appearance of this work. Changing work priorities and the recession in the U.K. were contributory factors but these excuses are of little interest to the average reader and I apologise to the authors for the late appearance of their chapters.

March 1992

Wendy A. Warr
6 Berwick Court
Holmes Chapel
Cheshire
CW4 7HZ
England

CONTENTS

KEYNOTE ADDRESS

CHEMISTRY IN THREE DIMENSIONS

Ernest L. Eliel
W. R. Kenan, Jr. Laboratories, Department of Chemistry University of North Carolina, Chapel Hill, NC 27599–3290, U.S.A.

ABSTRACT

Prior to van't Hoff and Le Bel, chemistry was two-dimensional. Since 1874, however, we have had to deal with the third dimension in molecular models, projection formulae, configurational descriptors and, most recently, computer algorithms used to describe and specify configuration. The problem is complicated because chirality, an important aspect of three-dimensional structure, is an attribute of the molecule as a whole whereas the commonly used Cahn-Ingold-Prelog configurational descriptors require factorisation of chirality into individual chiral elements. This paper deals with the history of chirality and the present status of describing it.

CHEMISTRY IN THREE DIMENSIONS

When I was asked to give the keynote address at this meeting, I was, at first, surprised because, despite my long service on the Publications Committee of the American Chemical Society, I am not by any stretch of the imagination an information expert. My contribution to chemical information is that of a generator and consumer whereas most of the participants in this Conference are in the business of systematically archiving and retrieving chemical information. But it became obvious that there is considerable current interest in the third dimension among information specialists. And with good reason: both the pharmaceutical industry[1,2] and the agricultural chemical industry[3] are getting increasingly interested in stereochemistry. And when the practising chemist is interested in stereoisomerism, the information scientist cannot afford to be far behind. Moreover, 'there's money in them thar hills' as we say in the American vernacular. Recently, an advertisement fell into my hands: 'Business Opportunities: Stereoselective Chemical Technology: Emphasising Chiral Manufacturing Technology, Products and Applications'. After extolling the importance of 'chiral technology' in the pharmaceutical, agrichemical and many other industries this ad offers a comprehensive report of the field for a price of no less than $2,850!

V. Prelog, who shared the 1975 Nobel prize in chemistry for his work in stereochemistry, has pointed out that stereochemistry is not a branch of chemistry but a point of view. Most molecules are non-planar. Whether one in fact looks at them in 3-D depends on one's needs and one's knowledge and background. In the sequel, I shall explore the three-dimensional point of view in stereochemical history, in stereochemical concepts and finally in information storage and retrieval.

W. A. Warr (Ed.)
Chemical Structures 2

It was Louis Pasteur[4,5] who, 130 years ago, first recognised that optical activity, discovered half a century earlier,[6,7] was due to molecular dissymmetry (not asymmetry, for a molecule can have a symmetry axis or axes and still display optical activity). Following Lord Kelvin[8] we now call molecular dissymmetry 'chirality', from Greek 'cheir' = hand. It was J.H. van't Hoff[5,9] and J.A. Le Bel[10] who, in 1874 made the connection between chirality and molecular structure; A. Werner some years later, recognised[11] that chirality is not confined to organic compounds but extends to inorganic including co-ordination compounds as well.

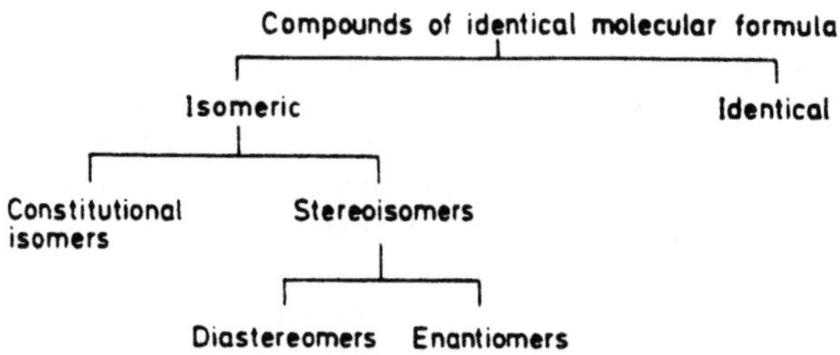

Figure 1. Isomers and homomers

Let me follow this bird's eye view of stereochemical history with an equally brief overview of stereochemical principles. As is shown in Figure 1, molecules of the same molecular formula may be either identical or isomeric. If isomeric, they may or may not differ in 'constitution' or connectivity (a concept well known to information scientists); if they so do, they are 'constitutional isomers'. If they have the same constitution and yet are not identical, they can only differ in their three-dimensional architecture: they are 'stereoisomers'. Stereoisomers, in turn, may be classified as 'enantiomers' (molecules whose structure is represented by non-superposable mirror images) or 'diastereomers' (molecules differing in spatial

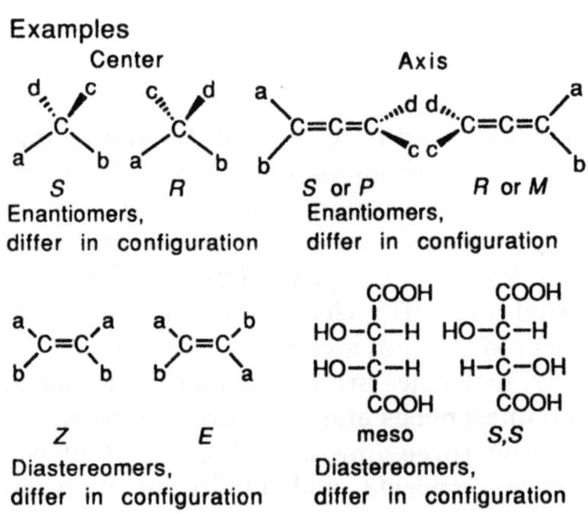

Figure 2. Enantiomers and diastereomers

array but not as objects and mirror images). Examples of enantiomers are given in the top part of Figure 2, whereas diastereomers are shown in the bottom part. Enantiomers are identical in most physical and chemical properties, such as boiling point, melting point, density, refractive index, spectra (UV, IR, NMR), free energy, etc. (so-called 'scalar properties'); they are also identical in their chemical behaviour toward achiral reagents. An analogy might be made with feet and socks: right and left feet, clearly enantiomeric, interact identically with a given sock, in view of the fact that socks are achiral (they have a plane of symmetry). This great similarity of enantiomers stems from the fact that they are 'isometric',[12] i.e., the intramolecular distances between corresponding atoms are the same in the two enantiomers, just as the distance between the big toe and the heel is the same in a right and left foot. (Given these circumstances it is perhaps unfortunate that enantiomers have unrelated and often quite different CAS Registry Numbers.) However, enantiomers differ in sign in such ('chiroptical') properties as optical rotation, optical rotatory dispersion, circular dichroism: so-called 'pseudoscalar properties' whose sign depends on absolute orientation in space. More important, enantiomers also differ in their reactivity toward other chiral molecules. An analogy is the relation of a right and left foot toward a right shoe: clearly there is a fit in one case and a misfit in the other. This relationship, of stereochemical fits and misfits, is at the heart of modern stereochemistry. It extends to the fit of an enzyme with its substrate (only one enantiomer of a substrate will fit a given enzyme and be acted on by it) or of a receptor with a drug. An example involving the enzyme lactate dehydrogenase[13] is shown in Figure 3. The enzyme (with its relatively rigid peptide chain) can accommodate (*S*)-lactic acid through a 'three-point contact':[14–16] when the acid function is juxtaposed with the guanidine moiety of arginine-171 and the hydroxyl function is hydrogen-bonded to the imidazole moiety of histidine-195, the NAD$^+$coenzyme (which is also bound specifically to the enzyme) is properly placed to abstract hydride from the carbinol function to oxidise it to the corresponding ketone (pyruvic acid). (The OH-proton is taken up by histidine-195 acting as a base.) If (*R*)-lactic acid is presented to the enzyme, it can enter the active site but will juxtapose the CH_3 group (rather than the hydrogen of the *C*HOH group) with the coenzyme and oxidation is then not possible.

Already in the middle of the 19th century, Eilhardt Mitscherlich wrote[17]

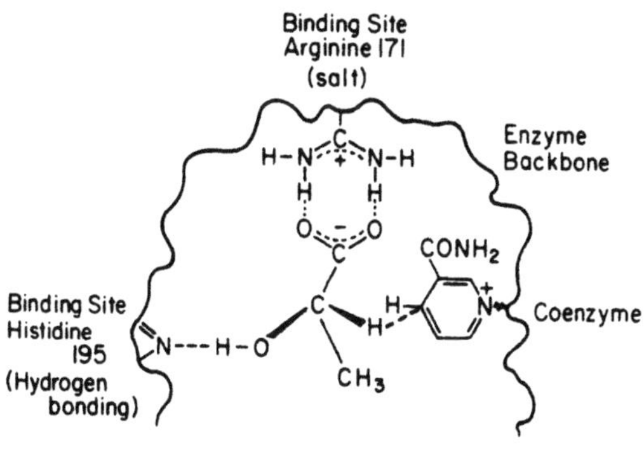

Figure 3. Active site of lactate dehydrogenase

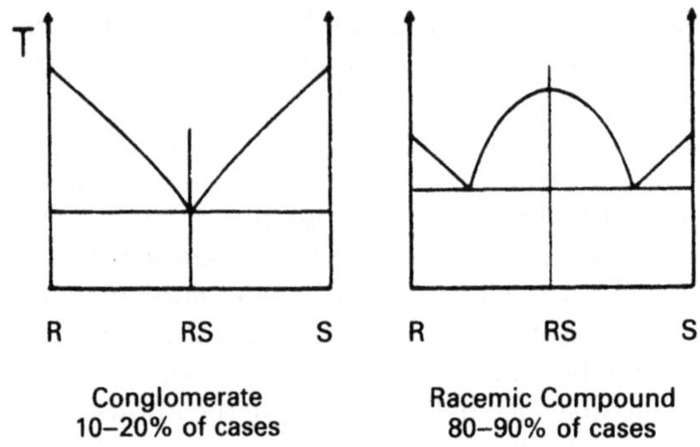

Figure 4. Melting point diagrams for conglomerates and racemic compounds

(comparing racemic and optically active sodium ammonium tartrate) 'The nature, the number, the arrangement of the atoms and the distances between them are the same'. This is, of course, what we mean by 'isometry' (*vide supra*). But Mitscherlich was comparing optically active compounds with racemates, not one enantiomer with another. What, then, is involved here? To understand Mitscherlich's statement we must understand the nature of racemates.[18] Racemates (Figure 4) can occur as mechanical mixtures of crystals of the individual enantiomers (so-called 'conglomerates'), as true 'racemic compounds' (in which enantiomeric molecules of opposite configuration are paired in the unit cell of the crystal) and (rarely) as racemic solid solutions in which the enantiomers are miscible in both the solid and liquid state. Conglomerates are a stereochemist's delight since they sometimes allow segregation of the enantiomers by mechanical separation (à la Pasteur). Unfortunately conglomerates constitute only about 10% of all racemates; racemic compounds are much more common and do not permit mechanical separation of enantiomers. In solution or melt, however, all racemates are usually mixtures and therefore it is appropriate (though not informative) that their Registry Numbers differ from those of the constituent enantiomers.

In addition to configuration, its meaning and its specification,[19,20] stereochemistry also deals with the concept of conformation.[21–23] The examples shown in Figure 5 show that different conformations of a species are produced by (usually

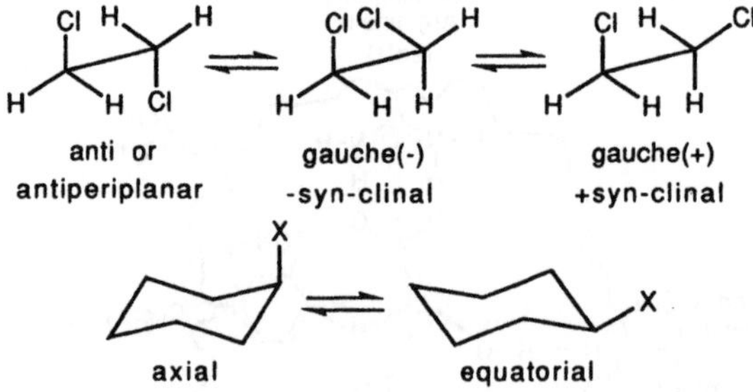

Figure 5. Different conformations of 1,2-dihaloethanes and substituted cyclohexanes

facile) rotations about single bonds. In principle, different conformational isomers (corresponding to energy minima in the energy profile of conformational change) may differ in reactivity. The rate constant k of a given species in a given reaction may be expressed[24,25] as $k = \sum_i n_i k_i$ where the n's refer to the populations (in terms of mole fractions) of the various conformations and the k's to their individual rate constants. For a substituted cyclohexane, with its two conformations, axial (a) and equatorial (e) (cf. Figure 6), the equation reduces to $k = n_e k_e + n_a k_a$.

It is important, however, to recognise that the most stable conformation of a molecule is not always the reactive one. An example is shown in Figure 6. The stablest conformation of cyclohexyl bromide (X = Br) or p-toluenesulphonate (X = $pCH_3C_6H_4SO_2O$) is the equatorial one. However, because of the requirement for an antiparallel arrangement of the leaving groups (here X and H in bimolecular elimination), the equatorial conformer cannot undergo the E_2 reaction, only the axial one can (17). Hence $k = n_a k_a$, i.e. the cyclohexyl-X species reacts exclusively in the less stable axial conformation. Similar principles apply to enzymes and their substrates and to pharmacophores and their receptors. The most stable conformation of the substrate or of the pharmacophore (drug) may not be the active one. (This is also true of the enzyme; enzymes may be activated by external compounds which produce a change in conformation, the so-called 'allosteric effect', or they may undergo a change in conformation when in contact with the substrate.)

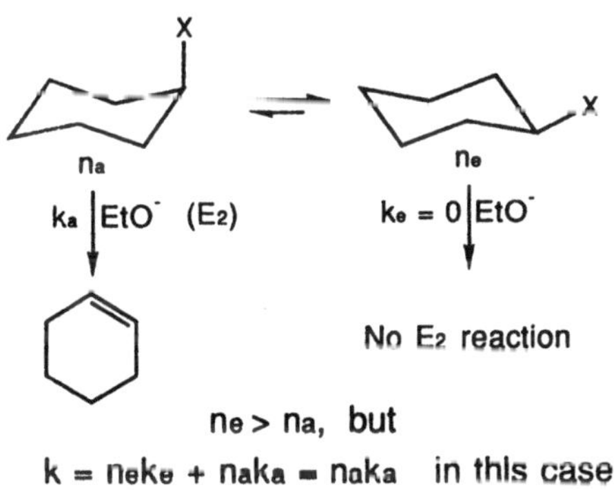

Figure 6. Bimolecular, base-catalysed elimination of cyclohexyl bromide or tosylate

This fact leads me to the third and last point of my lecture, namely the impact of stereochemistry on information storage and retrieval. Clearly it would be valuable to store chemical formulae in 3-D form. Such 3-D formulae can be extracted from the Cambridge Crystallographic Database for those compounds for which three-dimensional structures are experimentally available, usually from X-ray crystallography. However, for most known chemical compounds in the CA Registry file, precise three-dimensional structure has not been determined experimentally. It is therefore fortunate that a collaborative approach has been developed between CAS and Tripos to generate three-dimensional co-ordinate data from the connection tables in the CA Registry File by means of appropriate computer programs (e.g., CONCORD). So far about 4.2 million of the 10 million entries in the Registry file have had 3-D co-ordinate data added. The selection principle used so far has been to pick pure compounds either without stereogenic (i.e. stereoisomer-producing) elements or with unspecified configuration.

Clearly this can be only the beginning. Many organic compounds have specified chiral centres; such compounds are prevalent among drugs (57% of all drugs are chiral). Configuration may affect drug activity, as well as toxicity (side effects). In

Darvon (Analgesic)

Novrad (Antitussive)

Propoxyphene

Figure 7. Enantiomers with different pharmacological properties

fact, there are compounds, such as Darvon and Novrad (Figure 7, note the mirror-image names) where enantiomers have entirely different (in this case desirable) pharmacological effects. (In the case of the infamous drug thalidomide, it appears that the desired sedative and the catastrophic teratogenic effect reside in opposite enantiomers.) Thus it is important to include 3-D data for chiral compounds (including the specification of their sense of chirality at each stereogenic element) in the CAS Registry File. Efforts under way to accomplish this will be discussed by others in this conference.

A second potential difficulty (and one more difficult to overcome than the absence of specification of chirality mentioned above) has to do with the fact that the 3-D co-ordinates relate to a single conformation (one hopes the most stable one). If one retrieves this conformation one can then further study its energy minimisation by one of several molecular modelling programs presently available. These programs can also lead one to conformations corresponding to energetically low-lying secondary minima, which may be of chemical and pharmacological importance. Special attention is drawn to Allinger's recently developed MM3 program.[26] This program, in addition to containing improved parameters to calculate E (potential energy) also permits the calculation of vibrational energy levels which are important in estimating H from E (one must add zero-point and, in some instances, higher vibrational energy) and S and hence the experimentally important G ($G = H - TS$).

If one does not retrieve the structure but chooses to search in the file for certain geometrical (or other attributes), for example, the proper dimension of a pharmacophore to make three-point contact with a receptor (using a program such as MACCS marketed by Molecular Design Limited), the availability of but a single conformation in the CAS Registry (or any other file of substances) imposes a serious limitation, for the reasons stated above. However, it may well be contended that retrieving some structures (those having appropriate ground-state conformations) is much better than not being able to do the job at all! Nonetheless, the desirability of retrieving accessible conformations other than the energetically most low-lying one remains a far from trivial intellectual challenge.[27]

In summary, it is fair to say that Chemistry, thanks to the possibility of organising much information by structure, is the science with the best retrieval system. The system is constantly being improved, in addition to the database being enlarged. With the current interest in non-racemic (optically active) chiral compounds in the pharmaceutical and agrochemical industries, representation of molecules in three-dimensional space has become increasingly important. In other words, the 'stereochemical point of view' has now touched the information community. The steps already taken, and additional ones being discussed in this symposium, make it clear that information scientists have accepted the challenge and are 'running with it'.

REFERENCES

1. *Stereochemistry and Biological Activity of Drugs*; Ariëns, E.J.; Soudijn, W.; Blackwell: Cambridge MA, 1983.
2. *Drug Stereochemistry: Analytical Methods and Pharmacology*; Wainer, I.W.; Drayer, D.E., Eds.; Marcel Dekker: New York, 1988.
3. *Chemicals in Agriculture: Volume 1: Stereoselectivity of Pesticides: Biological and Chemical Problems*; Ariëns, E.J.; Van Rensen, J.J.S.; Welling, W., Eds.; Elsevier: Amsterdam, 1988.
4. Pasteur, L. 'Researches in Molecular Asymmetry of Natural Organic Products' (1860) Alembic Club Reprint No. 14: Edinburgh, 1905. (Note the erroneous translation from 'Recherches sur la Dissymétrie Moléculaire'. . . .)
5. *Foundations of Stereochemistry*; Richardson, G.M., Ed.; American Book Company: New York, 1901.
6. Biot, J.B. *Bull. Soc. Philomath. Paris* **1815**, 190.
7. Biot, J.B. *Bull. Soc. Philomath. Paris* **1816**, 125.
8. *Baltimore Lectures on Molecular Dynamics and the Wave Theory of Light*; Thomson, W. (Lord Kelvin); C.J. Clay and Sons: London 1904.
9. van't Hoff, J.H. 'Voorsteel tot uidbreiding der tegenwoordig in de scheikunde gebruikte structuur-formules in de ruimte'. *Arch. Nerl. Sci. Exact et Nat.* **1874**, *9*, 445–454.
10. Le Bel, J.A. 'Sur les relations qui existent entre les formules atomiques des corps organiques et le pouvoir rotatoire de leurs dissolutions'. *Bull. Soc. Chim. Fr.* **1874**, (2) *22*, 337–347.
11. Werner, A. 'Zur Kenntnis des asymmetrischen Kobalt atoms I.' *Ber. Dtsch. Chem. Ges.* 1911, *44*, 1887–1898.
12. Mislow, K. 'On the Classification of Pairwise Relations Between Isomeric Structures'. *Bull. Soc. Chim. Belg.* **1977**, *86*, 595–601.
13. Vennesland, B. 'Stereospecificity in Biology'. *Top. Curr. Chem.* **1974**, *48*, 39–65.
14. Easson, L.H.; Stedman, E. 'Studies on the Relationship Between the Chemical Constitution and Physiological Action. V. Molecular Dissymetry and Physiological Activity'. *Biochem. J.* **1933**, *27*, 1257–1266.
15. Bergmann, M. 'Synthesis and Degradation of Proteins in The Laboratory and in Metabolism'. *Science* **1934**, *79*, 439–445.

16. Ogston, A.G. 'Interpretation of Experiments on Metabolic Processes Using Isotopic Tracer Elements'. *Nature* **1948**, *162*, 963.
17. Mitscherlich, E.A. *C.R. Séances Acad. Sci.* **1844**, **19**, 720.
18. *Enantiomers, Racemates, and Resolutions*; Jacques, J.; Collet A.; Wilen, S.H.; Wiley-Interscience: New York, 1981.
19. Cahn, R.S.; Ingold, C. K.; Prelog, V. 'Specification of Molecular Chirality'. *Angew. Chem. Int. Ed. Engl.* **1966**, *5*, 385–415.
20. Prelog, V., Helmchen, G. 'Basic Principles of the CIP-System and Proposals for a Revision'. *Angew. Chem. Int. Ed. Engl.*, **1982**, *21*, 567–583.
21. Barton, D.H.R. 'The Conformation of the Steroid Nucleus'. *Experientia* **1950**, *6*, 316–320.
22. Barton, D.H.R. 'The Conformation of the Steroid Nucleus'. *Top. Stereochem.* **1971**, *6*, 1–10.
23. *Conformational Analysis*; Eliel, E.L.; Allinger, N.L.; Angyal, S.J.; Morrison, G.A.; Wiley-Interscience: New York, 1965.
24. Winstein, S.; Holness, N.J. 'Neighboring Carbon and Hydrogen. XIX. t-Butylcyclohexyl Derivatives. Quantitative Conformational Analysis'. *J. Am. Chem. Soc.* **1955**, *77*, 5562–5578.
25. Eliel, E.L.; Ro, R.S. 'Conformational Effects in S_N2 Reactions'. *Chem. Ind.* **1956**, 251–252.
26. Allinger, N.L.; Yuh, Y.H.; Lii, J.-H. 'Molecular Mechanics. The MM3 Force Field for Hydrocarbons. 1.' *J. Am. Chem. Soc.* **1989**, *111*, 8551–8566.
27. After the present conference ended, the author learned (Ricketts, D.M. personal communication, June 7, 1990) that program COBRA, marketed by Oxford Molecular, can generate multiple low-energy conformations from two-dimensional constitutional formula. The problem remains that if all low-energy conformations are stored, the database may become excessively large; if, on the other hand, the multiple conformation search is carried out 'on the fly' (i.e. the conformations are generated as the search proceeds) the demand for CPU capacity may be excessive.

A HIERARCHY OF THE STRUCTURE OF MATTER FROM THE VIEWPOINT OF INFORMATION RETRIEVAL AND STRUCTURE–PROPERTY CORRELATIONS

Sandor Barcza
Sandoz Research Institute, Sandoz Pharmaceuticals Corporation,
East Hanover, New Jersey 07936, U.S.A.

ABSTRACT

A (regrettably only) two-dimensional chart is presented, systematising and summarising the structure of matter. Within this, we zoom with higher magnification at the areas relevant to chemical information retrieval, structure-property correlations, especially related to the design of bioactive molecules. The chart covers small to large scale and low to higher resolution. It points out some of the parallels between man-made (e.g., drug, transistor) and natural (hormone, receptor, ribosome) assemblies of matter. The need to create integrated systems of this type is specially indicated by newer developments in molecular biology, materials science, computer systems and systems analysis. The scientist needs to find information and correlations over a sliding scale, with continuously variable window and magnification.

PURPOSE

The purposes of this note are:

1. To recommend that the chemical documentation community broaden its scope of attention from chemical documentation and retrieval to documentation and retrieval of the structure of matter.
2. To outline an integrated documentation and retrieval system ranging (for the sake of completeness) over all scales of matter[1,2] leaving no gaps and without redundancies, and in which the user could choose and continuously adjust the target range, aperture and resolution.

The major capabilities of chemical information retrieval today are: formula searches, substructure searches, exact match searches, and sequence match, homology, and subsequence searches. (Name, textual, and data searches are ignored in this note, as essentially belonging to the realm of text and data retrieval.)

Recent developments are: generic (Markush) structure storage and retrieval,[3] reaction storage and retrieval,[4] three-dimensional structure searches,[5–8] and various extensions of documentation in molecular biology.

W. A. Warr (Ed.)
Chemical Structures 2

Extension and integration are needed because:

1. Chemistry makes inroads into larger and larger structures.[9–11]
2. A near-continuum exists and should be well recognised.
3. A system is desired without gaps or redundancies.
4. Molecular biology opens new vistas rapidly.
5. Rapid progress occurs in man-made assemblies of matter.
6. More examples of parallelism between man-made and natural structures occur.
7. Structure-property correlations are being increasingly used and over a wider scale of structures.

Multilevel descriptions are often desirable even for the same entities: for example, the description of a virus[12] at the levels of covalent structure diagrams, sequences,[13] domains and macromolecular sub-units.

The chart in this note is an initial attempt to summarise description of the widest scale of the structure of matter, in as far as two dimensions can illustrate it.

THE CHART

The large scale (organism to galaxy) part of the chart is included to illustrate the principle and to encourage adaptation to chemistry of any techniques for storing, cataloguing, searching, graphing, etc. that are known in astronomy, geography, geology, architecture etc.

Commonalities abound:

1. Atoms *versus* planets in space.
2. Molecular surfaces *versus* geological layers.
3. Mapping and displaying these.
4. Transforms between data domains.
5. The use of Fourier and other transforms.
6. Distance geometry.
7. Image reconstruction from measurements.
8. Image compression and decompression.[14]
9. Three-dimensional graphics for buildings *versus* molecules.
10. Walk around, walk into, measure, section, etc.

It has become accepted to describe living systems at the different levels of ecosystems, anatomy, physiology, biochemistry, biophysics, etc. Modern information technology should make these descriptions easier to store, search, and display. As in other areas, one should be able to correlate better the different levels, and to be able to slide from one level to another.

As man attempts to study, emulate and control Nature, parallels between natural and man-made constructs must be documented, retrieved and displayed. Some of these are shown under 'Machines'.

In the realm of three-dimensional structure searching, one searches not only the distribution of matter (atoms or electrons and nuclei) but the distribution of any property over space. This includes lipophilicity, charge separation, polarisability etc.

While we portray the structure of matter in an hierarchical scheme, we also recognise that searching the structures is done at various levels of precision: exact match, non-exact match with false positives allowed, non-exact match with false negatives allowed, fuzzy searches, heuristic searches, and similarity searches. Similarity searches themselves can be done at many adjustable levels, and similarity itself is multidimensional.

In storing, searching and displaying the structure of matter, the elements of time, change, motion, dynamics, etc. have to be included. These can be spontaneous (natural), caused by man, or potential. Important examples include life cycles of stars, cultures, organisms, cells and interconversions of matter, such as in metabolic pathways, chemical reactions, conformational changes, and so on.

REFERENCES

1. *Powers of Ten*; Morrison, P., Morrison, P., Freeman. W.H.: San Francisco, 1982. (A book about the relative size of things in the Universe and the effect of adding another zero.)
2. *Cosmic View: The Universe in 40 Jumps*; Boeke, K.; John Day: New York, 1957.
3. Gillet, V.J., Downs, G.M., Holliday, J.D., Lynch, M.F., Dethlefsen, W. 'Searching a Full Generics Database'. In these Proceedings.
4. Christie, B., Moock, T. 'Multi-step Reaction Schemes in the Reaction Access System (REACCS)'. In these Proceedings.
5. Artymiuk, P.J., Grindley, H.M., Mitchell, E.M., Rice, D.W., Ujah, E.C., Willett, P. 'Representation and Searching of 3-D Protein Structures'. In these Proceedings.
6. Martin, Y.C., van Drie, J.H. 'Identifying Unique Core Molecules from the Output of a 3-D Database Search'. In these Proceedings.
7. Guner, O.F., Henry, D.A., Christie, B.D., Hughes, D.W., Miller, M.A. 'Integrated Approach to Three-Dimensional Information Management with MACCS-3D'. Paper presented at the 199th National Meeting of the American Chemical Society, Boston, April 1990.
8. Andrews, J.D., Guner, O.F., Hughes, D.W., Miller, M.A., Rock, C.A. 'Using a Three-dimensional Chemical Database Management System to Store and Analyse the results of Molecular Dynamics Studies'. Paper presented at the 199th National Meeting of the American Chemical Society, Boston, April 1990.
9. Cram, D.J. 'The Design of Molecular Hosts, Guests and their Complexes (Nobel Lecture)'. *Angew. Chem. Int. Ed. Engl.* **1988**, 27, 1009–1020.
10. Lehn, J-M. 'Supramolecular Chemistry – Scope and perspectives. Molecules, Supermolecules, and Molecular Devices (Nobel Lecture)'. *Angew. Chem. Int. Ed. Engl.* **1988**, *27*, 89–112.
11. Ringsdorf, H., Schlarb, R., Venzmer, J. 'Molecular Architecture and Function of Polymeric Oriented Systems: Models for the Study of Organisation, Surface Recognition, and Dynamics of Biomembranes'. *Angew. Chem. Int. Ed. Engl.* **1988**, *27*, 113–158.
12. Klug, A. 'From Macromolecules to Biological Assemblies (Nobel Lecture)'. *Angew. Chem. Int. Ed. Engl.* **1983**, *22*, 565–582.
13. Rice, C.M., Lenches, E.M., Eddy, S.R., Shin S.J., Sheets, R.L., Strauss, J.H. 'Nucleotide Sequence of Yellow Fever Virus'. *Science* **1985**, *229*, 726–733.
14. *Fractals Everywhere*; Barnsley, M., Academic Press: London, 1988.

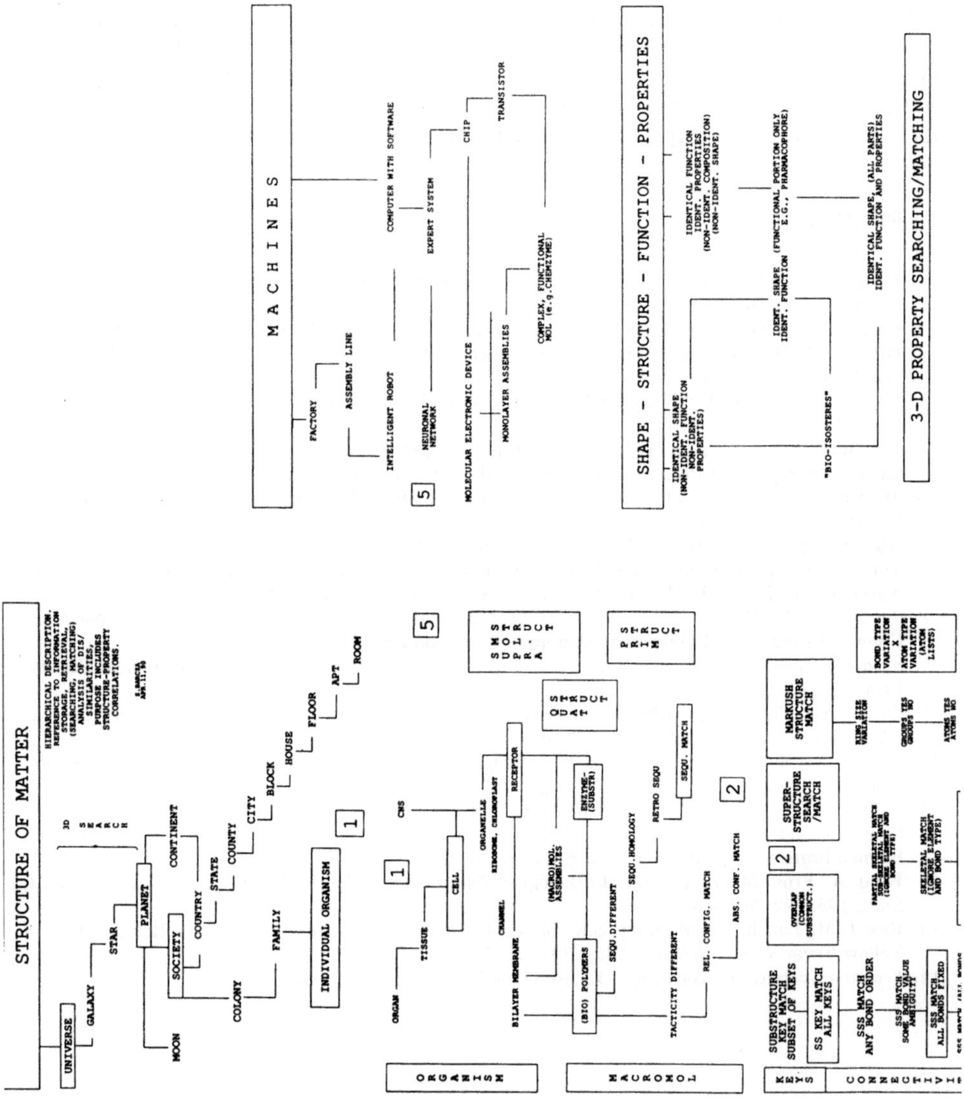
STRUCTURE OF MATTER
M A C H I N E S
SHAPE - STRUCTURE - FUNCTION - PROPERTIES
3-D PROPERTY SEARCHING/MATCHING

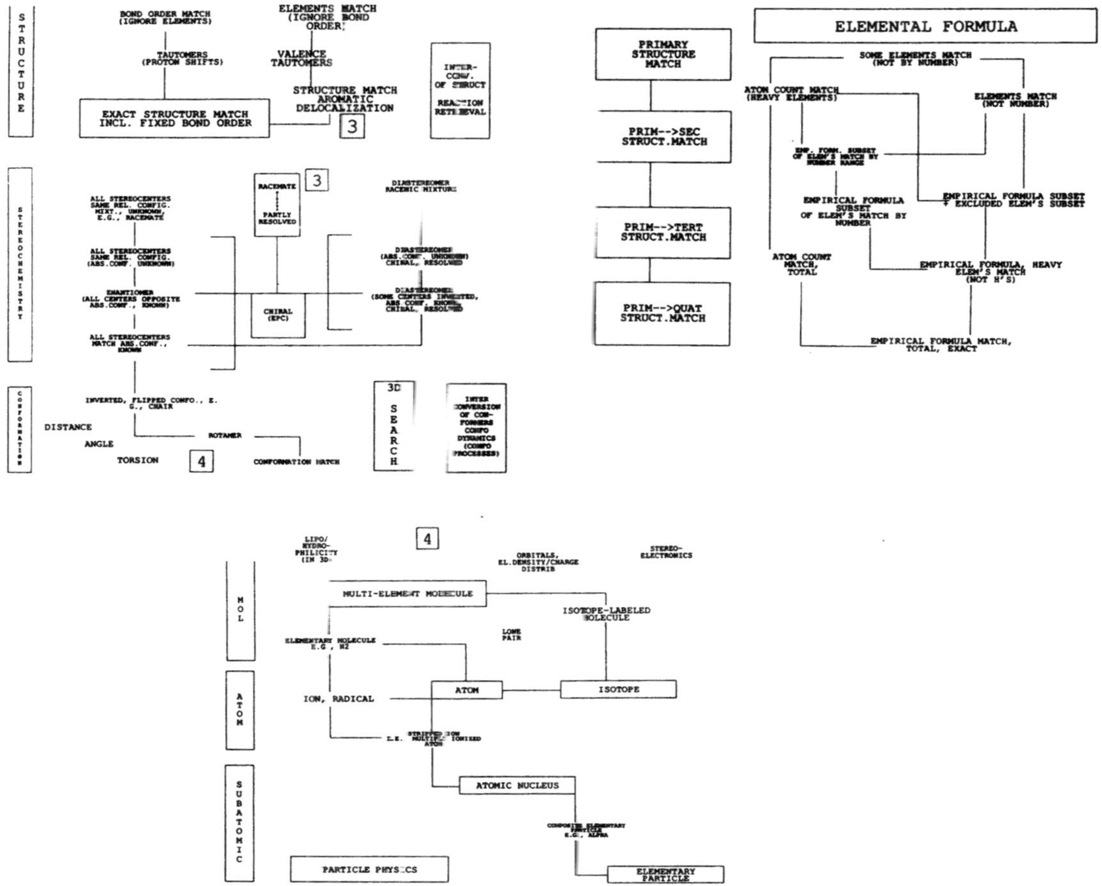
STRUCTURE
BOND ORDER MATCH (IGNORE ELEMENTS)
ELEMENTS MATCH (IGNORE BOND ORDER)
TAUTOMERS (PROTON SHIFTS)
VALENCE TAUTOMERS
STRUCTURE MATCH AROMATIC DELOCALIZATION
EXACT STRUCTURE MATCH INCL. FIXED BOND ORDER
3
REACTION RETRIEVAL
PRIMARY STRUCTURE MATCH
PRIM-->SEC STRUCT.MATCH
PRIM-->TERT STRUCT.MATCH
PRIM-->QUAT STRUCT.MATCH
ELEMENTAL FORMULA
SOME ELEMENTS MATCH (NOT BY NUMBER)
ATOM COUNT MATCH (HEAVY ELEMENTS)
ELEMENTS MATCH (NOT NUMBER)
EMP. FORM. SUBSET OF ELEM'S MATCH BY NUMBER RANGE
EMPIRICAL FORMULA SUBSET OF ELEM'S MATCH BY NUMBER
EMPIRICAL FORMULA SUBSET + EXCLUDED ELEM'S SUBSET
ATOM COUNT MATCH, TOTAL
EMPIRICAL FORMULA, HEAVY ELEM'S MATCH (NOT H'S)
EMPIRICAL FORMULA MATCH, TOTAL, EXACT
STEREOCHEMISTRY
ALL STEREOCENTERS SAME REL. CONFIG. MIXT., UNKNOWN, E.G., RACEMATE
ALL STEREOCENTERS SAME REL. CONFIG. (ABS.CONF. UNKNOWN)
ENANTIOMER (ALL CENTERS OPPOSITE ABS.CONF., KNOWN)
ALL STEREOCENTERS MATCH ABS.CONF., KNOWN
RACEMATE
PARTLY RESOLVED
CHIRAL (EPC)
DIASTEREOMER RACEMIC MIXTURE
CONFORMATION
INVERTED, FLIPPED CONFO., E. G., CHAIR
DISTANCE
ANGLE
TORSION
ROTAMER
CONFORMATION MATCH
4
3D SEARCH
INTER CONVERSION OF CONFORMERS CONFO DYNAMICS (CONFO PROCESSES)
LIPO/ HYDRO- PHILICITY (IN 3D)
ORBITALS, EL.DENSITY/CHARGE DISTRIB
STEREO- ELECTRONICS
MOL
MULTI-ELEMENT MOLECULE
ISOTOPE-LABELED MOLECULE
ELEMENTARY MOLECULE E.G, N2
LONE PAIR
ATOM
ION, RADICAL
ISOTOPE
ATOMIC NUCLEUS
SUBATOMIC
PARTICLE PHYSICS
ELEMENTARY PARTICLE

A PERSONAL COMPUTER PROGRAM SYSTEM FOR NMR DATABASE CONSTRUCTION

Hidetsugu Abe, Tohru Okuyama, Kouji Yoshida, Eiji Kouno, and Shin-ichi Sasaki Toyohashi University of Technology, 1-1, Hibarigaoka, Tempaku, Toyohashi 440, Japan

ABSTRACT

During the construction of a proton NMR database, several utility programs have been developed for improving efficiency of the compilation. The personal computer program system presented here is one of such utilities. This program serves three major functions. The first function is to input structural formulae interactively by drawing them on a CRT screen. The second is to convert and process the raw spectral data from a spectrometer into a proper set of information. The last function is to make assignment of signal groups in the given spectral data to hydrogen-containing substructures in the corresponding structural formula. These functions are executed interactively with graphical representation of spectrum and structures drawn on a CRT screen.

INTRODUCTION

Various spectrometric methods have been utilised in organic chemistry. Among them, nuclear magnetic resonance (NMR) spectrometry is the most useful for structural interpretation in various situations. Additionally, various computer program systems for structural elucidation of organic compounds have been developed.

There are two kinds of computer-aided structure elucidation methods: logic-oriented and database-oriented approaches. The computer program systems adopted for either approach require a high quality spectral database. The authors have been constructing proton and carbon-13 NMR spectral databases applicable to the computer systems for both approaches for more than 10 years.[1]

Figure 1 shows the procedures for compiling the two kinds of spectral data. As shown here, carbon-13 spectral data have been collected from various books and literature sources. Recently, the source of carbon-13 spectral data has been changed to a data collection in computer-readable form.[2]

On the other hand, the authors have worked from the very beginning on compiling 90 MHz proton spectra; gathering a variety of organic compounds, measuring their spectra, making their signal assignments, coding them, and so on.

Since these procedures are very tedious and time-consuming, several personal computer programs have been developed, to make the process more efficient.

DATA FORMATS FOR THE DATABASE

Figure 2 shows a set of 90 MHz proton NMR spectral data in exchange format. The first and the second lines contain serial number, name, molecular weight,

W. A. Warr (Ed.)
Chemical Structures 2

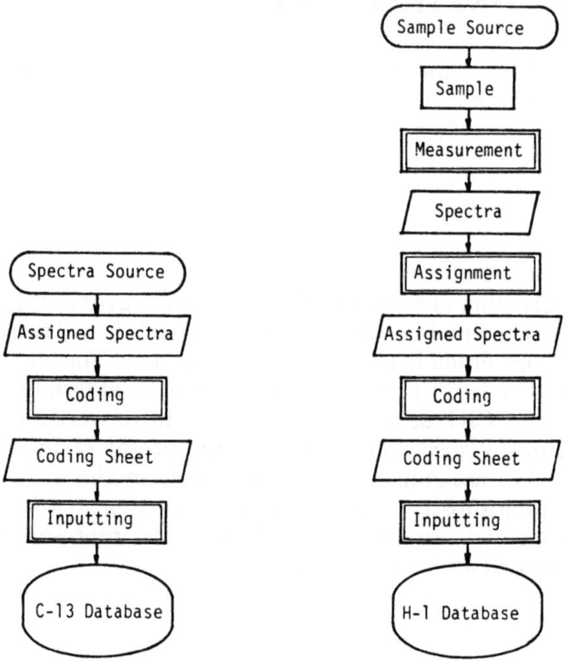

Figure 1. General procedure for compiling NMR spectral data

```
 1  15540 2-Ethyl-3-methoxypyrazine;
 2          4708 S  138.17  C  7  H 10  N  2  0  1;
 3  25680-58-4  TO      sp  CDCL3            ___._-___._/_ __._-___._/____;
 4 Y1  Y1  Y   Y   Q   NR  NR  C3  C3  C2;
 5   1   2   7   3   4   5   8   0   1   6   4   0   3  10   9;
 6  3  1.28-          ___  9
 7  2  2.84-          ___ 10
 8  3  3.98-          ___  8
 9  1  7.91-          ___  1 OR  2
10  1  8.03-          ___  2 OR  1;
```

Figure 2. A proton NMR data set in exchange format

molecular formula, CA registry number, if available, name of sample supplier, and melting and/or boiling point, if available. The third and the fourth lines show the structural formula in CANOST notation corresponding to the compound number 15540 shown in Figure 3. The details of the notation have been reported elsewhere.[3]

The remaining lines show positions, area intensities and assignments of the signal groups in the spectrum. A database containing more than 8,000 compounds has been constructed based on this format. At present, spectral data are being compiled in three additional formats: fully digitized, full peak, and patterned.

The spectral data recorded in the first format, fully digitized format, are to be used for spectral data display only.

Both second and third formats record data to be used for detailed search purposes. In the second format, all the peaks in a spectrum are recorded with their area intensities, heights and positions. Allocated hydrogen numbers and assigned hydrogen-containing substructures for all signal groups are also recorded.

In the third format, signal positions, splitting patterns, observed coupling constants and assignments are presented.

Examples of the formats for the compound number 15540 are shown in Figure 4. Fully digitized data are not shown here.

In the full peak format, allocated hydrogens and assignments of hydrogen-containing substructures in the compound for all signal groups are recorded as shown in the figure.

The spectrum in the patterned format is also shown in the figure. The splitting patterns and observed coupling constants for all signals are recorded as shown here. The terms S, D, T and Q mean singlet, doublet, triplet and quartet, respectively.

The basic idea of the present study is to automate, as far as possible, the construction of the data set. For this purpose, the personal computer program

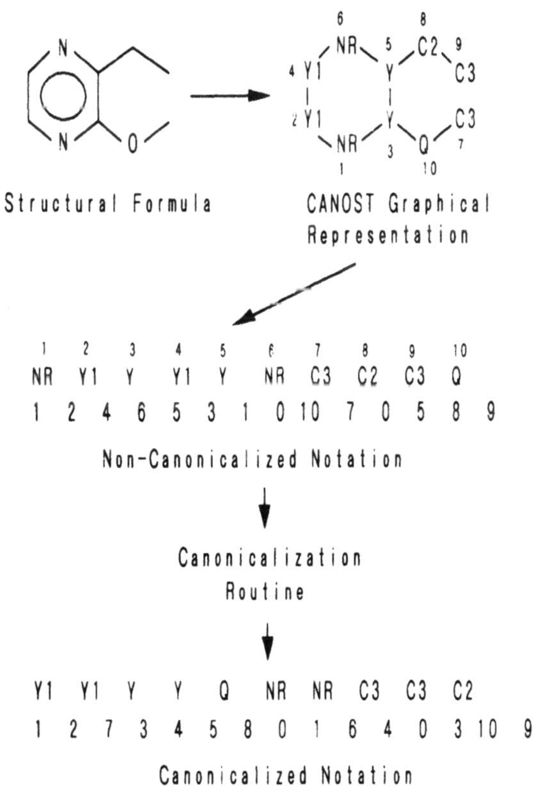

Figure 3. An example of CANOST notation

1. Full Peak Data Set

NO.	GROUP	AREA	HEIGHT	PPM
1	1	1250	800	8.04
2	1	1974	1446	8.00
3	1	2139	1109	7.93
4	1	1220	602	7.90
5	2	9183	9321	3.97
6	3	677	345	2.95
7	3	2312	1126	2.87
8	3	2649	1264	2.79
9	3	996	477	2.70
10	4	3011	1833	1.35
11	4	5148	3606	1.27
12	4	2420	1371	1.18

GROUP= 1 H= 2 NODE= 2 4
GROUP= 2 H= 3 NODE= 7
GROUP= 3 H= 2 NODE= 8
GROUP= 4 H= 3 NODE= 9

2. Patterned Data Set

PPM	SPLITTING		AREA	ASSIGN
8.02	DR	2.8	1	1 OR 2
7.92	DL	2.8	1	2 OR 1
3.97	S		3	7
2.82	Q	7.4	2	8
1.27	T	7.4	3	9

3. Conventional Data Set

PPM	AREA	ASSIGN
8.02	1	1 OR 2
7.92	1	2 OR 1
3.97	3	7
2.82	2	8
1.27	3	9

Figure 4. Three kinds of data formats for compound number 15540

system which is the subject of this paper is now being developed. The major programs and their functions are summarised in Table 1.

The first of the three is a structure input program allowing the user to draw various structural formulae easily on a CRT screen by means of a mouse. The second program performs several procedures for data format conversion and data reduction. The last program is used for signal assignments. The present paper will describe the second and the third programs in detail.

The general idea of the usage of the program system is shown in Figure 5. First, the structural formula of a compound is input by means of the first program. Next, the raw spectral data from a NMR spectrometer is processed with the input structural formula by means of the second program. In the last step, the signal assignment task for the spectrum is performed by means of the third program and a complete data set is obtained.

Table 1. PC programs for assisting NMR database construction

1. Structure Input
    - Graphical Structural Formula Input
    - Formula Conversion (to CANOST Notation)
2. Spectral Data Processing
    - Data Format Conversion (pseudo-IBM to MS-DOS)
    - Data Reduction
3. Signal Assignment
    - Signal Assignment
    - Splitting Pattern Search

Another function, a nomenclature program is also shown in the figure. Although systematic names of organic compounds are still very useful for information retrieval tasks in organic chemistry, the time available for teaching and learning nomenclature systems is getting shorter and shorter. So, the authors are now developing a computer-assisted nomenclature system.[4]

DATA PROCESSING PROGRAM

The major function of this program is getting peak positions and their area intensities from given raw data.

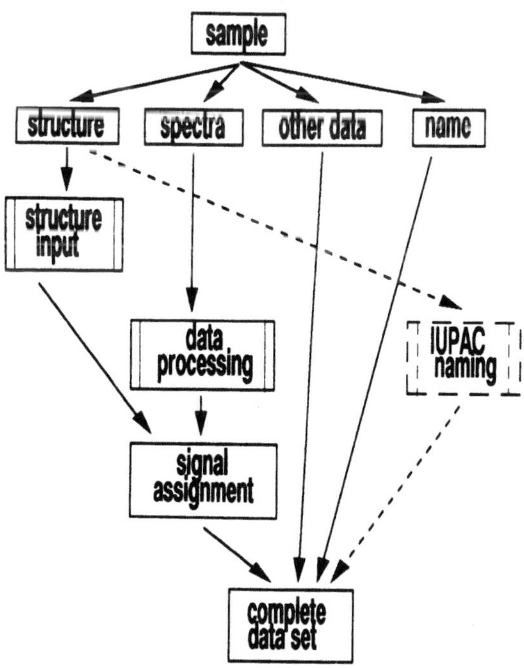

Figure 5. Required processes for compiling one record of proton NMR database

Table 2. Functions of the data processing program

1. Conversion of Data Format from pseudo-IBM to MS-DOS
2. Smoothing of Input Raw Data
3. Correction of Base Line
4. Detection of Peaks
5. Designation of Solvent Peak(s)
6. Designation of Impurity Peak(s)
7. Grouping of Peaks
8. Allocation of Hydrogens to Peak Groups

To obtain peak positions and their area intensities, several processes have to be carried out on the raw data as shown in Table 2. In principle, these processes are executed automatically as far as possible. However, since all of these processes could not be performed automatically for all variations of the raw data, interactive treatments are required on occasion.

Automated Processes

Smoothing. The Fourier-transformed raw data obtained from a spectrometer oonsist of 8192 data points for 1845 Hz spectral range. It corresponds to about 20.5 ppm. The data are first smoothed by means of an adoptive smoothing method.[5] The procedure of this method is as follows:

It is assumed that the observed wave function X(i) consists of the summation of signal S(i) and noise N(i) as in Equation 1.

$$X(i) = S(i) + N(i) \tag{1}$$

and the noise N(i) is irregular continuous noise with mean value E[N(i)] = 0 and variance $E[N^2(i)] = Sn^2$. Then, the assumed value of signal P(i) is given by Equation 2:

$$P(i) = a(i)X(i) + b(i) \tag{2}$$

The most appropriate values for a(i) and b(i) are calculated by means of the least square method and Equation 3 is given as follows:

$$P(i) = a(i)[X(i) - \bar{X}(i)] + \bar{X}(i) \tag{3}$$

$$\text{where } a(i) = [s_1^2(i) - S_n^2]/S_x^2(i),$$

$$\bar{x}(i) = E[x(i)] \quad \text{and} \quad S_x2(i) = E[\{x(i) - \bar{x}(i)\}^2]$$

The variance of noise S_2 is calculated from the first 300 points and the last 300 points of the given data. The actually adopted smoothing points for Equation 3 are 15.

Base Line Correction. After the data smoothing process is finished, a base line correction procedure is performed as follows. Two alternative situations are considered here, one is in the neighbourhood of the base line and the other is in the neighbourhood of a peak.

For the first situation, the beginning point P1 is defined in the neighbourhood of the base line as shown in Figure 6. Then the end point P2 is searched. It is the lowest one among 5 points which are in the neighbourhood of the point whose intensity is greater than the standard deviation of noise. Usually the point P1 is located at the right side of a peak or the very beginning of a spectrum and the point P2 is located at the left side of a peak. Then the lowest point between P1 and P2 is defined as P3. If the difference between the intensities of P1 and P2 or P2 and P3 is greater than 20, the straight lines between P1 and P3, and P3 and P2 are set as new base lines as shown in case 2 of Figure 6. If the differences are both smaller than 20 (case 1), the line between P1 and P2 is defined as the base line.

For the second situation, the point P4 is defined as the lowest point among 5 points whose intensities are smaller than the standard deviation of noise at the right end in the neighbourhood of a peak. Then the straight line between P2 (which is previously defined) and P4 is set as the new base line (case 3).

Peak Detection. The following is the procedure for detection of peaks:

Step 1: All the spectral data are differentiated by means of a so-called interpolation formula.

Step 2: The peak detection limit is set at 20 times the standard deviation of noise. If the value is greater than 2000, the limit is set at 2000. These parameters have been determined empirically.

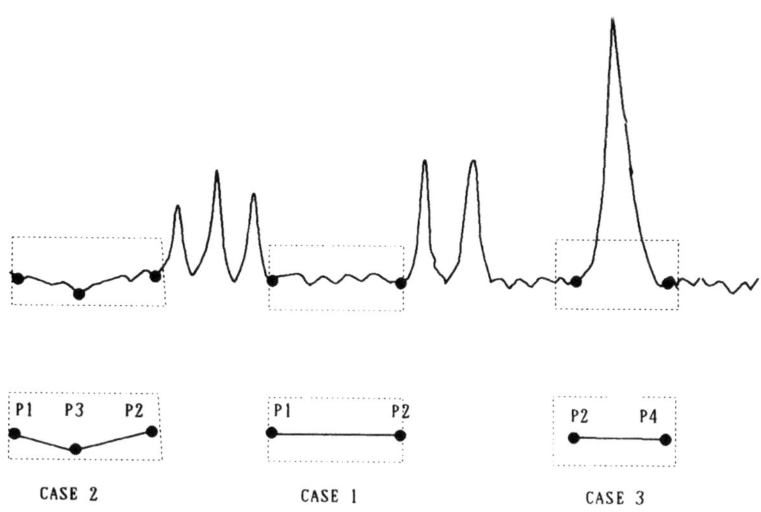

Figure 6. Base line determination procedures

Step 3: All points with derivative value zero and their left and right neighbours having respectively plus and minus derivative values are designated as candidates for peak tops.

Step 4: Among these candidates, those with heights greater than the previously defined detection limit are designated as peaks. If there is a pair of peaks neighbouring within 5 points, they are merged and the height of the new peak is set as that of the higher one.

Step 5: Area intensities of peaks are calculated by means of the trapezoidal method.

Grouping of Peaks. Generally speaking, a proton group in a given structure gives plural peaks in the spectrum. Therefore, grouping of peaks in a spectrum into a certain number of clusters is necessary for assignment tasks.

In the system under consideration, a pair of peaks existing contiguously within intervals of 100 points (*ca.* 22.5 Hz) is considered to be in the same group. The automatic grouping results can be modified interactively.

Designation of Reference Peak. The peak which is in the rightmost position is automatically designated as the reference substance peak. This designation can be changed easily in the interactive process.

Calculation of Relative Peak Positions. The relative positions (δ) of all peaks are then calculated by means of Equation 4:

$$\delta = \frac{f_s - f_r \times 106}{Z} \qquad (4)$$

where f_s is the frequency of a peak, f_r is the frequency of the reference substance peak and Z is the observation frequency.

Designation of Solvent Peaks. Next, those peak(s) which originate from residual hydrogens in deuterated solvent are found by reference to a correlation table (Table 3) which contains standard positions of those peaks for various solvents. These automatically designated peaks can be changed later in the interactive process.

Allocation of Hydrogens to Peak Groups. This procedure is performed as follows:

Step 1: The area intensity GAREAi for the i-th peak group is calculated by summarising the intensities of all peaks in the group.

Step 2: The total area intensity ISUM is calculated by summarising the intensities of all peaks except the reference peak and solvent peak(s).

Step 3: The allocated hydrogen number GHi for each peak group is calculated by Equation 5:

$$GH_i = ANINT[(GAREA_i/ISUM) \times MH] \qquad (5)$$

Table 3. Peak positions of solvent-originated signals

Solvent	Peak Postions(ppm)		
Acetic acid	2.05	11.53	
Acetone	2.05		
Acetonitrile	1.95		
Benzene	7.20		
Chloroform	7.25		
Cyclohexane	1.40		
Heavy water	4.75		
1,2-Dichloroethane	3.69		
Diethyl ether	1.16	3.36	
Dimethyl formamide	2.76	2.94	8.05
Dimethyl sulphoxide	2.50		
Dioxane	3.55		
Ethanol	1.17	3.59	2.60
Methanol	3.35	4.84	
Methyl cyclohexane	0.92	1.54	1.65
Methylene chloride	5.35		
Pyridine	7.20	7.58	8.70
Tetrahydrofuran	1.75	3.60	

where ANINT is the intrinsic function which gives the nearest integer for a given real number, and MH is the number of hydrogens in the given molecular formula.

Step 4: If there is a peak group whose GH_i becomes zero, the peaks in the group are omitted as impurity peaks and the process is repeated from the beginning.

Step 5: If the summation of GHi is not equal to MH, the following adjustment of allocated hydrogen number is executed. If the summation is smaller than MH, add one to the GH_i which has the smallest difference from RATIOi. If the summation is greater than MH, subtract one from the GHi which has the largest difference from $RATIO_i$. This procedure is repeated until the summation becomes equal to MH.

The result of the automatic processes described above is shown in Figure 7. This list shows the detected peaks and allocated hydrogens. Seventeen peaks are detected for the data. Area intensities, heights and positions for all peaks are shown under the column headings AREA, PEAK HEIGHT and PEAK POSITION, respectively. The group assignments are shown under the column heading GROUP. Peaks which have same group number are assigned to the same group. The numerals with negative signs indicate solvent, impurity and reference peaks. The numbers of hydrogens allocated to the groups are shown in the second paragraph.

Interactive Processes

The result of the automated processes described above is displayed on the CRT screen as shown in Figure 8. Then, the user can modify interactively any results by designating the menu items shown on the right side of the screen. Each menu item has subsidiary commands as shown in Table 4.

Base Line. If the user selects this function, the spectrum accompanied by the automatically defined base line is displayed. To correct the default base line, the

Detected Peaks

NO.	GROUP	AREA	PEAK HEIGHT	PEAK POSITION PPM	HZ
1	1	1235	801	8.04	934.8
2	1	1993	1447	8.00	937.8
3	1	2119	1110	7.93	944.5
4	1	1221	603	7.90	947.2
5	-2	185	306	7.28	1002.6
6	3	9167	9322	3.97	1299.2
7	3	1147	462	3.95	1300.6
8	-3	346	427	3.38	1351.9
9	5	649	345	2.95	1390.4
10	5	2328	1125	2.87	1397.9
11	5	2590	1263	2.79	1405.3
12	5	992	477	2.70	1412.7
13	-3	501	300	1.86	1488.2
14	7	2984	1836	1.35	1533.7
15	7	5183	3608	1.27	1541.1
16	7	2423	1372	1.18	1548.8
17	-1	323	463	0.00	1654.8

Allocated Number of Hydrogens

GROUP= 1 NUMBER OF HYDROGEN = 2
GROUP= 2 NUMBER OF HYDROGEN = 0
GROUP= 3 NUMBER OF HYDROGEN = 3
GROUP= 4 NUMBER OF HYDROGEN = 0
GROUP= 5 NUMBER OF HYDROGEN = 2
GROUP= 6 NUMBER OF HYDROGEN = 0
GROUP= 7 NUMBER OF HYDROGEN = 3

Figure 7. Detected peaks and allocated hydrogens

user may simply pick the'START-END' subcommand and designate two new points as the desirable start and end points of the new base line on an arbitrary region. This newly settled base line is saved by picking 'SAVE' subcommand.

Two additional subcommands 'EXPAND' and 'RESET' are used for expansion and resetting of the arbitrary spectral region. The 'EXIT' subcommand is for returning to the main screen.

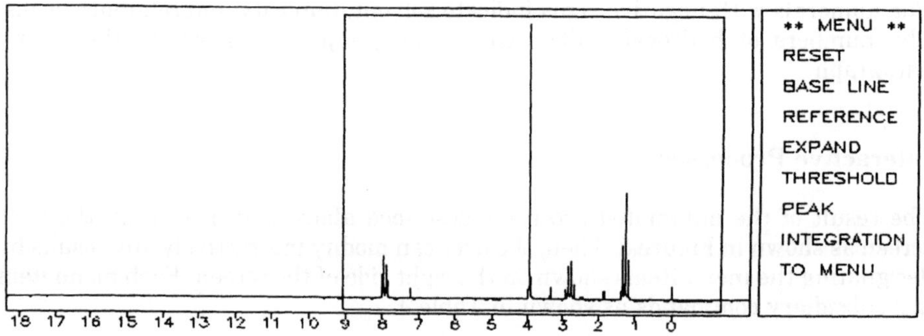

Figure 8. Initial CRT screen for the interactive process

Table 4. Command menu for the present system

Main Command	Subcommand
RESET	
BASE LINE	RESET EXPAND START-END SAVE EXIT
REFERENCE	RESET CHANGE EXPAND SAVE EXIT
EXPAND	
THRESHOLD	RESET CHANGE EXPAND SAVE EXIT
PEAK	RESET ADD DELETE EXPAND PRINT IMPURITY SOLVENT EXIT
INTEGRATION	RESET EXPAND PRINT EXIT
EXIT	

Reference. If the user selects this function, the default reference peak is shown as an overlapping vertical line. The 'CHANGE' subcommand is used for changing the reference peak. The user may designate any peak as the new reference peak. The functions of other subcommands are the same as above.

Threshold. The default threshold line is shown by selecting this command. Change of the level may be done by 'CHANGE' and 'SAVE' subcommands.

Peak. If the user selects this function, the screen is changed as shown in Figure 9. All peaks which are detected previously are shown by vertical lines. The impurity peak(s) and solvent peak(s) which have been designated in the previous step are also shown and they may be changed by means of 'IMPURITY' and 'SOLVENT' subcommands. The two subcommands 'ADD' and 'DELETE' are used for adding and deleting peaks arbitrarily.

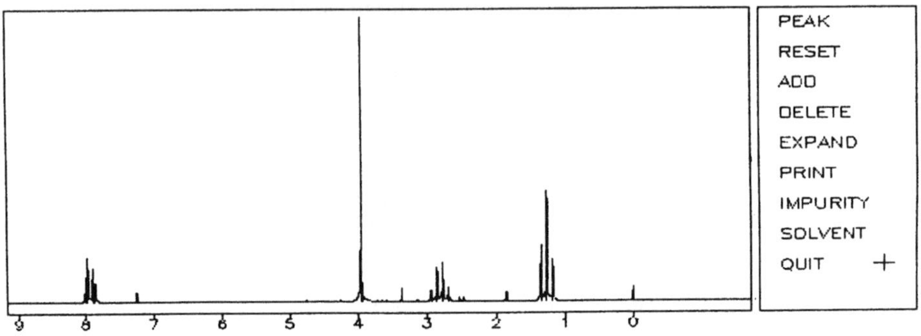

Figure 9. Detected peaks shown on CRT screen

Integration. This command is used for displaying an integration curve on the spectrum for the user's convenience.

SIGNAL ASSIGNMENT PROGRAM

This program will assist a user in assigning signal groups of a given spectrum to every hydrogen-containing atomic group in the corresponding structural formula.

The assignment task is performed primarily by using signal position versus partial structure correlation table. The table has been made from 8,000 spectral data which were measured and compiled in the authors' laboratory.[1]

These spectral data are stored with corresponding structural formulae represented by CANOST notations and appropriate assignment data as mentioned previously.

All hydrogen-containing atomic groups and their corresponding signal positions were abstracted from the data file. The atomic groups are specified with their alpha and beta circumstances as shown in Figure 10.

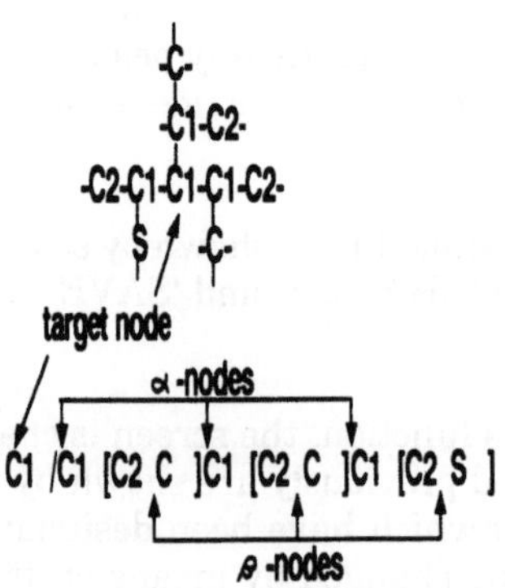

Figure 10. An example of substructure representation

A total of 47,690 signal position data was abstracted for 6,632 kinds of such atomic groups. As shown in Table 5, average value, standard deviation, and maximum and minimum values of atomic groups are then calculated. The atomic groups in the table are represented by means of CANOST codes which are used in CANOST notation system employed for representing structural formulae in the data file.

Table 5. Part of substructure-chemical shift correlations

	substructure	ave.	s. d.	min.	max.
1	C3 /C2 [C2]	.91	.421E-01	.43	2.15
2	C3 /C2 [C1]	.93	.824E-01	.45	1.20
3	C3 /C2 [C]	.85	.130E+00	.85	1.20
4	C3 /C2 [D1]	.99	.238E-06	.99	.99
5	C3 /C2 [D1Z]	.98	.392E-01	.96	1.13
6	C3 /C2 [DS]	1.15	.178E+00	.70	1.42
7	C3 /C2 [DSZ]	1.07	.000E+00	1.07	1.07
8	C3 /C2 [T]	1.15	.459E-01	1.12	1.24
9	C3 /C2 [Y]	1.22	.934E-01	.99	1.41
10	C3 /C2 [V1]	1.11	.000E+00	1.11	1.11
	:	:	:	:	:
6631	GE1/C2 [C3]C2 [C3]C2 [C3]	3.65	.000E+00	3.65	3.66
6632	SI1/C2 [C3]C2 [C3]C2 [C3]	3.61	.000E+00	3.61	3.61

The general idea of automatic assignment is illustrated in Figure 11. As shown in this figure:

1. All hydrogen-containing atomic groups in the given structural formula are searched in the table to find out the maximum and minimum values of signal positions.
2. Those signal groups in the given spectrum which are within the maximum and minimum values and have sufficient allocated hydrogens are assigned to the atomic group concerned.

The result is displayed on the CRT screen as shown in Figure 12. The grouping results obtained by means of the previous routine are shown over the spectrum and the structural formula is on the right. The primary assignment results mentioned above are shown in the lower part of the screen. The calculated hydrogen number RH, rounded hydrogen number GH and atomic group(s) which could be assigned to the signal group are also shown.

This example indicates that the atomic group numbered 1 may be assigned to the signal group 1, the atomic group 2 to the signal groups 3 and/or 5, the atomic group 3 to the signal group 5, and the atomic group 4 to the signal group 7, respectively. The automatically presented assignments can be modified interactively by the user.

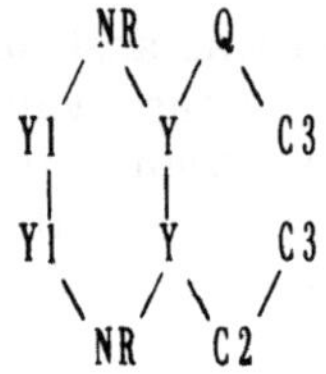


```
Y1 /Y1 [NR ]NR [Y  ]
C3 /Q  [Y  ]
C2 /C3 [   ]Y  [Y  NR ]
C3 /C2 [Y  ]
```

```
   :     :                              :       :          :     :
   9 C3 /C2 [Y  ]                     1.22  .934E-01     .99  1.41
   :     :                              :       :          :     :
 451 C3 /Q  [Y  ]                     3.87  .926E-01    3.32  4.30
   :     :                              :       :          :     :
 639 C2 /C3 [   ]Y  [Y  NR ]          2.39  .169E+00    2.22  2.84
   :     :                              :       :          :     :
5410 Y1 /Y1 [NR ]NR [Y  ]             8.39  .397E+00    7.91  8.96
   :     :                              :       :          :     :
```

Figure 11. Search procedures for candidate signal assignments for substructures

```
 GROUP= 1   RH= 1.926  GH= 2   H= 2    NODE=   1(1)   1(1)
 GROUP= 3   RH= 3.037  GH= 3   H= 3    NODE=   2(3)
 GROUP= 5   RH= 1.941  GH= 2   H= 5    NODE=   2(3)   3(2)
 GROUP= 7   RH= 3.095  GH= 3   H= 3    NODE=   4(3)
COMMAND: COPY
```

Figure 12. CRT screen showing allocation of hydrogens

Splitting Pattern Search

If the user wants to make a more precise examination of the results obtained by means of the above-mentioned signal assignment program, first-order splitting pattern search will be executed.

As shown in Figure 13, standard coupling constant values (Cij) for all possible combinations of nodes (atomic groups) which would be coupled each other in the given structure are obtained by referring to the table previously established.[6] The element Cij of the matrix is the standard coupling constant between nodes i and j. The numbering of each node corresponds to the non-canonical notation in Figure 3.

node	1	2	3	4	5	6	7	8	9	10
1 :	0.0	0.0	0.0	0.0	0.0	0.0	0.0	0.0	0.0	0.0
2 :	0.0	0.0	0.0	4.8	0.0	0.0	0.0	0.0	0.0	0.0
3 :	0.0	0.0	0.0	0.0	0.0	0.0	0.0	0.0	0.0	0.0
4 :	0.0	4.8	0.0	0.0	0.0	0.0	0.0	0.0	0.0	0.0
5 :	0.0	0.0	0.0	0.0	0.0	0.0	0.0	0.0	0.0	0.0
6 :	0.0	0.0	0.0	0.0	0.0	0.0	0.0	0.0	0.0	0.0
7 :	0.0	0.0	0.0	0.0	0.0	0.0	0.0	0.0	0.0	0.0
8 :	0.0	0.0	0.0	0.0	0.0	0.0	0.0	0.0	7.6	0.0
9 :	0.0	0.0	0.0	0.0	0.0	0.0	0.0	7.6	0.0	0.0
10 :	0.0	0.0	0.0	0.0	0.0	0.0	0.0	0.0	0.0	0.0

Figure 13. Expected coupling constants among substructures

Next, the splitting pattern for all nodes is calculated by a simple binary dividing method. A simple example is shown in Figure 14, where the signal of Ha is presumed as single at the beginning and it is divided into a doublet by He with the interval of Jae, and then each of them is again divided into two by Hd with the interval of Jad and so on. The intensities of all splitting peaks are also calculated as shown in the figure.

Since the splitting patterns calculated by the above mentioned procedure are based on pre-defined fixed coupling constant values, there may be some deviations between calculated and observed patterns. Therefore a stepwise perturbation procedure is employed for comparing these two patterns as follows:

Step 1: It is assumed that the calculated splitting peak positions for a proton group are cp1, cp2, and cp3, and the peak group to which the proton group is assigned consists of peaks positioned gp1, gp2, . . ., gp6 as shown below.

cp1 = −7.0	gp1 = 1517.7
cp2 = 0.0	gp2 = 1524.3
cp3 = 7.0	gp3 = 1526.1
	gp4 = 1530.8
	gp5 = 1533.5
	gp6 = 1535.8

Step 2: Distances di between adjacent peaks in the calculated splitting pattern are calculated.

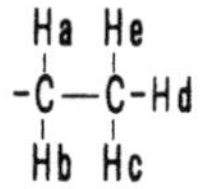

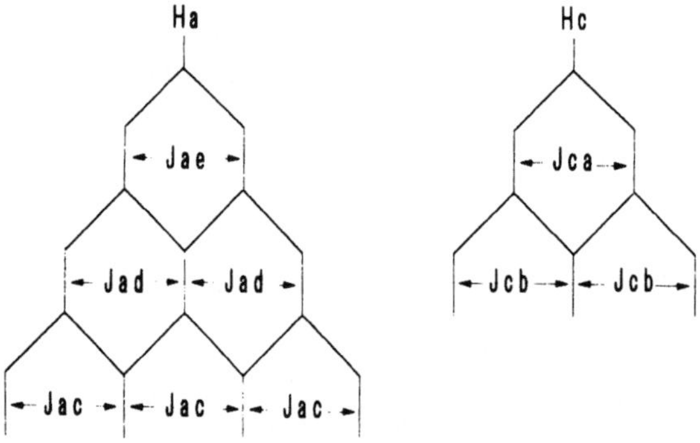

Figure 14. An example of splitting pattern determination

$$di = cpi+1 - cpi$$

In this case d1 = d2 = 7.0

Step 3: The peak detection allowance is set at 2.0. The actual initial value used in the procedure is 0.2, and it will be incremented by 0.2 when the pattern search is failed. The value 2.0 is employed here to make the explanation easier.

Step 4: The peak gp1 is assumed as a candidate for the first member of the triplet peaks. Then, those peaks whose positions are within 7.0 + 2.0 from it are searched for. In this case, those peaks positioned between 1522.7 and 1526.7 are searched and gp2 and gp3 are found as the candidates for the second member of the triplet.

Step 5: From gp2(1524.3 Hz), those peaks whose positions are within 7.0 + 2.0 from it are searched and gp4 is found as the candidate of the third member. Similarly, the peak gp5 is found from gp3. Therefore, two peak sets [gp1 gp2 gp4] and [gp1 gp3 gp5] are found as the candidate for the triplet.

Step 6: The search procedure is repeated similarly from the peak gp2, and an additional peak set [gp2 gp4 gp6] is found.

Step 7: No additional candidate sets are found from gp3, gp4 and so on.

An example for the result of the pattern search is shown in Figure 15. The first 4 lines mean that three possible candidate doublets which respectively consist of

```
------------------------------------------------------------------
SUBSTRUCTURE NO.=  1    MEMBER =  2   4
      I=  1  POSI.=     8.02   DIFF=     1.87  SET=  2  1
      I=  2  POSI.=     7.96   DIFF=     1.96  SET=  3  2
      I=  3  POSI.=     7.92   DIFF=     2.10  SET=  4  3
SUBSTRUCTURE NO.=  2    MEMBER =  7
      I=  1  POSI.=     3.97   DIFF=     0.00  SET=  6
SUBSTRUCTURE NO.=  3    MEMBER =  8
      I=  1  POSI.=     2.82   DIFF=     0.50  SET= 12 11 10  9
SUBSTRUCTURE NO.=  4    MEMBER =  9
      I=  1  POSI.=     1.27   DIFF=     0.22  SET= 16 15 14
-------- NEW CANDIDATE SET -----
SUBSTRUCTURE NO.=  1    MEMBER =  2   4
      I=  1  POSI.=     8.02   DIFF=     1.87  SET=  2  1
      I=  2  POSI.=     7.92   DIFF=     2.10  SET=  4  3
------------------------------------------------------------------
```

Figure 15. Results of splitting pattern search

peaks 1 and 2, 2 and 3, and 3 and 4 could be assigned to the substructures numbered 2 and 4. The signal positions (POSI. =) mean weighted average positions of the pairs of peaks. The numerals following the term 'DIFF=' are the differences from standard coupling constant values previously established.

For the first signal group (consisting of peaks numbered 1, 2, 3 and 4), three candidate splitting patterns are presented and the user has selected the first and the third combinations. Then the corrected patterns are listed in the second paragraph.

As shown in Figure 16, each of the candidate patterns is superimposed on the spectrum displayed on the CRT screen, and the user can select his preference.

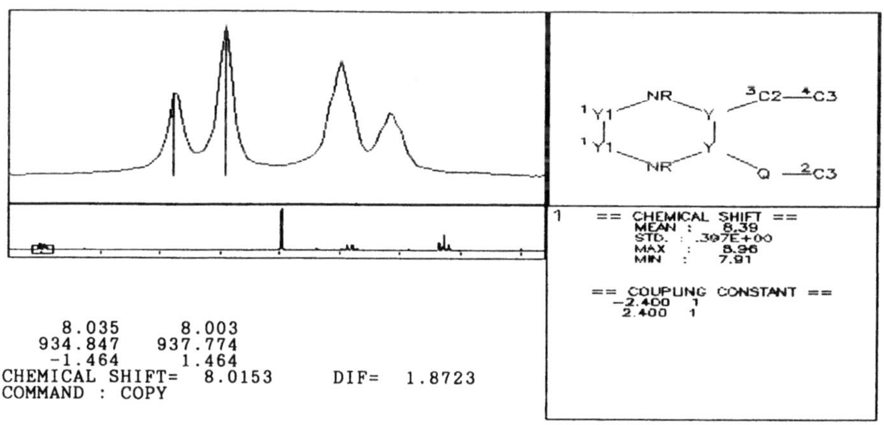

Figure 16. CRT screen showing the procedure for assessment of supposed splitting pattern

CONCLUSION

Development of the described personal computer-aided program system has improved the efficiency of database construction being done in the authors' laboratory. Moreover, the program should become a useful tool for interpretation of proton NMR spectral data. Generally speaking, the spectra measured by superconductive magnet spectrometers have much simpler splitting patterns than those obtained from lower magnetic field spectrometers. Therefore the splitting pattern search procedures presented here will become very helpful for interpretation of such spectra.

REFERENCES

1. Sasaki, S., Yamasaki, T., Tanaka, K., Abe, H. 'C-13 and H-1-NMR spectra collections as a base for the retrieval system SPIRES'. In *Computer-supported Spectroscopic Databases*, Zupan, J., Ed., Ellis Horwood: Chichester, 1986, pp. 133–154.
2. The C-13 NMR Spectra Collection of FIZ-Karlsruhe, Germany.
3. Abe, H., Kudo, Y., Yamasaki, T., Tanaka, K., Sasaki, M., Sasaki, S. 'A Convenient Notation System for Organic Structure on the Basis of Connectivity Stack'. *J. Chem. Inf. Comput. Sci.* **1984**, *24*, 212–216.
4. Abe, H., Takahashi, S., Sasaki, S. Paper submitted to *J. Math. Chem.*
5. Kawata, S., Minami, S. 'Adaptive Smoothing of Spectroscopic Data by a Linear Mean-Square Estimation'. *Appl. Spectrosc.* **1984**, *38*, 49–58.
6. Funatsu, K., Acharya, B.P., Sasaki, S. Unpublished work.

'B-BASE' – A STRUCTURE-ORIENTED, NUMERICAL, FACTUAL DATABASE FOR ^{11}B-NMR SPECTROSCOPY AND RELATED INFORMATION ABOUT OTHER NUCLEI

Heinrich Nöth and Erich S. Striedl
Institut für Anorganische Chemie, Ludwig-Maximilians-Universität München, Meiserstrasse 1, 8000 München 2, Germany

ABSTRACT

The structure-oriented, numerical, factual database, B-Base, is described as an application of the database management system ChemBase. The constructive work within the last two years resulted in a database, B-Base, which at present incorporates approximately 9,400 document units. Today the database covers primary literature up to the middle of 1988 and will be updated continuously. In general, data concerning boron compounds with 1–4 B atoms of co-ordination number 2–4 are stored. A concept for Lewis adduct compounds has been developed. In addition, this concept has been expanded for metal complexes and cluster compounds. Data fields for dividing B-Base into structure related file segments and for indexing multistructural graphs have been newly added. Evaluation of structure-shift relations by means of ChemBase software and the suitable PlotIT program for statistics resulted in about 60 basic graphs, visualising shift ranges of indicated substructures. These graphs are accessible within the application B-Base by using an alternative form when viewing a document unit.

INTRODUCTION

The structure-oriented, factual database B-Base represents to a certain extent an extended up-to-date edition of several monographs or reviews on ^{11}B-NMR spectroscopy which have been published within the last ten years.[1–4] This paper is written based on a poster presented at the Second International Meeting on Chemical Structures in Noordwijkerhout in June 1990. The scope of the poster is to give a summary of the topical contents of B-Base[5–7] and the possibilities of using this database at present. The main aspect is the evaluation of qualitative structure-shift relations and visualisation of results via structure related graphs of shift ranges within the application B-Base.

THE B-BASE DATABASE

Each document unit within B-Base includes chemical name, CA Registry Number (both only if handled within an article), authors, bibliographic reference, ^{11}B chemical shifts, line width, solvent and coupling constants as shown in Figure 1 and basically described earlier.[5,6] Structures with atom-related ^{11}B and hetero-atom chemical shifts are interactively searchable via ChemBase software, enabling the input and output of (sub)structures with atom-related shift values or

W. A. Warr (Ed.)
Chemical Structures 2

4,5-Diethyl-2,5-dihydro-2,2,3-trimethyl-1-natrio-1, 2,5-azasilaborol	B-BASE	955
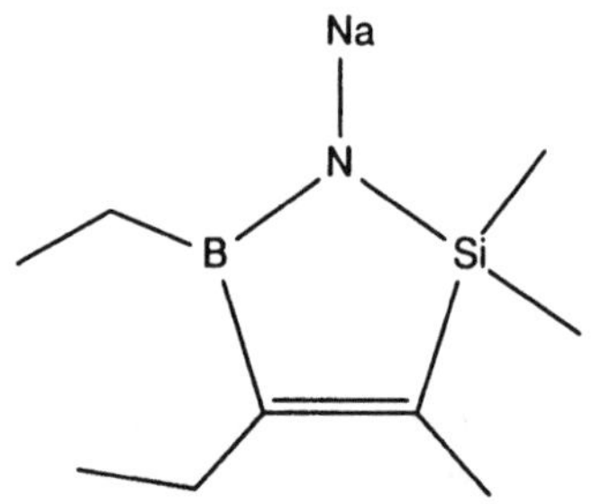	CAS	79483-03-7
	d-11B [ppm]	48.0
	LW [Hz]	950
	Solvent thf	
	BC2N	BSSC
	LIST	3
	Solvents: 14N, 23Na (C6D6); 29Si (toluene- d8); LW(23Na) = 2000 Hz	

3BCN-X.MET	,12		
Formula	C9 H19 B N Na Si	MW	203.15
AU	Koester, R., Seidel, G., Boese, R., Wrackmeyer, B.:		
SO	Chem. Ber. 120 (1987) 669	Ref.	87CH06

New Datafield **LIST** - explanations see text below

New Datafields **MET, MET-I** and **MET-2, MET-2-I** for referencing and indexing related graphs (**Metafiles**).

Figure 1. Form for one document unit within B-Base

ranges. Evaluation of structure-shift relations brought up the evidence of a new datafield 'LIST', which enables quick dividing of the whole database into file segments according to particular classes of compounds. An explanation of the classification schedule is given below.

2 co-ordination number of boron atom = 2
3 co-ordination number of boron atom = 3
4 co-ordination number of boron atom = 4
M mixtures of compounds
B boron hydrides, carbaboranes
C cluster compounds
D co-ordination compounds/transition metal complexes

In practice the user will choose a stored list and operate on it after setting it as the search domain. Boolean combinations ('AND', 'OR', 'NOT') on different lists are available with ChemBase software.[8,9] This procedure is comparable to the organisation of Derwent's Central Patents Index (CPI).[10] According to a new feature of ChemBase software,[11] the shift ranges of discrete substructures can be obtained directly *via* display of their graphs. This new enhancement of B-Base is demonstrated by Figure 2. In practice it is accessible by using a specially designed

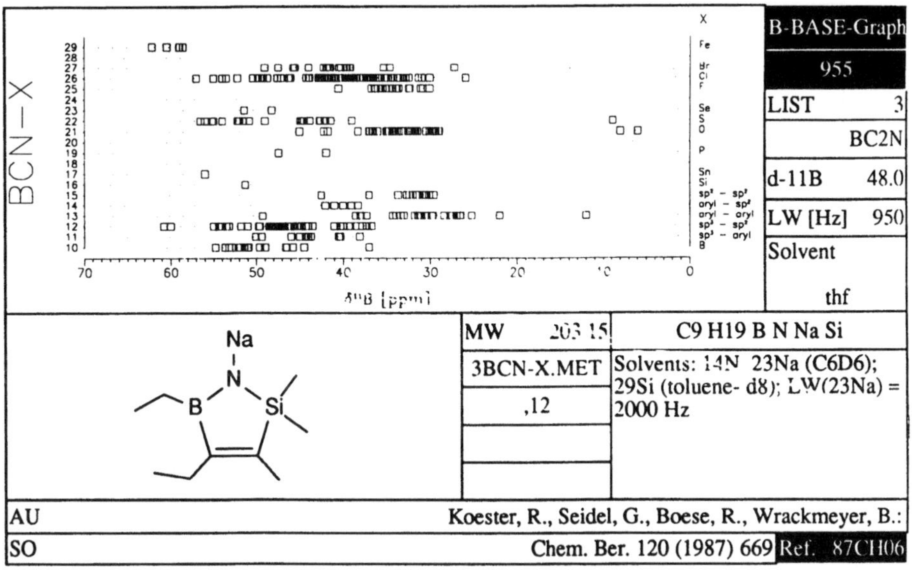

Figure 2. B-Base-Graph links each documentation unit with a graph, illustrating the shift range of the compound's boron environment

form, named B-Base-Graph (BB-GRAPH.FRM). The form incorporates one database field, defined for displaying metafiles, one box for the name of the related metafile (MET or MET-2) and a data field (MET-I or MET-2-I) which asigns the Y-position of the structure within the graph.

In the same way storage and display of ORTEP plots from single-crystal X-ray structural analysis is possible and being considered (Figure 3). To prevent casual access, the metafile reference field is hidden by the label box B-BASE-Ortep.

ATOM-RELATED SHIFT VALUES

Every heteroatomic NMR shift value mentioned within a primary publication is assigned to the discrete related atom as depicted in Figure 4. The values disappear on demand according to a toggle option within the Molecule Display Settings.[11]

More than 1400 document units of B-Base refer to multinuclear studies on boron compounds. Figure 5 displays the frequency of investigated nuclei other than B.

STRUCTURAL REPRESENTATION OF SPECIAL BOND TYPES

Lewis Adduct Compounds

The following abridged description of the documentation of Lewis adduct compounds within the factual database B-Base shows a concept, being developed

4-tert-Butyl-1-oxo-3-(2',2',6',6'-tetra- methyl-piperidino)-1,2,4,3-thioxazaboretidin		B-BASE-Ortep 4924
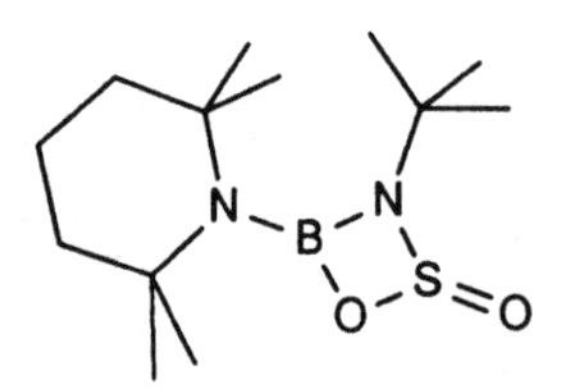	CAS-RN d-11B 28.0 LW [Hz] 170 Solvent benzene-d6 BSSC BN2O LIST 3 MW 286.25	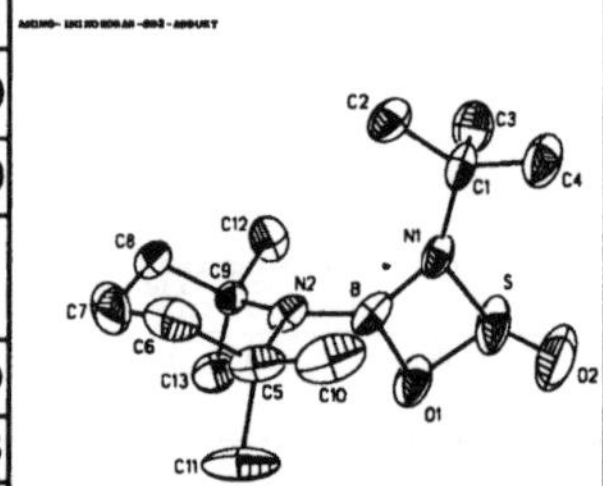
C13 H27 B N2 O2 S	14N: LW = 935Hz, X-Ray	
AU		Striedl, E.:
SO	Diploma Thesis, University of Muenchen (1988)	Ref. 88DT01

Figure 3. Storage and display of X-ray structural analysis ORTEP plots

with regard to one of the key questions of structure-NMR-shift-correlation in the field of boron chemistry, namely determining the co-ordination number of the central boron atom considered.

The formation of adduct compounds of boron containing compounds with the solvent leads to a change of the co-ordination number at the boron atom but structural descriptions of these compounds in reports vary from incorrect 3-co-ordinated boron compounds (only mentioning solvents) to mesomeric structures without any distinction between different bond types.

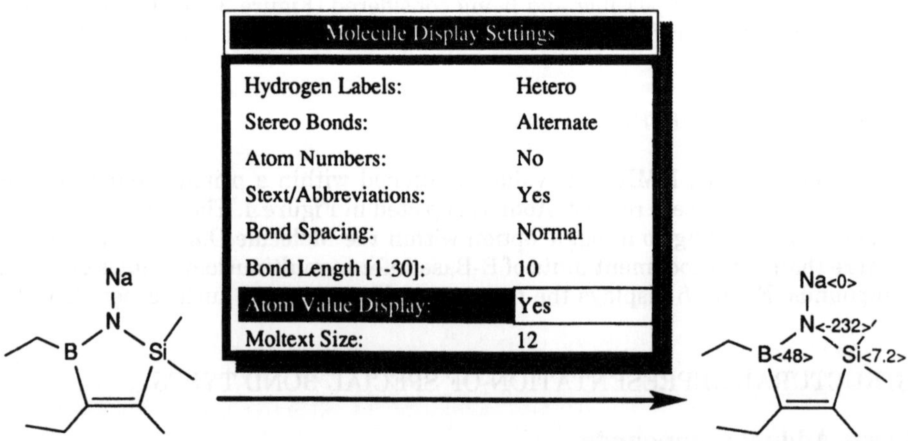

Figure 4. Heteroatomic NMR shift values are assigned to the discrete related atoms

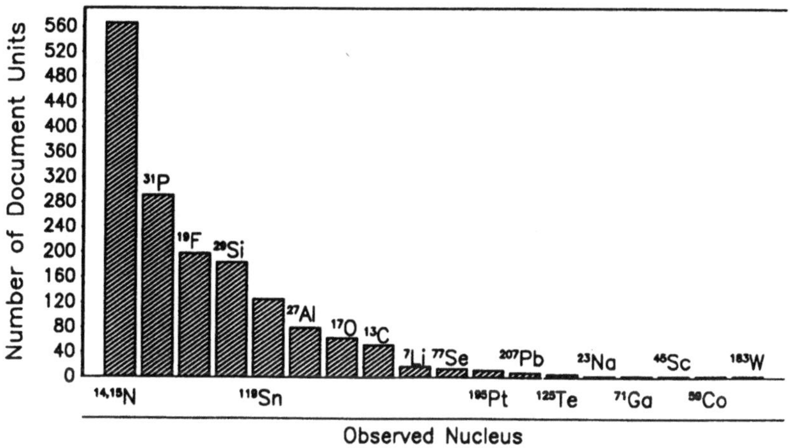

Figure 5. Frequency of documentation of NMR data other than ^{11}B within 'B-Base' (9,400 document units with ^{11}B-shift)

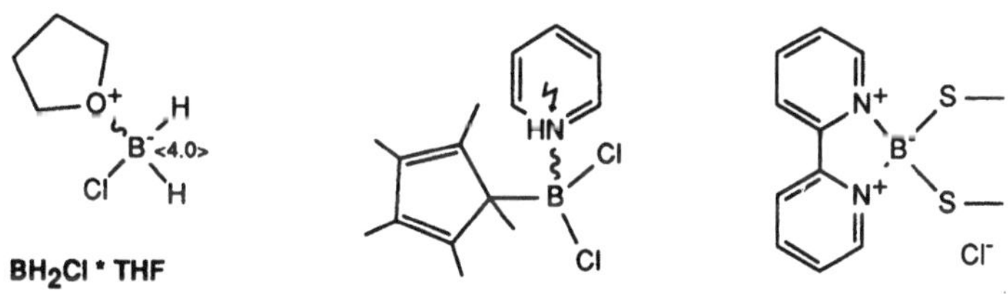

A B C

Figure 6. Documentation of Lewis adduct compounds in B-Base

As illustrated in Figure 6, in structure A, the bond between the two components, sometimes marked by an asterisk, must be represented as a single bond. Assigning a wavy 'either bond' emphasises the adduct character, but signifies no further difference. Formal charges must be applied to the related atoms according to the automatic supplying of hydrogen atoms, because structures like B in Figure 6, which would otherwise be wrong, will result. Mesomeric structures like C in Figure 6 are better documented without emphasis using the 'Either bond'.

D- or F- π-Bonds Within Transition Metal Complexes

Today Chemical Abstracts Service (CAS) has no structure conventions applied to transition metal complexes[12] which separate σ- from π-bonding interaction between metal atoms and ligands. Even the projected GMELIN-ONLINE will treat all metal to ligand connectivities as ordinary covalent bonds.[13]

Our approach to this problem is shown in Figure 7: within B-Base the user gets

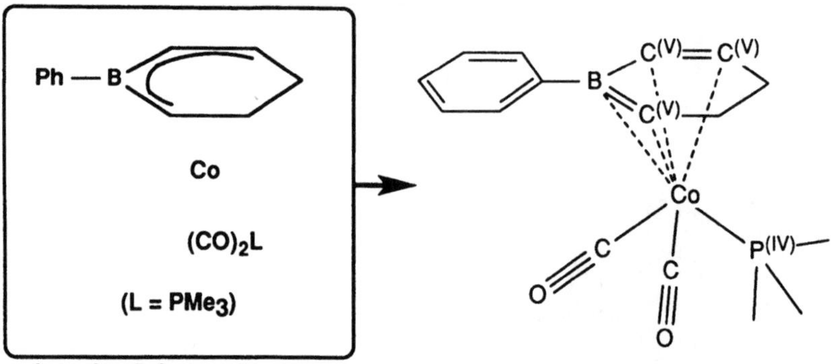

(redrawn from Ref. 14)

Figure 7. Redefining valences and assigning 'stereo down' bond types to metal-ligand π-bonds results in a graphical distinction between σ- and π-interaction

graphical information about where a π-bonding interaction is located. These dashed lines result from a 'stereo down bond' displayed at the stereo display mode 'Alternate'. Because they are simple optical attributes (coded as a single bond in the connection table), these 'π-bond types' cannot be searched for in the database. Therefore a user of B-Base executes a rough pre-selection by the program's search options and then selects the desired compounds *via* visual examination.

Cluster Compounds

In addition, this concept has been expanded for cluster compounds as depicted in Figure 8.

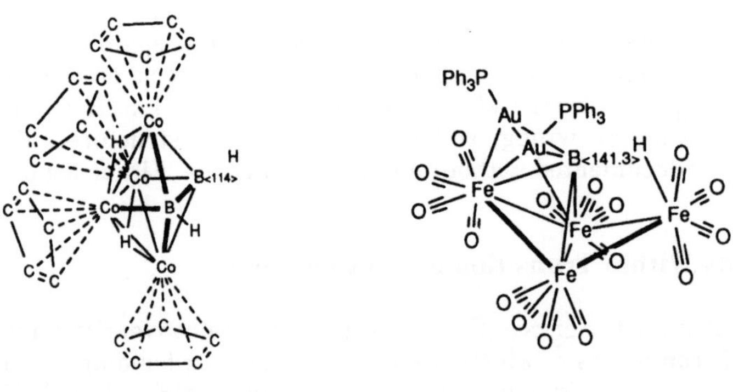

Figure 8. Representative sample structures for cluster compounds within B-Base

VISUALISATION OF STRUCTURE-SHIFT RELATIONS WITHIN 'B-BASE'

As shown, using B-Base simplifies a fast comparison between substructures and related shift values from NMR experiments. An important additional aspect is the possibility of a comparative overall access to published boron NMR data. Visualisation of shift ranges for various classes of boron compounds via bar graphs as illustrated by Figure 2 appears to be an obvious application in order to evaluate the contents of B-Base.

New bar graphs (or metafiles) on more specific questions may be generated easily by the following procedure:

1. Run a substructure search with B-Base.[5–7].
2. Save an X,Y-data file (ASCII) consisting of ^{11}B-NMR-shift and metafile index.
3. Generate a Hewlett Packard Graphics Language (HPGL) plot file from the data file by a suitable program. If a user has access to the PlotIT Interactive Graphics & Statistics Software,[15] he will be able to take advantage of using the already existing definition files, which contain the complete legends of each basic plot integrated in B-Base.
4. Transform the plot file with FROMHPGL.EXE[11] into a *.MET-File.

Figures 9a–c depict the possibilities of evaluating and visualising qualitative structure-shift relations with B-Base 'on demand'. Plot 9a represents a basic graph, visualising the shift values of all document units which refer to the Boron Substructure Code (BSSC) BO_2X. The BSSC implicates two oxygen atoms within the first co-ordination sphere of boron and a third ligand atom, indicated by the legend of the ordinate. The plots 9b and 9c show more precise shift ranges for particular substructures. Graphs like these will give useful hints about the explicit ^{11}B-NMR shift of a specific compound.

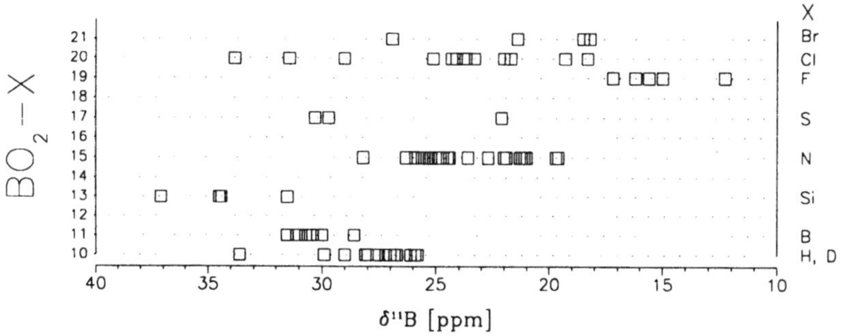

Figure 9a Basic plot for BSSC = $BO_2 - X$

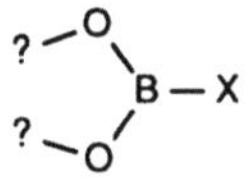

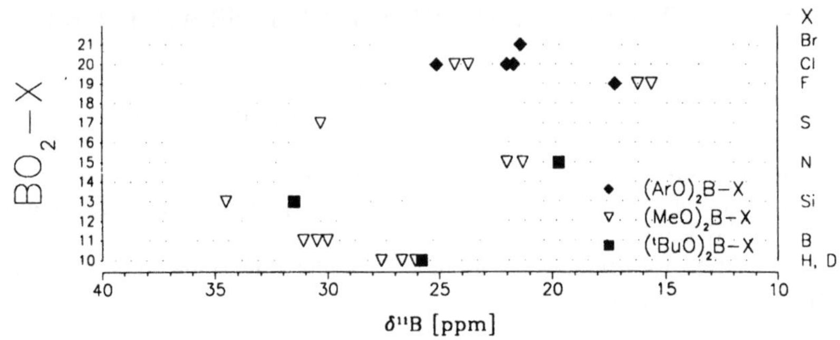

Figure 9b Plot 9a fixed with discrete substituents at both O-atoms

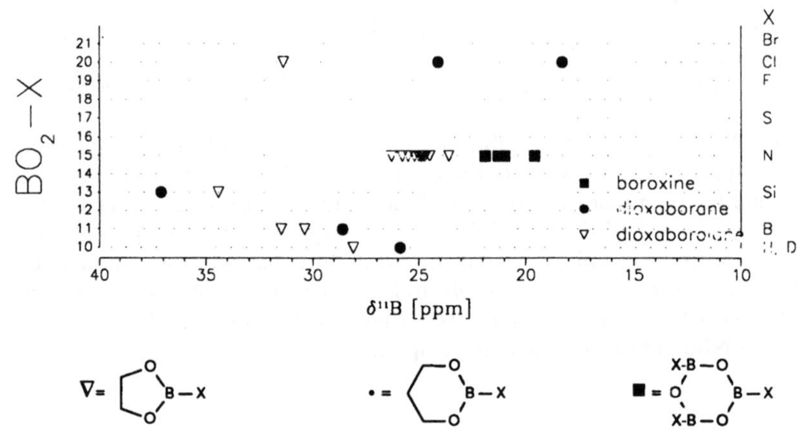

Figure 9c Different types of boron containing ring compounds

CONCLUSION

The concept, presented here, for the documentation of Lewis adduct compounds fills an essential gap within B-Base. The extension to metal complexes and other compounds with partial bond orders may be considered as an acceptable compromise, but additionally encodable bond types will be useful to the whole field of inorganic chemistry.

Graphical descriptions of shift ranges, well-known from ^{1}H- or ^{13}C-NMR spectroscopy, represent a further service for boron chemists, the more so because within B-Base they are integrated into each document unit. Integrating the PlotIT Graphics & Statistics software as part of the application B-Base leads to a powerful tool for discussing ^{11}B-NMR spectra on the broadest basis available today.

ACKNOWLEDGMENTS

We thank the Bundesminister für Forschung und Technologie and Chemical Concepts GmbH for support of this work.

REFERENCES

1. Nöth, H., Wrackmeyer, B. 'Nuclear Magnetic Resonance Spectroscopy of Boron Compounds'. In *NMR, Basic Principles and Progress*; Diehl, P., Fluck, E., Kosfeld, R., Eds.; Springer-Verlag: Berlin, 1978; Vol. 14.
2. Todd, L.J., Siedle, A.R. 'NMR studies of Boranes, Carboranes and Heteroatom Boranes'. *Prog. NMR Spectrosc.* **1979**, *13*, 87–176.
3. Siedle, A.R. 'Boron-11 NMR Spectroscopy'. *Ann. Rep. NMR Spectrosc.* **1982**, *12*, 177–261.
4. Wrackmeyer, B. 'Nuclear Magnetic Resonance Spectroscopy of Boron Compounds Containing Two-, Three- and Four-Co-ordinate Boron'. *Ann. Rep. NMR Spectrosc.* **1988**, *20*, 61–203.
5. Nöth, H., Striedl, E. 'B-Base: a Boron-11 NMR Factual Database for Personal Computers'. *Mitteilungsblatt Ges. Dtsch. Chem. Fachgruppe Chemie Information* **1989**, *15*, 3–10.
6. Nöth, H., Striedl, E. 'B-Base – eine ^{11}B-NMR-Faktendatenbank fur den Personal Computer'. In *Software Entwicklung in der Chemie 3*; Gauglitz, G., Ed.; Springer-Verlag: Berlin, 1989; pp. 101–108.
7. Striedl, E. Thesis, University of Munich, in preparation.
8. Meyer, D.E. 'Use of Microcomputer Software to Access and Handle Chemical Data'. In *Chemical Structures The International Language of Chemistry*; Warr, W.A., Ed.; Springer-Verlag: Berlin, 1988; pp. 251–260.
9. del Rey, D. 'Applications of Personal Computer Products for Chemical Data Management in the Chemist's Workstation'. In *Graphics for Chemical Structures; Integration with Text and Data*; Warr, W.A., Ed.; ACS Symposium Series 341; American Chemical Society: Washington, 1987; pp. 48–61.
10. Shenton, K.E., Norton, P., Ferns, E.A. 'Generic Searching of Patent Information'. In *Chemical Structures The International Language of Chemistry*; Warr, W.A., Ed.; Springer-Verlag, Berlin, 1988; pp. 169–178.
11. Supplied by Molecular Design Limited with Version 1.3 of ChemBase, August 1989.
12. Ryan, A.W., Stobaugh, R.E. 'The Chemical Abstracts Service Chemical Registry System IX. Input Structure Conventions'. *J. Chem. Inf. Comput. Sci.* **1982**, *22*, 22–28.
13. EXCERP Programm EXC3D; Input 3-dimensional Structure Conventions for GMELIN-ONLINE database; Manual for excerptors, August 1989.
14. Herberich, G.E., Raabe, E. 'Syntheses with 1-phenylbora-2,5-cyclohexadiene. Complexes of Iron, Cobalt and Nickel with Partially Hydrogenated Borabenzene Ligands'. *J. Organomet. Chem.* **1986**, *309*, 143–156.
15. PlotIT, Interactive Graphics & Statistics, Version 1.5, August 1989, by Scott P. Eisensmith, Scientific Programming Enterprises, Haslett, Mi 48840.

INTEGRATING CHEMICAL NOMENCLATURE INTERFACES TO STRUCTURE-BASED INFORMATION SYSTEMS

Graham H Kirby, Mark R Lord and John D Rayner
Department of Computer Science, University of Hull, Hull HU6 7RX, England

ABSTRACT

This paper derives from experience gained in using the Hull Chemical Nomenclature Translator as a front-end to various structure-based software packages. The benefits of nomenclature input facilities for chemical structure software present a need to link stand-alone packages into multiprocess systems running on PCs under MS-DOS, using connection tables for data exchange.

The attention of software developers and users is drawn to current problems in redirecting all data input to be read from files and in allowing alternative entry points which avoid repeated initialisation. They are urged to make more use of the Standard Molecular Data (SMD) format for transfer of structure data.

Techniques are described that have aided the production of multiprocess systems capable of running within the memory and single process limitations of MS-DOS. The advantages of more powerful desktop computers with multitasking operating systems are outlined for chemical information software tailored to users' needs.

NOMENCLATURE TRANSLATION

The computer translation of systematic chemical nomenclature into chemical structures is now possible. At least three groups have software which demonstrates this. Chemical Abstracts Service has for a long time used software running on mainframe computers to translate names in its style of nomenclature into connection tables.[1] More recently, the Beilstein Institute has developed the VICA name interpreter for an IBM 3090 mainframe. This translates with a high success rate a wide range of names into a connection table using a dictionary of some 3000 name fragments.[2]

The Hull University Nomenclature Translator (HUNT), unlike the above, runs on IBM PC and compatible microcomputers. It uses a formal and linguistic approach to the translation of the language of IUPAC systematic organic chemical nomenclature (hereafter shortened to organic nomenclature) into a connection table.[3] A formal grammar has been developed to describe many constructs in organic nomenclature by a context-free grammar.[4] Approximately 250 rules manipulate about 350 morphemes, or name fragments, which are associated with some 120 syntactic classes. The grammar covers hydrocarbons, acids, aldehydes, alcohols, ketones, ethers and steroids and is predominantly based on substitutive nomenclature but with some conjunctive and radicofunctional nomenclature. Some trivial names and terms allowed in IUPAC nomenclature are also included.

From the grammar, a name parser and associated dictionary of morphemes have been generated.[5] Semantic information is associated with the morphemes in the

W. A. Warr (Ed.)
Chemical Structures 2

dictionary so that, following a successful parse of a given name, the chemical validity of the name is checked and, where no semantic errors are detected, a Concise Connection Table (CCT) is generated. The CCT was designed specifically to represent the structure of nomenclatural fragments and the way they are connected to each other.[6] It reflects the hierarchy present in a name and is concise in that it does not explicitly represent every atom. Software has been written to generate the display of a structure from the CCT.

Figure 1 shows the structure of HUNT. In addition to the main phases of syntactic and semantic analysis, several other tasks are carried out. Simple lexical

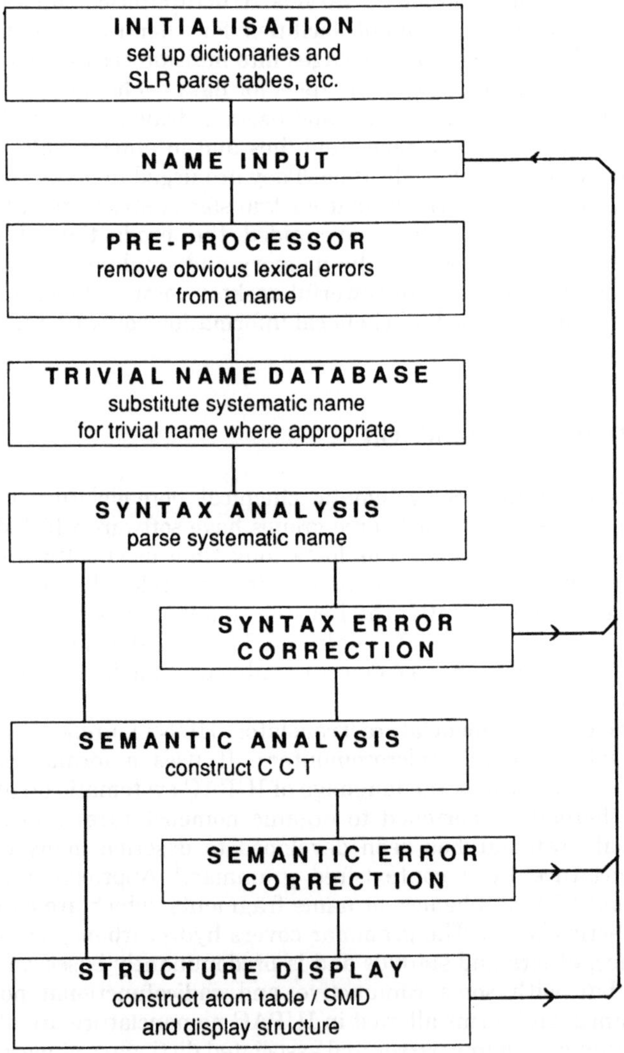

Figure 1. Hull University nomenclature translator

errors such as non-compulsory capital letters and multiple adjacent punctuation are detected and corrected by a pre-processor before syntax analysis. A trivial name database is provided to supplement the grammar rules. Every input name is checked against entries in the database, and any matching trivial name is replaced by its corresponding systematic name, prior to syntax analysis.

Given then that computer translation of organic nomenclature into structural information is now a practical proposition, what use is it? We see nomenclature translation widening access to a variety of information systems that deal with chemical structures, their properties and reactions. Many of these systems are structure-based, requiring structure input to be provided either graphically or by a notation such as a connection table or WLN. However, in some circumstances, nomenclature input has advantages.

Firstly, nomenclature input is simple to perform, requiring only keyboard skills. It can be carried out by chemically unskilled operators. Graphical input, in particular, requires both training and chemical knowledge and can take appreciable time. Nomenclature as conventionally printed includes characters and italic typeface not found on computer keyboards. Nomenclature translators, though, adopt conventions to overcome such problems. For example, Greek letters can be spelled out in full, function keys can be programmed to display special characters, fragments that should be italicised can be recognised from their context, and many typing errors can be detected and corrected by dialogue with the operator.

Secondly, many of the world's trade lists and inventories of chemicals, both in printed form and as computer databases, primarily use nomenclature and do not contain structures. Examples include the European Customs Inventory of Chemicals and the European Inventory of Existing Commercial Chemical Substances (EINECS), both produced by the Commission of the European Communities. The lists of recommended International Non-proprietary Names (INN) for pharmaceutical substances, produced by the World Health Organisation, give the IUPAC name with the INN. Lists of hazardous substances also tend to use nomenclature, for instance the UN list of Recommendations on the Transport of Dangerous Goods. Users of these lists, such as the UK Laboratory of the Government Chemist, who need to pursue investigation of particular entries in other chemical information systems, are limited to nomenclature input.

The potential use of nomenclature translation therefore is as an interface to a variety of structure-based information systems, and we wish now to address the issue of software interfacing. This is necessary in order to provide HUNT as a front-end software module for other existing and future structure-based systems.

INTERFACING ISSUES

There are two fundamental problems to be addressed in combining a number of separate software modules into a coherent package which should appear to its users as a single system. Firstly, the data to be handled must be transferable from one software module to another, and secondly it must be possible for run-time control to be switched within the whole system from one module to the next (see Figure 2). Ideally all of this should happen with a minimum of user intervention other than relevant application level selections.

Where the component software modules originate from the same authoring organisation, integration at both data and systems levels may be quite straight-

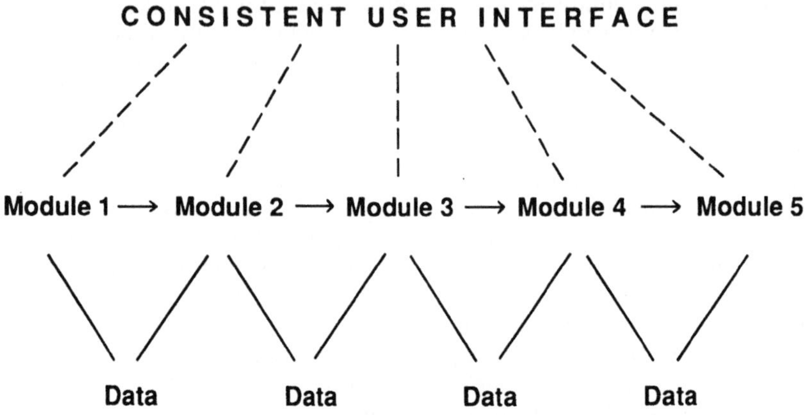

Figure 2. Multi-module software architecture

forward, subject only to questions of good overall systems design. The recently announced WHAT IF molecular modelling and drug design program[7] illustrates this approach by integrating most functions into one large FORTRAN program, supplemented by a variety of specially programmed interfaces to permit use of certain popular external programs in that field.

WHAT IF is an example of what can be achieved by one group working in a single language with detailed access to low-level program code. However, when a third party is considering the production of a facility based on a variety of software, involving both in-house and external proprietary systems, each level of interfacing can involve significant problems of compatibility in data formatting and systems support.

Our appreciation of these problems arises from experience with the provision of a variety of alternative display facilities for HUNT. In addition to our original character-graphics display method, we have linked HUNT with several different molecular structure display and modelling packages, including PsiBase, MolIdea, PCMODEL and MOLGRAF (see Figure 3). This has been accomplished by writing a module to convert from CCT to SMD, which opens up nomenclature input to any system accepting SMD (see Figure 4).

The issue of transferring data between separate modules should become less of a problem as the SMD file format is refined[8] and reaches wider acceptance. It is essential that software developers should allow for their programs to be used in other than stand-alone fashion, by allowing for input or output in this common structural representation, at least as one alternative. A whole range of processing modules can be envisaged (such as those in Figure 5) which could be combined at will, and which should be in the public domain.

Nevertheless, the physical transfer of data between component software modules still relies on operating system support as well as the modules themselves having a common understanding of the data transfer format. The construction of efficient multiprocess software systems on IBM PCs remains a major problem. The various possible approaches to creating such a hybrid system are shown in Figure 6.

Most proprietary chemical-oriented packages are designed as stand-alone systems, and MS-DOS, the most common PC operating system, allows only one

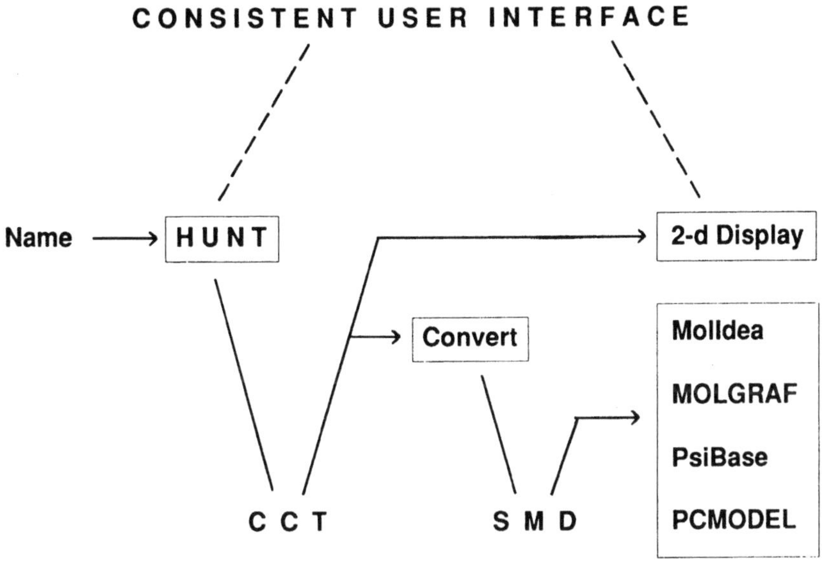

Figure 3. HUNT module interfaces

Structure Graphic Input *or* **Nomenclature Input** etc.

leading to

Connection Table or other Structure Representation

converted to

SMD

for use in

Database Query *or* **Structure Modelling** *or* **Reaction Analysis** etc.

Figure 4. Example modules and data formats

Nomenclature	**------->**	**S M D**		
Structure Input	**------->**	**S M D**		
W L N	**------->**	**S M D**		
S M D	**------>**	**2-d Screen Display - Various Conventions**		
S M D	**------>**	**3-d Screen Display - Various Conventions**		
S M D	**------>**	**Postscript printer page description language**		
S M D	**------>**	**Co-ordinate generation**	**------>**	**S M D**
S M D	**------>**	**Database Query**		
S M D	**------>**	**Subgraph Analysis / Reaction analysis**		

Figure 5. Desirable public domain software modules

1.	**Object Code Linking**	**- requires code compatibility and restricted by process max. size.**
2.	**Master Control and Processes**	**- reloads and re-initialises each process whenever used.**
3.	**Multi-tasking Environments**	**- allow processes to be suspended and restarted without repetition of start - up instructions.**

Figure 6. Operating and system support regimes

such process to be run at a time. Our experience suggests that it is generally not feasible to produce a hybrid software system that appears as a single MS-DOS program, by linking object code modules and libraries from different authors using a variety of languages. Not only will the resulting package almost certainly be too large for the 640Kb RAM limit of MS-DOS, but the issues of proprietary rights, compatibility of object code and file formats must be addressed.

The second, and more versatile, approach is to construct a system in which a small master control program invokes the required application processes in an appropriate sequence, each process terminating before the next starts, and each providing data for the next by means of temporary files. Under MS-DOS this may be achieved through the EXEC function in a small program, which remains in memory. This program regains control after each process terminates, in order to load and initiate execution of the next.

In this way we have successfully interfaced HUNT with the MolIdea and PsiBase systems, but our experience has shown another drawback in the association of separately designed modules. While some data may be transferred between processes by means of simple data files, the initialisation and menu set-up choices for each component module will have to be repeated on each re-load, with menu selections typically coming from the keyboard.

This results in a system which the user is encouraged to see as a single coherent package, but which requires the user continually to restate initialisation details which do not change during the session. In the worst case, many key-presses may be needed to re-navigate the menu front-end of a component module. Furthermore, significant delays may occur as a restarted component re-initialises its data structures (see Figure 7). This can be the situation for HUNT, which has a large volume of syntax tables and dictionary data to absorb each time it is re-loaded.

The first problem can be avoided only by a more sophisticated design of application modules, to provide file-based, non-keyboard entry of common set-up sequences, so that the user environment can be saved and reinstated from one use of the module to the next. If this is combined with a system of alternative entry points for execution of the software, a much greater freedom to mix and match software modules will be gained. Ideally, operating environments are needed which allow processes to be suspended and resumed instead of being closed down and restarted (the third point in Figure 6).

MULTITASKING OPERATING ENVIRONMENTS

With the arrival of the 80×86 series of microprocessors, particularly the Intel 80386 and the more recent 80486, multitasking facilities are now available on PCs. The increased addressing range of these chips allows more processes to be resident in memory at a time, and more sophisticated memory management and access control prevent separate processes from corrupting each other.

PCs based on the 80386 microprocessor, with the capability of full 32-bit memory addressing, started to become more available towards the end of 1987. This chip, like the 80286 (used in PC ATs), can use conventional memory (up to 640Kb) in real mode when emulating 8086/MS-DOS systems. In addition, the protected mode can be used to harness the real power of the machine, without losing the MS-DOS compatibility. The processor is capable of operating in the protected mode of the 80286 (virtual 286 mode) or its own protected mode (native

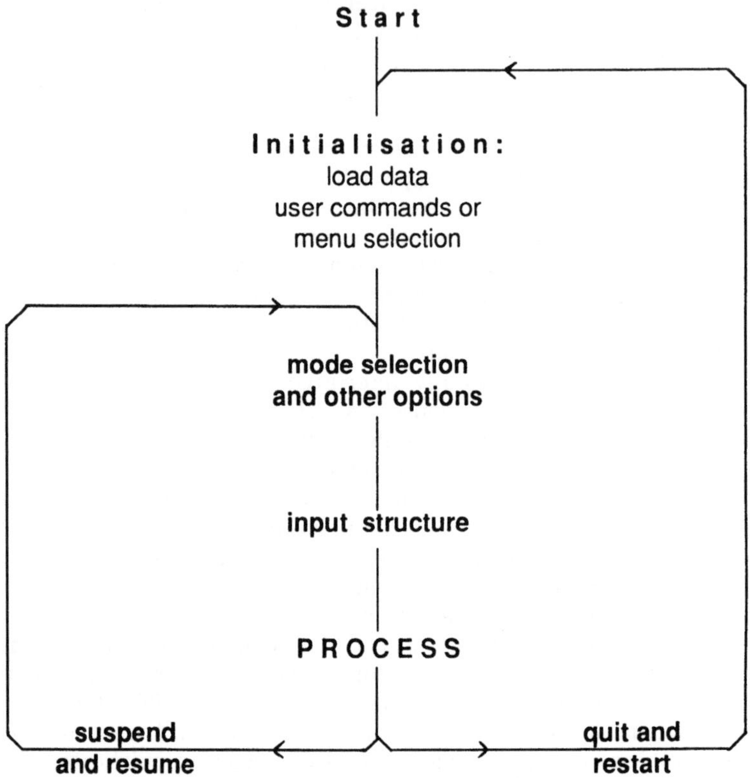

Figure 7. Process interface

mode) using 16-bit and 32-bit code, respectively. Using the new protected mode, the 80386 has a facility called virtual 8086 mode, in which the processor appears to operate as a series of discrete 8086 processors, each having its own memory space. These virtual machines are insulated from each other's address space by the protected working mode of the chip, preventing potentially fatal conflicts or crashes.

An extender, such as DESQview 386 or Microsoft Windows/386, is required in order to run an 80386-based PC under virtual 8086 mode. The use of such a system is the most simple multitasking option for allowing HUNT to be linked to other chemical packages. This is because the separate executable programs can each be given a virtual 8086 machine to themselves, with the interprocess communications being performed by the extender system. We have run the HUNT system under both the DESQview 386 and Microsoft Windows/386 environments. The former more easily allows the system designer to create the user interface under system control, so that the user needs less knowledge of operating system techniques.

DESQview 386 is essentially a windows-based resource manager for a PC. It comprises two main parts, namely DESQview and an expanded memory manager (QEMM-386). DESQview itself can run on any PC architecture. However, QEMM-386 is the control program responsible for administering the 386-specific features.

HUNT and the various structure display packages that accept SMD, or similar, input are each loaded into a separate 640Kb MS-DOS window, from where the various initialisation stages are executed. HUNT is shown in the foreground window (usually full screen) and is set to run in the foreground, whilst the display systems are set to run in the background and suspend when their initialisations have completed, awaiting input. This operation is performed by means of a special keyboard macro, known as a start-up script. In addition, each of these programs has a corresponding information file (DVP), which DESQview uses to administer them properly.

Once the above initialisation stage is complete, the user is invited to enter a name in HUNT for translation. If this name is valid, the corresponding SMD format is constructed, from the CCT, and dropped into a pre-determined file. Next, the user is asked to select from the available alternative display modules, using function keys. This key-press invokes a further script, unique to the current selection, which brings the appropriate display window to the foreground, sets it running in the foreground, shows the corresponding structure and initiates the eventual return to HUNT when the user is ready to input a new name. Selecting the quit option within HUNT results in the termination of all the processes.

Microsoft Windows/386 operates in a similar way to the DESQview system described above. However, it has a more graphical interface and does not have a built-in macro facility. Also, non-standard applications that display graphics (i.e., those graphics applications not specifically written for Windows) need to be run full screen.

The operation of HUNT and associated display modules within Windows/386 firstly requires the user to load each process manually by clicking with the mouse on the appropriate program information file (PIF, similar to the DESQview DVP file). Once loaded, the desired process, initially HUNT, is brought to the foreground either by a combination of key-presses, or by clicking in the title bar. Any process not currently in use can either be left in the background, or reduced to an icon. At the end, each process needs to be individually terminated by the user.

With the above extender systems the 640Kb MS-DOS restriction still applies within each virtual 8086 system used. Thus, a move to a completely new operating system, with appropriate hardware, may be a better solution for the future.

OS/2 and UNIX seem to be the two main rivals that are competing to replace MS-DOS as the dominant desktop operating system for the 1990s. OS/2 was jointly designed by IBM and Microsoft to replace MS-DOS with a system that is claimed to be capable of supporting the type of application and environment that will prevail in the future. It provides multitasking and direct addressing of 16Mb for 80×86-based machines, with the availability of a powerful windowing interface called Presentation Manager. In addition, it provides an MS-DOS compatibility box, which can run a single MS-DOS application. The current versions of OS/2 are designed to run on the 80286 processor, although a version specifically for the 80386 is expected in the near future, with a facility to multitask MS-DOS applications.

UNIX is an older operating system than OS/2, the First Edition system being released towards the end of 1971. The UNIX system has since been considerably enhanced and has been implemented on many different computers ranging from micros to mainframe systems. It has both multitasking and, unlike OS/2, multi-user capabilities. With the addition of a graphical interface, it has become the predominant operating system for high powered graphics workstations. As desktop computers increase in power and RAM, UNIX becomes a more attractive operating system for microcomputers. However, there are several competing groups producing slightly different versions, which currently hampers the portability of applications over various ranges of hardware.

As with OS/2, certain PC-based UNIX systems allow users to run MS-DOS applications by use of a special MS-DOS window. By using the virtual 8086 mode of the 80386, it will be possible with some UNIX systems to run simultaneously several MS-DOS programs as well as UNIX ones. Whatever operating system is used for providing structure-based systems with nomenclature interfaces, the use of this facility would be beneficial.

OS/2 and Presentation Manager are the natural progression for MS-DOS and Windows users, whilst UNIX has the additional multi-user facility and is scalable between a range of different architectures. Although OS/2 is currently an 80286 operating system, it has to be remembered that it is much younger than UNIX and a longer period of time may be required before a true operating system standard is established.

CONCLUSION

The availability of software providing translation of nomenclature input into connection tables serves to illustrate the problems caused by the insular design of much structure-based software, and the limitations of the MS-DOS operating system on what has been the standard PC used in the chemical information field. We would echo the comments made here three years ago on the need for open software architectures.[9] Software developers should consider more fully the wider consequences of their systems interface designs, to encourage greater flexibility for users in building software systems from modules tailored to their individual needs.

Integration means more than being able to couple together modules developed within the same organisation using a common interface. Greater use must be made of the only international standard format for data exchange of structural information, namely SMD. Several modules using or producing SMD have been identified that should be in the public domain so as to establish a basic set of software tools for building systems.

Developers of structure-based software must prepare for the multitasking capabilities offered by the combination of more powerful desktop computers and different operating systems by learning some lessons from current software. The continuing increase in processing power will create a demand from users for the building of customised systems by mixing and matching modules from a variety of sources, with the restarting and data interchange between modules being as transparent as possible.

Whatever multitasking operating environment is adopted for a particular implementation, the problem remains that, for example, a display module may still run through its menu selection of display styles, with the user repeating earlier specifications. To avoid this, a number of alternative entry points to modules would allow restarts after suspension to be handled differently from after re-load. Modules must also be written using logical input/output classes to allow for redirection.

For the past five years or so, MS-DOS has, in Europe at least, provided a stable and pseudo-standard environment for PC software. This situation is ending, and there is uncertainty amongst software developers about the future. They should be aware of the benefits to the chemical information community of portability, standard interfaces and user-customised systems. Multiple versions of the same software for differing operating environments should be avoided.

REFERENCES

1. Vander Stouw, G.G., Elliott, P.M., Isenberg, A.C. 'Automated Conversion of Chemical Substance Names to Atom-Bond Connection Tables'. *J. Chem. Doc.* **1974**, *14*, 185–193.
2. Goebels, L., Lawson, A. 'Organic Chemical Nomenclature: The Role of Beilstein Today and Tomorrow'. Presented at the conference on 'Chemical Nomenclature into the Next Millennium – Has it a Role?'; London 12/13th November 1987.
3. Cooke-Fox, D.I., Kirby, G.H., Rayner, J.D. 'Computer Translation of IUPAC Systematic Organic Chemical Nomenclature. 1. Introduction and Background to a Grammar-Based Approach'. *J. Chem. Inf. Comput. Sci.* **1989**, *29*, 101–105.
4. Cooke-Fox, D.I., Kirby, G.H., Rayner, J.D. 'Computer Translation of IUPAC Systematic Organic Chemical Nomenclature. 2. Development of a Formal Grammar'. *J. Chem. Inf. Comput. Sci.* **1989**, *29*, 106–112.
5. Cooke-Fox, D.I., Kirby, G.H., Rayner, J.D. 'Computer Translation of IUPAC Systematic Organic Chemical Nomenclature. 3. Syntax Analysis and Semantic Processing'. *J. Chem. Inf. Comput. Sci.* **1989**, *29*, 112–118.
6. Rayner, J.D. 'A Concise Connection Table Based on Systematic Nomenclatural Terms'. *J. Chem. Inf. Comput. Sci.* **1985**, *25*, 108–111.
7. Vriend, G. 'WHAT IF: A Molecular Modelling and Drug Design Program'. *J. Mol. Graphics* **1990**, *8*, 52–56.
8. Barnard, J.M. 'Draft Specification for Revised Version of the Standard Molecular Data (SMD) Format'. *J. Chem. Inf. Comput. Sci.* **1990**, *30*, 81–96.
9. Williams, M., Franklin, G. 'Future Directions in Integrated Information Systems: Is there a Strategic Advantage?'. In *Chemical Structures The International Language of Chemistry*; Warr, W.A., Ed.; Springer-Verlag: Berlin, 1988; pp.11–21.

AUTONOM – A CHEMIST'S DREAM: SYSTEM FOR (MICRO)COMPUTER GENERATION OF IUPAC-COMPATIBLE NAMES FROM STRUCTURAL INPUT

Janusz L Wisniewski
Beilstein Institute, Varrentrappstrasse 40–42,
D-6000 Frankfurt/Main 90, Germany

ABSTRACT

This paper describes the general design and state-of-the-art of AUTONOM, a computer expert system for unambiguous automatic translation of structural diagrams of organic compounds into chemical names. The algorithm designed for AUTONOM operates on structure diagrams entered *via* a structure drawing package and generates names exclusively on the basis of the connection tables of the input structures. The IUPAC rules have been adopted as the basis of naming rationales, however, some 'respectable' nomenclature practices, which lead to succinct and intelligible names but have not yet been sanctioned or codified by IUPAC are also followed. It is expected that the system, when fully implemented will provide the chemical community with a valuable tool.

INTRODUCTION

A complete, comprehensive grammar for systematic chemical nomenclature does not exist to date. The system of recommendations[1] developed by the IUPAC Commission on Nomenclature of Organic Chemistry (hereafter referred to as 'IUPAC') has not become a truly universal standard, presumably because of the complexity of the recommended rules, but also because of the frequent alternatives allowed in name assignment. It must also be said that a certain reluctance on the part of chemical industry to recognise the need for fully systematic nomenclature can be detected, and this leads in practice to the use of much quasi-systematic and inconsistent nomenclature.

Both of the leading producers of chemical information reference works (Beilstein, and the Chemical Abstracts Service (CAS)) follow IUPAC recommendations, but only to a certain extent. In order to obtain consistency in the selection of preferred index names both CAS and Beilstein have developed their own nomenclature rules or practices. These 'dialects' of IUPAC are nowhere clearly defined for the average user. This has not contributed to solving the difficulties in establishing a unambiguous nomenclature standard. As long as such a standard does not exist, the practising chemist will find himself alienated from systematic nomenclature. We believe that a truly unambiguous computer-based 'structure-to-name' translator could help the emergence of such a nomenclature standard.

The Research and Development Division of the Beilstein Institute has been working on this problem for several years, and some of the preliminary results have been announced at the most influential nomenclature conferences since 1987,

W. A. Warr (Ed.)
Chemical Structures 2

in particular the conference 'Chemical Nomenclature into the Next Millenium – Has It a Rôle ?'.[2] Our answer is undoubtedly 'Yes', and the purpose of this paper is to describe some of the progress which leads us to this conclusion.

To date there are no reports of a general purpose computer program which uniquely translates structural diagrams into chemical names. The first naming program for a very limited class of compounds (addition compounds, alloys, copolymers, and mixtures) was reported by Van der Stouw *et al.* in 1976.[3] Five years later, Mockus *et al.* reported4 on CAS plans to introduce a computer program for automatic structure-to-name translation in the generation of CA Index Names. The paper presented in detail the design of the algorithm, however, no subsequent report of the implementation of such an algorithm has so far followed. The most recent reports in this area[5] confirm the feasibility of a microcomputer-based program that would accept structure input and generate IUPAC-like systematic names, but the level of sophistication of the solutions appears to be still relatively low.

The present paper describes the current state of AUTONOM (AUTOmatic NOMenclature), a computer system which has been developed at the Beilstein Institute. The system, when fully implemented, should provide the chemical community with a completely automated 'structure-to-name' translator.

Although the following discussion will concentrate on a personal computer (PC) version of the program, it should be stressed that the concepts and solutions presented here apply also to the mainframe version of AUTONOM.

THE ALGORITHM

The algorithm applied in AUTONOM analyses a structure diagram of an organic compound entered *via* a graphic package and generates a chemical name purely on the basis of the connection table of the input structure. The communication between the structure drawing package (user-operated) and the naming part of the algorithm is realised by the interface module of AUTONOM. Once the complete structure has been entered, this interface fills the input buffer with data. The internal construction of the buffer is static and the interface is responsible for handling the differences in coding of the input structure by various drawing packages.

The text name generated by AUTONOM can be directly displayed on a monitor, stored as a database query term, or re-routed to another application for further processing.

The algorithm operates on a systematic step-by-step basis; while a chemist can often apply nomenclature rules without conscious effort by just 'seeing' the structure, the algorithm of AUTONOM has no such ability and must systematically collect, order and store numerous items of information about the structure during the processing cycle. The general approach adopted for the processing of a structure is based on the concept of localising (and possibly selecting) the smallest structural entities. These entities are nothing more than skeletal units i.e., catenated or cyclically connected atoms forming discrete nameable objects such as functional groups, chains or ring systems.

The naming cycle of a structure is conceptually divided into the following six phases:

1. Ring system perception.
2. Functional group recognition.
3. Ring system identification.
4. Parent structure selection.
5. Name tree creation.
6. Name assembly.

The selection and identification of objects takes place during the first three phases of the naming cycle. On completion of the third phase all atoms of the input structure are then distributed among the identified objects. So far no hierarchy among the objects has been established. This is now done using nomenclature rules during the fourth and fifth steps (parent structure selection and name tree creation). The descriptive term 'name tree' used here is by no means abstract. The objects and their mutual relations are, on completion of the fifth phase, represented by a data structure having a form of tree with the parent structure as the root and substituents as branches of the tree. During the last phase of the naming cycle hierarchically ordered objects are related with corresponding text name fragments which are then compiled into the complete name of the input structure.

RING SYSTEM PERCEPTION

The ring system identification currently implemented in AUTONOM is a two-step process initiated by the ring system perception. The reliability of the ring perception routine is the crux of the naming algorithm.

The model adopted for AUTONOM can be considered as representative for the 'smallest set of smallest rings (SSSR)' category of algorithms. From the point of view of nomenclature it was concluded that the perception of such a set of rings is the most suitable one for successfully naming the cyclic entities occurring in the structure. Once the SSSR for the structure is generated the algorithm finds all rings that share at least one atom with one or more other rings and so builds a list of nameable ring systems. On completion of the process all the atoms involved in cyclic closures are marked and distributed among the ring systems.

Once all the ring systems have been localised and their atoms marked, the algorithm generates for each ring system a corresponding hash code, which contains, in a very compact form, information on atom characteristics and atom interconnections of the ring system. The simple non-reversible hashing scheme used for the coding is unique in the sense that the same ring system is always tagged with the same hash string, but it does not mean that each system has a unique code. The main purpose of the hashing, as will be explained in greater detail later in this paper, was to relate all the rings in the input structure with classes of rings of similar structure.

FUNCTIONAL GROUP RECOGNITION

Acyclic portions of the structure are scanned to select substructure units bearing functional group characteristics, such as heteroatom compositions with unsaturated bonds. In the subsequent process of searching in a dictionary, each selected unit is compared with entries of pre-defined functional groups. Each entry in the dictionary is associated with a pre-determined priority rank and, in the case of a match, the priority rank is used to generate a sorted list of all functional groups present in the structure. Additionally, each selected and ranked group from the list is characterised by a 'base' atom, i.e., the atom through which it is connected to parent candidate skeletal units, such as ring systems or chains. The highest ranking entries from the list are then used in the next phase (parent structure selection).

The term 'highest ranking' derives from the IUPAC recommendations[1] and is supposed to set clear conditions on selection of the principal functional group, which later (in the completed name), is cited as the suffix and is the main criterion for choosing the parent structure skeletal unit. In many cases, particularly in the area of acids and their derivatives, the recommendations formulated by IUPAC might lead to ambiguous names. AUTONOM follows the recommended group ranking, but additionally, (when necessary) intersects the rank principle with the 'maximum multiplicity' rule, mainly in order to simplify and shorten the name.

Functional group recognition, although here presented as the sovereign phase of the naming cycle, is not a single run process which retires after the ranked list of functional groups has been determined. The list must be frequently updated automatically in the course of the naming of a single structure. The functional groups have been identified with preferred assignment of 'direction' with reference to the expected localisation of the parent structural skeleton. Since in this phase of the naming cycle no topological hierarchy of structural objects is yet established the preferred 'direction' is imposed by the established natural topology of atoms belonging to a functional group, e.g.:

Expected parent skeleton $\longleftarrow$ R_1–C(=O)–O–R_2

If it later proves that this assignment of 'direction' of the carboxylic acid happens to be correct then no changes to the group are necessary. However, if the parent skeleton is finally determined to be localised in the opposite direction relative to the group, i.e.:

R_1–C(=O)–O–R_2 $\longrightarrow$ to parent skeleton

then the initially identified acid must be split into two other groups, namely (–O) with the prefix descriptor 'oxy' and (=O) with the prefix descriptor 'oxo'. The splitting takes place some time later, during the name tree creation, after the mutual topological relation between the parent structure skeleton and the skeleton of the group has been determined.

RING SYSTEM RECOGNITION

The ring perception phase discussed earlier in this paper delivers a complete list of all localised ring systems. Each entry in the list contains a data record of the atoms constituting the system, the array of pointers pointing to positions of its sub-rings in the previously generated SSSR, and the hash code. Each such a record is then processed by the ring recognition routines.

In the first step (obligatory for each perceived ring system), a dictionary scan for ring systems with trivial names is carried out *via* an atom-by-atom search mechanism. The atom-by-atom matching is conducted in a special dictionary containing connection tables together with prescribed numbering and names. Much attention has been paid to optimising the organisation and construction of the dictionary in terms of efficient computer processing demands.

Each ring system chosen for inclusion in the dictionary is coded with exactly the same hashing algorithm used for localising the ring systems of the input structure during the ring perception phase. Thus the complete set of trivial-name ring systems is grouped into small sets with a common hash code as an address. At the time of writing, the measured average rings/set ratio was approximately 7. Although the statistical distribution of rings among sets is far from being uniform, the maximum registered number of rings in a single set is less than 25. This means, in practice, that in the worst case each localised ring system must be matched, atom-by-atom, against a maximum of 25 dictionary entries. Since the size of the dictionary will increase by the time the AUTONOM project is completed, it is planned, in order to reduce storage requirements significantly, to apply an effective data compression algorithm.[6]

During the dictionary search each perceived ring system is compared with the entries in a single dictionary ring set until an exact match is obtained. Then the algorithm, using the array of fixed locants stored together with each dictionary entry, continues with the complex routine of ring-atom numbering.

For non-symmetrical ring systems, the only possible composition of locants is the one stored in the array of the matched dictionary entry. For symmetrical rings this is not the case. Depending on the number of symmetry axes, a single atom can be numbered with many equivalent locants. The idea of storing all possible permutations of locants together with the dictionary entry was rejected as introducing unacceptable degradation in system performance. Instead a routine based on look-ahead techniques has been designed. The routine is able to regenerate all possible permutations of locant compositions on the basis of the one composition stored in the dictionary and the connection table of the matched ring system. Once all the permutations are available, the final ring-atom enumeration is decided by applying the following criteria, in the order given:

1. Principal functional groups.
2. 'Indicated' hydrogens.
3. Multiple bonds in compounds whose names indicate partial hydrogenation (cycloalkenes, pyrazolines, and the like).
4. Number of substituents.
5. Lowest locants for substituents named as prefixes.
6. Lowest locants for substituents named as prefixes in alphabetical order of itation.

For a wide spectrum of ring system classes dictionary access has been replaced with purely algorithmic ring identification. The dictionary does not contain any entry that falls into one of these classes; AUTONOM generates the names of such ring systems exclusively on the basis of the analysis of their internal composition. The policy strictly followed during programming AUTONOM is to cover the greatest possible number of ring system classes using algorithmic identification, and reference to the dictionary is only carried out if the latter proves to be distinctly more efficient from a system performance point of view. The classes of ring systems which are recognised purely by an algorithm include:

- Monocyclic hydrocarbons.
- Bicyclic alkanes.
- Monospirocyclic alkanes.
- Dispirocyclic alkanes.
- Tricyclic alkanes,
- Heteromonocyclics named by the Hantzsch-Widman method.[7]
- Replacement ('a' terms) nomenclature heteromonocyclics.
- Replacement ('a' terms) nomenclature heterobicyclics.
- Replacement ('a' terms) nomenclature monospiroheterocyclics.
- Replacement ('a' terms) nomenclature dispiroheterocyclics.
- Replacement ('a' terms) nomenclature heterotricyclics.

There are also groups of ring systems for which a combined dictionary and algorithmic scheme of identification is desirable. This combined approach is applied for all fused polycyclic hydrocarbons and heterocyclics with monospiro or dispiro connections to the ring systems classified in the above list.

Ring assemblies localised in the input structure are treated in a special way. According to IUPAC a ring assembly is a linear composition of two or more identical ring systems joined by acyclic single or double bonds, not necessarily at equivalent positions. The part of ring assembly identification conducted in this phase consists of finding all ring systems involved in assembly connections and the recognition of these systems not as isolated rings but as members of a bigger cyclic configuration. In nomenclature practice, this means inclusion of specific nomenclature criteria in the process of atom enumeration (such as special priority status for the atoms at the point of attachment, or the assignment of non-primed, primed or multi-primed locants to atoms constituting particular ring systems of the assembly).

PARENT STRUCTURE SELECTION

This phase of the naming cycle begins with identification of potential parent chain(s). The chain identification process has been built around a single recursive routine using techniques common for graph transition methods.

Once candidate parent chains have been identified, the algorithm ranks the selected candidate structural units (ring(s) and functional parent(s), if any, have already been identified in the previous phases) according to the relevant nomenclature principles. If more than one candidate structural unit of the same rank competes for selection as a parent, the following sequence of principles, in the order given, is applied until a decision is reached:

1. Greatest number of the principal functional groups cited as a suffix.
2. Preferred hetero atom content.
3. Rings preferred to chains.
4. Seniority of rings (if ring candidates present).
5. Seniority of chains (if chain candidates present).
6. The greatest number of substituents.

7. Lowest locants for the substituents.
8. Lowest locants for the substituents cited first as prefixes in alphabetical order.

Seniority of ring systems is decided by applying 12 criteria which are, in general, in agreement with the IUPAC recommendation C.0.14 (p. 101 in[1]). The senior chain, i.e., the chain upon which the nomenclature and enumeration of all atoms in the structure will be based, is chosen by applying successively 10 criteria formulated by the IUPAC recommendation C.0.13 (p. 97 in[1]).

The structural objects which cannot be assigned the status of parent automatically become substituents on the selected parent unit.

NAME TREE CREATION

The hierarchic principle underlying the approach to a construction of a chemical name (parent, substituent, substituent-on-substituent, etc.) was the decisive factor in the design of the data format for effective name generation analysis. The data structure finally implemented has been built around the concept of an ordered binary tree.[8]

The parent fragment selected in the preceding phase is established as the root of the name tree, while the other structural units become branches of the name tree. Starting at the root, the tree is traversed in an upward direction and mutual relationships among the branches (e.g., type of bonding, locants of connections, indicated hydrogen locants for ring sytems, locants of multiple bonds in chains, etc.) are established and data concerning these relationships are added to the branch descriptors. Concurrently, if necessary, new structural objects are generated and corresponding branches added to the tree. The formation of new objects is encountered either in the case of newly generated functional groups resulting from the operation of the group splitting or in the case of newly identified substituent chains. In the case of multiple candidate substituent chains starting at the same atom, the usual sequence of IUPAC rules is applied to find the senior one.

The structural objects no longer needed are immediately discarded and the memory segments they have occupied are deallocated and returned for further use. This is the case with the original functional groups which need to be split or with ring systems involved in an assembly junction. In the latter case one catenated ring assembly structural unit and corresponding single common assembly data record are generated while the individual assembly members are discarded.

The full description of all the possible objects and their mutual relationships is complete once the tree has been completely traversed in an upward direction. The structure is now fully mapped onto hierarchically ordered computer data records containing complete information on nameable structural fragments present in the structure. The data records are related to one another by a system of pointers organised as an easily manageable ordered binary tree.

NAME ASSEMBLY

Once the complete name tree is created, the algorithm begins to traverse the tree downward starting at most distant branch and moving to the root. The data records indicated by the pointers of visited branches are processed and the resulting textual name fragments are generated. Track is kept of the sequence of visited branches in the form of labels organised as a dedicated stack. In the next step, the stack is successively emptied and the name fragments indicated by the current top item of the stack are taken and combined into longer fragments. The job of combining smaller name fragments into longer fragments, and finally into a complete chemical name, is executed by a set of sophisticated routines. These handle such complex operations as proper alphabetisation, multiplication, punctuation, redundant vowel elision, suppression of unnecessary locants, insertion of superscript or italic strings, etc.

The whole name assembly process is constantly monitored by an intelligent 'triviality' controller. The controller is a set of complex routines responsible for tracking and replacing (when necessary), the sort of overdone systematic pedantry which computers are prone to, and replacing this with concise IUPAC-sanctioned traditional nomenclature. The example from Figure 1 well illustrates the pithiness which has been introduced into the name generated by AUTONOM (for an extremely complex and fictional compound !).

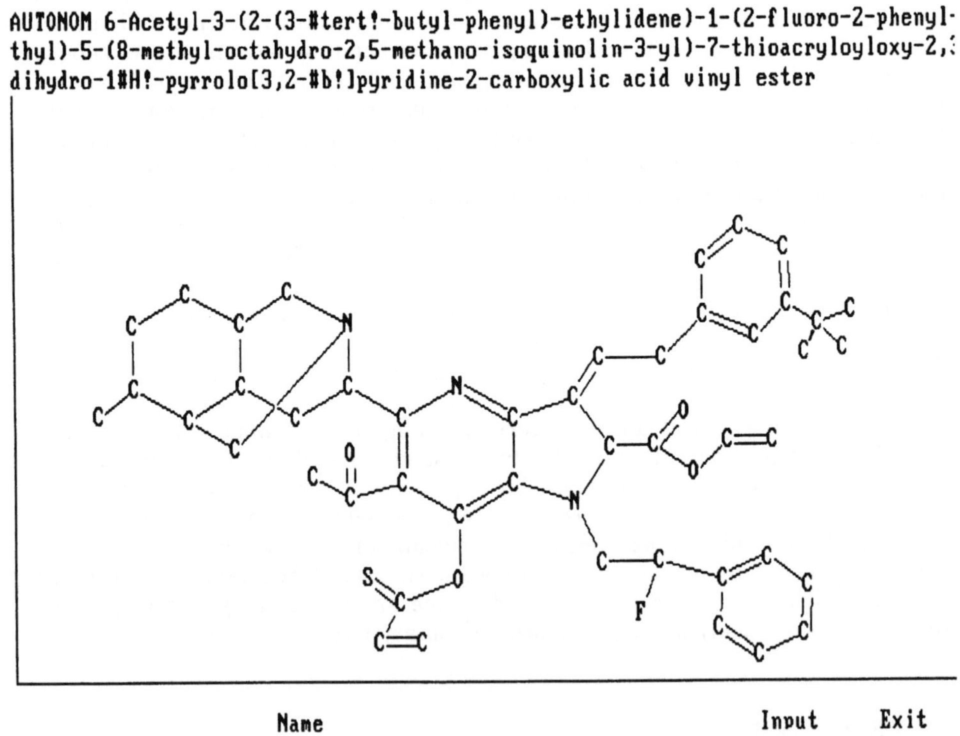

Figure 1. Name and structure output screen from AUTONOM

SUMMARY

Figure 1 demonstrates (original output screen) the use of AUTONOM in a PC-based implementation. The structure was drawn using a mouse and the program returned the name (in the form shown) in 19 seconds. The name fragments surrounded by #..! appear italicised when printed. Tests conducted on random samples from the Beilstein database indicate that the program (at the time of writing) is capable of achieving expert status in *ca.* 61% of the structures tested, where the expert status is defined as identical output compared to the Nomenclature Department of Beilstein. At the moment AUTONOM does not deal with sterochemistry, which accounts for the great majority of the 'missing' 39% of truly expert status. The remainder of the 39% consisted of a polite refusal by the program to generate a name at all. This is one advantage a computer system has over humans: it has no problems in admitting the limits of its own capabilities!

In the PC-based version of the program, input structures are limited to 255 heavy (non-hydrogen) atoms. On average, depending on the complexity of a structure, AUTONOM names up to 17 compounds per minute (measured on 80386-based Compaq Deskpro 386).

The program runs on an IBM-AT (or compatible) with a minimum of 640 kB of RAM and a hard disk with the minimum of 2.0 MB free for storage of the dictionaries.

Programming the system to its current state was a substantial task resulting in 99 different routines and functions and in 19,700 lines of code.

ACKNOWLEDGMENTS

The development of the AUTONOM project is generously supported by the Bundesministerium für Forschung und Technologie (BMFT). The author and the Beilstein Institute gratefully acknowledge this support. Contributions from Dr. L. Goebels, who is responsible for nomenclature aspects of the AUTONOM project and Prof. A. Lawson, who leads the project, are also gratefully acknowledged.

REFERENCES

1. International Union of Pure and Applied Chemistry. Nomenclature of Organic Chemistry, Sections A–F and H; Pergamon: Oxford, 1979.
2. 'Chemical Nomenclature into The Next Millennium – Has It a Rôle?', Conference on Chemical Nomenclature, London, England, Nov. 1987.
3. Vander Stouw, G.G., Gustafson, C., Rule, J.D., Watson, C.E. 'The Chemical Abstracts Service Chemical Registry System. IV. Use of the Registry System to Support the Preparation of Index Nomenclature'. *J. Chem. Inf. Comput. Sci.* **1976**, *16*, 213–218.
4. Mockus, J., Isenberg, A.C., Vander Stouw, G.G. 'Algorithmic Generation of Chemical Abstracts Index Names. 1. General Design'. *J. Chem. Inf. Comput. Sci.* **1981**, *21*, 183–195.
5. Meyer, D.E., Gould, S.R. 'Microcomputer Generation of Chemical Nomenclature from Graphic Structure Input'. *Am. Lab. (Fairfield), Conn.* **1988**, *20*(11), 92–96.
6. Wisniewski, J.L. 'Effective Text Compression with Simultaneous Digram and Trigram Encoding'. *J. Inf. Sci.* **1987**, *13*, 159–164.
7. International Union of Pure and Applied Chemistry. 'Revision of the Extended Hantzsch-Widman System of Nomenclature for Heteromonocycles'. *Pure Appl. Chem.* **1983**, *55*(2), 409–416.
8. Tenenbaum, A.M., Augenstein, M.J. '*Data Structures Using Pascal*'; Prentice-Hall: Englewood Cliffs, New Jersey, 1981, p. 318.

MACROMOLECULES: STRUCTURE REPRESENTATION AND NOMENCLATURE

W. Val Metanomski and Joy E. Merritt
Chemical Abstracts Service, Columbus, Ohio 43210, U.S.A.

Kurt L. Loening
Topterm, North American Division, Columbus, Ohio 43221, U.S.A.

ABSTRACT

Macromolecules, including biopolymers, represent a challenge in generating meaningful structure representations and corresponding names, because frequently complete structural information is not known. Illustrations are shown from existing recommendations of the International Union of Pure and Applied Chemistry (IUPAC), and the International Union of Biochemistry (IUB). Examples from drafts under consideration by the IUPAC Commission on Macromolecular Nomenclature are also presented. Current handling of macromolecules by Chemical Abstracts Service (CAS) as well as some enhancements under development are illustrated.

INTRODUCTION

Traditionally, polymers (macromolecules) have been described in terms of component monomers (source) from which they are prepared, e.g., polystyrene, poly(vinyl acetate), poly(methyl methacrylate), or styrene-acrylonitrile copolymer. Occasionally, even an assumed or hypothetical monomer is used, e.g., poly(vinyl alcohol) or ethylene-vinyl alcohol copolymer. Unlike well defined and well characterised low-molecular-weight compounds, polymers do not consist of identical molecules, but are mixtures of molecules of different molecular weights, and different sequences and shapes.

As the structures of polymers were elucidated and the authors started to report them, the need arose for a structure-based nomenclature. Still, it has to be remembered that in many cases structural representations are idealised. Deviations from ideality have to be ignored in constructing a useful nomenclature which still applies to the predominant structure of real polymer molecules.

IUPAC COMMISSION ON MACROMOLECULAR NOMENCLATURE

One of the goals of the International Union of Pure and Applied Chemistry (IUPAC) is to study topics of international importance to pure and applied chemistry which need regulation, standardisation, or codification. The mission of the IUPAC Commission on Macromolecular Nomenclature is to provide syste-

W. A. Warr (Ed.)
Chemical Structures 2

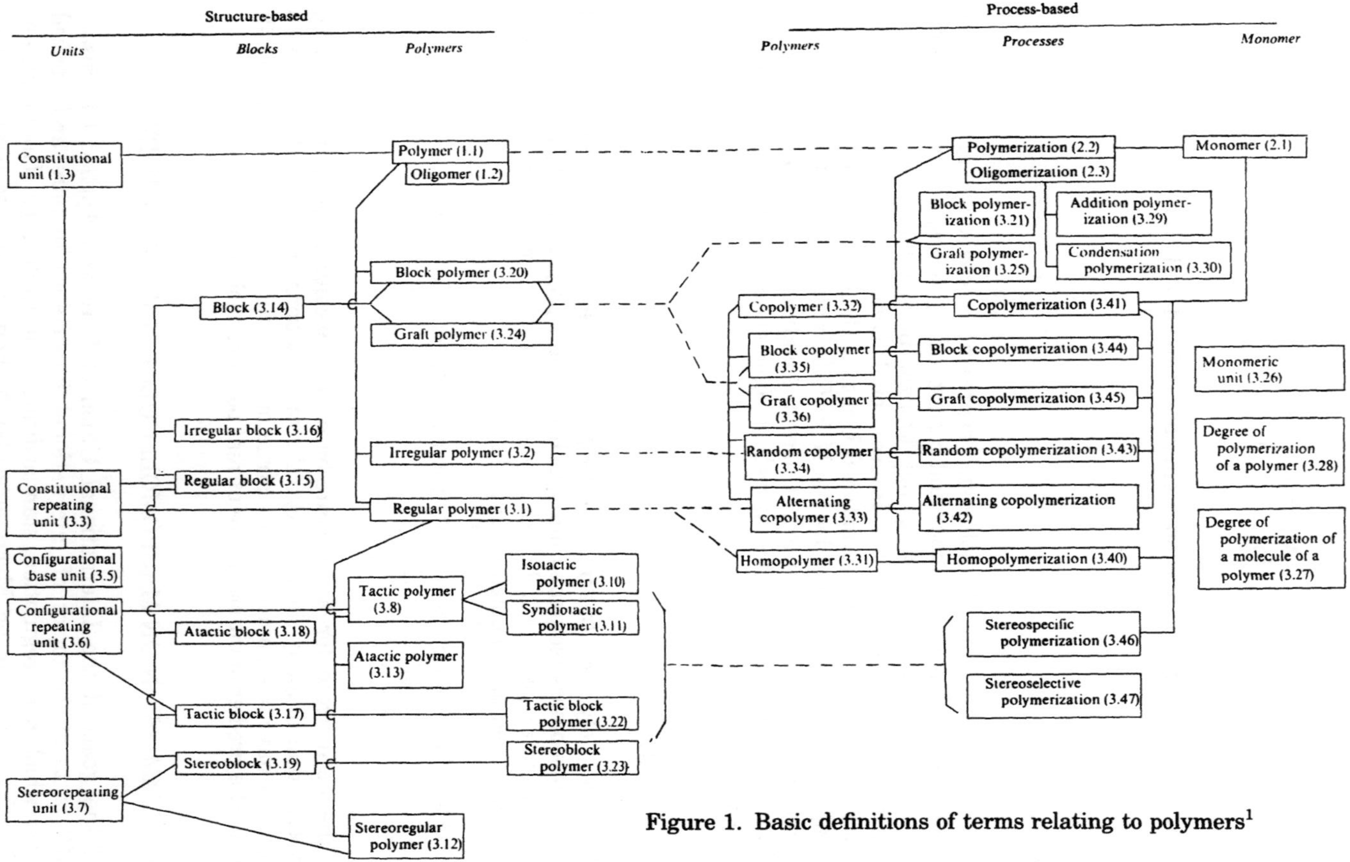

Figure 1. Basic definitions of terms relating to polymers[1]

matic and unambiguous terminology and nomenclature in the field of macromolecules acceptable to scientists throughout the world, and consistent with usage in other disciplines.

The Commission has developed sets of definitions for many of the basic terms dealing with polymer molecules and has published recommendations on structure-based and source-based nomenclature for organic, inorganic, and co-ordination polymers and copolymers. The Commission is currently working on the extension of nomenclature to more complicated cases such as irregular polymers, cross-linked polymers, double-strand polymers, interpenetrating networks, and multi-component polymer systems.

Figure 1 lists all the terms defined by the Commission. Two broad sets of definitions are presented. One is based on the structure of polymer molecules and the other on the processes by which polymeric substances come into being.

Structure-Based Nomenclature for Regular Polymers

For polymers that are regular, that is, have only one species of constitutional repeating unit (CRU) in a single sequential arrangement, structural representations and corresponding names identify polymer structures most accurately.

The system was first presented by the Polymer Nomenclature Committee of the American Chemical Society[2].

The graphical representations of polymer structures (chemical diagrams and formulae) on this and following pages have been drawn according to the IUPAC Commission's proposed recommendations, now in the final stages of the review[3].

A. *Regular Single-Strand Organic Polymers*[4]

a. $\text{-}\!\!\left[OCH_2CH_2 \right]\!\!\text{-}_n$

poly(oxyethylene)

b.

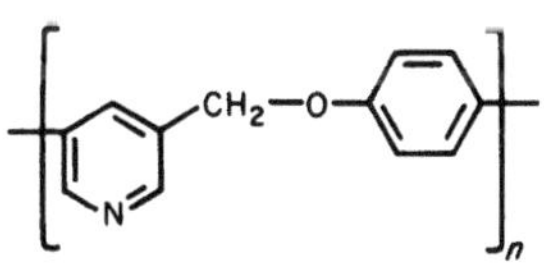

poly(3,5-pyridinediylmethyleneoxy-1,4-phenylene)

c.

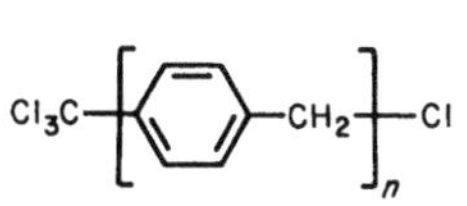

α-(trichloromethyl)-ω-chloropoly(1,4-phenylenemethylene)

Regular single-strand polymer chains can be represented as multiples of a preferred bivalent constitutional repeating unit (CRU) which can itself be named according to the rules for the nomenclature of organic chemistry[5]. The name of the polymer is 'poly(CRU)'. End groups are specified by prefixes placed ahead of the name of the polymer.

B. *Regular Ladder Organic Polymer*[6]

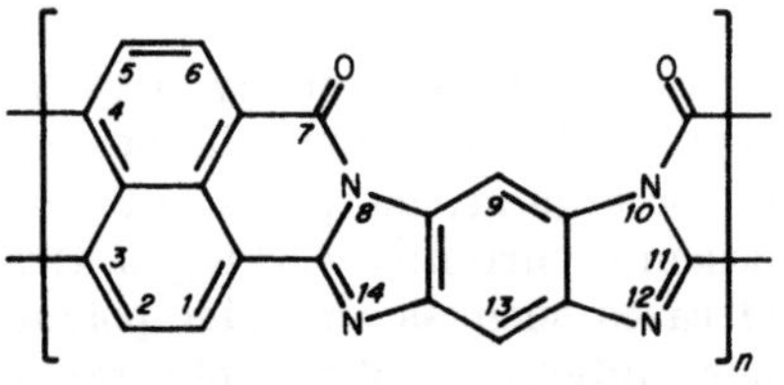

poly[(7-oxo-7*H*,10*H*-benz[*de*]imidazo[4′,5′:5,6]benzimidazo – [2,1-a]isoquinoline-3,4:10,11-tetrayl)-10-carbonyl]

This regular ladder polymer is an example of a double-strand polymer whose molecules are formed by an uninterrupted sequence of rings having two or more atoms in common. Nomenclature rests upon the selection, orientation, and naming of a preferred CRU, usually a tetravalent group.

C. *Regular Spiro Organic Polymer*[6]

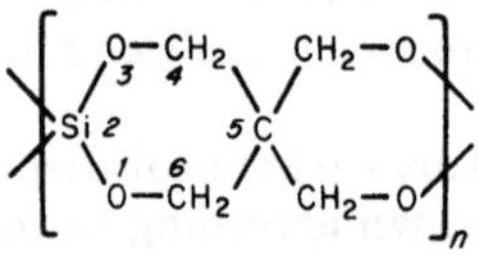

poly[1,3-dioxa-2-silacyclohexane-2,2:5,5-tetrayl-5,5-bis-(methyleneoxy)]

This regular spiro polymer is an example of a double-strand polymer consisting of an uninterrupted sequence of rings with adjacent rings having one atom in common.

D. *Regular Single-Strand Co-ordination Polymers*[7]

a.

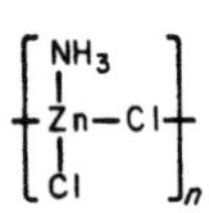

catena-poly[(amminechlorozinc)-μ-chloro]

b.

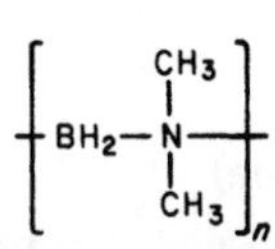

catena-poly[(dihydroboron)-μ-(dimethylamido)]

Similar to the single-strand organic polymers, a regular single-strand co-ordination polymer is represented by a preferred CRU in which both terminal constituent units are connected through single atoms to other identical CRUs or to an end group. The name of the CRU is formed by citing the names of its constituent units according to the established principles of inorganic and co-ordination nomenclature[8]. The prefix '*catena*' designates a linear polymer.

E. *Regular Quasi-Single-Strand Co-ordination Polymers*[7]

a.

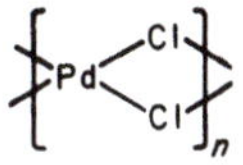

catena-poly[palladium-di-μ-chloro]

b.

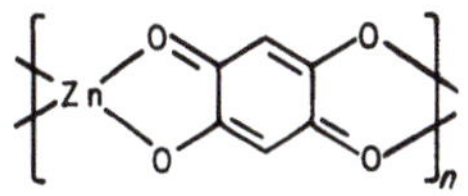

catena-poly[zinc-μ-[2,5-dihydroxy-*p*-benzoquinonato(2−)-$O^1,O^2{:}O^4,O^5$]]

A regular quasi-single-strand co-ordination polymer is represented by a preferred CRU in which only one terminal constituent unit is connected through a single atom to the other identical CRU or to an end group. It is named in the same manner as a regular single-strand polymer.

Stereospecific Polymers[9]

Structure-based nomenclature of regular polymers can denote stereochemical features if the CRU has one or more sites of defined stereoisomerism in the main chain of the polymer molecule.

A.

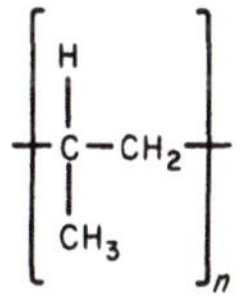

isotactic poly(propylene)

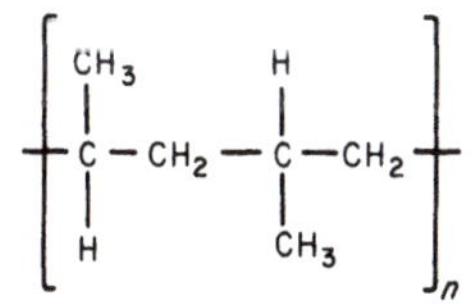

syndiotactic poly(propylene)

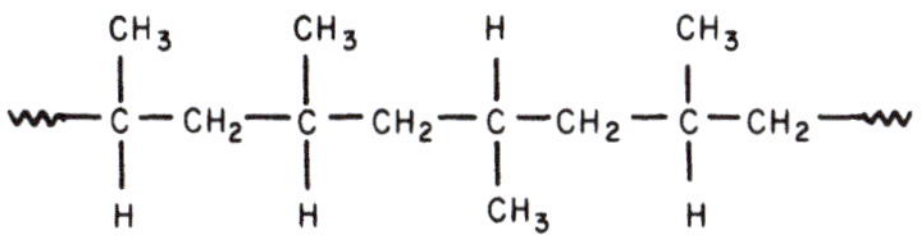

atactic poly(propylene)

Three steric arrangements are shown as Fischer projections for poly(propylene) with the corresponding descriptors 'isotactic', 'syndiotactic', and 'atactic'.

B.

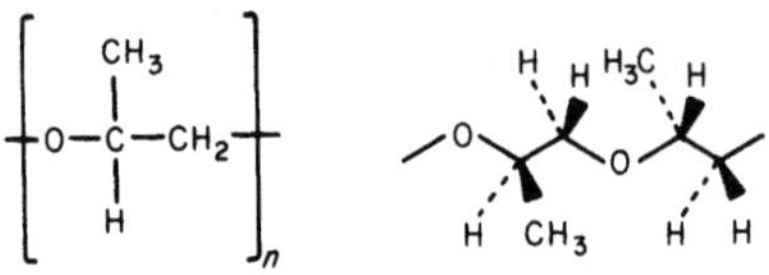

isotactic poly(oxypropylene)

C.

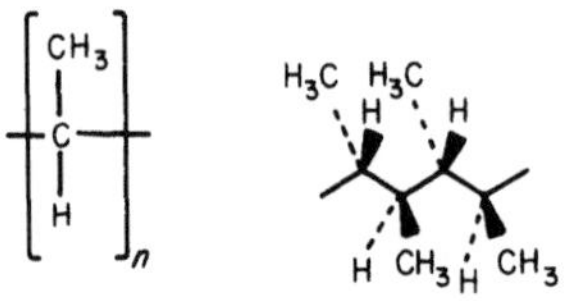

isotactic poly(ethylidene)

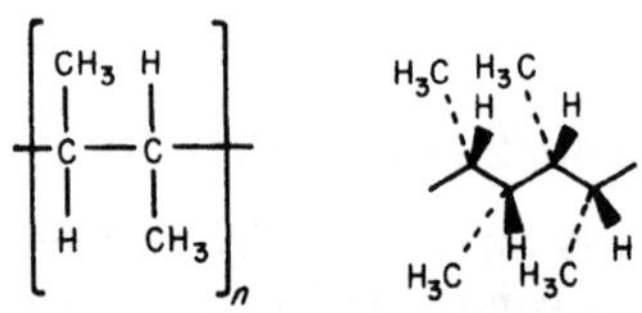

syndiotactic poly(ethylidene)

D.

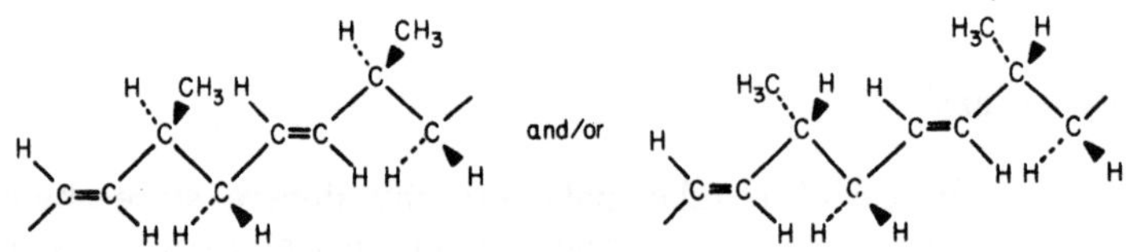

isotactic poly(3-methyl-*trans*-1-butenylene) or
transisotactic poly(3-methyl-1-butenylene)

E.

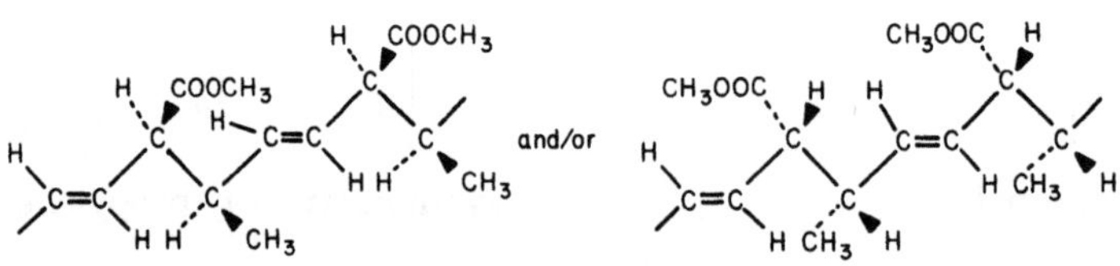

diisotactic poly[*threo*-3-(methoxycarbonyl)-4-methyl-*trans*-1-butenylene]
or
transthreodiisotactic poly[3-(methoxycarbonyl)-4-methyl-1-butenylene]

A diisotactic polymer is an isotactic polymer that contains two chiral or prochiral atoms with defined stereochemistry in the main chain of the CRU. The relative configuration at two contiguous carbon atoms in the main chain bearing substituents a and b is designated by the prefix 'erythro' or 'threo', as appropriate, by analogy with the terminology for carbohydrate systems[10].

Source-Based Nomenclature for Copolymers[11]

Although the use of a comprehensive structure-based system of naming copolymers is preferred, the structural identity of all the constitutional units as well as their sequential arrangement are rarely available. Therefore, a source-based nomenclature system is more practical. The names of monomers, trivial, semi-systematic, or systematic are used to specify monomeric units.

A. *Unspecified Copolymer*

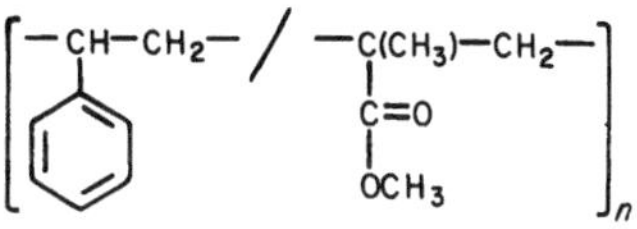

poly[styrene-*co*-(methyl methacrylate)]
copoly(styrene/methyl methacrylate)

The first name reflects the main system which meets the requirement of providing a name for any copolymer, however complex. The second name reflects a more concise alternative system.

B. *Statistical Copolymer*

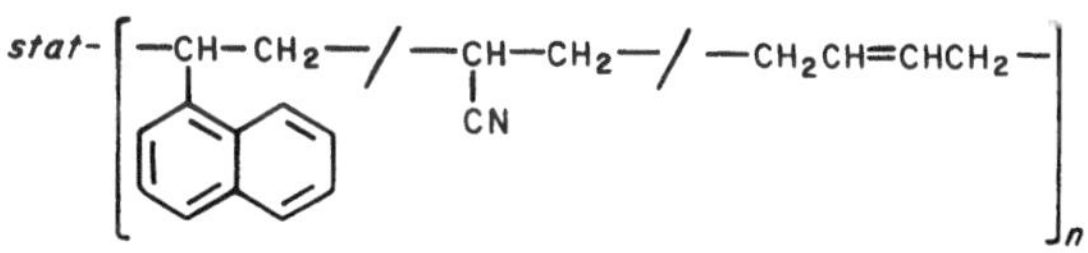

poly(1-vinylnaphthalene-*stat*-acrylonitrile-*stat*-butadiene)
stat-copoly(1-vinylnaphthalene/acrylonitrile/butadiene)

The term '*stat*' denotes a statistical sequence arrangement of monomeric units.

C. *Block Copolymer*

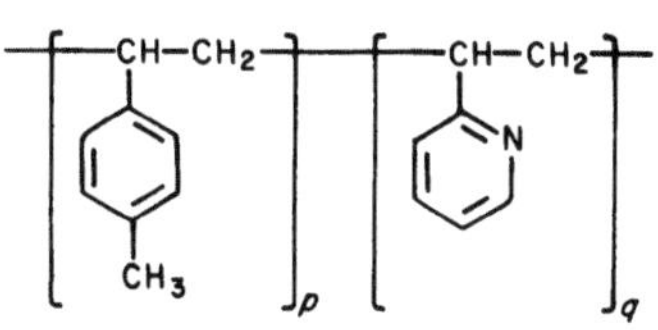

poly(*p*-methylstyrene)-*block*-poly(2-vinylpyridine)
block-copoly(*p*-methylstyrene/2-vinylpyridine)

The term '*block*' indicates the existence of a linear arrangement of blocks, each of which comprises units derived from a different species of monomer.

D. *Graft Copolymer*

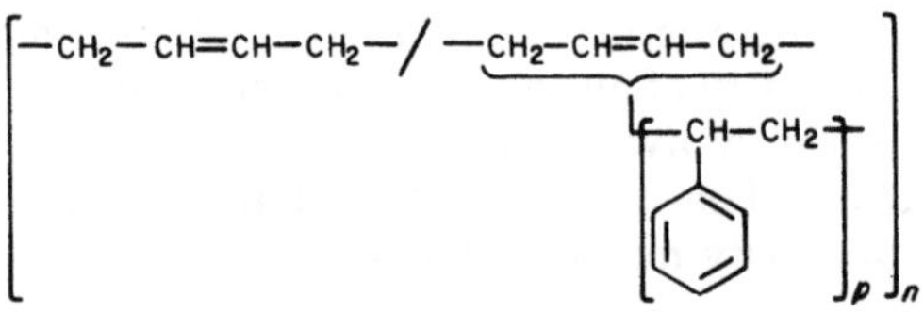

polybutadiene-*graft*-polystyrene
graft-copoly(butadiene/styrene)

The term '*graft*' denotes a polymer comprising molecules with one or more species of block connected to the main chain as side chains.

Structure-Based Nomenclature for Irregular Polymers[12]

Irregular polymers are polymers which cannot be described by only one species of CRU in a single sequential arrangement. This nomenclature system, now in the final stages of review by the Commission, takes care of the irregular polymers which have been partially modified, have specified arrangements of monomeric units, or reflect other cases for which the source-based system is inadequate.

A. *Styrene-vinyl chloride polymer joined head to tail*

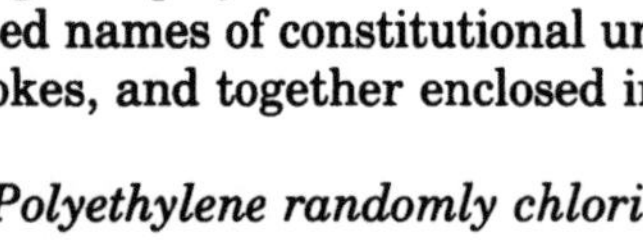

poly(1-phenylethylene/1-chloroethylene)

Irregular polymers are named by placing the prefix 'poly' before the structure-based names of constitutional units. The individual units are separated by oblique strokes, and together enclosed in parentheses or brackets.

B. *Polyethylene randomly chlorinated*

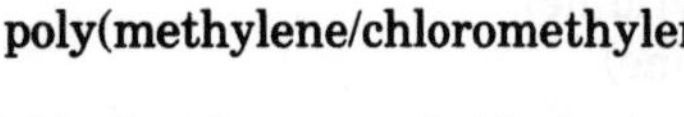

poly(methylene/chloromethylene/dichloromethylene)

C. *Biblock polymer with blocks joined by a specific group*

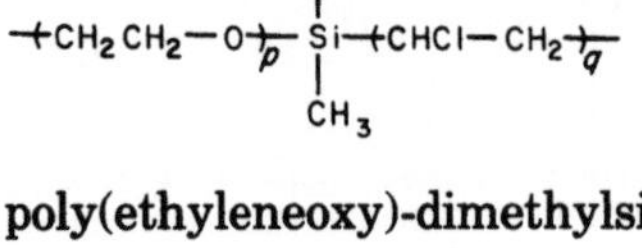

poly(ethyleneoxy)-dimethylsilylene-poly(1-chloroethylene)

D. *Star copolymer*

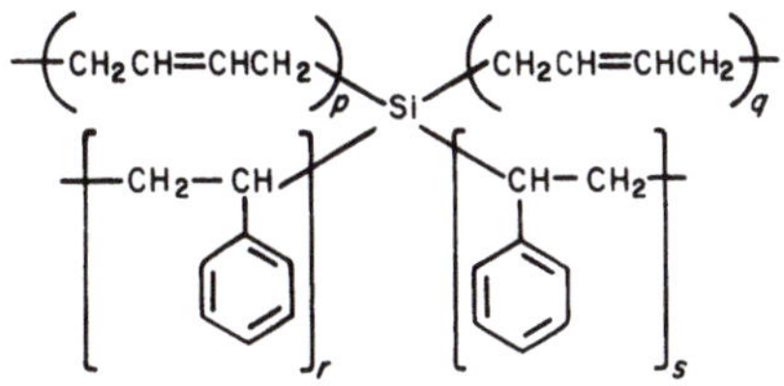

bis[poly(2-butenylene)]bis[poly(1-phenylethylene)]silane

Compendium of Macromolecular Nomenclature[13]

The Compendium (The Purple Book) is now in press and is expected to be published in early 1991. It contains nine previously published recommendations including those listed in the References Section of this paper[1,4,7,9,11,14,15], a history of the Commission since 1968, a list of members of the Commission, an introduction to macromolecular nomenclature, and a bibliography of biopolymer nomenclature recommendations.

INTERNATIONAL UNION OF PURE AND APPLIED CHEMISTRY – INTERNATIONAL UNION OF BIOCHEMISTRY (IUPAC-IUB)

The International Union of Biochemistry (IUB) was founded for the purpose of advancing biochemistry internationally by supporting the International Congress of Biochemistry, by co-ordinating research, discussion, and publications in the field of biochemistry, and by establishing co-operation among biochemical societies in adhering countries.

Responding to the concern for nomenclature, an IUPAC-IUB Combined Commission on Biochemical Nomenclature (CBN) was created in the 1960s which later became the IUPAC-IUB Joint Commission on Biochemical Nomenclature (JCBN). The goal of that Commission and of the IUB Nomenclature Committee (NC-IUB) is to facilitate communication in biochemistry by making recommendations on the nomenclature and symbolic representations of natural products, including biopolymers, and of related compounds such as synthetic analogues, and by ensuring that words and symbols mean the same to different users.

Biopolymers

One of the most important areas of biochemistry for which special symbols are essential is that of biopolymers. It is almost impossible to represent the name of even a simple protein, polynucleotide, or polysaccharide except by the use of logical and universally accepted abbreviations. There are three main series of symbols for monomeric units, *viz.* those for amino acids, mononucleosides, and monosaccharides, of which the amino acid series is the oldest. The monomeric units are generally designated by three-letter symbols: a capital letter followed by two lower-case letters.

A standard treatment has been devised for the three groups of macromolecules which are built up from these units. Where the sequence of residues is known, the symbols are written in order and joined by short lines (dashes, hyphens). Where the sequence of residues is not known, the group of symbols, separated by commas, is inclosed in parentheses. Example: Ala-Gly-(Met,Pro)-Lys means that the sequence of methionine and proline is unknown.

Macromolecules composed of repeating sequences may be represented by the prefix 'poly' or the subscript '*n*', both indicating 'polymer of'. The symbols for the monomeric units of the sequence are enclosed in parentheses.

In several cases it is thought desirable to recommend for the same substance two different forms of abbreviation or symbol, one or the other of which is more convenient for specific purposes.

Abbreviated Nomenclature of Synthetic Polypeptides[16]

A. *Homopolymer* (condensation)

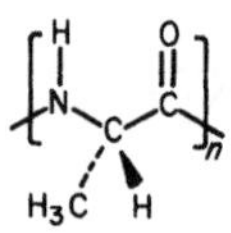

$(Ala)_n$
poly(Ala)
poly(L-alanine)

The prefix 'poly' or the subscript '*n*' indicates 'polymer of'. The L- is implied by the three-letter symbols for amino acids.

B. *Linear Copolymer* (unknown sequence)

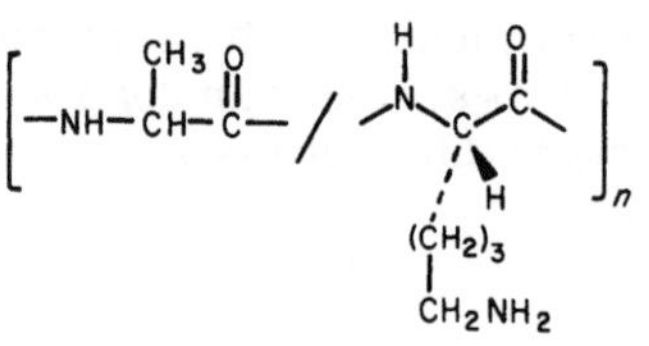

$(DLAla,Lys)_n$
poly(DLAla,Lys)
poly(DL-alanine,L-lysine)

The comma between the symbols of the amino acid residues indicates an unknown sequence. The relative composition of amino acids may be specified, if known, in terms of molar percentage[16].

C. *Linear Copolymer* (regular alternating sequence)

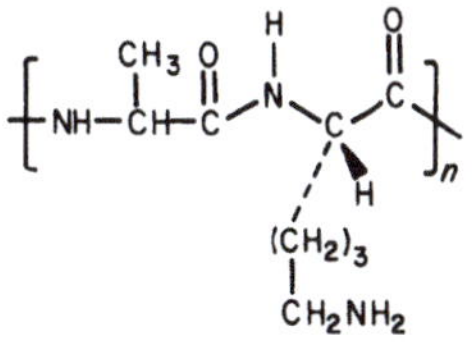

(DLAla-Lys)$_n$
poly(DLAla-Lys)
poly(DL-alanine-L-lysine)

The hyphen between the symbols for the residues indicates a known sequence.

D. *Block Copolymer*

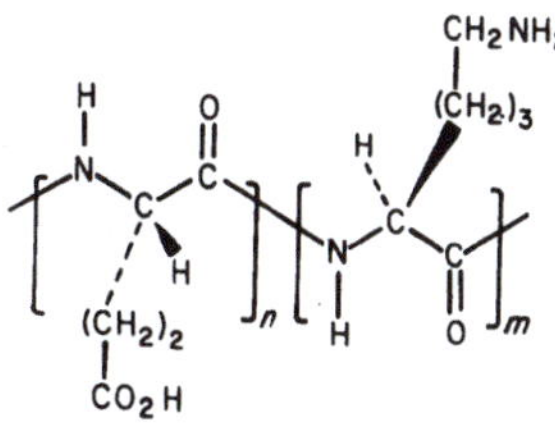

(Glu)$_n$-(Lys)$_m$
poly(Glu)$_n$-poly(Lys)$_m$
poly(L-glutamic acid)-poly(L-lysine)

This is a block polymer combined through the α-COOH terminus of poly(Glu) to the α-NH_2 terminus of poly(Lys). The percent composition of the blocks, if known, may be superscripted to the right of the residue symbols.

E. *Graft Copolymer*

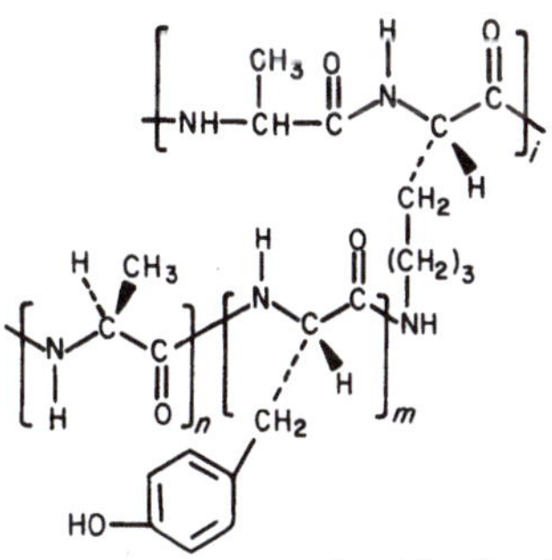

(DLAla-Lys)$_i$
(Ala)*n*-(Tyr)*m* ——┘

poly(DLAla-Lys)
poly(Ala)-poly(Tyr) ——┘

poly(DLAla-Lys) poly(Ala)-poly(Tyr) ——┐
(bracket joining Lys to the end)

poly(DLAla-Lys) — poly(Ala)-poly(Tyr)

This is a graft polymer with the main chain of a regular alternating sequence of DL-alanine and L-lysine connected through the ϵ-NH_2 group of lysine to the α-COOH group of L-tyrosine in the side chain, which consists of a block polymer of L-tyrosine and L-alanine. Branches (side chains) connected to the main chain can be designated in one of several ways: by a vertical line joining the main chain and the branch (side chain); by an extended bond joining the appropriate residues with the main chain written first. If the origin is unknown, the line originates at the 'p' in 'poly', if 'poly' is used, or at the first parenthesis (bracket), if the subscript '*n*' is used. The use and limitations of the double dash and further refinements in conventions are given in the IUB rules[16,17].

Table 1. Symbols for Amino Acids and Peptides[17]

One-Letter	Three-letter	Amino acid	One-Letter	Three-letter	Amino acid
A	Ala	alanine	N	Asn	asparagine
B	Asx	aspartic acid	P	Pro	proline
		or asparagine	Q	Gln	glutamine
C	Cys	cysteine	R	Arg	arginine
D	Asp	aspartic acid	S	Ser	serine
E	Glu	glutamic acid	T	Thr	threonine
F	Phe	phenylalanine	V	Val	valine
G	Gly	glycine	W	Trp	tryptophan
H	His	histidine	X	Xaa	unknown or
I	Ile	isoleucine			'other'
K	Lys	lysine			amino acid
L	Leu	leucine	Y	Tyr	tyrosine
M	Met	methionine	Z	Glx	glutamic acid
					or glutamine

Abbreviations and Symbols for Nucleic Acids, Polynucleotides, and Polymerised Nucleotides[18]

A. *Copolymer* (alternating)

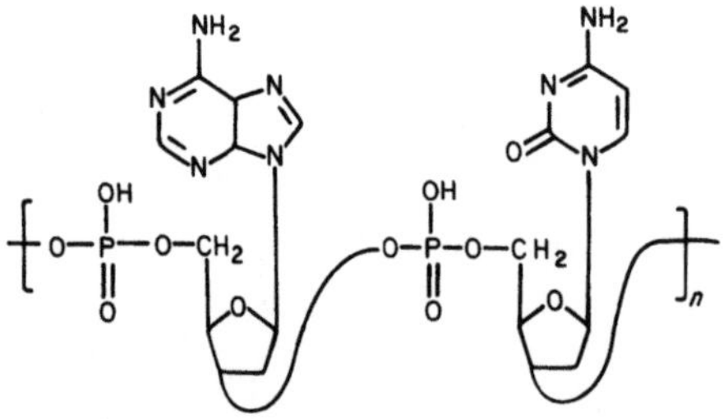

poly[d(A-C)]
poly(dA-dC)
poly(deoxyadenylate-deoxycytidylate)

The hyphen used in known sequences is a contraction of the arrow (→) that is understood to indicate 3' to 5' linkages. Subscripts denoting length, e.g., a number, an average or range may be added if desired. The use of the centre dot · to indicate a non-covalent association between two or more polynucleotide chains, such as that ascribed to hydrogen-bonding, and the use of the comma, indicating unknown association, are documented in the IUB rules[19].

Table 2. Single-Letter Code Recommendations for Nucleic Acid Sequences[19]

Symbol	Meaning	Origin	Symbol	Meaning	Origin
G	G	Guanine	W	A or T	Weak interaction (2 H bonds)
A	A	Adenine			
T	T	Thymine	H	A or C or T	not-G, H follows G in the alphabet
C	C	Cytosine			
R	G or A	puRine	B	G or T or C	not A, B follows A
Y	T or C	pYrimidine	V	G or C or A	not-T (not-U), V follows U
M	A or C	aMino			
K	G or T	Keto	D	G or A or T	not-C, D follows C
S	G or C	Strong interactions (3 H bonds)	N	G or A or T or C	aNy

The symbols are intended to be applicable to both deoxyribonucleic and ribonucleic acids. In circumstances where confusion between DNA and RNA is likely the sequence may be prefixed with the lower-case letter d or r. Similarly, no distinction will be made in the symbols between base, nucleoside, and nucleotide. There are also corresponding three-letter symbols for the purines, pyrimidines, and nucleosides[19]. The one-letter symbols were designed for polymer representation. The three-letter symbols should be used whenever chemical changes involving nucleosides or nucleotides are being discussed.

Polysaccharide Nomenclature[20]

A. *Homopolysaccharide* (Homoglycan)

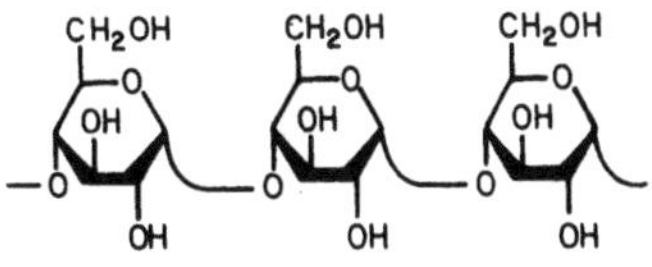

(1→4)-α-D-Glc*p*-(1→4)-α-D-Glc*p*-(1→4)-α-D-Glcp-(1→
(1→4)-α-D-Glucopyranan (amylose)

Polysaccharides composed of only one kind of monosaccharide are homopolysaccharides (homoglycans). In general homoglycans will be named, as a general term, by replacing the ending 'ose' of the sugar by the suffix 'an'. The linkage designation shows the carbon atoms involved in the glycosidic bonds, and precedes the symbols designating the configuration of the sugar. Further conventions for naming branched, cyclic and heteroglycans are given in the IUPAC-IUB recommendations for oligosaccharides and polysaccharides[20,21].

Table 3. Abbreviated Names for Use for Oligosaccharides and Polysaccharides[21]

Symbol	Name	Symbol	Name
All	Allose	Lyx	Lyxose
Alt	Altrose	Rib	Ribose
Gal	Galactose	Xyl	Xylose
Glc	Glucose	Rha	Rhamnose
Gul	Gulose	Fuc	Fucose
Ido	Idose	Fru	Fructose
Man	Manose	Neu	Neuraminic acid
Tal	Talose	Mur	Muramic acid
Ara	Arabinose		

The symbols represent the individual sugars or their residues. The use of the symbol to represent the free sugar is not recommended in textual material. The ring form is indicated by using the first letter of furanose, pyranose, or septanose, italicised and uncapitalised, added after the abbreviated name of the monosaccharide. No abbreviations are recommended for deoxy sugars, except for deoxyribose which is abbreviated dRib.

Glycoproteins

A glycoprotein is a compound containing carbohydrate covalently linked to protein. IUPAC-IUB recommendations for representing glycoproteins, glycopeptides, and peptidoglycans have been issued[22].

CHEMICAL ABSTRACTS SERVICE

Chemical Abstracts Service (CAS) since 1967 has represented polymers in the CAS Registry System, and provided corresponding CA Index names[23-26], in two ways:

1. in terms of the component monomer(s) from which they are prepared (called monomer-based registrations); and
2. in terms of the final structure of the resulting polymer (referred to as structural repeating unit (SRU) which is equivalent to the term constitutional repeating unit (CRU) preferred by the IUPAC Commission).

The SRU representations exist only for those relatively few cases where the polymer structure is well documented by the author, is known, or can safely be assumed. The cases for which SRUs are assumed are limited to such condensation polymers as polyamides, polyesters, polyurethanes, and polycarbonates, prepared from a difunctional monomer or two symmetrical comonomers. SRUs are not assumed for condensation polymers prepared from unsymmetrical monomers or from more than two monomers, or for addition polymers.

Current CAS Practices

Each example below contains the following information:

1. Type of polymer.
2. Trivial name as encountered in the literature.
3. CAS Registry Number of the monomer-based registration.
4. Structural diagram (monomer(s) enclosed in brackets with a subscript '*x*').
5. CA Index name (in terms of component monomer(s)).

Examples A, C, and D contain in addition:

6. CAS Registry Number of the corresponding SRU.
7. Structural diagram of the SRU (enclosed in brackets with a subscript '*n*').
8. CA Index name of the SRU.

A. *Homopolymer* (condensation): Nylon 11

a. 25587-80-8

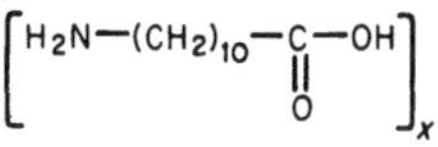

Undecanoic acid, 11-amino-, homopolymer
and
b. 25035-04-5

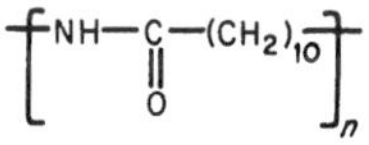

Poly[imino(1-oxo-1,11-undecanediyl)]

B. *Homopolymer* (addition): poly(*p*-chlorostyrene)

24991-47-7

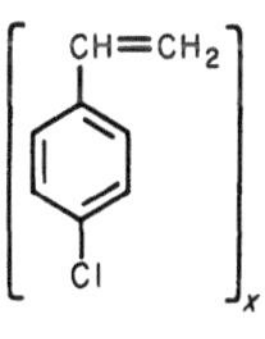

Benzene, 1-chloro-4-ethenyl-, homopolymer

C. *Copolymer* (condensation, two symmetrical monomers): poly(neopentylene adipate)

a. 27925-07-1

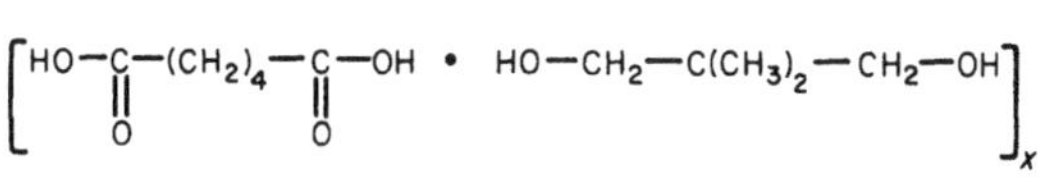

Hexanedioic acid, polymer with 2,2-dimethyl-1,3-propanediol
and

b. 28039-87-3

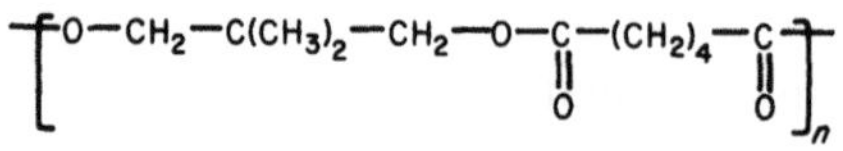

Poly[oxy(2,2-dimethyl-1,3-propanediyl)oxy(1,6-dioxo-1,6-hexanediyl)]

D. *Copolymer* (condensation, two symmetrical monomers):
poly(ether-ether-ketone) (PEEK)

a. 29658-26-2

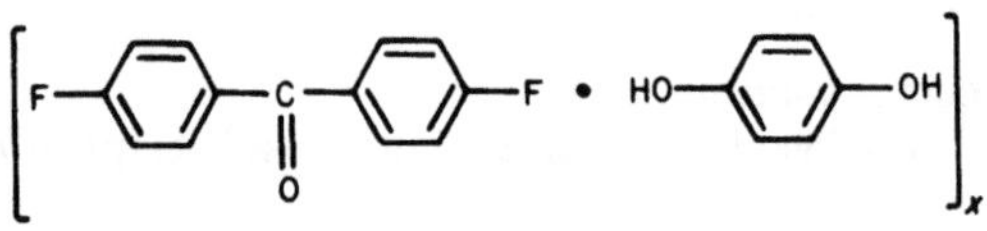

Methanone, bis(4-fluorophenyl)-, polymer with 1,4-benzenediol and

b. 31694-16-3

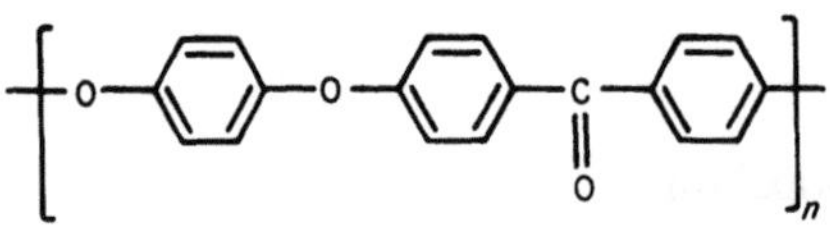

Poly(oxy-1,4-phenyleneoxy-1,4-phenylenecarbonyl-1,4-phenylene)

E. *Copolymer* (condensation, three monomers):
poly(butylene adipate sebacate)

63411-59-6

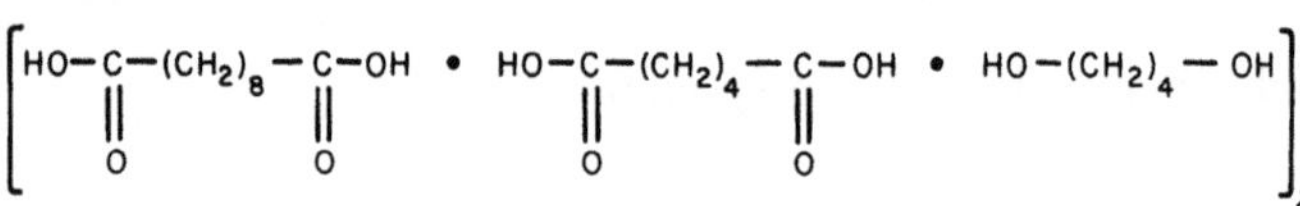

Decanedioic acid, polymer with 1,4-butanediol and hexanedioic acid

F. *Copolymer* (addition): styrene-vinyl acetate copolymer

25213-29-0

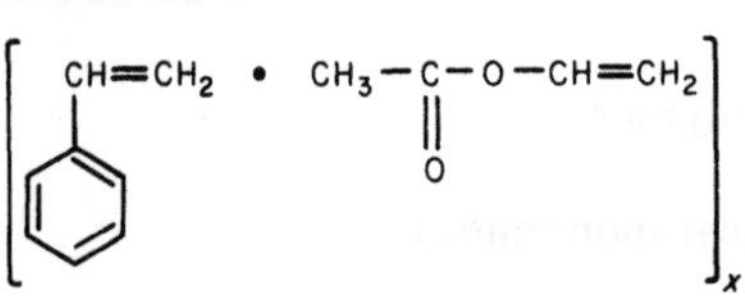

Acetic acid, ethenyl ester, polymer with ethenylbenzene

Enhancements Under Development at CAS

Organic Polymers. Work is in progress on the algorithmic generation of

1. Polymer class term.

2. Monomeric unit(s).
3. Structural (constitutional) repeating unit (SRU).

Each example below illustrates the assignment of polymer class term to the monomer-based registration, derivation of corresponding monomeric units, and generation of the complete SRU whenever applicable.

A. poly(*p*-phenylene succinamide)

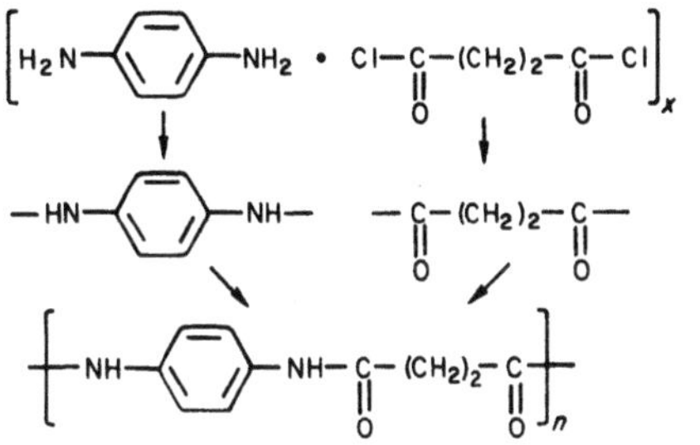

polyamide

monomeric units

SRU

B. poly(ethylene hexamethylene sebacate terephthalate)

polyester

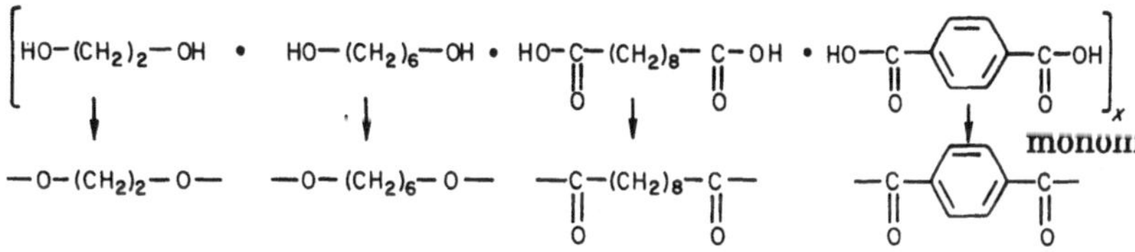

monomeric units

C. tetrafluoroethylene-trifluoroethylene-vinylidene fluoride copolymer

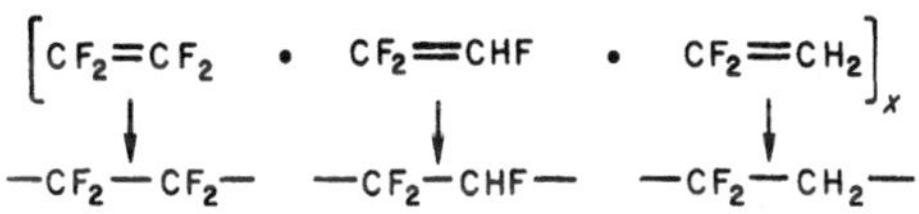

fluoropolymer

monomeric units

Biosequences

Searchable biosequence structures are being added to the CAS Registry File on STN International. Novel search and display techniques including annotations have been developed to permit online sequence searching.[27]

A. *Proteins* (without annotation)

```
    115537-36-5

    Amylase, α- (Schwanniomyces castellii clone pYcl-α reduced)

SEQ*      1 MRFSTEGFTS KVVAAILAFS RLVSAQPIIF DMRDVSSSAD KWKDQSIYQI
         51 VTDRFARSDG STTADCLVSD RKYCGGSYKG IIDKLDYIQG MGFTAIWISP
```

* IUPAC-IUB one-letter symbols for amino acids.

101 VVEQIPDNTA YGYAYHGYWM KNIDELNTNF GTADELKQLA SELHFRSMYD
151 GRRCYNHYAW NGDGSSVDYS SFTPFNQQSY FHDYCLITNY NDQTNVEDCW
201 EGDTEVSLPD LSTEDNEVIG VFQTWVSDFV QNYSIDGLRI DSAKHVDTAS
251 LTKFEDASGV YNLGEVYQGD PTYTCPYQNY MKGVTNYPLY YPVYRFFSDT
301 SATSSELTSM ISTLQSSCSD VSLLGNFIEN HDQVRFPSVT SDTSLIKNDM
351 AFIILGDGIP IIYYGQEQGL NGGSDPANRE ALWLSGYNTD SEYYELISKL
401 NQIRNEALWL SGYNTDSEYY ELISKLNQIR NQAIKKDSAY STYKSSVVSS
451 SDHYIATRKG SDANQLISIF NNLGSNGSQD ITVSNTGYSS GDKVIDIISC
501 NSVLAGDSGS LSVSISGGMP QVYAPSSVLS GSGICNG

SEQ3 **

1 Met-Arg-Phe-Ser-Thr-Glu-Gly-Phe-Thr-Ser-Lys-Val-Val-Ala-Ala-
16 Ile-Leu-Ala-Phe-Ser-Arg-Leu-Val-Ser-Ala-Gln-Pro-Ile-Ile-Phe-
31 Asp-Met-Arg-Asp-Val-Ser-Ser-Ser-Ala-Asp-Lys-Trp-Lys-Asp-Gln-
46 Ser-Ile-Tyr-Gln-Ile-Val-Thr-Asp-Arg-Phe-Ala-Arg-Ser-Asp-Gly-
61 Ser-Thr-Thr-Ala-Asp-Cys-Leu-Val-Ser-Asp-Arg-Lys-Tyr-Cys-Gly-
76 Gly-Ser-Tyr-Lys-Gly-Ile-Ile-Asp-Lys-Leu-Asp-Tyr-Ile-Gln-Gly-
91 Met-Gly-Phe-Thr-Ala-Ile-Trp-Ile-Ser-Pro-Val-Val-Glu-Gln-Ile-
106 Pro-Asp-Asn-Thr-Ala-Tyr-Gly-Tyr-Ala-Tyr-His-Gly-Tyr-Trp-Met-
121 Lys-Asn-Ile-Asp-Glu-Leu-Asn-Thr-Asn-Phe-Gly-Thr-Ala-Asp-Glu-
136 Leu-Lys-Gln-Leu-Ala-Ser-Glu-Leu-His-Phe-Arg-Ser-Met-Tyr-Asp-
151 Gly-Arg-Arg-Cys-Tyr-Asn-His-Tyr-Ala-Trp-Asn-Gly-Asp-Gly-Ser-
166 Ser-Val-Asp-Tyr-Ser-Ser-Phe-Thr-Pro-Phe-Asn-Gln-Gln-Ser-Tyr-
181 Phe-His-Asp-Tyr-Cys-Leu-Ile-Thr-Asn-Tyr-Asn-Asp-Gln-Thr-Asn-
196 Val-Glu-Asp-Cys-Trp-Glu-Gly-Asp-Thr-Glu-Val-Ser-Leu-Pro-Asp-
211 Leu-Ser-Thr-Glu-Asp-Asn-Glu-Val-Ile-Gly-Val-Phe-Gln-Thr-Trp-
226 Val-Ser-Asp-Phe-Val-Gln-Asn-Tyr-Ser-Ile-Asp-Gly-Leu-Arg-Ile-
241 Asp-Ser-Ala-Lys-His-Val-Asp-Thr-Ala-Ser-Leu-Thr-Lys-Phe-Glu-
256 Asp-Ala-Ser-Gly-Val-Tyr-Asn-Leu-Gly-Glu-Val-Tyr-Gln-Gly-Asp-
271 Pro-Thr-Tyr-Thr-Cys-Pro-Tyr-Gln-Asn-Tyr-Met-Lys-Gly-Val-Thr-
286 Asn-Tyr-Pro-Leu-Tyr-Tyr-Pro-Val-Tyr-Arg-Phe-Phe-Ser-Asp-Thr-
301 Ser-Ala-Thr-Ser-Ser-Glu-Leu-Thr-Ser-Met-Ile-Ser-Thr-Leu-Gln-
316 Ser-Ser-Cys-Ser-Asp-Val-Ser-Leu-Leu-Gly-Asn-Phe-Ile-Glu-Asn-
331 His-Asp-Gln-Val-Arg-Phe-Pro-Ser-Val-Thr-Ser-Asp-Thr-Ser-Leu-
346 Ile-Lys-Asn-Asp-Met-Ala-Phe-Ile-Ile-Leu-Gly-Asp-Gly-Ile-Pro-
361 Ile-Ile-Tyr-Tyr-Gly-Gln-Glu-Gln-Gly-Leu-Asn-Gly-Gly-Ser-Asp-
376 Pro-Ala-Asn-Arg-Glu-Ala-Leu-Trp-Leu-Ser-Gly-Tyr-Asn-Thr-Asp-
391 Ser-Glu-Tyr-Tyr-Glu-Leu-Ile-Ser-Lys-Leu-Asn-Gln-Ile-Arg-Asn-
406 Glu-Ala-Leu-Trp-Leu-Ser-Gly-Tyr-Asn-Thr-Asp-Ser-Glu-Tyr-Tyr-
421 Glu-Leu-Ile-Ser-Lys-Leu-Asn-Gln-Ile-Arg-Asn-Gln-Ala-Ile-Lys-
436 Lys-Asp-Ser-Ala-Tyr-Ser-Thr-Tyr-Lys-Ser-Ser-Val-Val-Ser-Ser-
451 Ser-Asp-His-Tyr-Ile-Ala-Thr-Arg-Lys-Gly-Ser-Asp-Ala-Asn-Gln-
466 Leu-Ile-Ser-Ile-Phe-Asn-Asn-Leu-Gly-Ser-Asn-Gly-Ser-Gln-Asp-
481 Ile-Thr-Val-Ser-Asn-Thr-Gly-Tyr-Ser-Ser-Gly-Asp-Lys-Val-Ile-
496 Asp-Ile-Ile-Ser-Cys-Asn-Ser-Val-Leu-Ala-Gly-Asp-Ser-Gly-Ser-
511 Leu-Ser-Val-Ser-Ile-Ser-Gly-Gly-Met-Pro-Gln-Val-Tyr-Ala-Pro-
526 Ser-Ser-Val-Leu-Ser-Gly-Ser-Gly-Ile-Cys-Asn-Gly

** IUPAC-IUB three-letter symbols for amino acids.

B. *Proteins* (with annotation)

RN	Registry Number	120249-58-3
FS	File Segment	Protein Sequences
PN	Protein Name	Insulin
OR	Organism Name	ox
CL	Clone Designation	none
SH	Sulfhydryl Group	– (indicates whether Cys residues are free or as disulfides)
GN	Gene Name	– (for nucleic acids)
SN	Substitution	8A-L-threonine-10A-L-isoleucine-21A-L-homoserine-27B-L-arginine-30B-L-threoninamide
SQL	Sequence Length	30,21
*NTE	Annotation	multichain modified (linear chains are assumed, cyclic peptides are so indicated)

type	from	to/span	description
terminal mod.	Thr-30		C-terminal amide
bridge	Cys-7	Cys-7′	disulfide bridge
bridge	Cys-19	Cys-20′	disulfide bridge
bridge	Cys-6′	Cys-11′	disulfide bridge
bridge	Ala-14	Ser-12′	dicarba bridge
uncommon	Hse-21′	–	
stereo	Pro-28	–	D
modification	Phe-1	–	boc
replacement	Ser-9	–	phospha

SEQ	1	FVDQHLCGSH	LVEALYLVCG	ERGFFYRPLT
SEQ	2	GIVEQCCTSI	CSLYQLEDYC	X

*This is a mock-up example to show various styles of NTE data. It does not represent a real sequence.

C. *Nucleic Acids* (current CA index name)

102036-84-0

Deoxyribonucleic acid,
*** d(A-G-A-C-C-A-A-T-C-C-T-A-A-C-C-A-A-T-G-T-C-T-G-G-T-T-A-A-G-A-T-G-G-T-C-C-A-A-T-C-C-C-G-A-A-A-C-T-T-C-T-A-G-T-T-G-C-G-G-T-T-C-G-A-A-G-A-A-G-C-C-A-G-A-A-T-G-T-T-T-C-T-G-A-A-A-G-T-T-T-C-A-G-A-A-A-A-T-T-C-T-A-G-T-T-T-T-G-A-G-A-T-T-T-T-C-A-G-A-A-G-T-A-C-G-G-C-A-T-G-A-T-G-A-T-G-C-A-T-A-A-C-A-A-G-G-A-C-T-T-T-C-T-C-G-A-A-A-G-T-A-C-T-A-T-A-T-T-G-C-T-C-C-T-C-T-A-C-A-T-C-A-T-T-T-T-A-A-A-T-A-C-C-C-C-A-T-G-T-G-T-C-C-T-T-T-G-A-A-G-A-C-A-C-A-T-C-A-C-A-G-A-A-A-G-A-A-G-T-G-A-A-G-G-C-A-T-C-G-T-T-A-G-C-A-G-T-T-T-T-G-T-A-G-A-T-T-C-A-A-C-C-T-C-A-A-T-T-T-G-C-A-G-A-G-T-T-A-C-G-T-T-C-T-A-A-T-A-T-A-T-T-T-A-C-A-C-A-A-G-A-C-T-G-A-C-C-C)

*** IUPAC-IUB single-letter code for nucleic acid sequences.

REFERENCES

1. IUPAC Commission on Macromolecular Nomenclature. 'Basic Definitions of Terms Relating to Polymers (Recommendations 1974)'. *Pure Appl. Chem.* **1974**, *40*, 477–491.
2. American Chemical Society Division of Polymer Chemistry, Committee on Nomenclature. 'A Structure-Based Nomenclature for Linear Polymers'. *Macromolecules* **1968**, *1*, 193–198.
3. IUPAC Commission on Macromolecular Nomenclature. 'Graphic Representation of Polymer Structures'. In preparation.
4. IUPAC Commission on Macromolecular Nomenclature. 'Nomenclature of Regular Single-Strand Organic Polymers (Recommendations 1975)'. *Pure Appl. Chem.* **1976**, *48*, 373–385.
5. IUPAC Commission on Nomenclature of Organic Chemistry. *Nomenclature of Organic Chemistry. Sections A, B, C, D, E, F and H*; Pergamon Press: Oxford, 1979. (The Blue Book).
6. IUPAC Commission on Macromolecular Nomenclature. 'Nomenclature of Regular Double-Strand (Ladder and Spiro) Organic Polymers'. In preparation.
7. IUPAC Commission on Macromolecular Nomenclature and Commission on Nomenclature of Inorganic Chemistry. 'Nomenclature for Regular Single-Strand and Quasi Single-Strand Inorganic and Coordination Polymers (Recommendations 1984)'. *Pure Appl. Chem.* **1985**, *57*, 149–168.
8. IUPAC Commission on Nomenclature of Inorganic Chemistry. *Nomenclature of Inorganic Chemistry. Recommendations 1990*; Blackwell: Oxford, 1990. (The Red Book).
9. IUPAC Commission on Macromolecular Nomenclature. 'Stereochemical Definitions and Notations Relating to Polymers (Recommendations 1980)'. *Pure Appl. Chem.* **1981**, *53*, 733–752.
10. IUPAC Commission on Nomenclature of Organic Chemistry, and IUPAC-IUB Commission on Biochemical Nomenclature. 'Tentative Rules for Carbohydrate Nomenclature, Pt. 1'. *Biochemistry* **1971**, *10*, 3983–4004; errata: 4995.
11. IUPAC Commission on Macromolecular Nomenclature. 'Source-Based Nomenclature for Copolymers (Recommendations 1985)'. *Pure Appl. Chem.* **1985**, *57*, 1427–1440.
12. IUPAC Commission on Macromolecular Nomenclature. 'Structure-Based Nomenclature for Irregular Polymers'. In preparation.
13. IUPAC Commission on Macromolecular Nomenclature. *Compendium of Macromolecular Nomenclature*; Blackwell: Oxford, 1991. In press. (The Purple Book).
14. IUPAC Commission on Macromolecular Nomenclature. 'Use of Abbreviations for Names of Polymeric Substances (Recommendations 1986)'. *Pure Appl. Chem.* **1987**, *59*, 691–693.
15. IUPAC Commission on Macromolecular Nomenclature. 'A Classification of Linear Single-Strand Polymers (Recommendations 1988)'. *Pure Appl. Chem.* **1989**, *61*, 243–254.
16. IUPAC-IUB Commission on Biochemical Nomenclature. 'Abbreviated Nomenclature of Synthetic Polypeptides (Polymerised Amino Acids) (Revised Recommendations 1971)'. *Eur. J. Biochem.* **1972**, *26*, 301–304; *Pure Appl. Chem.* **1973**, *33*, 437–444.
17. IUPAC-IUB Joint Commission on Biochemical Nomenclature. 'Nomenclature and Symbolism for Amino Acids and Peptides (Recommendations 1983)'. *Eur. J. Biochem.* **1984**, *138*, 9–37; correction: **1985**, *152*, 1; *Pure Appl. Chem.* **1984**, *56*, 595–624.
18. IUPAC-IUB Commission on Biochemical Nomenclature. 'Abbreviations and Symbols for Nucleic Acids, Polynucleotides and Their Constituents (Recommendations 1970)'. *Eur. J. Biochem.* **1970**, *15*, 203–208; corrections: **1971**, *18*, 558; **1972**, *25*, 1; *Pure Appl. Chem.* **1974**, *40*, 277–290.
19. IUB Nomenclature Commission. 'Nomenclature for Incompletely Specified Bases in Nucleic Acid Sequences (Recommendations 1984)'. *Eur. J. Biochem.* **1985**, *150*, 1–5.
20. IUPAC-IUB Joint Commission on Biochemical Nomenclature. 'Polysaccharide Nomenclature (Recommendations 1980)'. *Eur. J. Biochem.* **1982**, *126*, 439–441; *Pure Appl. Chem.* **1982**, *54*, 1523–1526.
21. IUPAC-IUB Joint Commission on Biochemical Nomenclature. 'Abbreviated Terminology of Oligosaccharide Chains (Recommendations 1980)'. *Eur. J. Biochem.* **1982**, *126*, 433–437; *Pure Appl. Chem.* **1982**, *54*, 1517–1522.

22. IUPAC-IUB Joint Commission on Biochemical Nomenclature. 'Nomenclature of Glycoproteins, Glycopeptides and Peptidoglycans (Recommendations 1985)'. *Eur. J. Biochem.* **1986**, *159*, 1–6; correction: **1989**, *185*, 485; *Pure Appl. Chem.* **1988**, *60*, 1389–1394.
23. Loening, K. L.; Metanomski, W.; Powell, W. H. 'Indexing of Polymers in Chemical Abstracts'. *J. Chem. Doc.* **1969**, *9*, 248–251.
24. Langstaff, E. M.; Ostrum, G. K. 'Access to Polymer Information in Chemical Abstracts'. *J. Chem. Inf. Comput. Sci.* **1979**, *19*, 60–64.
25. *Chemical Abstracts Index Guide*; Chemical Abstracts Service: Columbus, Ohio, 1990; Appendix IV, 222.
26. *Searching for Polymer Information in CAS ONLINE*; Chemical Abstracts Service: Columbus, Ohio, 1990.
27. *Protein Sequence Searching in the Registry File*; Chemical Abstracts Service: Columbus, Ohio, 1990.

SEARCHING A FULL GENERICS DATABASE

Valerie J. Gillet, Geoffrey M. Downs, John D. Holliday, Michael F. Lynch and Winfried Dethlefsen
Department of Information Studies, University of Sheffield, Western Bank, Sheffield, S10 2TN, England

ABSTRACT

A hierarchy of screening methods applicable to full generic structures, including generic radical terms, is described. The most general level of description is fragment screening, which includes atom and bond centred fragments, 'bubbled-up' from full generic structures, where the logical relationships between fragments are retained in MUST and POSS screens. Ring systems are characterised and represented as a section of the bit string in a similar logical procedure.

A second screening method which retains the topology of a generic structure as an AND/OR tree is the reduced graph method, where the ring and non-ring components of structures are distinguished. Non-ring components are further distinguished as aggregates of carbon atoms and non-ring nodes consisting of connected heteroatoms. The nodes of a reduced graph are further characterised by a hierarchy of descriptors, ranging from an indication of the status of structural features within nodes to the most detailed description, i.e., the constituent atoms for nodes derived from specific partial structures and parameter lists for nodes derived from generic partial structures.

Finally the integration of these two screening methods is described, i.e., the inclusion of fragment and ring screens within the nodes of a reduced graph.

INTRODUCTION

During the last decade much effort has been directed towards solving the problems associated with the storage and retrieval of generic chemical structures particularly at Sheffield and within other organisations including Derwent Publications Ltd. and Chemical Abstracts Service[1–12]. This paper discusses recent progress at Sheffield towards a complete solution to full structure searching and includes details of: improved screening techniques, by the integration of several different approaches (including fragment screening, ring screening and reduced graphs); the refined level of search which allows structure matches to be determined; and the implementation of user-defined levels of searching.

The difficulties posed by generic structures have been identified many times and are included here for the sake of completeness. A typical generic structure is characterised by an invariant part together with associated variable groups. The variables are given as lists of alternatives expressed as either partial structure diagrams, line formulae, specific or generic radical names, or textual non-structural information such as *electron withdrawing group*. The variables may be attached at fixed or variable positions and may be substituted to any level of

W. A. Warr (Ed.)
Chemical Structures 2

nesting. Groups may occur with varying incidences, e.g., repeated units, and groups may have independent existence or combine together to form rings.

The different types of variation which characterise generic structures can be summarised as:

- Substituent variation. Variation amongst a list of alternative groups, any of which is possible at that position, e.g., R1 = methyl, ethyl or propyl.
- Position variation. Variation in the mutual positions of attachment of partial structures, e.g., monochlorophenyl.
- Frequency variation. Variation of the possible frequencies of partial structures, e.g., $-(CH_2)_{1-3}Cl$.
- Homology variation. Variation within a series of specific structures which are homologues to one another. The series may be finite or infinite and is defined by a set of one or more structural properties common to all specifics contained in the series, e.g., alkyl 1-4C, halogen.

The search challenges of generics arise from their size and great variability, (it is easy to find patents which include numbers of substances which exceed the number currently stored in the CA Registry System), and from the use of generic nomenclature, such as alkyl, where descriptions are given in intensional form, e.g., as those structural features which may or may not be present. These factors lead to the requirement for novel techniques, particularly for the handling of generic nomenclatural expressions.

The aim of the GENSAL project from the outset has been to allow the distinction between specifically and generically described radicals to be removed so that the correspondence between a generic radical and one of the members of the class can be identified. For example, a searcher may be interested in structures where a substituent is any straight chain or branched alkyl radical with between one and three atoms. The equivalence of this expression with the substructures *methyl, ethyl, propyl* and *isopropyl* should be recognised by the search system without need for class enumeration, either manually or automatically. In addition, there is also a requirement for more complex levels of search, for instance, a searcher may be interested in retrieving all file structures up to the level of generalisation of *t-alkyl* in response to *t-butyl* in the query structure, and is not at all interested in the more general *alkyl* or *hydrocarbyl*.

Table 1. Parameters

A	Total atom count
C	Total carbon count
T	Acyclic ternary branch count
Q	Acyclic quaternary branch count
E	Alkene unsaturation count
Y	Alkyne unsaturation count
RC	Ring count
RN	Ring atom count
RS	Ring substitution count
RF	Ring fusion count
RA	Normalised ring count
RZ	Ring heteroatom count
Z	Total heteroatom count

The form of representation of generic nomenclatural expressions is crucial in providing solutions at this level of complexity as will be seen later. At Sheffield, generic partial structures are represented internally as a series of parameters where each parameter has an associated range of possible values, with default values which can be overridden at input. The parameters were initially defined as shown in Table 1 and as an example, the default values for alkyl are:

$$A\langle 1\text{-}\rangle C\langle 1\text{-}\rangle T\langle 0\text{-}\rangle Q\langle 0\text{-}\rangle E\langle 0\rangle Y\langle 0\rangle RC\langle 0\rangle \ldots RZ\langle 0\rangle Z\langle 0\rangle .$$

The searching algorithms discard the nomenclatural term and perform operations on the parameter values alone. User-defined levels of search can be implemented by simple operations on these integer ranges as will be described later. These parameter values are referred to as *Enumerative Parameter Values (EPV's)*.

The first requirement, then, of a system for generics, is a flexible language for handling the structures. This is provided by the structure input language GENSAL which represents generic structures in a complete and unambiguous topological form, and incorporates all of the features described above. GENSAL uses a combination of partial structure diagrams, specific nomenclatural terms, and generic radical terms in which those structural attributes which are given in terms of numbers or ranges of structural features are recorded as parameters. A GENSAL description is shown in Figure 1. The GENSAL interpreter program

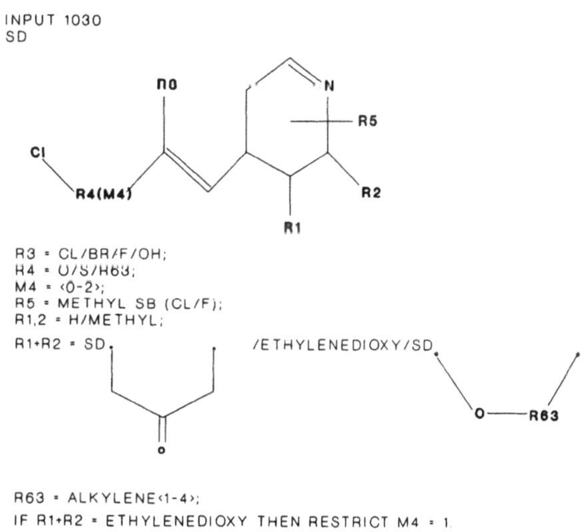

Figure 1. GENSAL description of a generic structure

analyses the input syntactically and semantically and generates the internal representation, the Extended Connection Table Representation (ECTR). The ECTR preserves the structural and logical relationships represented in GENSAL as an AND/OR tree with the leaf nodes of the tree representing partial structures as connection tables, or parameter lists, or textual non-structural information, and the branches in the tree reflecting the logical relations between the partial structures. The ECTR of a generic structure is given in Figure 2 and the information information contained within the ECTR is summarised as:

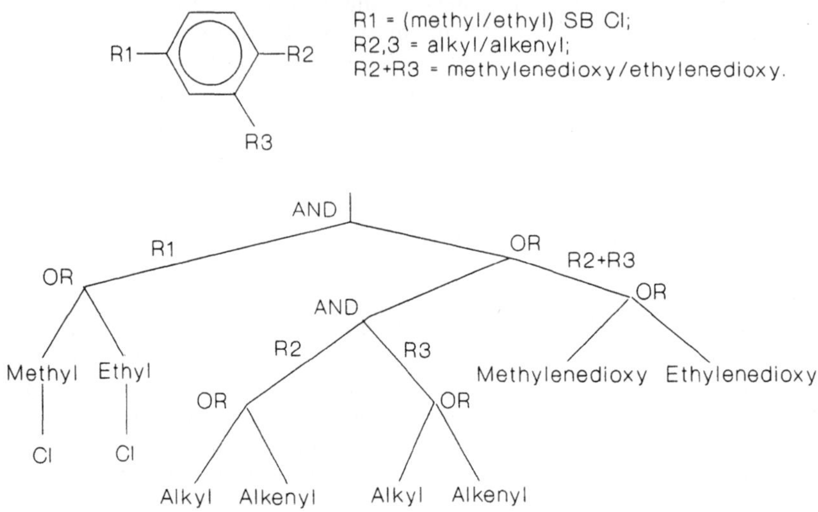

Figure 2. The ECTR as an AND/OR tree

- Structural Information. This is the information contained in the partial structures, in the form of partial connection tables and parameters representing specific and generic radicals respectively.
- Positional/Multiplicity Information. Fixed and variable attachment positions between two partial structures, and the frequency of occurrence of partial structures, are represented without enumeration.
- Logical Information. The logical relationships between partial structures are fully reflected in the ECTR with partial structures that may occur together given in AND relationship, and partial structures which are alternative to one another given in OR relationship.

This flexible ECTR provides a powerful structure both for representation of complex features of generics, such as combined substituents, and for development of searching algorithms and general utilities, e.g., the bubble-up process.

THE BUBBLE-UP

The bubble-up process is a general method of accumulating information within the logical framework of the ECTR. It is a method that can be applied regardless of the nature of the information required and is used to accumulate ring and fragment screens to represent entire structures as well as for the descriptors which represent the nodes of a reduced graph. Bubble-up is a two stage process whereby local information is determined within a partial structure as the ECTR is traversed from top to bottom, and then the information is accumulated by successive operations of union and intersection, at AND and OR nodes respectively, on the way back up. The actual operations vary according to how the information is represented, e.g., the union of bit strings is achieved by the logical inclusive OR whereas the union of integer ranges is achieved by the addition of the individual elements across the ranges. Bubble-up over the whole ECTR accumulates information in a way which is independent of the partitioning of the generic structure (where partitioning refers to the way in which a structure is split into partial structures). The bubble-up process is illustrated in Figure 3 for the accumulation of a molecular formulae range for a generic structure.

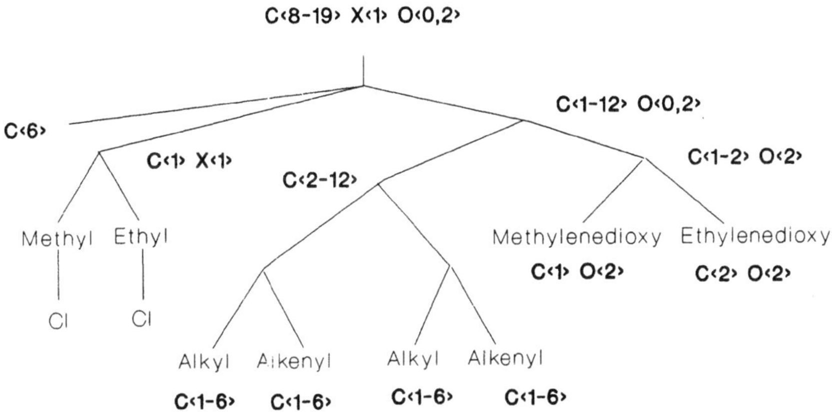

Figure 3. Bubble-up of molecular formulae ranges for a generic structure

SCREENING TECHNIQUES

The size and complexity of generic structures leads to the requirement for more elaborate screening systems than for specific structures. The different approaches which have been implemented at Sheffield include fragment and ring screens and also reduced graphs of varying discriminatory power.

The fragment generator records the presence or absence of certain discriminatory atom and bond sequences and augmented atoms. Ring screens complement the fragment screens and provide information specifically about the cyclic parts of a structure. A new approach to ring screening has been developed to accommodate the complex environment of generic structures[13]. The ring screens categorise the rings in the structure according to size and composition and include the number and types of heteroatoms and the number of fusion points for each ring.

In a generic structure there is usually some part which is common to all the alternative structures which are covered by the generic and this is referred to as the invariant part, whereas the parts which are not common to all the alternatives are the variant parts. Fragment and ring descriptors are therefore split into those wholly occurring within the invariant part, i.e., the descriptors which *must* be present, and those occurring at least partly within the variant part, i.e., the descriptors which *may* be present[5]. Two bit vectors are used to reflect this differentiation. The first vector represents the MUST descriptors and the second represents the POSS descriptors (i.e., the union of the must and may descriptors). The bubble-up process is used to accumulate ring and fragment descriptors with the correct logic in order that they are correctly assigned as MUST or POSS.

Reduced graphs represent a different kind of screen from the more conventional fragment screens and so provide an orthogonal search strategy. They represent a gross simplification of structures which retains the topology between components, whereas fragment screens indicate the presence or absence of detailed atom and bond fragments without any indication of the topology between them. Ring screens, like reduced graphs, are concerned with larger structural units, and characterise particular rings within a system, but again do not represent the topology between ring systems.

Several types of reduction have been considered for specific structures, for example, reduction to ring and non-ring components, reduction according to aggregates of carbon and heteroatoms, reduction on the basis of unsaturations within a structure, or reduction on a more graph theoretical basis, e.g., homeomorphically reduced graphs. Reduction to ring and non-ring components is most appropriate for generic structures particularly for generically described structures and this type of reduction has a foundation in systems for chemical nomenclature and notation.

In the first instance, graph reduction was performed on the basis of the ring (R) and non-ring (N) components of a chemical graph. Such a reduction is trivial for specific structures, but for generics requires complex path tracing methods through the ECTR, the details of which are not appropriate for this presentation. In general terms, it can be seen that nodes may extend across partial structure boundaries in such a way that atoms from different alternative partial structures collapse to the same node (Figure 4). Nodes may also span partial structures which

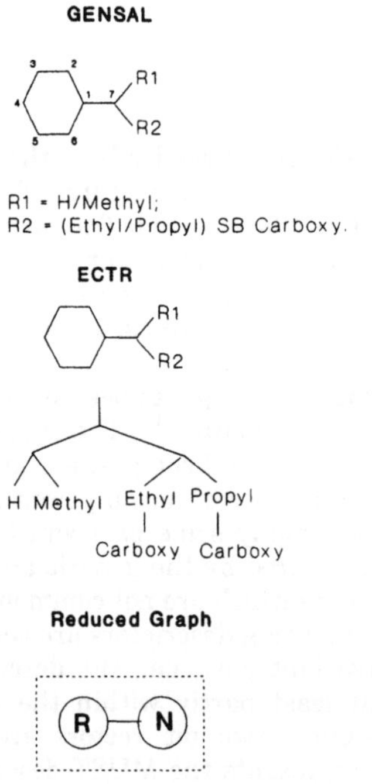

Figure 4. The GENSAL, ECTR and ring/non-ring reduced graph representations of a generic structure

are represented by different R-groups, e.g., the partial structures represented by R1 and R2 in Figure 4 all collapse to the same non-ring node. Where alternative partial structures reduce to nodes of different types, the logical relationship between these is retained in the reduced graph. Thus the reduced graph is also an AND/OR tree but has a much simplified structure compared to the ECTR, with the leaf nodes representing the ring and non-ring components and the branches indicating the logical relationships between them. Optional nodes arise from the occurrence of hydrogen with a list of alternative substituents. Some nodes in the reduced graph are invariant in the same way as a generic structure has invariant atoms, and these nodes are enclosed by the dotted area of Figure 4. Nodes in AND relationship are indicated by solid lines or by branches labelled by arcs; nodes in OR relationship are indicated by unlabelled branches, and optional connections are indicated by broken lines. The reduced graphs of generic structures represent classes of simpler graphs in much the same way as a generic structure represents a class of specific structures, as can be seen in Figure 5.

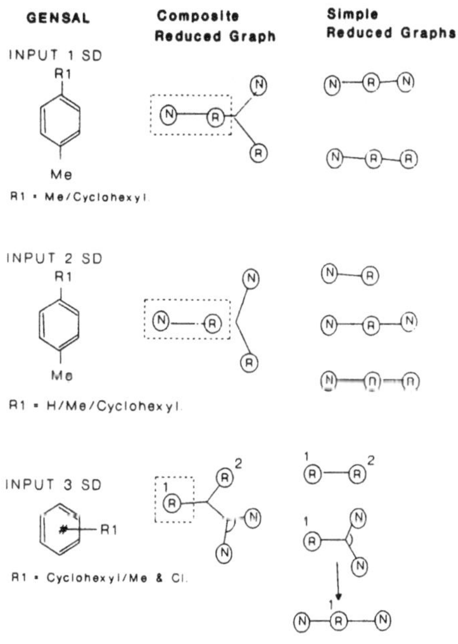

Figure 5. Reduced graphs represent classes of simpler graphs

The processing of generic partial structures requires an analysis of the parameter list to determine whether the partial structure is ring or non-ring and it is possible to extend nodes from a specific to a generic partial structure, e.g., 'carboxy SB alkyl' reduces to a single non-ring node. However, not all generic radicals can be equivocally decided as ring or non-ring, e.g., 'hydrocarbyl', and provision has been made for these both in representation and matching criteria. While the features of generic structures give rise to substantial complexities of processing, they are now fully represented in our software.

The differentiation of the reduced graph is increased by the inclusion of structural information within the nodes. This is achieved by combining information from the partial structures which constitute a node and relies on the universal applicability of parameters. The parameters which are used to represent generic partial structures can be applied to other representations. Thus, specific partial structures may also be represented by a parameter list by reducing their specificity. The parameter list can be derived from the partial connection table, and is termed a *specific* or *specific-derived* parameter list to differentiate it from the usual 'generic' parameter list. For instance, the specific parameter list representing '-CH = C(CH_3)-CH = CH_2' contains the enumerative parameter values C⟨5⟩E⟨2⟩T⟨1⟩Z⟨0⟩.

Structural information is accumulated for the nodes of a reduced graph by conflating the parameter values derived by the collapse of two or more partial structures which constitute that node. For instance, the non-ring node which is generated by the collapse of the two alternative substituents 'C1-7 alkoxy' (C⟨1-7⟩E⟨0⟩T⟨0-⟩Z⟨1⟩) and '-CH = C(CH3)-CH = CH2' (C⟨5⟩E⟨2⟩T⟨1⟩Z⟨0⟩) is specified by an accumulated 'node' parameter list with parameter values of C⟨1-7⟩E⟨0,2⟩T⟨0-⟩ Z⟨0-1⟩. These are derived from the parameter values of the respective generic and specific parameter list, by the bubble-up process.

Node parameter lists can be compared by simple examination of the corresponding parameter values during matching operations. Consider the following example of node-node comparison of the node (a) for 'C4-8 alkadiene' in a reduced query graph with the collapsed node (b) for 'C1-7 alkoxy/-CH = C(CH_3) -CH = CH_2' in a reduced file graph, showing that there is a match for each pair of parameter values, because there is intersection for each pair of integer ranges:

node (a)	C⟨4-8⟩	E⟨2⟩	T⟨0-⟩	Z⟨0⟩	
compared with node (b)	⟨1-7⟩	⟨0,2⟩	⟨0-⟩	⟨0-1⟩	yields the
intersection	⟨4-7⟩	⟨2⟩	⟨0-⟩	⟨0⟩	

Instead of this exact comparison of parameter lists i.e., comparison of EPVs, the comparison can also be performed at a further reduced level of description, and implemented as a rapid bit-string operation on parameter lists, with bit combinations representing *Reduced Parameter Values* (*RPVs*). An indication of the status of these structural features is given by a simple bit combination which indicates whether this feature is obligatorily present, possibly present, or obligatorily absent.

Search Results

The queries used in this study represent whole structures and include specific queries, each of which has a generic parent structure in the database and also a sample of generic queries. Each of these generic queries is derived from a parent structure in the database, and consists of the sets of specific structures exemplified in each of the patents, i.e., all variables are described as specific partial structures and there is no homology variation. The searches are performed by using reduced graph representations of both query and database structures. The nodes of the reduced graphs are characterised by counts or ranges of carbon and heteroatoms i.e., represented as EPVs, and by RPVs representing branching and unsaturation characteristics derived either from specific partial structures or from the generic radicals of the database structures. Each query should then retrieve at least its associated parent structure and may in addition retrieve others. Searches using these queries have recently been performed on our test database of about 2000 structures in GENSAL.

Using 200 specific queries, each of which is known to retrieve at least its parent, the mean number of database structures retrieved is five; 88 queries retrieve only the parent structures, while the maximum number of database structures retrieved is only 36. Using the smaller set of 60 structurally explicit generic queries, the mean number of database structures retrieved is 13; 16 of these queries retrieve only their parent structures, while the worst case query retrieves 135 structures. These results are displayed graphically in Figures 6 and 7.

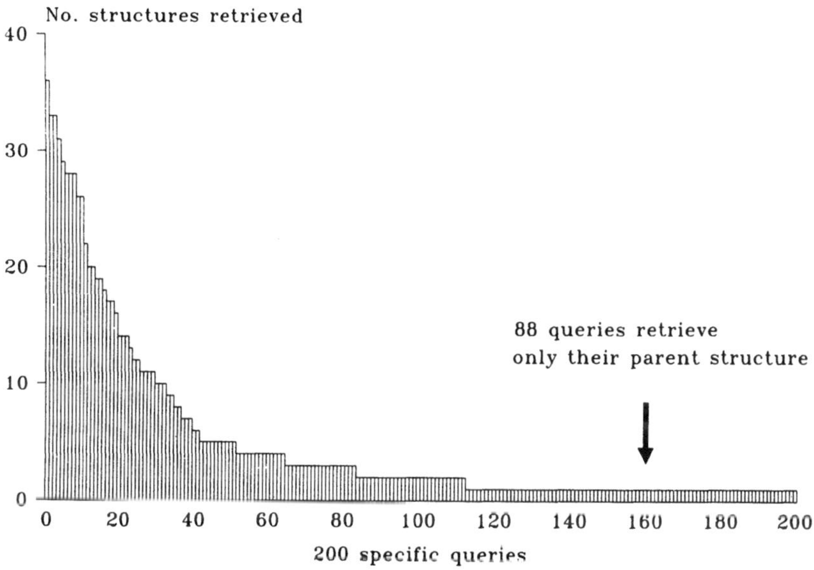

Figure 6. Results of searching 200 specific queries against the full generics database

The main conclusions to be drawn from these investigations are that whereas the basic approach demonstrates the feasibility of the method as a screen, enhancements are necessary to achieve more effective results. Possible enhancements include increasing the discriminating power of the nodes, particularly where heteroatoms are involved, and also performing less severe reduction on the non-ring components.

Two additional levels of reduction have now been implemented. The first involves increasing the information content of the nodes where the discrimination of the ring nodes is increased by the inclusion of the ring screens described above within each node, as MUST and POSS screens. Thus each ring node is now characterised by its component rings and includes information describing the size of each component ring, its fusion points and the heteroatoms contained within the ring. This has required minor modifications to the ring generation programs in order that screens are bubbled-up at the node level rather than the structure level. The discrimination of the non-ring nodes is increased by subdividing the heteroatoms into N, O, P, S, X and Q where X represents any halogen atom and Q

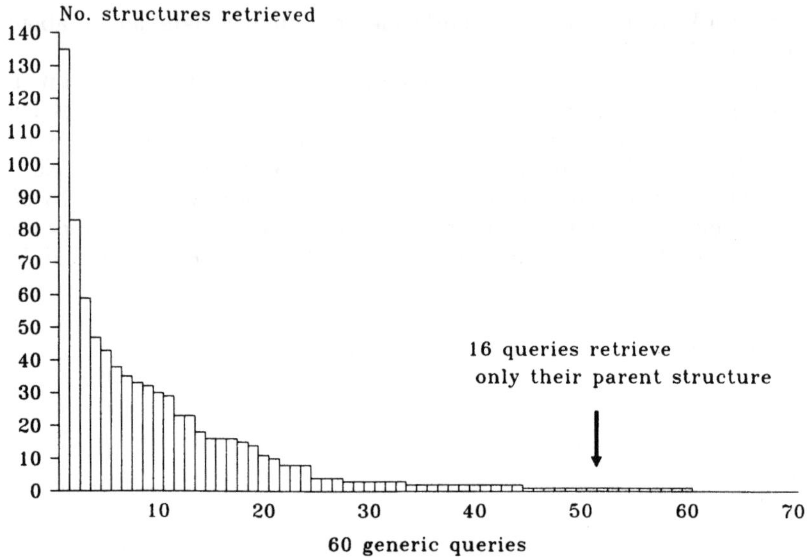

Figure 7. Results of searching 60 generic queries against the full generics database

represents any other heteroatom. The searches for the generic queries were repeated with this additional information and the results are displayed in Figure 8.

The second level of reduction involves characterising an additional node type. The non-ring nodes are subdivided into aggregates of non-ring carbon atoms (C) and aggregates of non-ring heteroatoms (Z). Figure 9 shows the results of

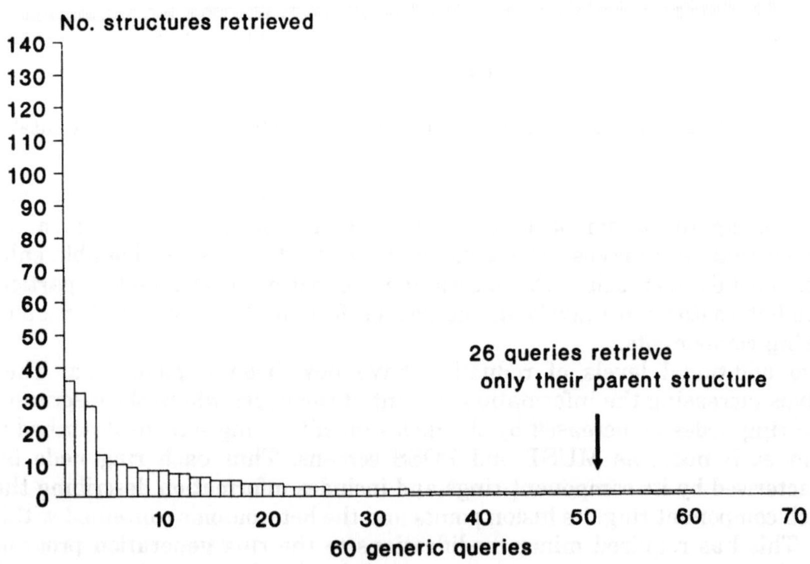

Figure 8. Results of searching 60 generic queries against the full generics database including ring screens and heteroactom differentiation

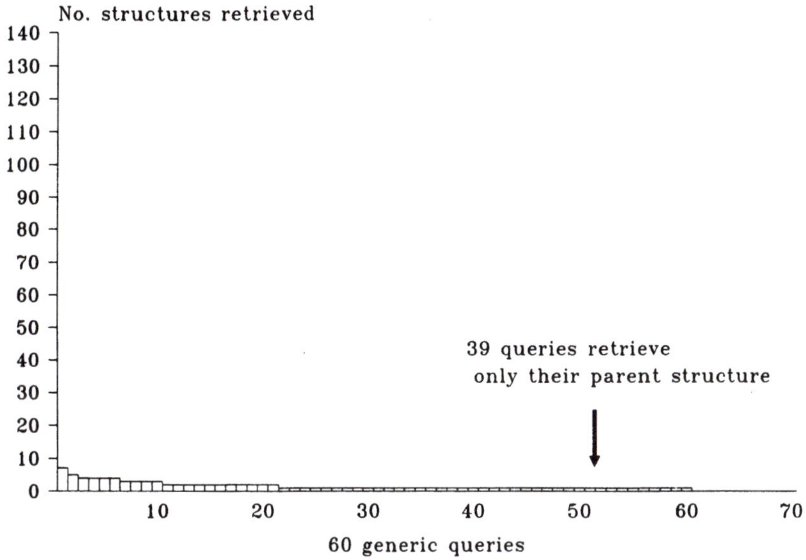

Figure 9. Results of searching 60 generic queries against the full generics database with ring/non-ring-carbon/non-ring-hetero reduction

successive screening operations where the output of the previous results is then matched using the additional node type. The maximum number of structures retrieved is now seven with 39 of the 60 queries retrieving only their parent structure. It can be seen that these methods result in a considerable increase in the effectiveness of reduced graphs as screens for generic structures.

This series of graphs represents a gradation of reductions beginning with the most severe and gradually increasing the information content. Naturally, there is a trade off between the severity of thc reduction and performance, hence the implementation of a series of screens. The most severe reduction has the advantages of rapid matching and smaller storage requirements but least effective results, whereas the less severe reductions give greater precision but at the expense of greater storage and computation.

THE REFINED SEARCH

The previous section described reduced graphs as screens for generic structures where the candidate structures which are retrieved match at the level of nodes in the reduced graph (i.e., they are node matches). Node matches, however, do not necessarily correspond with structure matches because of the collapse of information, e.g., the node representing alkoxy C⟨1-6⟩, in R/N reduction, matches with the node representing $-CH_2-CH_2-O-CH_3$ but this does not correspond with a structure match. Of course, in some cases node matches may correspond with structure matches, e.g., the node representing alkoxy C⟨1-6⟩ matches with the node representing $-O-CH_2-CH_2-CH_3$ and this does correspond with a structure match. The refined search is required to determine which of the node matches, if any, are also structure matches, and the starting point is the correspondence between the query and the candidate which is already provided as pairs of matching nodes.

In this discussion, a *match* exists between structures which are related to each other by community, i.e., they have at least one specific structure in common. It should be noted that a structure match is a match in terms of the GENSAL system, i.e., any structural differentiations which are not represented in GENSAL, e.g., stereochemistry, cannot be distinguished in the refined search.

Ultimately, a structure match can be determined only by resolving a node into its constituent partial structures, and performing an atom-by-atom search where these are represented by connection tables, or matching by parameters where one of the partial structures in a pair represents a generic radical.

The stages involved in proceeding from a node match to the refined search can be broadly summarised as:

- Identification of a node match for the whole reduced graph.
- Identification of a match for those RPVs which are used as node descriptors (using the complete parameter list or a part of it).
- Comparison of EPVs of complete parameter lists.
- Evaluating the origins of matching nodes to identify:

  node matches which correspond to structure matches;

  and node matches which fail to satisfy this condition and must be submitted to a refined search.
- Resolving reduced graph nodes which arise from the superposition of alternative partial structures into alternative reduced graph nodes so that nodes represent either single partial structures, or partial structures combined by AND logic only. The operations of matching RPVs and EPVs and evaluating the nodes for coincidence of a node match with a structure match can now be repeated.
- Refined searching, i.e., the comparison of parameter lists or connection tables, accounting for the mutual attachment of partial structures.

The first two stages were described in the section on screening techniques. Comparison of EPVs within reduced graph nodes provides a more detailed level of description and is also used to identify structure matches and non-matches as early as possible in the succession of search operations. For example, there are cases where a structure match can already be identified during the node search. The conditions required for the coincidence of a node match with a structure match for a given query node/file node pair are:

- Neither of the two nodes in a pair is derived by the collapse of any partial structures which are alternative to each other.
- At least one of the nodes must be derived from one simple generic expression, i.e., from one parameter list in the ECTR.
- EPVs for all parameters are used as node descriptors.

The possibility of a node match being coincident with a structure match requires, therefore, that the origin of each of the nodes is known, i.e., the logical relationships between the partial structures which constitute a node, and also how those partial structures are represented: by parameter lists or connection tables or a combination of both. One of the nodes must be derived from a genuine generic parameter list, i.e., one that is derived from a generic radical term within GENSAL; its counterpart is either a genuine generic parameter or a derivative parameter list but must not be derived from alternative partial structures. The derivative parameter list may be any of the following:

- Specific derived *via* the connection table of one simple specific expression.
- Specific derived in a wider sense, i.e., from the interconnection of two or more simple specific expressions e.g., from t-butyl——CH = CH——n- propyl given as one of several alternatives in R1——CH = CH——R2. The parameter list is derived by the addition of specific-derived parameter lists for each connection table by the bubble-up process.
- Derived by the addition of genuine parameter lists, e.g., alkyl—— alkenylene——.
- Derived by the addition of one or more genuine and one or more specific derived parameter list, e.g., t-butyl——alkenylene.

The coincidence of a node match with a structure match can be evaluated in the original reduced graph. If a mismatch between nodes is identified then the candidate structure is rejected, otherwise the reduced graph is resolved, i.e., nodes representing alternative partial structures are resolved, and any structure matches which can be identified at this level are determined; the remaining node pairs are then subjected to further steps of the refined search. The exact order of implementing these steps, and whether all steps are necessary, e.g., the comparison of RPVs as a step preceding the comparison of EPVs, has yet to be determined.

The Mutual Correspondence of Node-pairs

Where both nodes in a node pair resolve to a combination of parameter lists and connection tables, (e.g.,alkenyl C⟨2-5⟩—COO—$CH_2CH(CH_3)_2$ matching with –alkylene C⟨1-3⟩—CH – CH COO—CH_2—alkyl C⟨3-6⟩), then the refined search is no longer a simple case of performing either an atom-by-atom search or the comparison of parameter lists, and the proper correspondence between the partial structures within the nodes must be found. This problem is alleviated, to some extent, by the increased differentiation of the nodes within the reduced graph, e.g., by focussing on the heteroatoms within a partial structure, since these often provide a natural division within partial structures.

Several different levels of reduction are illustrated, in increasing levels of differentiation, in Figure 10. Firstly, all of the structures reduce to a single non-ring node, if the reduction is on the basis of only ring and non-ring components, and ranges of carbon and heteroatoms. Column I reduces on the basis of aggregates of carbon atoms and aggregates of heteroatoms. Column II is based on a concept of heterofunctionality[14] used in the GREMAS system as degree of 'heteroorientation'. In a slightly modified application of this principle, in a reduced graph of the type shown in column II, each carbon atom having a heterovalency of two or more is treated as a separate carbon area, i.e., as a separate node, with the exception only of carbon atoms which are substituted exclusively by two or exclusively by three halogen atoms. In column III, the carbon atoms with two, three or four heterovalencies are distinguished from aliphatic carbon chains with one or more carbon atoms, and additionally the bonds between the nodes are given.

The reduction according to heterofunctionality as given in column II assists in determining the proper correspondence between partial structures; the processes of the refined search are now explained using the examples in Figure 10 where structure A is the query structure. The modified node search now screens out structures E, F and I by a comparison of node types; and D, G and H are screened out by the node descriptors. The refined search is then necessary only for structures B and C. The final step of this refined search is now attained quite easily, because the paired nodes in the graph of the query and file structures describe precisely the proper mutual correspondence of the heterofunctions and they prescribe precisely the required segregations and/or additions within carbon chains.

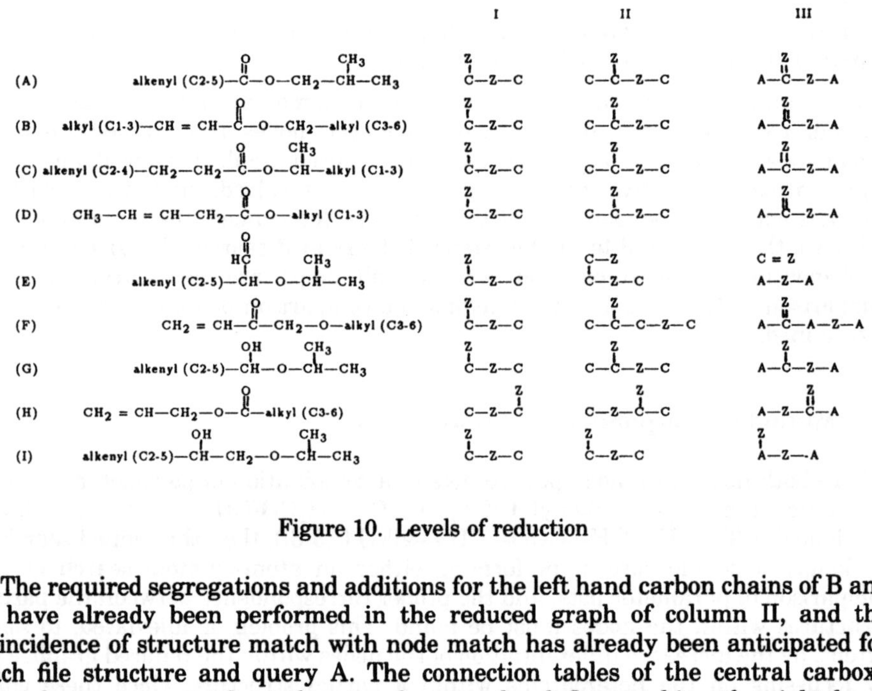

Figure 10. Levels of reduction

The required segregations and additions for the left hand carbon chains of B and C have already been performed in the reduced graph of column II, and the coincidence of structure match with node match has already been anticipated for each file structure and query A. The connection tables of the central carboxy partial structure match at the atom-by-atom level. In matching the right hand chains of B and C, the —CH_2- and -$CH(CH_3)$- are searched as substructures in the partial structure —CH_2—$CH(CH_3)_2$, taking into account the connections to the —O— of the carboxy group in all cases. This substructure search fails for C. For B the final comparison is of the genuine parameter list of alkyl C3-6with the specific derived parameter list of the remaining segregated partial structure —$CH(CH_3)_2$ in A, and yields a structure match.

USER DEFINED LEVELS OF STRUCTURAL GENERALISATION

The most useful search strategy for a generic query against a file of generic structures is community, i.e., the query and the file structure should have at least one specific structure in common. Thus the query shown in Figure 11 retrieves generic file structures having the formula indicated with the values of R as indicated in Table 2.

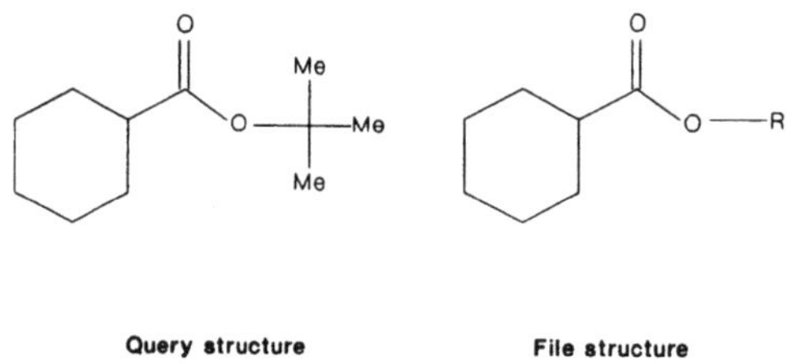

Figure 11. Query and file structures retrieved (see Table 2 for values of R)

Table 2 Values of R for the file structures of Figure 11

(a)	R = t-butyl/i-propyl
(b)	R = alkyl ⟨4-8⟩T⟨1-⟩
(c)	R = alkyl T⟨1-⟩
(d)	R = alkyl
(e)	R = hydrocarbyl

It may be the case that the searcher is interested in structures (a), (b) and (c), whereas he is not at all interested in (d) and (e) because they are structurally too general. The differences in structural generalisation are clearly revealed within the partial structure representations of the ECTR. The alternative partial structures of (a) are represented as partial connection tables whereas (b) to (e) are represented by parameter lists where the different levels of generalisation are indicated by differences in the integer ranges of particular parameters as shown in Table 3. Table 3 also contains the specific-derived parameter lists of the corresponding query structure (g) and the two alternative specifics within (a), t-butyl (a1) and i-propyl (a2).

Thus, whereas the matching relation for full structures is that of community, other matching relations may be required and can be applied to partial structures within the overall strategy. The identity of two partial structures which are specific, i.e., given as connection tables, is finally determined only by an atom-by-atom search (whereas the non-identity may be determined already by comparison of specific-derived parameter lists, as discussed in the section on screening techniques). The relations between two structure parts expressed by parameters can be determined by examining the corresponding parameter values of both lists. The matching relations which can be applied to integer ranges are: identity, strict inclusion, subsumption (i.e., non-strict inclusion, which may be strict inclusion or identity), intersection, or community which is any of the above.

When whole parameter lists are considered, identity exists if, for every parameter, the corresponding integer ranges are identical. The other matching relations can be extended similarly to whole parameter lists, e.g., the parameter list 'A' is a strict structural generalisation of the parameter list 'B' if the integer range of each parameter of 'B' is subsumed by the integer range of the corresponding parameter in 'A' *and* if this subsumption is a strict inclusion for at least one pair of integer ranges so that the whole 'B' is strictly included by 'A'. Thus, as Table 3 shows, strict inclusion between generic nomenclatural expressions is exemplified by the following series of broadening generalisations: 't-butyl' is strictly included by 'C⟨4-8⟩ t-alkyl' which is strictly included by 't-alkyl' which is strictly included by 'alkyl' which is strictly included by 'hydrocarbyl'.

Table 3. Parameters and levels of generalisation

	A	C	E	Y	T	Q	RC	..	RZ	Z
(a1)	4	4	0	0	1	0	0	..	0	0
(a2)	3	3	0	0	0	0	0	..	0	0
(b)	4-8	1-	0	0	1-	0-	0	..	0	0
(c)	1-	1-	0	0	1-	0-	0	..	0	0
(d)	1-	1-	0	0	0-	0-	0	..	0	0
(e)	1-	1-	0-	0-	0-	0-	0-	..	0-	0
(g)	4	4	0	0	1	0	0	..	0	0

The relation between a specific partial structure and a parameter-expressed generic partial structure may be established by generating a specific-derived parameter list for the specific structure. In this case the parameter list has integer values consisting of single integers and can be compared with the parameter list representing the generic partial structure as described above. The relation is subsumption if each integer of the specific derived parameter list is subsumed by the corresponding integer range of the generic parameter list, otherwise the structures are strange, i.e., non-matching.

In practice, during a search such as that described in Figure 11 where the intention is to exclude such generalisations of 't-butyl' as are given in file structures by (d) or (e), the searcher must be able to express that (c) is the broadest level of generalisation which is desired. The search algorithm then retrieves every specific 't-butyl' and every parameter-expressed generic term which is, like (b) and (c), subsumed by 'alkyl T⟨1-⟩' and which strictly includes 't-butyl', but rejects file expressions like (d) and (e) and 'alkyl A⟨5-8⟩ T⟨1-⟩:' (d) is rejected because its T⟨0-⟩ is not subsumed by the T⟨1-⟩ of (c); (e) is rejected for the same reason and additionally because of the values of E,Y,RC...RZ; 'alkyl A⟨5-8⟩ T⟨1-⟩' is rejected because its A⟨5-8⟩ does not include the A⟨4⟩ of 't-butyl', although A⟨5-8⟩ is subsumed by the A⟨1-⟩ of (c).

CONCLUSION

The results presented here are based solely on reduced graphs without a preceding bit string search; however, ring screens and details of heteroatoms have been incorporated within the reduced graph resulting in a highly effective screening method for full structure search of generics. The efficiency of the method can be improved by applying a fast fragment screen prior to reduced graph matching and this is currently under investigation. We have also outlined solutions to the refined search and a method of providing generalised search strategies including user-defined levels of search which rely on the use of parameters. This work is now well in hand and we expect to report our results later in 1990.

ACKNOWLEDGMENTS

We gratefully acknowledge funds from International Documentation in Chemistry (IDC) mbH, Derwent Publications Ltd., Chemical Abstracts Services, and the British Library.

REFERENCES

1. Lynch, M.F.; Barnard, J.M.; Welford, S.M. 'Computer Storage and Retrieval of Generic Chemical Structures in Patents. 1. Introduction and General Strategy'. *J. Chem. Inf. Comput. Sci.* **1981**, *21*, 148–150.
2. Barnard, J.M.; Lynch, M.F.; Welford, S.M. 'Computer Storage and Retrieval of Generic Chemical Structures in Patents. 2. GENSAL. A Formal Language for the Description of Generic Chemical Structures'. *J. Chem. Inf. Comput. Sci.* **1981**, *21*, 151–161.
3. Welford, S.M.; Barnard, J.M.; Lynch, M.F. 'Computer Storage and Retrieval of Generic Chemical Structures in Patents. 3. Chemical Grammars and Their Role in the Manipulation of Generic Chemical Structures'. *J. Chem. Inf. Comput. Sci.* **1981**, *21*, 161–168.
4. Barnard, J.M.; Lynch, M.F.; Welford, S.M. 'Computer Storage and Retrieval of Generic Chemical Structures in Patents. 4. An Extended Connection Table Representation for Generic Structures'. *J. Chem. Inf. Comput. Sci.* **1982**, *22*, 160–164.
5. Welford, S.M.; Barnard, J.M.; Lynch, M.F. 'Computer Storage and Retrieval of Generic Chemical Structures in Patents. 5. Algorithmic Generation of Fragment Descriptors for Generic Structure Screening'. *J. Chem. Inf. Comput. Sci.* **1984**, *24*, 57–66.
6. Barnard, J.M.; Lynch, M.F., Welford, S.M. 'Computer Storage and Retrieval of Generic Chemical Structures in Patents. 6. An Interpreter Program for the Generic Structure Language GENSAL'. *J. Chem. Inf. Comput. Sci.* **1984**, *24*, 66–70.
7. Gillet, V.J.; Welford, S.M.; Lynch, M.F.; Willett, P.; Barnard, J.M.; Downs, G.M.; Manson, G.A.; Thompson, J. 'Computer Storage and Retrieval of Generic Chemical Structures in Patents. 7. Parallel Simulation of a Relaxation Algorithm for Chemical Substructure Search'. *J. Chem. Inf. Comput. Sci.* **1986**, *26*, 118–126.
8. Gillet, V.J.; Downs, G.M.; Ling, A.I.; Lynch, M.F.; Venkataram, P.; Wood, J.V.; Dethlefsen, W. 'Computer Storage and Retrieval of Generic Chemical Structures in Patents. 8. Reduced Chemical Graphs and Their Applications in Generic Chemical Structure Retrieval'. *J. Chem. Inf. Comput. Sci.* **1987**, *27*, 126–137.
9. Downs, G.M.; Gillet, V.J.; Holliday, J.H.; Lynch, M.F. 'Computer Storage and Retrieval of Generic Chemical Structures in Patents. 9. An Algorithm to Find the Extended Set of Smallest Rings (ESSR) in Structurally Explicit Generics'. *J. Chem. Inf. Comput. Sci.* **1989**, *29*, 207–214.
10. Downs, G.M.; Gillet, V.J.; Holliday, J.H.; Lynch, M.F. 'Computer Storage and Retrieval of Generic Chemical Structures in Patents. 10. The Generation and Logical Bubble-up of Ring-screens for Structurally Explicit Generics'. *J. Chem. Inf. Comput. Sci.* **1989**, *29*, 215–224.
11. Fisanick, W. Storage and Retrieval of Generic Chemical Structure Representations. US Patent 4 642 762, Feb. 10, 1987.
12. Shenton, K.; Norton, P.; Fearns, E.A. 'Generic Searching of Patent Information'. In *Chemical Structures: The International Language of Chemistry*; Warr, W.A., Ed.; Springer-Verlag: Berlin, 1988; 169–178.
13. Downs, G.M.; Gillet, V.J.; Holliday, J.R.; Lynch, M.F. 'Theoretical Aspects of Ring Perception and Development of the Extended Set of Smallest Rings (ESSR) Concept'. *J. Chem. Inf. Comput. Sci.* **1989**, *29*, 187–206.
14. Meyer, E. 'Topological Search for Classes of Compounds in Large Files – Even of Markush Formulas – at Reasonable Machine Cost'. In *Computer Representation and Manipulation of Chemical Information*; Wipke, W.T.; Heller, S.R.; Feldmann, R.J.; Hyde, E.; Eds.; John Wiley and Sons: New York, 1974; 105–122.

AUTOMATIC TRANSLATION OF GENSAL REPRESENTATIONS OF MARKUSH STRUCTURES INTO GREMAS FRAGMENT CODES AT IDC

Günter Stiegler, Berthold Maier, Holger Lenz
IDC Internationale Dokumentationsgesellschaft für Chemie mbH
Otto-Volger-Strasse 19, W-6231 Sulzbach, Germany

ABSTRACT

The IDC file of generic chemical structures and reactions in patents is based on the GREMAS search system. GREMAS allows very rapid search of structures, substructures and other information in this large database. Until now, the GREMAS fragment codes of generic chemical structures had to be derived manually. Using the formal language GENSAL for the description and graphical input of generic chemical structures, we are able to process them by a computer program. In addition to the research work on generic chemical structures by Professor Lynch's team in Sheffield, we have developed a method of translating the GENSAL expressions into GREMAS fragments automatically. So we can use the well established GREMAS search software and the GENSAL input is compatible with the large and valuable IDC file. In this paper, the way of processing Markush structures at IDC using GENSAL is outlined. The generation of GREMAS fragments from an extended connection table representation using graph reduction methods is described.

INTRODUCTION: WHY AUTOMATIC TRANSLATION OF GENSAL INTO GREMAS?

The GREMAS search system[1–5] has proved itself for storage and retrieval of generic chemical structures in patents for more than 30 years (see Figure 1). It allows very rapid search of structures and substructures in a large database. The IDC database of organics and polymers consists of about 1,200,000 documents with 9,000,000 structures since 1959. As the GREMAS code incorporates a very detailed description of organic chemical structures and is based on the classification principles in organic chemistry, the GREMAS system allows high precision in search and minimal loss of recall. With a highly sophisticated search strategy, the searcher using GREMAS is able to determine, whether he wants to retrieve more specifically given structures as for instance pyridyl, or if he also wants to find patents where only general information, such as 'nitrogen-containing heterocycle', or even a more general structure such as 'heterocyclic ring system', is given. He also has access to nonstructural information and to reaction indexes and is able to combine the search for all these types of information in a single query.

The main disadvantage until now was that the GREMAS fragments had to be evaluated manually in the case of Markush structures of any complexity. This is a

W. A. Warr (Ed.)
Chemical Structures 2

GREMAS Search System at IDC

- Storage and retrieval of Markush structures and of specific structures
- 1,200,000 documents with 9,000,000 structures since 1959
- Well established software for very rapid search of structures and substructures in a large database
- High precision of search and minimal incomplete recall
- GREMAS highly sophisticated search strategy allows searches in different levels of structural generalization
- Search for structures, reactions and nonstructural information with a single query

- Generation of the GREMAS fragment codes has to be done manually for queries and for input of the file structures

Figure 1. GREMAS search system at IDC

time-consuming, expensive task and needs an extensive training course both for the indexers and the searchers.

The first computer programs for GREMAS generation from topological input were developed in 1968 by Sens, Schilling and Vogt and were based on the work of E. Meyer of BASF. With the aid of these programs, which were installed on an IBM-system 360/65 with 512K memory, the automatic GREMAS code generation could be done for specific structures and for Markush structures to a certain extent. These programs were further developed and rewritten in some parts at IDC and they are still in use for other purposes. The limitations in handling generic structures with these old programs were that the Markush structure must not have more than three monovalent variable groups; each of them could have up to nine alternatives and the definitions of the alternatives had to be structurally explicit. These limitations are exceeded by almost every patent Markush structure nowadays and therefore these programs cannot be used for the GENSAL GREMAS translation.

Since we have the GENSAL language (see Figure 2) for description of Markush structures[6], and software for translation of them into machine internal representations, we are able to derive search representations automatically. Great

GENSAL

- Formal language for the description of Markush structures
- Close to the chemist's language in patents
- Easy to learn
- Enough expressive power for the description of generic structures
- Smart input program MicroGensip

- GENSAL search system not yet available

Figure 2. GENSAL

effort has been made in Sheffield since 1983 to develop a GENSAL search system, but the goal has not yet been reached. Therefore our strategy is to use GREMAS as a well established and effective search screen, which can be automatically derived from GENSAL. In that way, we combine the advantages of GREMAS, which have already been mentioned, with those of GENSAL, an easy to learn formal language for mechanised input, powerful enough for the description of Markush structures in chemical patents. In addition to that, we have a smart interactive input program called MicroGensip.

WAY OF PROCESSING MARKUSH STRUCTURES IN PATENTS

The chemist's language in chemical patents is a mixture of chemical nomenclature, structure diagrams and common language. The syntax is often choosen freely; the meaning must be evaluated from the context and for that a lot of chemical understanding is necessary. Quite often the meaning in some parts still remains unclear. For automatic processing, we need a formal language like GENSAL, which has a defined syntax, and enough expressive power, and which is very close to the language in patents. The first step of indexing patents is to get an understanding of their chemical background (see Figure 3). After that the Markush structures are translated from the chemist's language into the GENSAL language, which is used for MicroGensip interactive input. The GENSAL surface language is analysed by the pass1 interpreter program for correct syntax and is stored for display, editing and processing. Then the GENSAL is compiled by the second pass of the interpreter to a machine internal representation (ECTR)[7,8], which is used for further processing.

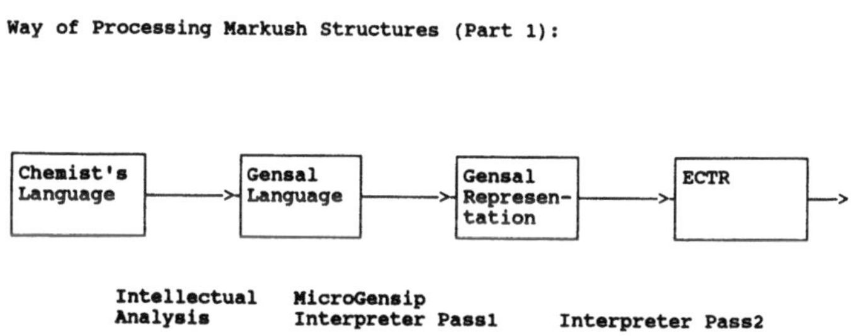

Figure 3. Way of processing Markush structures (Part 1)

The next steps are: determination of cyclic and acyclic regions, analysis of rings, and generation of the GREMAS fragment codes (see Figure 4). Finally the IDC file structure is constructed. It consists of GREMAS fragments representing Markush structures and specific structures, nonstructural information and the GREMAS reaction index.

THE MACHINE INTERNAL REPRESENTATION OF GENSAL USED AT IDC

The internal representation of GENSAL is called ECTR (Extended Connection Table Representation)[7]. It was described first by Barnard, Welford and Lynch in

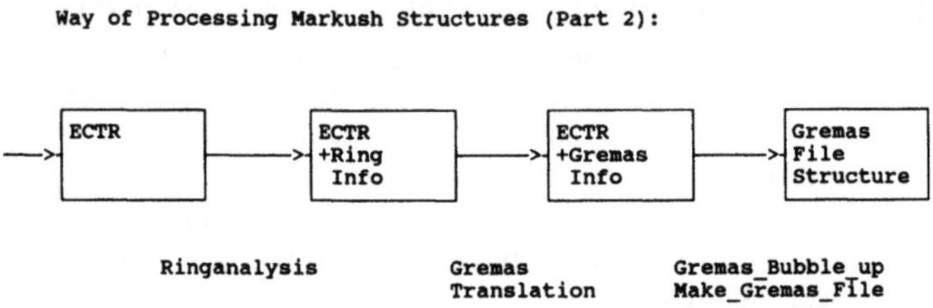

Figure 4. Way of processing Markush structures (Part 2)

1982 and was further developed by Barnard, U. Schwarz of BASF, the Sheffield research group (Prof. M. Lynch and co-workers)[6–10] and co-workers of IDC. The ECTR has the structure of a logical AND/OR tree, where the leaves are the partial structures and the branches show the logical relationship.

As in a patent Markush structure the substituents can be substituted further and further again, the ECTR structure is recursive. This means that every leaf in this tree can be the root of a new AND/OR tree itself. For this reason, the ECTR has an unlimited number of hierarchical levels of partial structures. The topmost partial structure is commonly the constant part of the Markush structure. The partial structure one level higher in the hierarchy than a partial partial structure is called the parent partial structure; the partial structure one level lower down is the child partial structure. The branches connecting these structures carry a lot of information. The partial structure can have one bond to another one: in that case we call it singly connected. Or it may have two connecting bonds, so we call it doubly connected. An important piece of information in the branches about the type of attachment of a partial structure to another one is what we call variable position substitution. We distinguish between three types of substitution: fixed position attachment to exactly one atom, variable position substitution on a ring, and variable position substitution on an arbitrary set of atoms. The tree-structure of the ECTR will be shown later with an example.

Specific partial structures such as phenyl are stored as connection tables. Generic partial structures such as aryl are stored as a set of parameters, which describe structural features, and their integer value ranges, which count the occurrence of these features. In the IDC system we use an extended and refined set of the structural parameters, built on those which were first suggested by the Sheffield research group in 1982 (see Figure 5). The classes of parameters are divided into two groups. The first group is used for specification of a generic partial structure itself; the second group gives information about its connection to other partial structures.

GENERATION OF GREMAS FRAGMENTS FROM THE ECTR

There are two types of GREMAS codes, three-letter codes and Y-codes, which have multiples of six characters. The three-letter codes are used for the description of carbon atoms, rings and ring systems, substitution patterns on rings and chains, and hetero-hetero atom bonds. The Y-codes are assigned to the GREMAS districts or regions (see Figure 6). There are four types of GREMAS districts, namely,

GENSAL Parameters at IDC

Classes of parameters:

Specification of generic partial structures

GREMAS districts
Atom counts
Unsaturation
Branching in carbon chains
Ring fusion
Charged atoms and radical electrons

Specification of the connection between partial structures

Connection to parent structure
Connection to child structure (substitution)

Figure 5. Classes of GENSAL parameters at IDC

carbon chain, alicyle, aromatic ring, and heterocyle. They are denoted by the letters R, S, T, U. The Y-code contains information about the types of atoms within the region, multiple bonds, connection to other regions and the logical relationship.

For GREMAS generation from the ECTR every partial structure in a Markush structure is processed separately. A program, which traces down the ECTR-tree, starting at the root and visiting every partial structure exactly once, is called tree-trace algorithm. This is the main program to all other programs for this purpose. The following steps of processing are done on every partial structure. The first step in a specific partial structure is a graph analysis, which determines whether the graph is connected or disconnected. Then every component of the graph is analysed in order to find out the GREMAS regions. This is a method of graph reduction[9], which produces a much simpler graph, which is built on the four types of GREMAS regions: carbon chains, alicycles, aromatic rings, and heterocycles. Then every

GREMAS districts:

Type of GREMAS district:	**GREMAS code:**
carbon chain	R
alicycle	S
aromatic ring	T
heterocycle	U

Figure 6. GREMAS districts

atom of each region is examined and its chemical neighbourhood is calculated. The main advantage of the reduction of the structure to single atoms and processing them separately is that combinations of variable groups are avoided as much as possible. Only in those cases where more than one variable group is in the necessary GREMAS neighbourhood around the atom do combinations need to be calculated. The GREMAS code of a carbon atom, for instance, is a function of all possible atoms between the first and the fourth sphere. In many cases the first sphere is sufficient.

The GREMAS translation of a carbon atom is done in two steps. The primary specification (see Figure 7) tells us what GREMAS district type it belongs to, how the heteroatom neighbourhood in terms of heteroorientation is, and how many hydrogen atoms it carries.

Primary specification of the carbon atoms

carbon atom specification	possible values
member of GREMAS district type	R, S, T, U, unknown
heteroorientation	0, 1, 2, 3, 4, unknown
hydrogen count	0, 1, 2, 3, 4, unknown

Figure 7. Primary specification of the carbon atoms

The heteroorientation of a carbon atom is defined as the sum of all congeners of the atom except hydrogen and carbon, which are bonded by an acyclic bond, multiplied by the bond order. The bond order can be 1, 2 or 3. So, for example, a carbon atom attached to carbon, hydrogen, and one hydroxy group has heteroorientation 1; a keto carbon has heteroorientation 2; and a carboxy carbon has heteroorientation 3.

The secondary specification (see Figure 8) gives the full description of the carbon atom according to the rules of the GREMAS system with a three-letter code.

The resulting GREMAS fragments are stored according to their logical relationship. They are either the same for all variable neighbourhoods of the carbon atom and therefore constant, or the different neighbourhoods contribute to different fragment codes, which are possibly present in this case.

In the case of generic partial structures such as aryl, the GREMAS regions and

Secondary specification of the carbon atoms

carbon atom specification	possible values
Full description of the carbon atom according to the GREMAS system,	three-letter code,
function of the neighbouring atoms in the n-th sphere of the C-atom, 1< =n< =4	according to the rules of the GREMAS system

Figure 8. Secondary specification of the carbon atoms

the codes of the C-atoms within every region are generated from the parameter values. Also in this case, the GREMAS codes depend on the neighbouring atoms, which are structurally explicit (specific) or again members of a generic partial structure. The processing of a partial structure results in locally constant and locally possible GREMAS fragments, which are stored together with this partial structure in the ECTR. Finally, a program called GREMAS bubble-up collects the GREMAS fragments and assigns the full logic as it is described in the Markush structure.

EXAMPLE OF GENSAL GREMAS TRANSLATION

Figure 9 shows a very simple example of a Markush structure in GENSAL. It contains some important features in order to show the principles. The constant part of the structure consists of a five-membered heterocycle and four variable groups R1, R2, R3, and R8. Three of them, namely R1, R2, R8, are singly connected groups, while the other one, R3, is doubly connected. The definition of the variable groups R1, R2, and R8 is done by nomenclatural terms, where alkyl and alkylene are generic partial structures, and hydrogen, methyl, ethyl, hydroxy, and amino are specific partial structures. The values for the doubly connected group are given by two explicit structure diagrams (SD), which are linear carbon chains with two and three atoms, respectively. The connecting bonds to the parent structure are marked with asterisks. The GREMAS neighbourhood of the carbon atom between the nitrogens depends on the alkyl and the alkylene groups. The neighbourhood of the charged nitrogen is a function of R2 and R3. So all combinations of the values for R2 and R3, which result in 2*3 = 6 possibilities, are calculated.

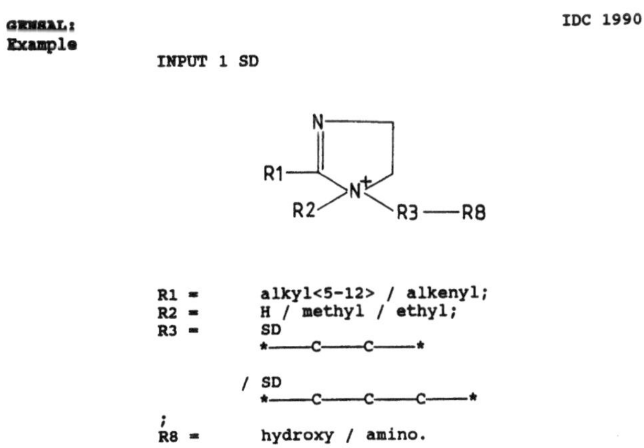

Figure 9. Simple example of Markush structure in GENSAL

The ECTR-tree for this example structure is illustrated in Figure 10. The partial structures are enumerated in such a way that the constant part of the Markush structure is number 1. It is given by a structure diagram (SD). The four

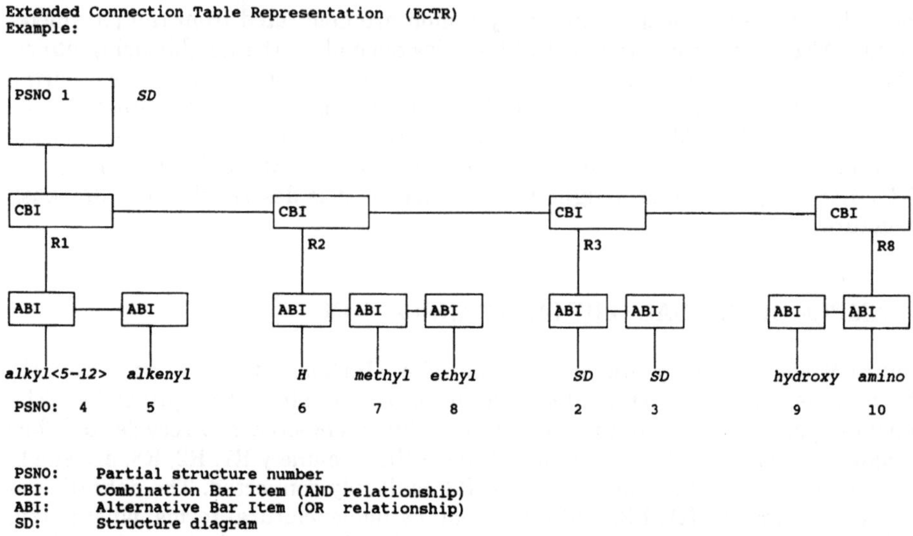

Figure 10. Extended Connection Table Representation (ECTR)

combination bars (CBIs) are related to the variable groups R1, R2, R3, R8 and point to the alternative definitions of the variable groups. As the partial structures 2 to 10 are not substituted further, the ECTR-tree terminates at this hierarchical level.

The resulting GREMAS codes of this example are shown in Figure 11. There are constant and possibly present three-letter codes and constant and possibly present Y-codes. A detailed description of the GREMAS codes would exceed the scope of this paper and will not be given here.

HARDWARE AND SOFTWARE REQUIREMENTS

Parts of the IDC GENSAL GREMAS System are implemented on a MS-DOS microcomputer, for instance the graphical input and the syntax analysis of

GREMAS codes,
example:

three-letter codes, constant	three-letter codes, possibly present
CIA CI7 RZU SDT WGB WGD WPZ WZA XC2	BAA BA7 CI8 EAA EA7 RAB RAU RBJ ROU WPA WPB WPD WPW WPX WPY WPO WSA WTA
Y-code (district code), constant	**Y-codes (district codes), possibly present**
YUQR2- YRU/2-	YRC---$A1--- YRCB--$B1--- YRCE--$B2---

Figure 11. GREMAS codes

GENSAL. All other steps of processing are done on a mainframe computer. Most of the programs are written in the high level programming languages PASCAL and FORTRAN.

EXPERIENCES AND LIMITS OF AUTOMATIC GREMAS GENERATION

In some cases, the automatic processing fails or the results are inaccurate. This happens when the ring set cannot be found, when the structural conventions do not allow a proper description of a patent in GENSAL, when the expressive power of GENSAL is not sufficient or when the logic or the chemical graph is so complex that the possible neighbourhoods of the atoms cannot be calculated. Sometimes these patents cause severe problems in manual GREMAS encoding too; in other cases the manual encoding can easily be done, while the solution by a program would be extremely difficult and uneconomical (see Figures 12 and 13).

Difficult Markush structure, Example 1:

Condensation at a specific ring
Example: OD 87/151377

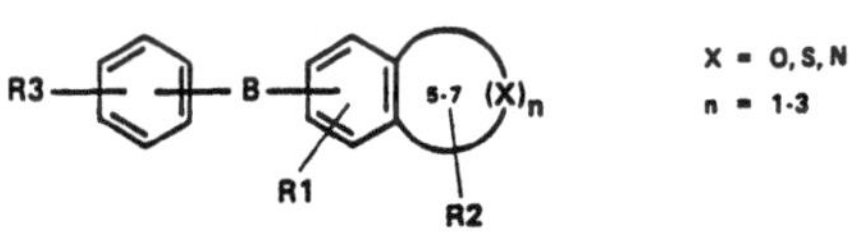

Figure 12. Difficult example 1

Difficult Markush Structure, Example 2:

Condensation at a generic ring
Example: OD 87/043481

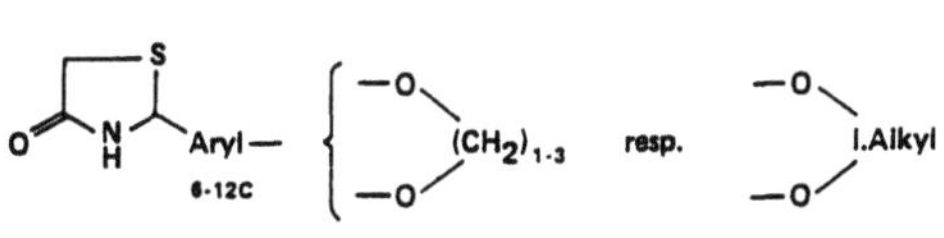

"6-12 C Aryl opt. subst. by an alkylenedioxy-group "

Figure 13. Difficult example 2

ACKNOWLEDGMENTS

We would like to give our grateful acknowledgments to Professor Mike Lynch and his co-workers at the University of Sheffield, who designed the GENSAL system and contributed several programs in their original version. Thanks are due to J. Barnard, G. Downs and Hampden Data Services Ltd. for developing MicroGensip and additional programs in co-operation with IDC. Thanks are also due to W. Dethlefsen of BASF, who made fundamental contributions to handling undecidedness in chemical patents, to GENSAL, and to the GENSAL GREMAS conversion. We also thank our colleagues at IDC, namely A. Jungblut, G. Most, and H.S. Neubert for programming work and technical implementation, and V. Batroff, K.W. Kempf, and M. Köhne for their assistance in GREMAS coding problems.

REFERENCES

1. Fugmann, R.; Braun, W.; Vaupel, W. 'Zur Dokumentation Chemischer Forschungsergebnisse'. *Angew. Chem.* **1961**, *73 (23)*, 745–751.
2. Fugmann, R.; Braun, W.; Vaupel, W. 'GREMAS – ein Weg zur Klassifikation und Dokumentation in der Organischen Chemie'. *Nachr. Dok.* **1963**, *14 (4)*, 179–190.
3. Fugmann, R; Dolling, U.; Nickelsen, H. 'Das Problem Chemischer Ringstrukturen aus Topologischer Sicht'. *Angew. Chem.* **1967**, *79 (17/18)*, 802–813.
4. Rossler, S.; Kolb, A. 'The GREMAS System, an Integral Part of the IDC System for Chemical Documentation'. *J. Chem. Doc.* **1970**, *10*, 128–134.
5. Suhr, C; von Harsdorf, E.; Dethlefsen, W. 'Derwent's CPI and IDC's GREMAS: Remarks on their Relative Retrieval Power with Regard to Markush Structures'. In Proceedings of a Conference organised by the Chemical Structure Association at the University of Sheffield, England, 26–29 March 1984.
6. Barnard, J.M.; Lynch, M.F.; Welford, S.M. 'Computer Storage and Retrieval of Generic Chemical Structures in Patents. 2. GENSAL: a Formal Language for the Description of Generic Chemical Structures'. *J. Chem. Inf. Comput. Sci.* **1981**, *21*, 151–161.
7. Barnard, J.M.; Lynch, M.F.; Welford, S.M. 'Computer Storage and Retrieval of Generic Chemical Structures in Patents. 4. An Extended Connection Table Representation (ECTR) for Generic Structures'. *J. Chem. Inf. Comput. Sci.* **1982**, *22*, 160–164.
8. Barnard, J.M.; Lynch, M.F.; Welford, S.M. 'Computer Storage and Retrieval of Generic Chemical Structures in Patents. 6. An Interpreter Program for the Generic Structure Language GENSAL'. *J. Chem. Inf. Comput. Sci.* **1984**, *24*, 66–71.
9. Gillet, V.J.; Downs, G.M.; Ling, A.; Lynch, M.F.; Venkataram, P.; Wood, J.V.; Dethlefsen, W. 'Computer Storage and Retrieval of Generic Chemical Structures in Patents. 8. Reduced Chemical Graphs and their Applications in Generic Chemical Structure Retrieval'. *J. Chem. Inf. Comput. Sci.* **1987**, *27*, 126–137.
10. Downs, G.M.; Gillet, V.J.; Holliday J.D.; Lynch, M.F. 'Computer Storage and Retrieval of Generic Chemical Structures in Patents. 9. An Algorithm To Find the Extended Set of Smallest Rings in Structurally Explicit Generics'. *J. Chem. Inf. Comput. Sci.* **1989**, *29*, 207–214.

CHIRAL STRUCTURE DATABASE FOR SACCHARIDES

Nancy L. Porter
Maxwell Online, Inc., 8000 Westpark Drive, McLean, VA 22102, U.S.A.

ABSTRACT

Locating a group of organic compounds sharing a common string of chiral centres with the same stereochemistry is currently a tedious and time-consuming exercise. After a substructure search has been performed, dictionary terms and three-dimensional structural diagrams must be used to eliminate compounds with the incorrect stereochemistry. The more chiral centres there are in the substructure, the more of the retrieved compounds must be eliminated from the final answer set. This low precision is due to the absence of absolute and relative orientation information in connection tables, even where three-dimensional structures can be displayed. In order to achieve chiral substructure searching capability, atom-pair information for chiral centres must be included in connection tables. A small chemical structure database using monosaccharides and disaccharides was created to develop and test this concept further. The results of this project may ultimately be applicable for searching any compound with one or more chiral centres.

INTRODUCTION

Locating a group of organic compounds sharing a common string of chiral centres with the same stereochemistry is currently a tedious and time-consuming exercise. If the user attempts to find the desired compounds through a substructure search, a large number of the compounds retrieved are not what he had in mind. Nomenclature-based searching, however, is of limited value if the chiral centre desired is located within a named moiety without being identified (as is frequently the case with sugars). Even if the stereochemistry is included in the nomenclature, the user must frequently draw the structure to verify that the designated orientation does indeed match the required substructure.

In order to address the problem of performing substructure searches for chiral moieties, a method of designating both absolute and relative configuration in molecular structures was developed. Then a small chemical structure database containing 25 mono and disaccharides was created in order to test the concept. Results from tests show this method to be viable and expandable to other types of compounds.

CURRENT DIFFICULTIES IN ACCURATE RETRIEVAL

Different problems for substructure searching on chiral compounds arise in two-dimensional structure databases compared with three-dimensional databases. The two-dimensional case is the worse of the two, because two-dimensional databases

W. A. Warr (Ed.)
Chemical Structures 2

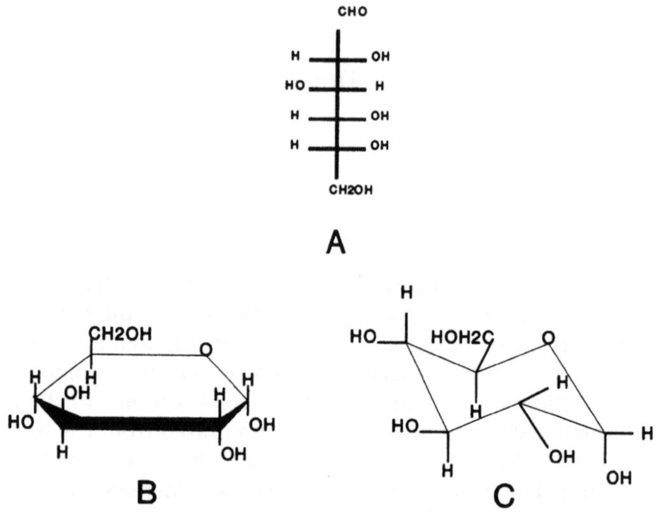

Figure 1. (A) Fischer representatation of D-glucose
(B) Haworth representation of α-D-glucopyranose
(C) Stable conformation of α-D-glucopyranose

generally do not store the three-dimensional information necessary to designate chiral centres. Figure 1 shows three different and common representations of the structure for D-glucose. Presenting any one of these structures as a query structure to a two-dimensional database would retrieve all aldohexoses, because there would be no way for the database to specify on which 'side' of the main carbon skeleton a substituent should appear. Narrowing the search by using nomenclature would further limit the retrieval to all D-aldohexoses, of which D-glucose would be one. The user would have to look up the absolute configurations of each chemical in the answer set to discern which of the retrieved compounds is really a match.

In three-dimensional databases, the necessary information to find a chiral centre is stored and accessible, but generally only searches for whole molecules or large moieties containing chiral centres result in precise retrieval. The user looking for compounds with a precise chiral configuration as a part of the substructure will frequently retrieve 'racemic' mixtures because the user has no way of telling the program which way is 'up' for the purpose of the search. Similarly, stored structures may not always be oriented the same way when the data is input. Thus substructure searches will retrieve compounds with the same relative configurations but with opposite absolute configurations. A search for the substructure shown in Figure 8 would retrieve all but four of the compounds in the database. While this is a significant improvement over the retrieval from a two-dimensional database, it is still not good enough for everyday use.

REQUIREMENTS FOR CHIRAL SUBSTRUCTURE SEARCHING

In order to achieve precise and accurate substructure searches for chiral moieties, at least two conditions must be met: there should be a standard method of designating the desired stereochemistry graphically, so that the user can input a specific configuration about the chiral centre of interest; and there should be a standard method of designating the molecular orientation and absolute configurations for storage in the structure database. In the best of all possible worlds, both standards should dovetail so that substructure searching can be easily achieved.

Achieving the first condition is fairly straightforward. A number of different ways of designating stereochemistry have evolved over the years, each with its own particular strengths and weaknesses. The method of representing the request graphically can therefore be tailored to the needs and knowledge base of the user. For instance, a user interested primarily in looking for a particular atom pair in a database might feel more comfortable with a display using Newman projections to specify the query substructure. However, even in using Newman projections, the user should know how the database designates 'top' of the molecule and whether the Newman projection looks at the molecule from the bottom up or from the top down.

Similarly, in creating a stereochemical designation for each chiral centre, every molecule in the database must first be oriented according to the same rules, and the same method of designating the attachments must be used for every chiral centre. In addition, the program must be able to access specific atomic information for every chiral substituent, allowing for more flexible substructure searching.

The method for designating the configuration at a chiral centre developed in this paper uses atom pairs: that is, every chiral centre is paired with the atoms on each side of it that connect it to the rest of the molecule. The substituents attached to both atoms are listed in a specific order that includes the chiral atom. Atom pairs were chosen for this method for two reasons: by anchoring the chiral centres firmly within the molecule, inversions leading to poor retrieval are avoided and redundant information is painlessly created; and both absolute and relative configurations are preserved. Atom pairs therefore allow a great deal of flexibility in designating stereochemistry on the part of the searcher, while reducing the risk of retrieving 'racemic mixtures' on the part of the user.

THE DATABASE

Sugars were deliberately selected to test the workability of this concept, for several reasons. The number of monosaccharides is small, the absolute stereochemistry of monosaccharides and of many disaccharides is well known, and sugars contain multiple chiral centres. Creating a database containing connection tables with atom-pair hyperstructure data for monosaccharide anomers and a few disaccharides would therefore present a 'worst case' scenario for retrieving the correct results to a query. The answers to a query could also be readily worked out by hand before submitting it to the database, allowing independent testing to be achieved. Finally, such a database could also be readily expanded at a later date to include other biological molecules, many of which have sugar chains as a base.

Once sugars were chosen as the basis of the database, the next consideration was to decide upon a standard graphical representation from which connection tables and the hyperstructure data could be derived. The Fischer representation was selected because monosaccharide units would then be represented the same way throughout, regardless of whether they occurred independently or as a part of a disaccharide.

The following sugars (I-XXV in Figure 2) were included in this database: D-glucose, L-glucose, D-mannose, L-mannose, D-ribose, L-ribose, D-xylose, L-xylose, D-gulose, L-gulose, D-fructose, L-fructose, D-fucose, L-fucose, D-arabinose, L-arabinose, D-allose, L-allose, D-galactose, L-galactose, D-lyxose, L-lyxose, D-sucrose, β-D-lactose, and 4-β-D-glucopyranosyl-β-D-glucopyranose.

The numbering rules were arbitarily established as follows: the main carbon chain was numbered from top to bottom (following the standard convention), with the left-hand moiety first if the compound was a disaccharide. Substituents attached to these carbon atoms with single bonds were regarded as single units for the purposes of the connection table, so that hydrogen atoms and hydroxide groups were each counted as single units, but carboxide oxygen atoms were included with their carbon atoms as one unit. For disaccharides, oxygen atoms in the sugar rings or connecting sugar units were counted after the main chain carbon atoms, top to bottom and left-hand moiety first, with the connecting oxygen atom being counted before the ring oxygen atom if both were attached to the same carbon atom. Hydrogen atoms were numbered next, from top to bottom and left-hand moiety first, followed by hydroxide groups, in the same manner.

Types were assigned to each bonding site as follows: A was given to non-chiral carbon atoms and their non-chain substituents, and to connecting oxygen atoms. D was assigned to chiral carbon atoms only. Bonding sites typed as E were the lower molecular weight non-chain substituents of chiral carbons, and G-type bonding sites were the higher molecular weight non-chain substituents of chiral carbons. Figure 3 shows the Fischer representation of D-glucose with its assigned bonding site numbers and types.

Hyperstructure data was generated using the following rules: only carbon atoms in the main sugar chain were included; each atom-pair entry in the hyperstructure table must contain at least one chiral carbon atom; the second atom of the pair was considered first in assigning configuration data; attachments to the chiral atoms

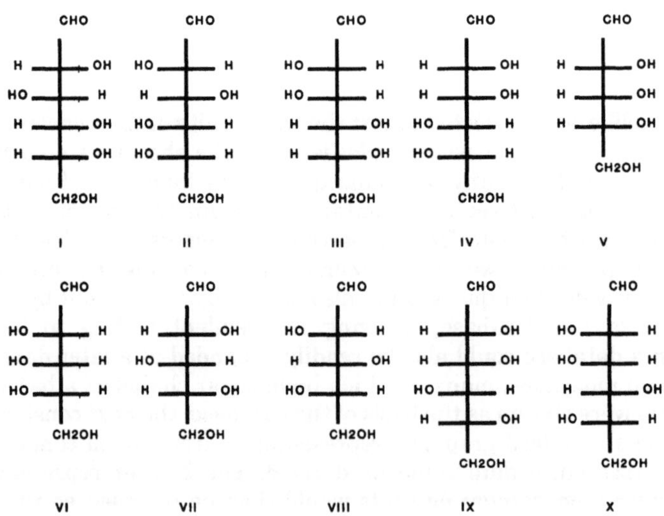

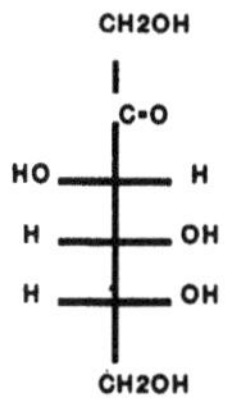

XI

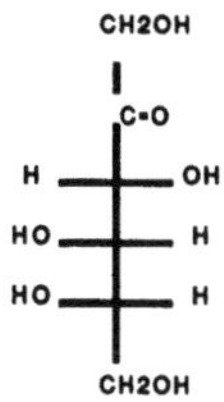

XII

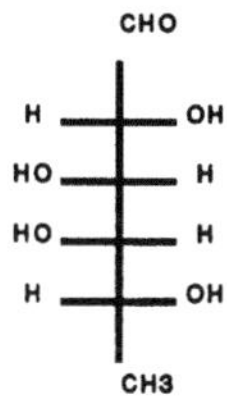

XIII

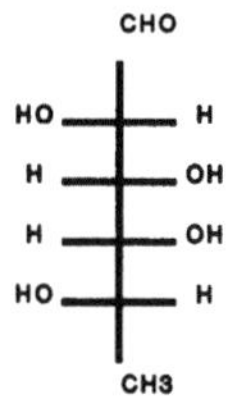

XIV

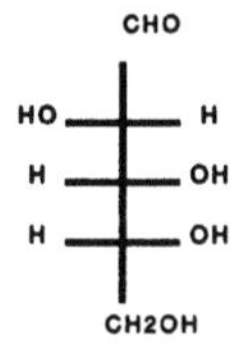

XV

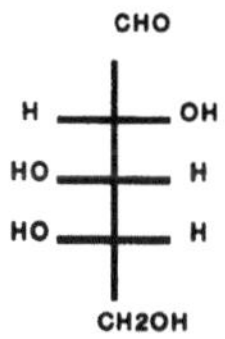

XVI

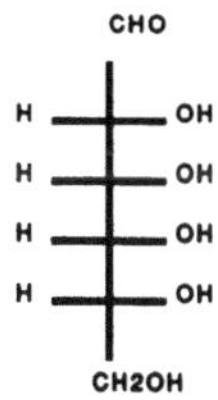

XVII

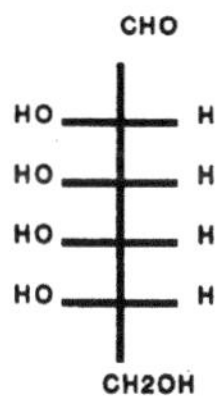

XVIII

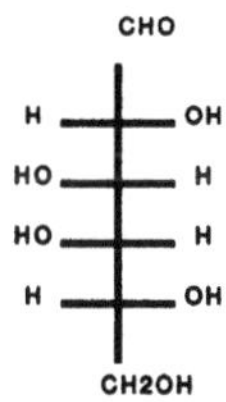

XIX

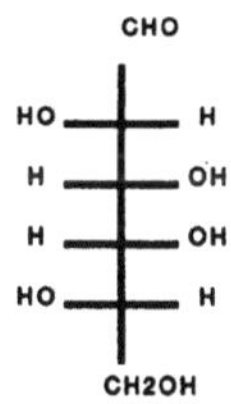

XX

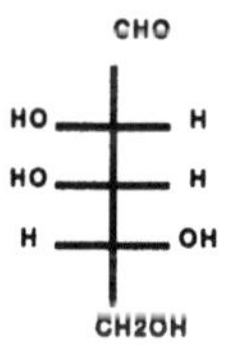

XXI

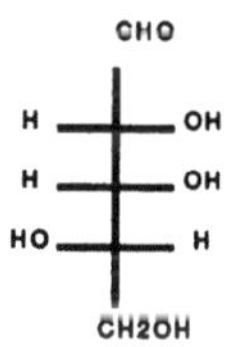

XXII

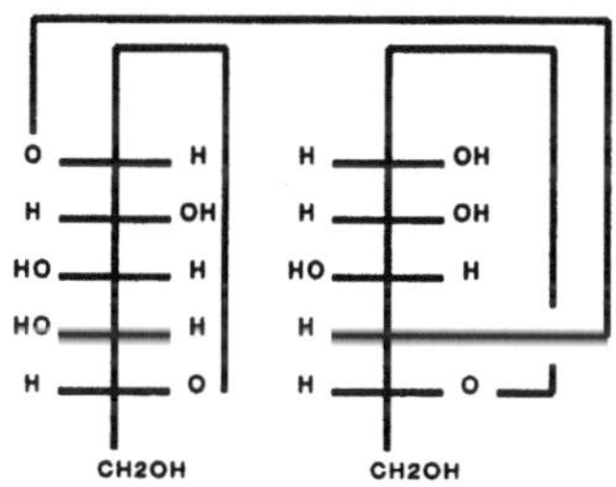

XXIII

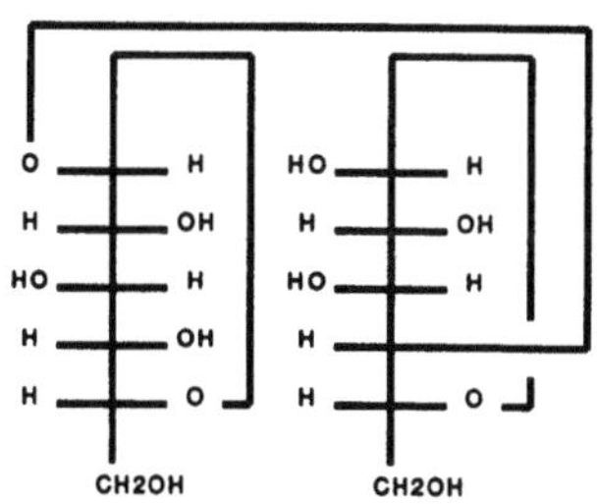

XXIV

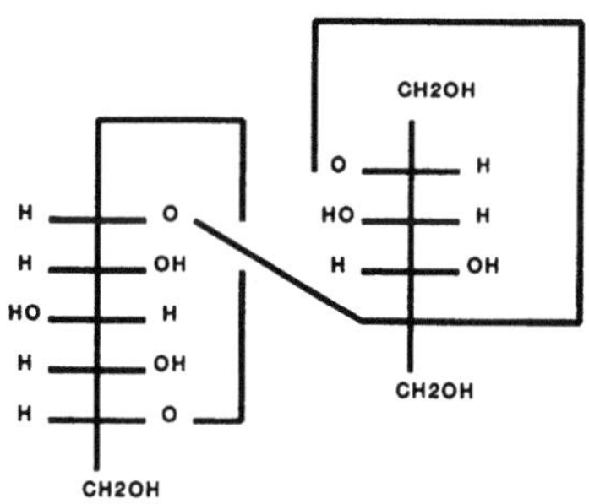

XXV

Figure 2. Sugars included in the database

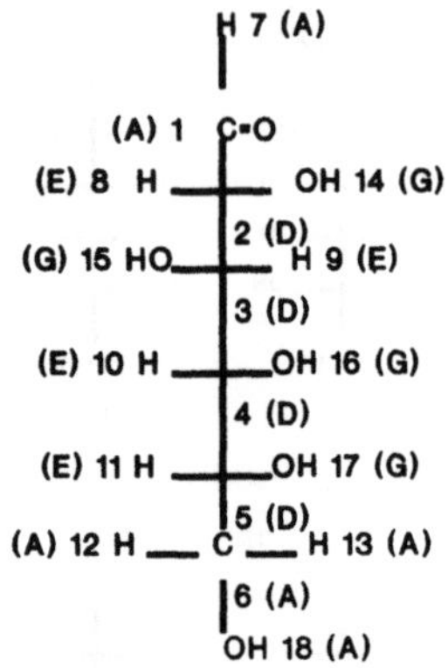

Figure 3. D-glucose with bonding sites numbered and types assigned

were numbered clockwise, staring with the atom connecting the pair to the rest of the molecule; the chiral atom under consideration was always listed last; and non-chiral carbon atoms were represented only by their bonding site numbers, excluding their attachments. Thus each atom pair is represented in the hyperstructure table by a number string that lists the attachments to chiral carbon atoms, in a standard order directly related to the absolute stereochemistry of the molecule. Figure 4 shows atom pair 2–3 of D-glucose and its corresponding configuration data. Figures 5, 6, and 7 show the connection tables and hyperstructure data for D-glucose, L-glucose, and D-sucrose, respectively. Please note that the connection tables for D-glucose and L-glucose are identical but that the hyperstructure data clearly shows that these two chemicals are enantiomers.

The search program used for the database was written in BASIC specifically for the database.

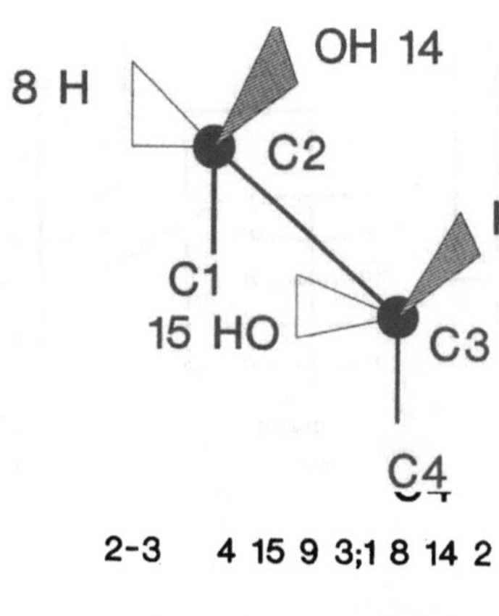

Figure 4. Atom pair 2–3 of D-glucose and corresponding configuration information

D-GLUCOSE

Type	Bonding site		Connections
A	1	CO	2 7
D	2	C	1 3 8 14
D	3	C	2 4 9 15
D	4	C	3 5 10 16
D	5	C	4 6 11 17
D	6	C	5 12 13 18
A	7	H	1
E	8	H	2
E	9	H	3
E	10	H	4
E	11	H	5
A	12	H	6
A	13	H	6
G	14	OH	2
G	15	OH	3
G	16	OH	4
G	17	OH	5
A	18	OH	6

Atom pair	Configuration
1-2	3 8 14 2;1
2-3	4 15 9 3;1 8 14 2
3-4	5 10 16 10 4;2 15 9 3
4-5	6 11 17 5;3 10 16 4
5-6	6;4 11 17 5

Figure 5. D-glucose structure data

L-GLUCOSE

Type	Bonding site		Connections
A	1	CO	2 7
D	2	C	1 3 8 14
D	3	C	2 4 9 15
D	4	C	3 5 10 16
D	5	C	4 6 11 17
D	6	C	5 12 13 18
A	7	H	1
E	8	H	2
E	9	H	3
E	10	H	4
E	11	H	5
A	12	H	6
A	13	H	6
G	14	OH	2
G	15	OH	3
G	16	OH	4
G	17	OH	5
A	18	OH	6

Atom pair	Configuration
1-2	3 14 8 2;1
2-3	4 9 15 3;1 14 8 2
3-4	5 16 10 4;2 9 15 3
4-5	6 17 11 5;3 16 10 4
5-6	6;4 17 11 5

Figure 6. L-glucose structure data

```
D-SUCROSE

Type Bonding site   Connections
D     1     C         2 13 14 16
D     2     C         1  3 17 30
D     3     C         2  4 18 31
D     4     C         3  5 19 32
D     5     C         4  6 14 20
A     6     C         5 21 22 33
A     7     C         8 23 24 34
D     8     C         7  9 15 25
D     9     C         8 10 26 35
D    10     C         9 11 27 36
D    11     C        10 12 13 15
A    12     C        11 28 29 37
A    13     O         1 11
A    14     O         1  5
A    15     O         8 11
E    16     H         1
E    17     H         2
E    18     H         3
E    19     H         4
E    20     H         5
A    21     H         6
A    22     H         6
A    23     H         7
A    24     H         7
E    25     H         8
E    26     H         9
E    27     H        10
A    28     H        12
A    29     H        12
G    30     OH        2
G    31     OH        3
G    32     OH        4
A    33     OH        6
A    34     OH        7
G    35     OH        9
G    36     OH       10
A    37     OH       12

=================================================================
Atom pair          Configuration
1-2                3 17 30 2;14 16 13 1
2-3                4 31 18 3;1 17 30 2
3-4                5 19 32 4;2 31 18 3
4-5                6 20 13 5;3 19 32 4
5-6                6;4 20 13 5
7-8                9 15 25 8;7
8-9                10 35 26 9;7 15 25 8
9-10               11 27 36 10;8 35 26 9
10-11              12 13 15 11;9 27 36 10
11-12              12;10 13 15 11
=================================================================
```

Figure 7. D-sucrose structure data

SAMPLE SEARCHES

For the first sample query, a detailed examination of the search routine against the records for D-glucose and L-glucose will be described in order to show how nicely the hyperstructure configurations allow differentiation between anomers. A query and the compounds retrieved are given for three test cases: a substructure search for two adjacent chiral centres, one for three adjacent chiral centres, and one for two chiral centres separated by one atom.

The user enters a query by indicating how many chiral centres are to be specified. The screen presents a Fischer skeleton with the requested number of branches. The user then enters the desired substituents in the appropriate places in the diagram. Non-specified sites are filled with question marks. The implicit assumption with this format of data input is that the user is orienting his query so that the 'top' of the query structure relates to the 'top' of the standard sugar representation.

Two Adjacent Chiral Centres

Suppose the user wishes to find those sugars containing the substructure illustrated in Figure 8. Notice that the user has only specified the stereochemistry for two atoms. The program would first translate the query structure into a configuration string based on types. The current query can be expressed as:

(A or D) G E D; (A or D) E G D

where (A or D) stipulates that both the lower atom and the upper atom in the retrieved structure must be attached either to another chiral chain atom or to a non-chiral chain atom; and E and G mean that the attachments to the lower carbon atom must be arranged in order of decreasing molecular weight going clockwise but are arranged in order of increasing molecular weight going clockwise for the upper carbon atom.

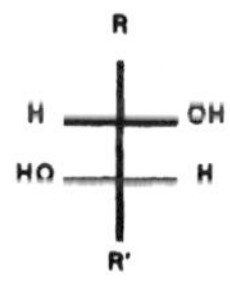

Figure 8. Test substructure 1

The program next examines the hypertext part of each structure record. Atom-pair entries containing too few elements are eliminated from consideration; thus atom pairs 1–2 and 5–6 in the D-glucose and L-glucose records are not compared against the query string. The first eligible atom-pair entry is compared against the query string by translating the listed bonding sites into types one at a time and performing a character match against the query string after each translation. The program stops the translation-and-match routine against an entry with the first non-matching character. Entries that match the query string are then matched against the actual query substructure for actual bonding site identities. If this database contained sugar derivatives, compounds matching the stereochemistry of the query structure but not the composition would therefore not be retrieved.

In the D-glucose connection table, the program looks first at the hypertext entry for atom pair 1–2, skips it because there are not enough elements in the configuration, and goes to atom pair 2–3. Atom pair 2–3 has an entry that translates to D G E D; A E G D. This character string matches the query string, so the substructure is compared against the part of the connection table that contains the data for this atom pair. The composition of the entry also matches the query substructure, so D-glucose is put out to the answer set.

In the L-glucose connection table, atom pair 1–2 is skipped and the examination of atom pair 2–3 begins. This entry's configuration string translates to D E G D; A G E D, but the program stops the translate-and-match process after the second character, which does not match the query string. However, atom pair 3–4 has a configuration string that translates to D G E D; D E G D, which is a match. The composition of this entry also matches the query substructure, so L-glucose is also put out to the answer set.

The compounds in this database matching the query structure are as follows: L-mannose, L-gulose, D-gulose, D-glucose, L-glucose, D-xylose, L-xylose, L-fructose, L-fucose, D-fucose, L-arabinose, D-galactose, L-galactose, L-lyxose, β-D-lactose, 4-β-D-glucopyranosyl-β-D-glucopyranose, and D-sucrose.

Please note that the implicit assumption was made that the user wanted to find sugars with the query substructure *relative to the standard sugar orientation for this database*. If the user really wanted the query substructure regardless of the orientation of the substructure relative to the rest of the molecule, the search would have to be repeated a second time with the query substructure rotated 180° perpendicular to the plane of the screen.

Three Adjacent Chiral Centres

Suppose the user wishes to find all the sugars in the database with the substructure shown in Figure 9. This substructure translates to the configuration strings:

D G E D; (A or D) E G D
(A or D) E G D; D G E D

Compounds to be retrieved must contain configuration strings matching both query strings in the order given and are attributed to adjacent atom-pair entries.

The compounds in this database retrieved as matches to the query substructure are: D-gulose, D-glucose, D-xylose, 4-β-D-glucopyranosyl-β-D-glucopyranose, and D-sucrose.

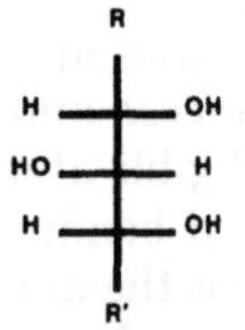

Figure 9. Test substructure 2

Two Chiral Centres Separated by One Atom

The third test substructure is shown in Figure 10. This substructure translates to the following configuration strings:

D (E or G or A) (E or G or A) D; (A or D) E G D
(A or D) G E D; D (E or G or A) (E or G or A) D

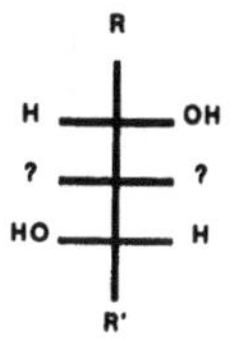

Figure 10. Test substructure 3

Compounds to be retrieved must contain both configuration strings in the order given and in adjacent atom pair entries.

The compounds in this database retrieved as matches to the query substructure are: L-mannose, L-glucose, L-fructose, L-fucose, D-fucose, L-arabinose, D-galactose, L-galactose, L-lyxose, and β-D-lactose.

CONCLUSION

The three test cases described above have demonstrated that the hypertext atom-pair concept appears to work well for the sample database. However, this database contained 25 similar compounds, so that the search strategy focused primarily on comparing stored configurations against query configurations. The applicability and practicality of the method employed here for a large database of chiral and non-chiral compounds remains to be determined. Certainly the aid of screen and modifications to the search program will be necessary to improve the search speed and to improve the usefulness of the method to larger databases. Additional modifications to make the request input more user-friendly and to allow greater flexibility in specifying query substructures will also have to be made.

Despite the changes outlined above, this method of defining structural conformations could be readily expanded to include rules for sugar derivatives, amino acids, and other organic and inorganic compounds containing chiral centres. Certainly it is flexible enough to be considered for inclusion in databases specialising in three-dimensional structure information as well as in those utilising standard two-dimensional information.

ACKNOWLEDGMENTS

The author offers sincere thanks to Maxwell Online Inc., for funding the trip to Noordwijkerhout and to the Biotech Programs Division of Biospherics Inc., whose questions led her to think about this problem.

BIBLIOGRAPHY

1. *Communication, Storage and Retrieval of Chemical Information;* Ash, J.E.; Chubb, P.; Ward, S.E.; Welford, S.M.; Willett, P., Eds; Ellis Horwood: Chichester, 1985.
2. Attias, R.; Dubois, J.-E. 'Substructure Search Systems: Concepts and Classification'. *J. Chem. Inf. Comput.* **1990**, *30*, 2–7.
3. Barnard, J.M. 'Online Graphical Searching of Markush Structures in Patents'. *Database* **1987**, *10(3)* 27–34.

4. Christie, B.D.; Munk, M.E. 'Structure Generation by Reduction: A New Strategy for Computer-assisted Structure Elucidation'. *J. Chem. Inf. Comput. Sci.* **1988**, *28*, 87–93.
5. Downs, G.M.; Lynch, M.F.; Willett, P.; Manson, G.A.; Wilson, G.A. 'Transputer Implementation of Chemical Substructure Searching Algorithms'. *Tetrahedron Comput. Methodol.* **1988**, *1(3)*, 207–217.
6. Dromey, R.E. 'A Linked-path Connection Table with Substructural Atom Ordering'. *J. Chem. Inf. Comput. Sci.* **1979**, *19*, 37–42.
7. Graf, W.; Kaindl, H.K.; Kniess, H.; Schmidt, B.; Warszawski, R. 'Retrieval by Means of the BASIC Fragment Search Dictionary Based on the Chemical Abstracts Service Registry III System'. *J. Chem. Inf. Comput. Sci.* **1979**, *19*, 51–55.
8. Hagadone, T.R.; Lajiness, M.S. 'Capturing Chemical Information in an Extended Relational Database System'. *Tetrahedron Comput. Methodol.* **1988**, *1(3)*, 219–230.
9. Rush, J.E. 'Status of Notation and Topological Systems and Potential Future Trends'. *J. Chem. Inf. Comput. Sci.* **1976**, *16*, 202–210.
10. Shelley, C.A. 'Heuristic Approach for Displaying Chemical Structures'. *J. Chem. Inf. Comput. Sci.* **1983**, *23*, 61–65.
11. Shelley, C.A.; Munk, M.E. 'An Approach to the Assignment of Canonical Connection Tables and Topological Symmetry Perception'. *J. Chem. Inf. Comput. Sci.* **1979**, *19*, 247–250.
12. Silk, J.A. 'Present and Future Prospects for Structural Searching of the Journal and Patent Literature'. *J. Chem. Inf. Comput. Sci.* **1979**, *19*, 195–198.
13. Walentowski, R. 'Unique, Unambiguous Representation of Chemical Structures by Computerisation of a Simple Notation'. *J. Chem. Inf. Comput. Sci.* **1980**, *20*, 181–192.
14. Wenger, J.C.; Smith, D.H. 'Deriving Three-dimensional Representations of Molecular Structure from Connection Tables Augmented with Configuration Designations Using Distance Geometry'. *J. Chem. Inf. Comput. Sci.* **1982**, *22*, 29–34.

ADDITION OF ATOM/BOND STEREOCHEMISTRY TO THE CAS REGISTRY FILE

Paul E. Blower, Jr., Dwight H. Lillie, Alan H. Lipkus, and Cheng Qian
Chemical Abstracts Service, P.O. Box 3012, Columbus, Ohio, 43210 U.S.A.

ABSTRACT

CAS registers stereoisomers using text descriptors derived from the corresponding chemical names. This system works well for the unique registration of stereoisomers but it is difficult to relate the text descriptor to the atoms and bonds of the connection table. This limits its usefulness for substructure search or display of stereochemistry in the structure diagram. CAS is currently preparing to augment the Registry connection table with atom/bond specific stereodescriptors. This paper focuses on two aspects of this work: the representation of stereochemistry and techniques for converting the Registry structure file to the stereo-augmented format.

INTRODUCTION

The Chemical Abstracts Service (CAS) Chemical Registry System was initially installed in 1965. The current version of the system[1], called Registry III, has been in operation since 1974. The Registry System now contains records for more than 10 million chemical substances, with 97% represented in the form of atom-bond connection tables.

The Registry System was originally begun to support the preparation of the indexes for *Chemical Abstracts* (CA). Since that time, the Registry System has evolved from a production tool for CAS publications and services to a system that provides a foundation for substance identification used by the scientific community worldwide. The Registry System is primarily known because of the CAS Registry File, available for online searching on the STN International network, and for CAS Registry Numbers. The CAS Registry Number, which links the structure with the CA index name and other data, is used for chemical substance identification by many governmental agencies and industrial organisations.

As these uses of the Registry System have increased in importance, it has become apparent that some changes in the way the Registry System represents various kinds of substances would benefit the users of Registry-based services. In addition, advances in chemical science over the past 25 years make it desirable that the Registry System be extended to handle information better in areas of rapid growth, particularly, in molecular biology and materials science. During the same period, computer technology has changed drastically, and changes in the Registry System should be made to make more efficient use of the computing equipment of the 1990s.

In view of these considerations, CAS has undertaken a multi-year effort, usually referred to as Registry Renovation, which will lead ultimately to the next

W. A. Warr (Ed.)
Chemical Structures 2

generation of the Registry System. During the course of this effort, users of Registry services will notice a number of changes. These will include modifications in the way the Registry System handles not only biological macromolecules and materials, but also alloys, polymers, and inorganic and co-ordination compounds. A more global change that will affect a large number of substances is in the handling of stereochemistry. Currently stereochemical data is only recorded in the chemical name and the controlled vocabulary known as the CAS Text Descriptor. This will be enhanced by a technique for recording specific atom and bond stereochemistry in the connection table.

REPRESENTATION

The stereochemical characteristics of chemical compounds have long been recognised as among the most important aspects of their structures. Since the 1960s, CAS has registered stereoisomers as unique chemical substances. The method for the stereoisomer registration is based on a nomenclature system[2,3]. While several different approaches are taken to describe the stereoisomers of different types of compounds, they are basically all *textual* descriptors, as distinguished from atom/bond specific descriptors. CAS stereodescriptors generally belong to three categories[4]: systematic descriptors, stereoparents, and stereodescriptors of co-ordination compounds.

This system works well for the unique registration of stereoisomers. However, it does not provide the information needed for structure display or substructure search of stereochemical detail. There are two basic complexities with the text descriptor method for describing stereochemistry. Firstly, the locants cited are derived from nomenclature atom numbers rather than the numbering system used in the connection table and there is no simple relationship between these two numbering systems. Secondly, the stereodescriptors are based on ligand priorities determined by the Cahn-Ingold-Prelog (CIP) Sequence Rule[5]. Applying this rule may involve examining a large portion of the molecule because the ligands cannot always be ranked by comparing the atoms directly attached to the stereocentre. Furthermore, changes in the structure remote from the stereocentre may change its descriptor. Thus, the system would be difficult to use for describing the stereochemistry of a substructure.

An alternative technique for representing the absolute configuration of tetrahedral stereocentres and geometric isomers at double bonds was devised by Petrarca, Lynch, and Rush[6]. It was adapted by Wipke and Dyott[7] in their Stereochemically Extended Morgan Algorithm (SEMA). The configuration is described by a *parity descriptor* which uses the atom numbers of the immediate neighbours. The parity of a tetrahedral stereocentre is determined by the following rule. Look down the bond from the lowest numbered neighbour to the central atom. If the ascending sequence of the remaining neighbours appears in clockwise order, the descriptor is *even*; otherwise, it is *odd*. This is illustrated in Figure 1, where the node numbers represent the relative atom numbers. Since the parity descriptor is associated with the connection table atom and uses only local information (i.e., the atoms numbers of its immediate neighbours) this method is more useful for computer processing.

In order to use this method in the CAS Registry System, we extended it to cover cases where stereochemistry is only partially known, such as the substance of

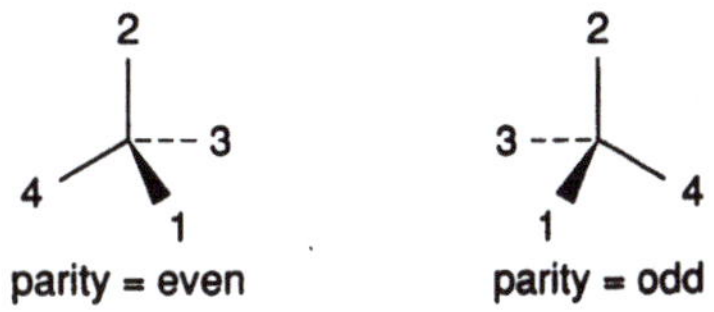

Figure 1. Determination of parity for tetrahedral centres

Figure 2. In this example, the author indicated that absolute stereochemistry was known for two centres on the butanone side chain, but the configurations of the four centres on the tetrahydrofuran ring were only known relative to each other. To handle cases like this, we define two levels of stereodescriptors, global and local. Global descriptors refer to the substance as a whole and indicate whether the stereochemistry is absolute or relative or whether the substance is a racemic mixture. The global descriptor also includes optical rotation data. Local descriptors are associated with specific, stereogenic atoms. They consist of a *configuration descriptor* and an optional *group descriptor*. Except for co-ordination centres (see below), the configuration descriptor is a parity descriptor with values *even*, *odd* or *unknown*. A group descriptor is used to identify sets of relative stereocentres like the four centres on the tetrahydrofuran ring in Figure 2. It indicates which stereocentres are in the group and whether the group is relative or racemic. This technique allows us to express all of the stereochemical detail provided in the original document.

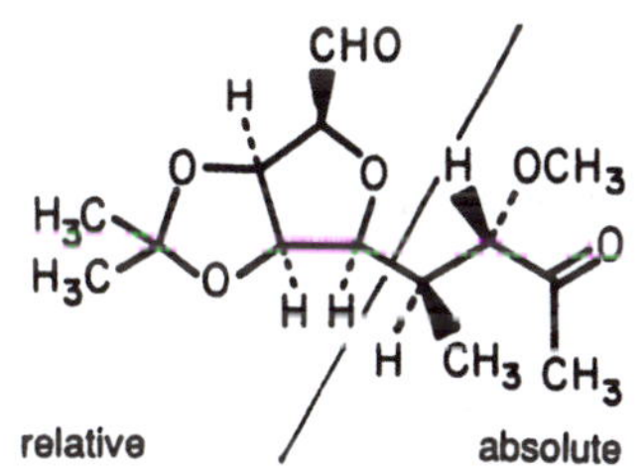

Figure 2. Substance with partial stereochemistry

We also extended the parity representation to cover all types of stereochemistry cited in the CA index name and localised to an atom or bond. In addition to tetrahedral atoms and double bonds, the representation includes allenes and the non-tetrahedral, atomic geometries important in co-ordination chemistry. For the more complex co-ordination geometries, our current plan is to base the configuration descriptor on an idealised model of the geometry around the co-ordination centre. Models would be defined for the 17 non-tetrahedral geometries[8] recognised by CAS. The model for an octahedral complex is illustrated in Figure 3a with the co-ordination sites labelled from 1 to 6. Figure 3b shows a schematic octahedral complex with ligating atoms $A_1, \ldots, A_6$, where the subscripts indicate the relative atom numbers in the connection table. The configuration descriptor associated with the octahedral centre in the connection table would be a list of the site numbers of the ligating atoms listed in ascending atom number order. For the schematic example in Figure 3b, the configuration descriptor would be 235614 because A_1 occupies site 2, A_2 occupies site 3, A_3 occupies site 5, etc. Because of the inherent symmetries of the co-ordination models, there are generally several equivalent descriptors for the same configuration. To account for equivalent representations, there will be a table of symmetry operations for constructing equivalent representations maintained for each co-ordination model. These tables will be used for comparing the stereochemistry of two co-ordination centres with the same geometry.

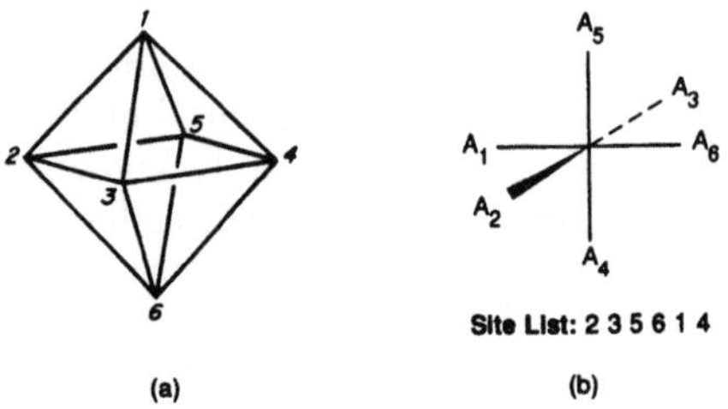

Figure 3. Model for an octahedral co-ordination centre

FILE CONVERSION

Stereochemistry is expressed in CA index names by three methods: stereoparents, systematic stereodescriptors, and descriptors for co-ordination compounds. We have developed a separate procedure for converting each of these three types of stereodescriptors to atom/bond specific descriptors in the Registry connection table. Table 1 gives statistics for the Registry File as of February, 1990. At that time, the total file contained over 10 million registered substances.

Table 1. Registry statistics for stereospecific substances

Type	Number	Percent
Systematic Nomenclature	1,527,000	15.3%
Stereoparents	747,000	7.5%
Coordination Compounds	259,000	2.6%
TOTAL	2,533,000	25.3%

Stereoparents

Stereoparents are derived from the common names of natural products. The following discussion focuses on stereoparents from the steroid, alkaloid or terpene classes. The stereoparent is the simplest member of a family of related, cyclic structures with complex stereochemistry. The stereochemistry is given by a stereospecific structure diagram which appears in the *CA Chemical Substance Index*; this is illustrated by cholestane (Figure 4). Stereochemistry is indicated in the diagram by dotted lines (α bonds), if the substituent is below the plane of the ring, and wedged lines (β bonds), if it is above the plane.

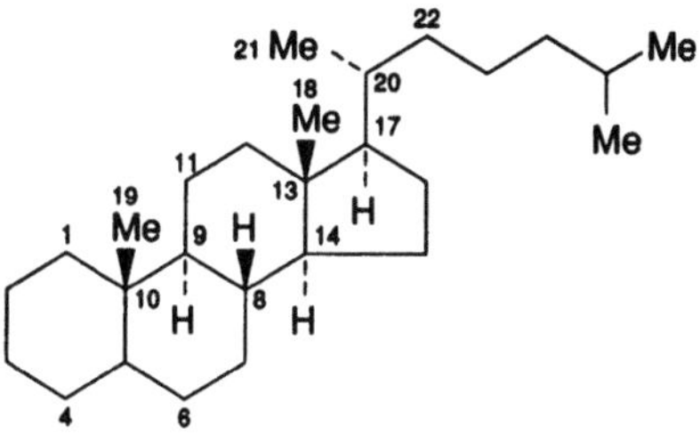

Figure 4. Stereoparent diagram for cholestane

In the diagram, the atoms are numbered so that the numbers can be used to indicate modifications of the parent structure. The naming conventions allow for specification of abnormal or additional stereochemistry and modifications in the molecular topology of the parent structure. Additional or abnormal stereochemistry on ring atoms of the stereoparent is designated by citing the atom number and direction (α or β) as illustrated in Figure 5. Four types of changes to the ring system can be indicated by keywords: *nor* (ring contraction), *hom* (ring expansion), *sec* (ring cleavage), and *cyc* (cyclisation). A structure derived from ring cleavage of the cholestane system is illustrated in Figure 6.

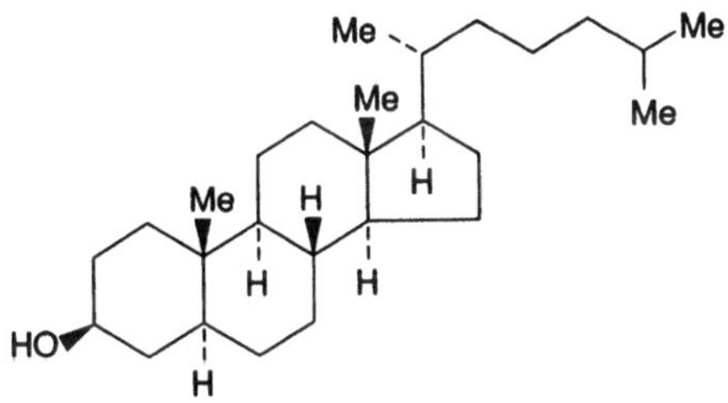

DES: 3B,5A.CHOLEST

Figure 5. Stereoparent with modified stereochemistry

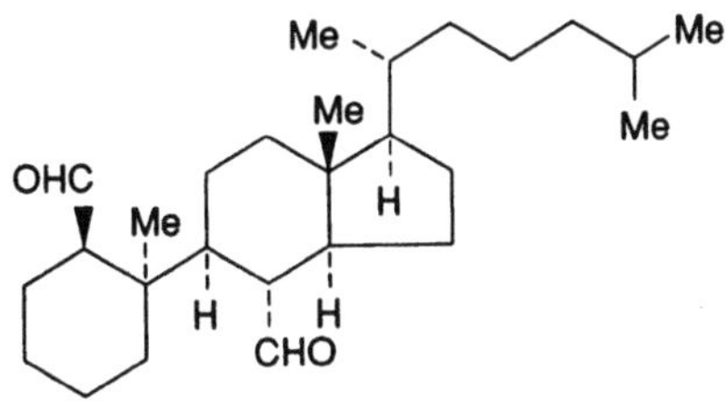

DES: 5B.CHOLEST..2,6,7-SEC

Figure 6. Stereoparent with modified topology

The procedure for converting substances described as modified stereoparents is based on the use of stereospecific templates, one for each stereoparent. The template will have nomenclature locant numbers and parity descriptors for all stereocentres. The procedure for adding atom/bond specific stereodescriptors to the connection table, in simplified form, consists of the following steps:

1. Retrieve the Registry File substance and parse its stereodescriptor into components that identify the stereoparent and describe the topological and stereochemical modifications.
2. Modify the topology of the template as described.
3. Perform a substructure match using the modified template to locate the stereoparent portion of the file structure.
4. Translate parity from the template to the file substance. Note that the template parity cannot be transferred directly because it depends on the two sequences of relative atom numbers which will, in general, be different.

Systematic Stereodescriptors

Stereodescriptors for systematically named substances use a combination of absolute and relative terms. An absolute term is assigned to a *reference* centre, and all other centres are related to it using relative terms. The absolute terms, *R* and *S*, follow the system developed by Cahn, Ingold, and Prelog. Various kinds of relative descriptors are used. The most readily interpreted are those that describe the stereochemical relationships of substituents on a ring system. In general terms[9], *cis* and *trans* are used for small rings with only two stereogenic atoms; *exo*, *endo*, *syn* and *anti* are used for bicyclo [X.Y.Z]anes; and α and β are used for other situations where the stereocentres are in the same ring system as the reference centre. The relative descriptors *R**/*S** are used for all other tetrahedral stereocentres, and *E*/*Z* are used for stereogenic double bonds. The use of these terms is illustrated in Figures 7 and 8.

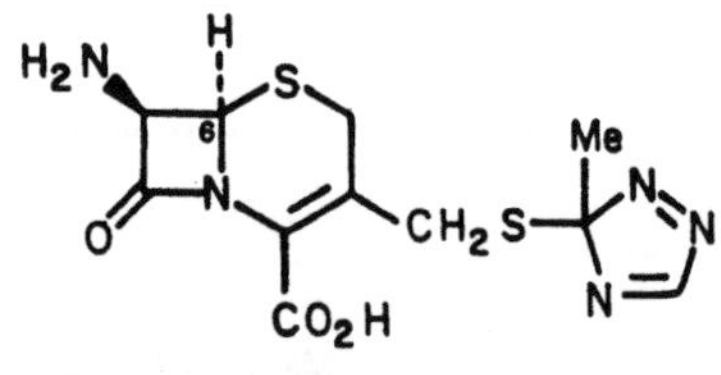

5-Thia-1-azabicyclo[4.2.0]oct-2-ene-2-carboxylic acid, 7-amino-3-[[(3-methyl-3H-1,2,4-triazol-3-yl) thio]methyl]-8-oxo-, (6R-*trans*)-[6R-(6α, 7β)]-

Figure 7. Example of systematic stereodescriptors

If there is any danger of confusion, descriptor terms are associated with locants. It is important to note, however, that locants do not refer to a single numbering system for the total substance. Instead, the locants for each parent fragment start with one (1). In Figure 8 for example, the locants in 1α,4*a*β, 8*a*α refer to the parent *naphthalenyl*, while the locants in 2*R**,3*S** refer to the parent *oxirane*.

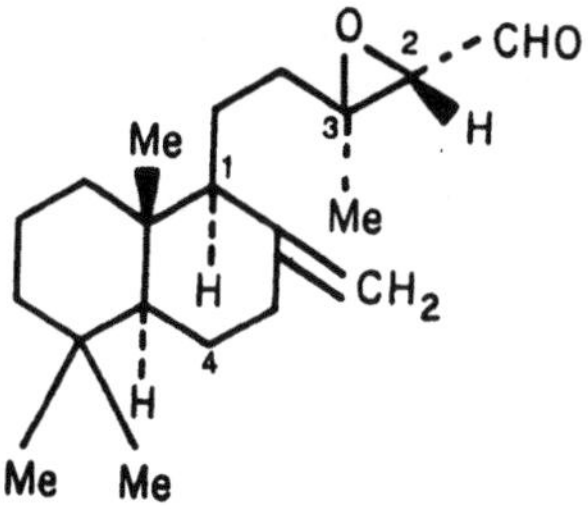

Oxiranecarboxaldehyde, 3-[2-(decahydro-5,5,8a-trimethyl-
2-methylene-1-naphthalenyl)ethyl]-3-methyl-,
[1S-[1α(2R*,3S*),4aβ,8aα]]-

Figure 8. Another example of systematic stereodescriptors

The general approach for converting Registry substances with systematic stereodescriptors consists of three stages. The first stage attempts to assign the nomenclature stereodescriptors to the appropriate atom or bond of the connection table. In the early 1970s, CAS developed a *nomenclature translation* program[10] to convert systematic names of organic compounds to connection tables. This program maintains the correspondence between nomenclature locants and atom numbers which we will use to map stereodescriptors into specific atoms.

The second stage will identify and resolve any problems which arise because of insufficient or ambiguous descriptors. Generally, we expect a one-to-one correspondence between the stereodescriptors in the name and the stereocentres in the connection table. There are, however, legitimate cases where there are more stereocentres than stereodescriptors. First, some structural configurations are implicitly understood to be *cis*, such as bridges in bicyclic systems and double bonds in small rings. The other situation involves cases of valid partial stereochemistry. These include derivatives of penicillins and cephalosporins and substances with unknown configuration at a heteroatom. The stereocentre with unknown configuration in the triazole ring of Figure 7 is an illustration of valid partial stereochemistry.

The third stage of conversion will translate the nomenclature stereodescriptor to an atom/bond specific parity descriptor. This process is straightforward for *R*/*S*, *R**/*S** and *E*/*Z* because only a single stereocentre is involved. Translating the relative ring descriptors (e.g., *cis*, α or *exo*) is somewhat more complex because it requires identifying the *ring plane* and the substituents referred to by the relative descriptor.

Co-ordination Compounds

Stereodescriptors for co-ordination centres[11] consist of four parts: a symmetry site term, a configuration number, a chirality symbol, and a ligand stereochemistry segment. The symmetry site term identifies the molecular geometry at the co-ordination centre, e.g., square planar (SP-4), trigonal bipyramid (TB-5), and octahedral (OC-6). The configuration number is a sequence of digits describing the geometrical arrangement of ligands about the centre in terms of their CIP rank.

The details of determining the configuration number vary somewhat from one geometry to another, but they can be illustrated for the OC-6 case (*Figure 9*). The configuration number is two digits. The first digit gives the CIP rank of the atom *trans* to the most preferred atom. The second digit gives the CIP rank of the atom *trans* to the most preferred of the remaining atoms: those in the plane perpendicular to the principal axis. In cases where there are sets of constitutionally equivalent atoms, a choice can arise in selecting the preferred atom. In such cases, preference is given to the atom *trans* to the least preferred atom; this is called the *trans maximum* rule. The chirality symbols *C* (clockwise) and *A* (anticlockwise) are used to distinguish enantiomeric configurations. These rules are illustrated in Figure 9.

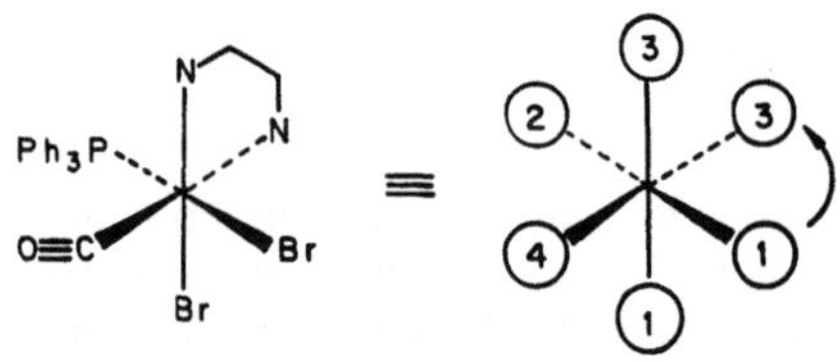

DES: OC-6-32-C

Figure 9. Example of co-ordination stereodescriptors

Constructing a configuration descriptor for a co-ordination centre can be accomplished by the following procedure:

1. Parse the stereodescriptor, select the appropriate model, and assign CIP rank numbers to the sites of the model.
2. Rank the atoms attached to the co-ordination centre by using the CIP procedure.
3. Find all possible ways to map the atoms to the sites of the model so that the CIP rank for the atom matches that of the site.
4. If the chirality of the structure thus constructed is opposite to the chirality symbol in the stereodescriptor, interchange any pair of *trans* ligands.
5. Compare the structure just constructed with any previous structures and eliminate redundancies. The procedure concludes successfully if one and only one structure can be constructed; otherwise, the substance would be flagged for manual review.

CONCLUSION

CAS has undertaken a multi-year effort to renovate the Registry System. As part of this effort, we are currently engaged in adding atom/bond specific stereodescriptors to the connection tables of stereospecific substances already on the Registry File. These extensions will, in turn, permit us to enhance structure display and substructure search to use accurately the stereochemical information available.

We have defined a representation that covers all types of stereochemistry cited in CA index names and localised to an atom or bond, including cases where stereochemistry is only partially known. In addition, we have developed algorithms for converting the stereodescriptor in the CA index name to atom/bond specific descriptors in the connection table. We expect that the 60 – 75% of the stereospecific substances on the Registry File can be converted algorithmically. Any substance that fails the algorithmic processing will be manually converted by entering the structure diagram with stereochemistry.

REFERENCES

1. Dittmar, P.G.; Stobaugh, R.E.; Watson. C.E. 'The Chemical Abstracts Service Chemical Registry System. I. General Design'. *J. Chem. Inf. Comput. Sci.*, **1976**, *16*, 111--121.
2. Blackwood, J.E.; Elliot, P.M.; Stobaugh, R.E.; Watson. C.E. 'The Chemical Abstracts Service Chemical Registry System. III. Stereochemistry'. *J. Chem. Inf. Comput. Sci.* **1977**, *17*, 3–8.
3. Blackwood, J.E.; Giles, P.M. Jr. '*Chemical Abstracts* Stereochemical Nomenclature of Organic Substances in the Ninth Collective Period (1972–1976)'. *J. Chem. Inf. Comput. Sci.* **1975**, *15*, 67–72.
4. *Chemical Abstracts Index Guide*, Appendix IV, 1989, pp. 180I–199I.
5. Cahn, R.S.; Ingold, C.; Prelog, V. 'Specification of Molecular Chirality'. *Angew. Chem. Int. Ed. Engl.* **1966**, *5*, 385–551.
6. Petrarca, A.E.; Lynch, M.F.; Rush, J.E. 'A Method for Generating Unique Computer Structural Representations of Stereoisomers'. *J. Chem. Doc.* **1967**, 7, 154–165.
7. Wipke, W.T.; Dyott, T.M. 'Stereochemically Unique Naming Algorithm'. *J. Am. Chem. Soc.* **1974**, *96*, 4834–4842.
8. *Chemical Abstracts Index Guide*, Appendix IV, 1989, pp. 184I–187I.
9. Detailed restrictions on the use of the various types of relative descriptors are given in *Chemical Abstracts Index Guide*, Appendix IV, 1989, pp. 181I–183I.
10. Vander Stouw, G.G.; Elliott, P.M.; Isenberg, A.C. 'Automated Conversion of Chemical Substance Names to Atom-bond Connection Tables'. *J. Chem. Doc.*, **1974**, *14*, 185–93.
11. Brown, M.F.; Cook, B.R.; Sloan, T.E. 'Stereochemical Notation in Co-ordination Chemistry. Mononuclear Complexes'. *Inorg. Chem.*, **1975**, *14*, 1273–1278.

RELATIVE CONFIGURATIONS IN STEREOCHEMICAL SUBSTRUCTURE SEARCH

Alan H. Lipkus and Paul E. Blower, Jr.
Chemical Abstracts Service, P.O. Box 3012, Columbus, Ohio, 43210 U.S.A.

ABSTRACT

Incorporation of stereochemically augmented connection tables into the CAS Registry File will make stereochemical substructure searching possible. A stereochemical search capability could be implemented by extending the present search system to determine the stereochemical validity of a topological substructure match. However, much of the stereochemistry in the Registry is described only in terms of relative configurations. When relative stereochemistry is involved, it may not be obvious whether a topological substructure match is stereochemically valid, since each relative set of stereoatoms may be viewed as having either of two chiralities. This type of problem can be represented as a weighted graph. By analysing the corresponding graph, one is able to determine whether a topological substructure match is stereochemically valid. A graph algorithm based upon a spin-glass analogy is described.

INTRODUCTION

The current version of the Chemical Abstracts Service (CAS) Chemical Registry System, called Registry III, was put into operation in 1974[1]. Since that time, many changes in the chemical sciences and computer technology have taken place. As a response to these changes, CAS is currently engaged in a multi-year effort to build the next generation of the Registry System. Enhancements in information handling for a number of areas of chemistry are planned. A significant portion of this effort concerns the handling of stereochemical information. At present, such information is recorded as an alphanumeric descriptor and in the chemical name. One of the enhancements planned for the Registry System is to interpret these textual descriptions of stereochemistry with specially designed algorithms and add the information directly to substance connection tables. This work is underway and has been described elsewhere[2].

The purpose of adding stereochemistry to Registry connection tables is to make this information available in a form more accessible for computer handling and manipulation. This will permit stereochemical information to be used to enhance structure display and substructure search, both of which currently operate at the topological level. In the case of substructure search, the addition of stereochemical capabilities implies that the user could specify a query with stereochemistry and be given an answer set of structures each of which contains the query with the desired stereochemistry. The incorporation of stereochemical information into substructure search should be, for the most part, a straightforward extension of topological search techniques. However, when that information involves relative, as opposed to absolute, stereochemistry, the search problem can be somewhat more

W. A. Warr (Ed.)
Chemical Structures 2

complicated. Provisions have been made to store relative stereochemistry in Registry connection tables, and the eventual stereochemical enhancement of substructure search is anticipated. We have therefore examined more closely the handling of relative stereochemistry in a stereochemical substructure search system.

STEREOCHEMICAL COMPARISON

A substructure search system operating at the topological level can serve as the basis for a system with stereochemical capabilities. This scheme assumes that a topological search, in which the stereochemistries of the query substructure and file structures are neglected, is performed first. Then, another step that involves a series of stereochemical comparisons is added. In this step, the stereochemistry of each structure in the topological answer set is compared with that of the query, and only those structures that contain the query with the desired stereochemistry are placed in a final answer set. This comparison uses the atom-to-atom map returned by the topological search to determine whether the configurations of each pair of topologically matching stereocentres are identical or opposite; if the configurations of each pair are identical, the structure is placed in the final answer set.

The method used to compare the configurations of the relevant stereocentres (we will consider here only centres with tetrahedral geometry) is based on the descriptors used for computer representation of these configurations. In the stereochemically enhanced Registry, such configurations will be represented by the parity descriptor method[3,4]. This is similar to the Cahn-Ingold-Prelog method[5] but uses only local information, i.e., the atom numbers of the immediate neighbours. The parity descriptor tells whether the three highest-numbered atoms run clockwise (parity = even) or counterclockwise (parity = odd) when viewed from the lowest-numbered atom. The parity descriptors assigned to topologically matching stereocentres in different structures cannot be literally compared when there are differences in atom numbers, but in that case they can be compared by a procedure that takes these differences into account.

When both query and structure have absolute (i.e., completely known) stereochemistry, their configurations can be compared as just described. There may, however, be cases in which some or all of the stereochemistry is relative. Relative stereochemistry involves sets of stereocentres whose configurations are known only in relation to one another. In the stereochemically enhanced Registry, each relative configuration will be identified as such, and configurations in the same relative set will be indicated by a common group number[2]. Relative configurations can complicate the stereochemical comparison of query and structure because each relative set has two equally valid representations due to incomplete knowledge about the stereochemistry. These are (1) the 'given' representation, in which the configurations show what is stored in the connection table, and (2) the 'reflected' representation, in which each configuration in the set is inverted.

Figure 1a shows an example of a stereochemical comparison involving absolute and relative stereochemistry. The query substructure and the structure, which is presumed to have been retrieved by a topological search, each have four stereocentres. It is assumed that in the query these stereocentres are grouped into

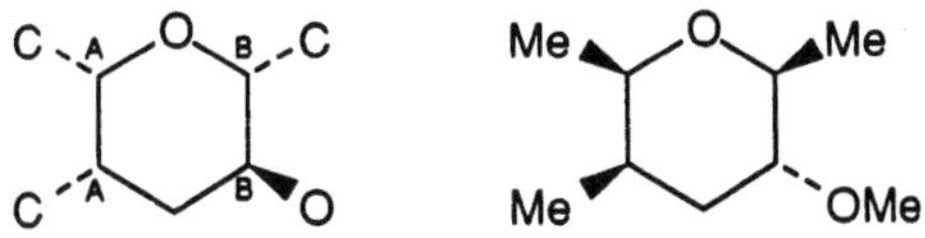

(a)

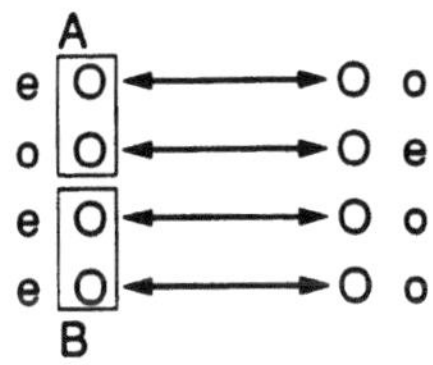

(b)

Figure 1. Stereochemical comparison: (a) substructure (left) and structure (right) to be compared; (b) schematic diagram of the comparison

two relative sets, A and B (the centres are labelled according to their set). A schematic diagram of this stereochemical comparison is shown in Figure 1b. The stereocentres are denoted by circles, and the arrows connecting pairs of circles indicate the mapping between query stereocentres (left) and structure stereocentres (right) as found by the initial topological search. The boxes, labelled A and B, enclose stereocentres in the same relative set; unenclosed circles denote centres with absolute stereochemistry. Each circle is labelled with the parity (e = even, o = odd) of the corresponding stereocentre, and we assume for illustrative purposes that the query and structure have similar atom numbers so that these parities can be literally compared. Since the reflection of both relative sets makes the configurations (or parities) of the query and structure identical, we conclude that this topological match is also a stereochemical match.

An algorithm is needed that can analyse this sort of comparison problem, involving relative stereochemistry, and determine whether or not a stereochemical match exists. It must be able to handle problems larger and more complex than the example just given even though such problems may occur infrequently in practice. A 'brute force' algorithm can be immediately identified: use both valid representations for each relative set to generate all possible stereochemistries for query and/or structure, and examine every combination for a stereochemical match. Such an algorithm is feasible, but further examination of the problem reveals a more subtle approach.

GRAPH REPRESENTATION

A compact way of representing a stereochemical comparison for the purpose of computer analysis is as a graph, i.e., a set of nodes connected by a set of edges. In this graph, each relative set in the problem is represented by a node. If absolute stereochemistry is present, the set of all absolute centres, whether from query or structure, is represented by an additional node. The graph is constructed by the following rules:

1. For each pair of topologically matching stereocentres at least one of which is a relative configuration, draw an edge connecting the two nodes representing their respective sets.
2. Assign to this edge a weight that depends on the relationship between the (given) configurations of the stereocentres; the weight is +1 if they are identical and −1 if they are opposite. Note that since at least one of the centres is relative, opposite configurations do not immediately imply that a stereochemical match is impossible.
3. If the latest edge would connect two nodes already connected, do not draw it, but its weight must be the same as the weight of the already existing edge (otherwise, a stereochemical match is impossible).

These rules explicitly neglect pairs of topologically matching stereocentres that both have absolute stereochemistry. It is assumed that such pairs have already been examined to make sure that their configurations are identical. If all topologically matching centres in the problem are of this type, meaning that there is no relative stereochemistry, the graph representation is not applicable.

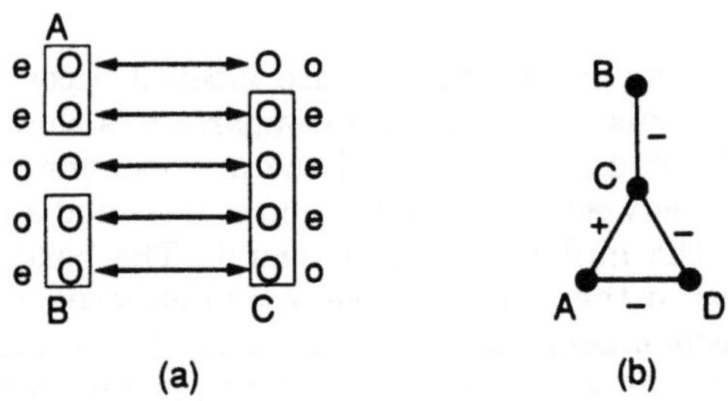

Figure 2. Graph representation of a stereochemical comparison: (a) schematic diagram; (b) corresponding graph

Figure 2 illustrates the derivation of the graph representation for a hypothetical stereochemical comparison problem. The problem, which is deliberately more complex than the problem in Figure 1, is shown schematically in Figure 2a. There are three relative sets of configurations, A, B, and C, as well as some absolute stereochemistry. As before, it is assumed that the parities shown can be literally compared. The graph derived by applying the above rules is given in Figure 2b. This is a connected graph (consisting of one component), but in some cases, the result could be a disconnected graph (consisting of two or more separate components). The nodes labelled A, B, and C correspond to the relative sets, while node D represents the two absolute centres in the problem. The edge weights are indicated only by their respective signs, '+' or '−', to simplify the graph. Note that if we were to change the parity of either one of the stereocentres in set B, no graph would be possible because the second edge between nodes B and C would have a weight opposite to the first, thus violating rule 3. Such a violation implies that a stereochemical match is impossible and is likely to occur in many real problems.

To determine whether or not a stereochemical match exists, it is necessary to determine whether or not some of the relative sets can be changed from their 'given' representation into their 'reflected' representation in a way that makes all the appropriate configurations (parities) identical. If we denote these two representations by a 0 and 1, the task can be viewed as determining whether it is possible to label each node of the graph with a 0 or 1 so as to satisfy two constraints. These constraints, based on the edge weights, are as follows:

1. If an edge has a weight of +1, the nodes it joins must be labelled similarly, i.e., both 0 or both 1.
2. If an edge has a weight of −1, the nodes it joins must be labelled differently, i.e., one a 0 and the other a 1.

The effect of the first constraint is to preserve the relationship between configurations that are identical, as given, by treating both relative sets in the same way (reflecting both sets or not reflecting either set). The effect of the second constraint is to make configurations that are opposite, as given, into identical configurations by treating both sets differently (reflecting one set and not the other). A binary labelling of nodes that satisfies these constraints everywhere in the graph will be called consistent.

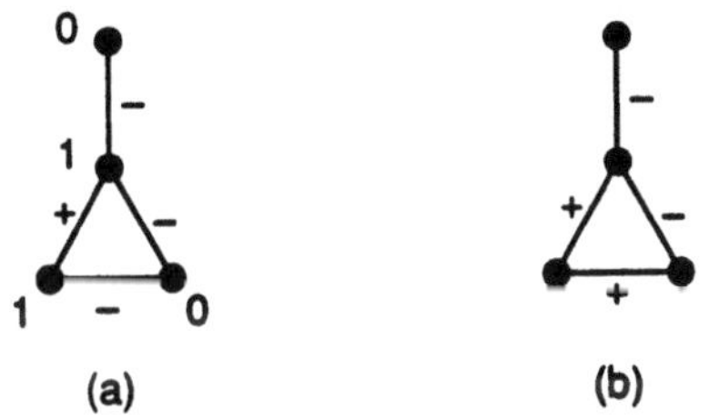

Figure 3. Graph labelling: (a) consistent labelling of the graph from Figure 2b; (b) graph with no consistent labelling possible

A consistent labelling for the graph of Figure 2b is shown in Figure 3a; another consistent labelling can be obtained by replacing the 0s with 1s and *vice versa*. The existence of a consistent labelling for this graph demonstrates that the topological match in Figure 2a is a stereochemical match. This conclusion can be verified by interpreting 0 as the 'given' representation and 1 as the 'reflected' representation. (The reverse is not acceptable because node D, labelled with a 0, represents absolute stereocentres which by definition cannot be reflected.) The interpretation implies that reflecting sets A and C, which correspond to the nodes labelled with a 1, makes all the appropriate parities identical. This is easily verified in Figure 2a. An example of a graph with no consistent labelling is Figure 3b, which differs from Figure 3a only in its edge weights. Any binary labelling of the nodes in this graph will somewhere violate one of the edge-weight constraints. Since a consistent labelling is impossible, any topological match described by this graph is not a stereochemical match.

ANALYSIS ALGORITHM

The existence of a stereochemical match has been shown to depend on a particular property of the graph representation for the problem, i.e., the existence of a consistent binary labelling. If this property of the graph can be appropriately characterised, it provides the basis for an algorithm. A useful observation in this regard is that the problem of finding a binary labelling of the nodes consistent with the edge-weight constraints can be related to the behaviour of physical systems known as spin glasses[6]. A spin glass is a collection of magnetic atoms randomly embedded in a non-magnetic host. The spins on the magnetic atoms experience different pairwise interactions that depend on the distance between them. Some pairs prefer to orient their spins in the same direction (ferromagnetic interaction). Others prefer to orient their spins in opposite directions (antiferromagnetic interaction). The remainder of the pairs do not experience any interaction. The spin glass attains its lowest energy when all of the preferred spin orientations are satisfied.

The analogy between the labelling problem and spin glasses is apparent. The nodes in the graph are like the magnetic atoms. The binary labels assigned to the nodes are like the 'up' or 'down' spin orientations, and the edge weights are like the interatomic distances in that they force certain pairs of binary labels (spin orientations) to be either the same or different. It can thus be seen that Figure 3a, which satisfies all of its edge-weight constraints, is analogous to a small spin glass in which the spins are all in their preferred pairwise orientations.

An important property of spin glasses is that it is sometimes impossible to satisfy simultaneously all preferred spin orientations. This property is referred to as 'frustration'. In a spin glass that is frustrated there always exist some spin pairs whose mutual orientation is not the preferred one. Frustration is an inherent property of the system and cannot be relieved by simply reorienting some of the spins. A graph such as Figure 3b exhibits a property quite similar to frustration in that it is impossible to satisfy simultaneously all of its edge-weight constraints. In this sense, Figure 3b is analogous to a frustrated spin glass.

A specific feature in spin glasses has been recognised as a necessary condition for frustration: the existence of a cycle having an odd number of antiferromagnetic interactions[7]. It can be seen how this leads to frustration by considering a walk that starts at any atom on such a cycle and goes around once. When the walk returns to the starting atom, the spin orientation has reversed itself an odd number of times and so is opposite to the original orientation. As an extension of our spin-glass analogy, this frustration condition can be used to characterise graphs (e.g., Figure 3b) for which a consistent labelling is impossible. Since an edge with a weight of -1 is analogous to an antiferromagnetic interaction, the necessary condition for there to be no consistent labelling is the existence of a cycle having an odd number of edges with a weight of -1. Putting this in more general terms: a stereochemical match exists only if the graph representation of the problem contains no cycles having an odd number of edges with a weight of -1. A useful corollary to this is the following: a stereochemical match always exists if the graph representation is acyclic.

An algorithm can now be outlined for analysing the stereochemical comparison problem when relative stereochemistry is present. For each topological match found by the initial topological substructure search, a graph representation is constructed using the previously described rules. The graph constructed may be disconnected and may thus have more than one component. (If no graph is possible, as noted earlier, it can be immediately concluded that the topological match is not a stereochemical match.) The following algorithm is then applied:

1. Find a component not already examined and go to step 2. If all components have been examined, stop: the topological match is a stereochemical match.
2. Calculate the number of nodes minus the number of edges for the current component. If it equals 1 (the component is a tree), go to step 1. Otherwise, go to step 3.
3. Use a cycle-detection routine to determine whether there exists in the current component a cycle having an odd number of edges with a weight of −1. If there is, stop: the topological match is not a stereochemical match. Otherwise, go to step 1.

The computational cost of cycle detection is comparatively high. The design of the algorithm is therefore intended to avoid unnecessary cycle detection by performing a simple test to determine whether the current component is a tree. The test is valid only for connected graphs, so it is applied only to components which, by definition, are connected. It is likely that many real stereochemical comparison problems will be simple enough to have an acyclic graph representation; the problem shown in Figure 1 is an example. In cases like this, the algorithm will completely avoid unnecessary cycle detection.

CONCLUSION

In practice, the problem of handling relative configurations in a stereochemical substructure search system may not be relevant to the majority of the searches conducted. Nevertheless, some algorithm for this problem is required if relative stereochemistry is to be handled at all. The algorithm developed in this paper is a viable alternative to the 'brute force' approach of generating all the possible stereochemistries implied by the relative configurations.

The analogy found to exist between the graph representation of the problem and a spin glass is unexpected and useful. The practical benefit of this analogy is being able to borrow the notion of frustration, a property exhibited by spin glasses. The property of frustration is found to provide a conceptually simple way of characterising whether or not there is a stereochemical match, and a test for this property forms the basis of the algorithm presented.

REFERENCES

1. Dittmar, P.G.; Stobaugh, R.E.; and Watson, C.E. 'The Chemical Abstracts Service Chemical Registry System. I. General Design'. *J. Chem. Inf. Comput. Sci.* **1976**, *16*, 111–121.
2. Blower, P.E., Jr.; Lillie, D.H.; Lipkus, A.H.; Qian, C. 'Addition of Atom/Bond Stereochemistry to the CAS Registry File'. In these Proceedings.
3. Petrarca, A.E.; Lynch, M.F.; Rush, J.E. 'A Method for Generating Unique Computer Structural Representations of Stereoisomers'. *J. Chem. Doc.* **1967**, *7*, 154–165.
4. Wipke, W.T.; Dyott, T.M. 'Simulation and Evaluation of Chemical Synthesis. Computer Representation and Manipulation of Stereochemistry'. *J. Am. Chem. Soc.* **1974**, *96*, 4825–4834.
5. Cahn, R.S.; Ingold, C.; Prelog, V. 'Specification of Molecular Chirality'. *Angew. Chem. Int. Ed. Engl.* **1966**, *5*, 385–551.

6. *Spin Glass Theory and Beyond*; Mezard, M.; Parisi, G.; Virasoro, M.A.; World Scientific: Singapore, 1987.
7. Toulouse, G. 'Theory of the Frustration Effect in Spin Glasses'. *Comm. Phys.* **1977**, *2*, 115–119.

THE DU PONT GLOBAL TECHNICAL INFORMATION SYSTEM

Jean G. Marcali, Florence H. Kvalnes, James A. Patterson, and Edward S. Wilks
E.I. du Pont de Nemours & Co., Wilmington, Delaware 19880–0014, U.S.A.

ABSTRACT

An integrated private database on STN International has been designed and implemented by Du Pont and Chemical Abstracts Service (CAS). The database, named SCION, consists of a chemical file and a document file. A unique feature of this system is that both Du Pont proprietary files and publicly available files on STN International can be searched using the same MESSENGER command language. The chemical file contains structures, Du Pont accession numbers (called C-Numbers), molecular formulae, systematic names, synonyms, descriptors, and CAS Registry Numbers (for equivalent organic, inorganic, and polymeric substances). The document file comprises bibliographic data, abstracts, subject indexing *via* controlled vocabulary terms, and an online hierarchical thesaurus of these controlled terms. Custom features, system capabilities, and preliminary reactions of users to the online database are discussed.

INTRODUCTION

Du Pont's proprietary technical information system (see Figure 1), established in 1964 and later enhanced for polymer storage and retrieval, was described in earlier papers[1–4]. This corporate system supports Du Pont's multidisciplinary scientific and technical efforts worldwide. It is the Company's only comprehensive index to, and collection of, documents covering Du Pont's technical efforts.

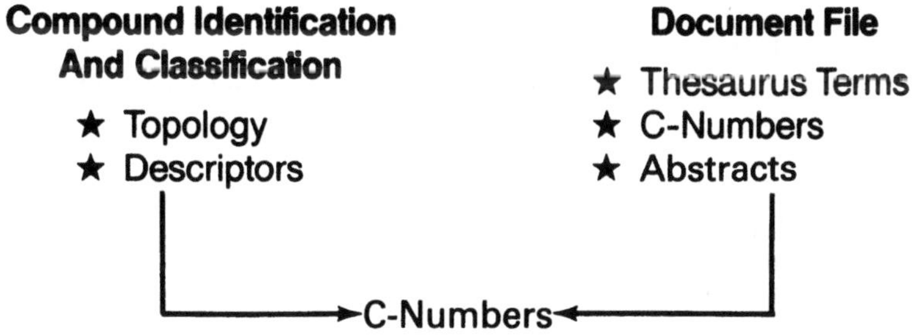

Figure 1. Du Pont proprietary technical information system

DU PONT PROPRIETARY TECHNICAL INFORMATION SYSTEM

Information scientists index and retrieve documents using general terms from the thesaurus and chemical terms. General terms represent concepts, e.g., POLY-

W. A. Warr (Ed.)
Chemical Structures 2

MERISATION. Chemical terms are seven-digit alphanumeric Du Pont accession numbers or C-Numbers, and each represents an individual chemical substance, including polymers. A C-Number is assigned sequentially to each unique substance entered into the system, and it links a series of files used for identification, document referencing, and classification of substances. Non-polymeric organic compounds are stored as compact connection tables based on topology; descriptors, which resemble chemical fragments, are also part of the record. Polymeric substance records consist of C-Numbers of component monomers as well as polymer class descriptors where applicable.

Du Pont's topology input procedure is an adaptation of the Registry system developed by CAS[5,6] from work done initially by Du Pont[7]. This work was shared with CAS and the American Chemical Society with the expectation that they would use it to benefit the chemical community. The CAS Registry System and Du Pont's chemical system are based on a chemical structure storage and search system[8] co-operatively developed by both organisations.

The Du Pont proprietary technical information system was state-of-the-art when it was developed 20 years ago. It is still operating in batch mode but has become increasingly difficult to support. In response to demand from the technical community for online access to the information, the abstract file was duplicated and converted to an interactive, free-text searchable database as an interim measure.

With the realisation that neither system was adequate to meet future requirements, an in-depth study of other major companies' proprietary information networks was initiated. Conclusions from the investigation showed that the basic principles of Du Pont's system were sound and sophisticated, but up-to-date software was needed.

A three-person team with extensive experience in Du Pont's proprietary technical information system and knowledge of the needs of the technical community was established. User requirements and specifications for the new system were developed and circulated to technical information staff and management for review.

The team decided that the system software should be selected from a source that was well-established in the field of technical information. In addition, if one software and one command language could be used for the entire system, i.e., both text and chemical structure, it would be advantageous to the technical community. Because the thesaurus, a hierachical list of controlled terms, was the key to the text or document file, thesaurus software was also necessary. Search software for chemicals had to have the capability to search by substructure or full structure, by name, by compound number, by molecular formula, and by class descriptor. Continuity in both systems support and staff was a very important consideration. Another criterion was that the system be kept up-to-date with enhancements resulting from ongoing research in information science.

MESSENGER software best met the criteria because it handled both text and chemical structure with one command language. Development of thesaurus software for STN International was well underway. Also, MESSENGER must be kept technologically competitive because it is the software used on STN, an international network currently consisting of close to 100 scientific and technical databases with nodes in Columbus, Ohio; Karlsruhe, Germany, and Tokyo, Japan. A totally integrated system for chemical substances and text would be in place and would be interactive globally; the system would be available over 99% of the advertised time. Chemical Abstracts Service would support the system, and the economics were competitive, if not attractive.

Du Pont submitted a Request For Proposal to CAS for consideration; by late 1986 CAS submitted a proposal. The Du Pont Plan and the proposal from CAS were circulated to Du Pont departments for review. They agreed that contract negotiations should proceed. Concurrent with contract negotiations, a security inspection of CAS facilities, systems, and procedures was made by a Du Pont team, which concluded that requirements for internal security standards would be met. Our contract with CAS to develop and maintain a Du Pont Private Registry was authorised in August 1987.

Since 1987, a team of six Du Pont staff (a system analyst and five technical information specialists) and twelve CAS staff (five chemists and seven system analysts) designed and built the database and converted 160,000 substances, abstracts and indexing records for over 112,000 documents, and a 26,000-term thesaurus to create SCION (Scientific Corporate Information Online). The database has the unique feature that both Du Pont proprietary files as well as publicly available files on STN International can be searched with the same command language, MESSENGER. SCION consists of a chemical file and a document file.

CHEMICAL FILE

The chemical file, formerly called the Chemical Information System (CIS), was designed to parallel the Chemical Abstracts Registry File as closely as possible. Standard Registry File fields were used and, with some exceptions, CAS structuring conventions were followed. This allows for 'crossover searching' of both the Du Pont Database and the Registry File using MESSENGER search commands. Additional fields were used for information relevant only to the Du Pont file. An example is shown in Figure 2.

CHEMICAL FILE DISPLAY

The CNUM represents the Du Pont C-Number. This is the tie-in to the bibliographic data in the document file. Each compound has a unique C-Number. In this paper C-Numbers are disguised for security reasons.

The RN is the CAS Registry Number and is included whenever the substance is in the CAS Registry File. About half of Du Pont's chemicals overlap structures in CAS Registry. The presence or absence of a CAS Registry Number for a Du Pont substance lets the scientist immediately know whether the compound has appeared in the public literature or any publicly available files on STN. The RN field can be used to sort an answer set for compounds that have a CAS registry number. The selected numbers can then be used to find references in STN files that contain CAS Registry Numbers, such as CA File.

The CAS field is the CAS name for the compound; the DUP field represents the Du Pont name. In some cases both names are the same. However, polymer names are different. We elected to retain our existing polymer nomenclature system[4], which meets our needs better than CAS polymer nomenclature.

The SYN field is synonyms, tradenames, codes, etc. This field is especially useful

L1 Answer 1 of 1 ★★★★ Du Pont Proprietary Information ★★★★

CNUM: 177001M
RN: 9003-68-3
CAS: 1,4-Benzenedicarboxylic acid, polymer with 1,2-ethanediol (9CI)
DUP: POLYESTER-ETHYLENE GLYCOL/TEREPHTHALIC
SYN: ARNITE®
SYN: DACRON®
SYN: HOSTADUR®
SYN: MYLAR®
SYN: POLY-ETHYLENE TEREPHTHALATE
SYN: PET POLYMER
SYN: P88®
SYN: REBAND®
CT: POLY-HOMO; POLYESTER
MF: $(C_8H_6O_4 \cdot C_2H_6O_2)_x$
CI: PMS

COM 1
CNUM 170056N
MF C_8H_6O
CI COM

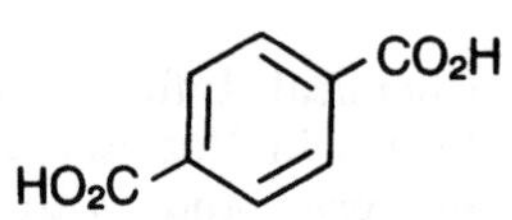

COM 2
CNUM 170111P
MF $C_2H_6O_2$
CI COM

$HOCH_2CH_2OH$

Figure 2. Chemical file display

in identifying tradenames and codes. The CT represents the controlled term field, which contains selected chemical class terms (e.g., polymer classes such as polyester and polyamide). This type of term is unique to the Du Pont database. The MF represents the molecular formula field. The CI is the class identifier. Broad classes, such as polymers and alloys are specified when appropriate. These classes are somewhat broader than controlled terms.

The FS (not shown) represents the File Segment field. It indicates the presence or absence of CONCORD 3-D co-ordinates for displaying and modelling structures using Alchemy software. Between 40% and 50% of the structures have 3-D co-ordinates.

The chemical structure is the most important part of the record. Structures are fully searchable using MESSENGER commands. All features used to search CAS Registry are available in this database.

CONVERSION

The price differential for manual structure input *versus* electronic input of existing connection tables, and the time required to build the file one compound at a time, provided much incentive to use existing CIS structure tables for input. However, problems had to be solved for input to be done electronically. New chemical input had to resemble standard Registry format. Inorganic compounds and polymers were not stored by connection table; also Du Pont and CAS used different structure conventions for certain chemical classes, for example:

DU PONT CONVENTION	*CAS CONVENTION*
$CH_3 - C(=O) - O - Na$	$CH_3 - C(=O) - OH.Na$
$CH_3 - NH_3 - Cl$	$CH_3 - NH_2.H - Cl$

The first task was to convert our connection tables to a format that was acceptable for CAS input. Figure 3 illustrates the Du Pont connection table for benzene on the left and the format for CAS input on the right.

Du Pont Compact Connection Table	**CAS Redundant Connection Table**
1 C	1, c, s2 d3
2 1 C★5	2, c, s1 d4
3 1 C★5	3, c, d1 s5
4 2 C★5	4, c, d2 s6
5 3 C★5	5, c, s3 d6
6 4 C★5	6, c, s4 d5

Ring Closure 005★5006

Figure 3. Connection tables

To register Du Pont chemicals in the new file, about 225 programs were needed to extract appropriate data from CIS and to convert them to CAS input format. This was no easy task. CIS was written before database management systems were available and, as a result, relevant data were scattered throughout the database in files linked by pointers and indexes, i.e., there was no single file to dump to retrieve relevant information.

Nevertheless, programs were written which gave us connection tables in a reorganised format that resembled CAS input. For compound classes whose structuring conventions matched those of CAS, searches were conducted to identify C-Numbers of related compounds. Connection tables for these C-Numbers were extracted and reformatted by conversion programs. Where structuring differences between CIS and CAS were minor, programs were written to convert electronically some simple classes (e.g., Group IA metal salts of carboxylic acids, hydrogen halide salts of simple amines, and quaternary amines with halogen anions). Other classes could have been converted electronically, but the small number of compounds involved did not justify the complex programming required.

Electronic conversion of polymer records was a challenge. CAS permits polymer input *via* combination of component registry numbers, molecular formula, and a special code called an expression value, which is used to group components that belong together. Because our polymer record included C-Numbers of component monomers, programs were written to extract the molecular formula for each component from CIS, combine this information with component C-Numbers and the expression value, and format these data according to CAS input criteria. Information in DUP, SYN, and CT fields was also added electronically from tapes created from similar files in CIS.

Approximately 123,000 non-polymeric organic substances, 25,000 polymers, and 7,000 inorganic substances had to be converted. Of these, about 93,000 organic compounds and 24,000 polymers were converted electronically.

Approximately 38,000 substances that were not converted by machine were converted manually (see Figure 4).

— Complexes
— Organometallic Compounds
— Inorganic Compounds/Doped Substances
— Ionomeric Polymers
— Indefinite-Aftertreatment Polymers
— "Apparent Duplicates" In Polymers
— Non-Polymeric Duplicates

Figure 4. Manually-converted substructure classes

MANUALLY-CONVERTED SUBSTANCE CLASSES

Classes of compounds converted strictly according to CAS conventions were complexes, most organometallic compounds, and inorganic compounds.

Exceptions to CAS conventions were applied. For doped substances and ionomeric, aftertreated and graft polymers, customised conventions were established.

DOPED SUBSTANCES

Inorganic dopes (pure substances deliberately adulterated with controlled amounts of impurities to impart special properties) and other fractional compounds were registered in CIS by ignoring real atomic ratios and registering the substance as if there were one atom of each kind present. For example, antimony-doped stannic oxide, SnO_2:Sb, was registered as Sb1Sn1O1, regardless of the real ratio of atoms. As inorganic substance registration began for the new chemical file, CAS released a timely registry enhancement: an innovative tabular representation for dopes that treated each ratio of dopant to carrier as a separate substance. We chose to register dopes with a nominal dopant content of 0.99, regardless of actual dopant concentration; the other atoms were registered in their correct ratios. Thus, SnO_2:xSb, where x is known or unknown, was registered as SnO_2:0.99Sb by the new tabular method, as shown in Figure 5.

Sb ·	Sn ·	O
	8:IN	
Sb		0.99
Sn		1
O		2

Figure 5. Doped substances

IONOMERIC POLYMERS

The difference in structuring conventions between CIS and CAS for salts of non–polymeric substances resulted in failure of machine conversion of ionomers. For example, sodium methacrylate is stored as $CH_2 = C(Me) - CO_2 - Na$ in CIS but $CH_2 = C(Me)CO_2H.Na$ at CAS; hence the ionomer:

POLYETHYLENE/METHACRYLIC ACID, AFTERTREATED TO METHACRYLIC ACID, SODIUM SALT (E/MAA/NaMAA)

failed machine conversion. This polymer is stored in CIS as:

(C = C	C = C(Me) – CO_2H	.	(C = C(Me) – CO_2H	. Na$))_x$
(E)	(MAA)		(NaMAA)	

whereas in CAS it is stored as:

$(C = C \ . \quad C = C(Me) - CO_2H \quad . \quad (C = C(Me) - CO_2H \quad . \quad Na))_x$

About 1,000 ionomers were converted manually.

REPLACEMENT OF INDEFINITE AFTERTREATMENT IN POLYMERS

After treated polymers with imprecisely-defined structures, e.g., chlorosulphonated poly-ethylene (below), were indexed in CIS by a C-Number representing POLY-ETHYLENE, AFTERTREATED AND A GENERAL TERM, e.g., POLYAFT-CHLOROSULPHONATED. Intersecting these terms gave some false correlation when searched.

Concurrently with creation of the new Du Pont document and chemical files, we adopted a new aftertreated-polymer system in which fragments created by the treatment were named and stored as part of the chemical structure. Thus, the polymer below was named: POLY-ETHYLENE, AFTERTREATED TO CHLORO-SULPHONYL. The chemical descriptors for this polymer are ethylene and a 'chlorosulphonyl-carbon atom' fragment:

$(C = C. \quad Cl - SO_2 - \dot{C}.)_x$

The artificially created 'tri-radical' carbon, which is not a true radical, symbolises a carbon atom in the base polymer backbone. Each new fragment is named starting from the furthest point from the 'tri-radical' carbon and working along the group to it. The carbon atom itself is not included in the name. Inclusion of 'tri-radical' carbon atoms in fragments facilitates distinction from free-radicals, e.g., ($.SO_2Cl$).

Names of multiply aftertreated polymers include all fragments plus intermediate ones. Fragments are alphabetised, regardless of actual reaction sequence. Thus, the polymer above, treated with aniline, is named:

POLY-ETHYLENE, AFTERTREATED TO ANILINOSULPHONYL/CHLORO-SULPHONYL

The chemical 'components' are:

$$(C = C \quad . \quad Ph - NH - SO_2 - \dot{C}. \quad . \quad Cl - SO - \dot{C}. \quad)x$$

Chemical-file word-descriptors POLY-AFTERTREATED, POLYAFT-ANILINO-SULPHONYL, and POLYAFT-CHLOROSULPHONYL, are also used. The command:

Search ETHYLENE and (POLYAFT-CHLOROSULPHONYL and POLYAFT-ANILINOSULPHONYL)/CT

retrieves this polyethylene polymer.

Search ETHYLENE and POLY-AFTERTREATED/CT

retrieves any aftertreated polymer containing the ethylene monomer component. This innovation incorporated the new groups at the chemical level and removed the need for crossover to the document file during a search.

'APPARENT DUPLICATES' IN POLYMERS

Intellectually, the three polymers:

1. POLY-ETHYLENE/PROPENE
2. POLY-ETHYLENE, GRAFT PROPENE
3. POLY-PROPENE, GRAFT ETHYLENE

are perceived as different, but in the new Du Pont chemical file the computer-stored chemical descriptors for all three are identical:

$$(C = C \quad . \quad C = C - C)_x$$

The word-descriptor POLY-GRAFT is also computer-stored for polymers 2 and 3 and is usable during search but it is not accessed by the computer as a distinguishing feature during registration. Thus, once any one of these three is registered, input of the others is rejected as a 'duplicate' unless uniqueness is assured by use of a customised input field, called a sequence descriptor.

Some sequence descriptors are shown in Figure 6. Note that all eight polymers were computer-stored as $(C = C \ . \ C = C - C)_x$, because a monomer may not be repeated within a polymer storage record unless 'bound', e.g., as a salt or complex; therefore $(C = C \ . \ C = C \ . \ C = C - C)_x$ is rejected. The sequence descriptor individualises each polymer.

Polymer Name	Sequence Description
1. Poly-Ethylene/Propene	NONE
2. Poly-Ethylene, Graft Propene	1G2
3. Poly-Propene, Graft Ethylene	2G1
4. Poly-Ethylene/Propene, Graft Ethylene	12G1
5. Poly-Ethylene/Propene, Graft Propene	12G2
6. Poly-Ethylene, Graft Ethylene/Propene	1G12
7. Poly-Propene, Graft Ethylene/Propene	2G12
8. Poly-Ethylene/Propene, Graft Ethylene/Propene	12G12

Figure 6. Sequence descriptors for ethylene-propene graft polymers

Because the special input field is not searchable by MESSENGER on STN, the sequence descriptor was also stored in the CT field for use in searching. Thus, a search for POLY-ETHYLENE/PROPENE, GRAFT ETHYLENE with the logic:

(182409W AND 168943R)/CRN AND 2/NC AND PMS/CI AND 12G1/CT

(where 182409W = C-Number for ethylene
168943R = C-Number for propene
CRN = Component Registry Number
NC = number of components
PMS/CI = Polymers/Class Identifier
CT = Controlled Term)

retrieved only Polymer No. 4 in Figure 6; all eight were retrieved when 12G1/CT was omitted from the search logic.

NON-POLYMERIC 'DUPLICATES'

Certain non-polymeric, organic molecules that are registered as different entities in CIS are really tautomeric forms of the same molecule. For example, after

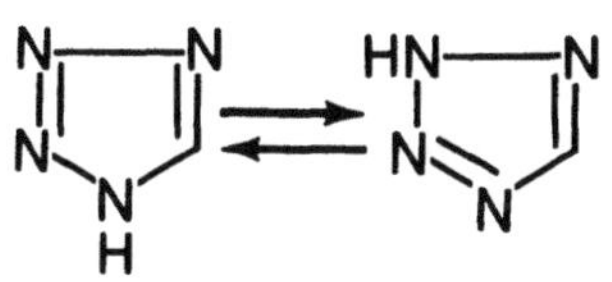

registration of either 1H-tetrazole or 2H-tetrazole in the new Du Pont chemical file as a normalised structure, attempted registration of the other resulted in rejection as a 'duplicate'. Document postings of the 'duplicate' were transferred to the C-Number for the accepted structure, and the 'duplicate' was deleted from CIS.

Another case of tautomerism involved thiono-acids, R – C(= S) – OH, versus thiolacids R – C(= O) – SH. CAS usually stores these acids in a normalised form. Thiono-acid forms of some dicarboxylic acids are used as condensation-polymer descriptors, so storing both forms in the new Du Pont chemical file by means of the CAS special input field 8:TO (TO = Tautomer Override) was sometimes necessary.

DOCUMENT FILE

The document segment of SCION also was designed to parallel files on the STN International Network, especially File CA, as closely as possible. This facilitates searching both in-house and public literature using one command language, MESSENGER. The same field names for bibliographic information were used where appropriate. The document file has two parts: one is a file of bibliographic information, abstract text, and the controlled terms used to index the document; the other is a thesaurus, a hierarchical list of controlled terms. Although File CA has no thesaurus, the capability is available in other databases on STN International.

DOCUMENT FILE DISPLAY

The standard bibliographic fields of accession number, author, title, date issued or publication date, and abstract text are present. Special fields include the home department number, which is assigned to the document by the issuing source, project codes, period covered (time during which the work was done), site at which the work was performed, and the security classification for the document. Date fields were designed to accommodate a four-digit year.

Because Du Pont is a global company, some of the documents are written, at least in part, in a language other than English. Following the practice of a number of databases, the language in which the document is written is provided.

Many of the documents in the database include the home department numbers of related reports to facilitate retrieval of earlier work on a particular project. This information has been incorporated in a cross reference field. Because all fields except the accession number and security class fields are searchable, the ability to do limited citation searching using this cross reference exists.

As part of a current awareness program, bibliographic data are added to the database frequently. To facilitate searching for newly added documents, a field for classification or category code is present. These codes cover major areas of interest to Du Pont. Typical examples are: Fibre Processing, Biochemistry, Engineering. By combining these category codes with either issue dates or the entry date field, the technical community can keep up to date with currently issued internal documents.

Indication of where a full copy of a document can be obtained is an important part of any database. Most documents in this database are available from a central source; some are available only from specific plant locations. The AV field indicates the location to be contacted when requesting a copy of the document.

The database contains research and development, manufacturing, and market research reports as well as correspondence and some publications. The DT field (document type) is used to select a specific type of document for further review or to restrict retrieval on a given subject to a specific document type.

Documents are indexed using concept co–ordination with a controlled vocabulary as well as links and rôle indicators for chemicals[9]. Substance C-Numbers with their rôle indicators and the controlled general terms are displayed in the link field. A link is an intellectual subdivision assigned by the indexer, which divides the document into discrete units to reduce false correlation at search. Each

Link A	Link B
Autoclaves/CT	Design/CT
Polymerization/CT	Extruders/CT

Figure 7. Use of links for intellectual division of a document

document has at least one link, which is designated by a letter of the alphabet. Usually the indexer follows the subdivisions or sections created by the document author. Figure 7 shows the concept of links. The document in Figure 7 discusses use of autoclaves for polymerisation of a polymer in one section and extruder design in another. Information on use of autoclaves for polymerisation was assigned to link A, and extruder design to link B. The command:

Search AUTOCLAVES/CT (L) DESIGN/CT

using the link operator L would not retrieve this document because the terms AUTOCLAVES and DESIGN are in different subdivisions or links. However, the command:

Search AUTOCLAVES/CT and DESIGN/CT

would retrieve it because the AND operator is not link-specific.

Common practice in indexing documents that discuss chemical reactions is to index each reaction in its own link. When rôle indicators are added, this technique simulates a reaction database. Rôles are used for every chemical, but experience has shown that an elaborate set of rôles leads to redundant indexing[10]. This database uses reactant, product, special agent, medium, and other (used when the aforementioned do not apply). A typical record is illustrated by Figure 8.

The abstract text for each document averages about 150 words and is more indicative than informative of document content. However, when convenient, results and data are included. Words in the abstract and title are part of the Basic Index and are searchable using free text searching techniques. Because each sentence in the abstract is a unit, proximity searching can be used for more specific retrieval during free text searching.

THESAURUS

The thesaurus is the cornerstone of searching by controlled terms. It is a dynamic tool that is constantly expanded and revised as technology, products, and processes change. Changes in thesaurus terms and their relationships are reviewed by a panel of experienced indexers with backgrounds in several scientific and technical disciplines.

The three types of relationships in the thesaurus are hierarchical, reference, and key terms. The command:

CRI	190743T	
DPT	F (FIBERS)	
DT	CLEARED FOR PUBLICATION	
SEC	PROPRIETARY INFORMATION PROTECTION CLASS	
CC	POLYMERS	
HDN	Z-52-112	
AU	Wittbecker, E L	
TI	Polycyclohexylene Ether	
ISS	19510401	
PER	195003 TO 195103	
PRJ	P8312-56789-90	
SS	Wilmington DE	
AV	CENTRAL REPORT INDEX	
CR	Z-51-9; QRA-50-21	
LA	ENGLISH	
AB	The ionic polymerization of 1,4-epoxycyclohexane to form high melting polycyclohexylene ether was studied. Catalyst concentration, temperature, and catalyst combination affected the rate and degree of polymerization. The best catalyst pairs identified were $FeCl_3$-$SOCl_2$, $FeCl_3$-succinyl chloride, and $SbCl_5$-succinic anhydride. Polymer properties included crystallinity which did not disappear up to 410°C and insolubility in many organic solvents. The polymer was soluble in a tetrachloroethane/phenol mixture. Copolymerizations with tetrahydrofuran and 1,4-epoxycyclohexane were studied in a variety of solvents.	
LINK	A	
	RCT	308294L (C-Number for 1,4-epoxycyclohexane)
	PRO	401825V (C-Number for the polymer)
	SAG	440678S; 432190B; 214532G; 284123F (C-Numbers for the catalysts)
	MED	195556E; 222394T; 385987C; 169012W (C-Numbers for the solvents
	OTH	401852V
CT	CONCENTRATIONS; CRYSTALLINITY; IONIC CATALYSTS; MELTING POINT; POLYMERIZATION; SOLUBILITY	
LINK	B	
	RCT	308294L; 398976Q
	PRO	412980Z
	SAG	440678S; 212345B
	MED	301987C; 192345G; 333456Y
	CT	IONIC CATALYSTS; POLYMERIZATION

Figure 8. Document file display

Expand CHEMICAL PROPERTIES + ALL/CT

results in the display shown in Figure 9, which illustrates a generic 'tree' by showing relationships between some chemical properties.

```
BT1 PROPERTIES
    → CHEMICAL PROPERTIES
    NT1 ACTIVITY COEFFICIENT
    NT1 COLOR STABILITY
    NT1 CORROSION STABILITY
        NT2 SALT SPRAY RESISTANCE
    NT1 MOLECULAR WEIGHT
        NT2 MOLECULAR WEIGHT DISTRIBUTION
            NT3 MONODISPERSE MOLECULAR WEIGHT
    RT CHEMICAL ANALYSIS
    KT CHEMICAL
    KT PROPERTIES
```

Figure 9. Thesaurus relationships

THESAURUS RELATIONSHIPS

BT, BT1, BT2 signify Broader Term, Broader Term level 1, and Broader Term level 2; the same applies to Narrower Terms, NT, NT1, NT2, NT3, etc. RT refers to Related Term and KT to Key Term. The hierarchical or generic relationships, Broader Term to Narrow Term, can result in five levels of specificity although the average number is three.

The reference terms are scope or explanatory notes, USE or SEE terms, and the standard RELATED terms. USE is an instruction to use the specified controlled term. SEE is the instruction to consult the list of controlled terms to determine which is more applicable. UF and SF are reciprocals for USE and SEE, respectively. RELATED (RT) terms are for consideration at search.

Key terms (KT) 'index' the thesaurus. Bound terms or phrases in the thesaurus are mostly natural word order rather than inverted. Therefore, key terms are vital. For example, to find all of the index or search terms that contain the word STABILITY, the instruction to list the key term and all of its related terms is given. The command, Expand STABILITY + KT/CT, lists (among other things) LIGHT STABILITY, STORAGE STABILITY, and THERMAL STABILITY.

The established hierarchies of broader to narrower terms permit searching generically. Generic searching, as practised in Du Pont's proprietary information database for many years, allows the collection, during search, of all of the document postings to narrower terms in a particular generic 'tree.' This avoids generic posting at indexing. In the new document file the thesaurus can be searched generically for any combination of broader, narrower, and related terms. Thus, the command:

Search CHEMICAL PROPERTIES + NT/CT

has the effect of searching the term CHEMICAL PROPERTIES and its narrower terms shown in Figure 9. The resulting set contains all documents posted to each of the individual terms. The command:

Search CHEMICAL PROPERTIES + BT/CT

retrieves all documents indexed individually to CHEMICAL PROPERTIES and to PROPERTIES. The command:

Search CHEMICAL PROPERTIES + RT/CT

retrieves all documents indexed to CHEMICAL PROPERTIES and to CHEMICAL ANALYSIS. The command:

Search CHEMICAL PROPERTIES + ALL/CT

retrieves all documents for all the broader, narrower, and related terms.

USER REACTION

A recent survey showed that users were enthusiastic about this new database. Cross reference to CAS Registry Numbers and the capability to search polymers by class and monomer component were perceived as advantages. Comprehensive retrieval at specific and generic levels with controlled terms as well as free text searching of the abstract text and bibliographic data afforded good flexibility. One command language to search both Du Pont proprietary and STN files was heartily welcomed.

IMPLEMENTATION

A training course, training tools, and menus are being developed. Additionally, an aggressive marketing program is in progress. Price schedules for users are being established, and costs will be billed electronically. An Information Help Line will be staffed by experienced technical information specialists.

SCION is in beta-test now; problems have been few. Release is expected in the fourth quarter of 1990. The next logical development is a properties data file.

CONCLUSION

Our system was in a very precarious state in 1985. But with the support of a skilled analyst and exemplary team effort, we have managed to keep the old system operational and convert about 75% of the files electronically and the remainder manually. Together, with the superb co-operation, flexibility, and expertise of the CAS staff, we have built SCION, with MESSENGER software and a proprietary security system.

KEY FEATURES

Key features of the system are interactive, global access with system reliability more than 99% of the advertised time (see Figure 10). One command language facilitates searching our proprietary database and public databases on STN International during one search session. Currency of data is improved. The system will be supported by CAS and will be upgraded as enhancements for STN are released. Our current awareness program is integrated into the system. Chemical substances are searchable by structure, names, including synonyms, class, and molecular formula and are displayable in 3-D with modelling capability. Polymers can be accessed by component C-Numbers and sequence descriptors. An online thesaurus of controlled terms for specific and generic searching is also a guide for indexing. Free text searching of abstract text and bibliographic data is available. These features, with menus and a private international telecommunications network, give the Du Pont scientific and technical community a powerful tool to find the right information to make timely, strategic decisions and maintain a competitive edge.

— INTERACTIVE, GLOBAL ACCESS

— SYSTEM RELIABILITY >99% OF ADVERTISED TIME

— ONE SEARCH, DU PONT & STN

— CURRENCY OF DATA

— SYSTEM SUPPORTED BY CAS

— CONTINUING ENHANCEMENTS

— CURRENT AWARENESS

— ONLINE THESAURUS

— MENUS

— SEARCHING
CHEMICAL STRUCTURE/SUBSTRUCTURE
CHEMICAL CLASS DESCRIPTORS
CHEMICAL NAMES/MOLECULAR FORMULAE
POLYMERS BY COMPONENT C-NUMBER
POLYMER SEQUENCE DESCRIPTORS
3D DISPLAY AND MODELING
CONTROLLED TERMS
FREE TEXT—ABSTRACT/BIBLIOGRAPHIC
ENHANCED GENERIC SEARCHING

Figure 10. Key features

ACKNOWLEDGMENTS

A project of this magnitude obviously involved many people. We gratefully acknowledge the contributions and dedication of the Du Pont/CAS team and respective support staffs, the support of other Central Report Index staff in maintaining operations, and the encouragement, support, and understanding given by management.

REFERENCES

1. Montague, B.A.; Schirmer, R.F. 'Du Pont Central Report Index: System Design, Operation, and Performance'. *J. Chem. Doc.* **1968**, *8*, 33–41.
2. Hoffman, W.S. 'Du Pont Information Flow System'. *J. Chem. Doc.* **1972**, *12*, 116–124.
3. Schultz, J.L. 'Handling Chemical Information in the Du Pont Central Report Index'. *J. Chem. Doc.* **1974**, *14*, 171–179.
4. Schultz, J.L. 'Polymer Nomenclature, Classification, and Retrieval in the Du Pont Central Report Index'. *J. Chem. Inf. Comput. Sci.* **1975**, *15*, 94–100.
5. Morgan, H.L. 'The Generation of a Unique Machine Description for Chemical Structures – A Technique Developed at Chemical Abstracts Service'. *J. Chem. Doc.* **1965**, *5*, 107–113.
6. Leiter, D.P.; Morgan, H.L.; Stobaugh, R.E. 'Installation and Operation of a Registry for Chemical Compounds'. *J. Chem. Doc.* **1965**, *5*, 238–242.
7. Gluck, D.J. 'A Chemical Structure Storage and Search System Developed at Du Pont'. *J. Chem. Doc.* **1965**, *5*, 43–51.
8. Hoffman, W. S. 'An Integrated Chemical Structure Storage and Search System Operating at Du Pont'. *J. Chem. Doc.* **1968**, *8*, 3–13.
9. Costello, J. C. Jr. 'Storage and Retrieval of Chemical Research and Patent Information by Links and Rôles in Du Pont'. *Am. Doc.* **1961**, *12*, 111–120.
10. Van Oot, J. G; Schultz, J. L; Mc Farlane, R. E; Kvalnes, F. H; Reister, A. W. 'Links and Roles in Co-ordinate Indexing and Searching: An Economic Study of Their Use, and an Evaluation of Their Effect on Relevance and Recall'. *J. Chem. Doc.* **1966**, *6*, 95–101.

CHEMICAL STRUCTURE REGISTRATION FOR BEILSTEIN ONLINE

Stephen Welford
Springer-Verlag, 8 Alexandra Road, Wimbledon SW19 7JZ, England

Clemens Jochum
Beilstein Institute, Varrentrappstr. 40–42, W–6000 Frankfurt, Germany

ABSTRACT

Beilstein Online is a comprehensive online database of organic chemical compounds and their reported chemical and physical properties. The database, which corresponds to the Beilstein Handbook of Organic Chemistry and covers the chemical literature from 1830, is available online on STN International and Dialog. This paper describes the structure registration system, which has been developed at the Beilstein Institute and used in the construction of this database. The paper concentrates on the treatment of tautomerism and stereochemistry, the connection table format in which structures are processed and delivered to online hosts, and the range of alternative delivery formats which are available for in-house use. Extension of the registration system and data structures for organometallic and inorganic substances in support of the Gmelin online database are also described.

INTRODUCTION

This paper describes the key aspects of chemical structure registration for the Beilstein database[1,2], and considers in particular the treatment of stereochemistry, charge mesomerism and tautomerism. We do not describe here the structure input conventions or input programs, MOLMOUSE and latterly MOLKICK, used for structure input to the Beilstein registry system, nor the administrative aspects of the structure input projects which, since 1984, have together resulted in the manual input of some 7 million structure records. Nor do we describe in detail the various related programs with which the registry software interacts, for example the chemical name to connection table (CT) conversion program VICA by which the majority of non-stereochemical structures in the Beilstein Handbook have been input automatically to the registry file, the tautomer generation program TAUFE which generates from a tautomer-normalised CT a preferred tautomeric form for publication in the Beilstein Handbook, and the range of computer-aided nomenclature programs, including a chemical name editor CAN-Edit and a fully automated CT to chemical name conversion program AUTONOM, which are being developed to support new methods of data processing and preparation at the Beilstein Institute[3].

W. A. Warr (Ed.)
Chemical Structures 2

Readers of this paper will be familiar with the general requirements of a structure registry system. The CAS Registry System provides a well-known precedent in this field, and a series of papers[4] introduces many of the concepts and terminology which are discussed here. The Beilstein registry system accepts structure input in a variety of forms; these include CTs in VICA and MOLMOUSE formats, and ROSDAL strings[5] generated by MOLKICK. Other input formats are also possible, including CTs from the Chemical Information System (CIS) and the C13NMR databank, but these will not be described further. The system interprets stereochemical information input graphically or entered in the form of conventional stereochemical descriptors[6], and applies the registry conventions regarding charge mesomerism and tautomerism. The structure is renumbered into a canonical form by a stereochemically-extended version of the algorithm originally described by Morgan[7] and formatted into a Beilstein Registry Connection Table, BRCT[8]. The BRCT is assembled together with related structural information, specifically chemical name and molecular formula, into a Structure Distribution File (SDF) record[9]. In the case of multicomponent compounds, for example salts, mixtures and addition compounds, the SDF record contains two or more BRCTs, one for each different structural fragment in the compound.

The SDF records are batched for adding to the registry file which, together with the compound property data files, is maintained at the Beilstein Institute on an IBM 3090 by the database management system ADABAS. If the SDF record is not present in the registry file then the compound has not been previously registered by Beilstein and a Beilstein Registry Number BRN is allocated and added to the SDF record. The BRN also becomes linked to the compound accession number for the purpose of loading the associated property data into the appropriate ADABAS data files.

The SDF structure file is used for delivery to hosts and in−house customers of the Beilstein structure file. Other delivery formats produced by the registry software include Télésystèmes-DARC and MDL MOLfile formats. Large DARC and MOLfile structure files have been created at the Beilstein Institute for the purpose of benchmarking the DARC-SMS and MDL MACCS-II substructure search systems against the S4 substructure search system developed jointly by the Beilstein Institute and the software house Softron GmbH[10].

KEY PROBLEMS IN STRUCTURE REGISTRATION

The key problems in structure registration have already been mentioned – the treatment of charge mesomerism, tautomerism and stereochemistry and the creation of a unique registry record for each compound. Software solutions to all of these problems have already been found, reported and put into practice in a variety of systems over the last two decades. Given the time constraints under which the Beilstein database creation software had to be completed and put into operation, it was not appropriate to embark on extensive research work to find new solutions to these basic problems; instead existing solutions were implemented in a combination which best suited Beilstein's needs. In this sense, the remainder of this paper does not report anything dramatically new, but rather serves to consolidate proven solutions, albeit in a new environment.

Although we concentrate now on the processing of structures by software, one other factor in the registration equation must be mentioned. No matter how elaborate the software, the quality of the registry file ultimately depends on the quality of the input structures. The end result of the registry software is essentially one of normalisation, that is the reduction of input variation by the forced selection or creation of a preferred form for each compound. Naturally, different systems emphasise and implement this process of normalisation differently. However, there exists in every system the danger that the special provisions in the registry software will be overwhelmed if the input variation is too high. There are basically two approaches to combat this problem. The first is to control rigidly the structure input by means of precise structuring conventions and a small experienced input team; the second is to equip the registry software to recognise and equate the variant input forms of each compound. A combination of both is probably optimal. In the case of Beilstein, the limited time frame for initial database creation and the huge number of structures to be input, and therefore the large number of input staff required, worked against the first approach and combined to force a greater emphasis on the registry software to manage the input variety. The area in which variation of input is greatest, as a result of a substantial lack of widely-accepted structuring standards, is in the graphic representation of stereochemistry[11]. This undoubtedly is the greatest source of difficulty for automated structure registry systems, and Beilstein is no exception. In mentioning this problem, we are perhaps making a plea for greater standardisation in this area, and the adherence to standards by authors, editors and publishers in the most general sense.

TAUTOMERISM

Structural tautomerism breaks down into several categories, which include the most common prototropic tautomerism, ring-chain and valence tautomerism, and charge mesomerism. Only the first and the last have been successfully treated in software – everyone is familiar with the CAS Registry III tautomer normalisation conventions[12]. It quickly became apparent, if not obvious, that in the case of a data handbook such as Beilstein, it was imperative that different tautomeric forms of a compound for which different data have been published must be registered differently; the situation is analogous to that of registering stereoisomers as separate compounds[13].

Given this criterion, and having already written tautomer normalisation software, we abandoned the idea of tautomer normalisation at the time of registration, except for the case of charge mesomerism which is discussed below. Different tautomeric forms of a compound, where these have been reported in the literature, are registered separately in Beilstein and have different Beilstein Registry Numbers. In the case of Beilstein on STN, the Beilstein structures are in any case reprocessed by a part of the CAS registry software at FIZ Karlsruhe in order to produce a structure search file which is searchable on STN under the CAS ONLINE conventions. Appropriate use of CAS tautomer bonds is therefore required on STN, which may lead to two or more BRNs being retrieved, each of which represents a separately registered tautomeric form. Since each of these will have the identical normalised CAS record, they will have identical displays on STN.

The ability to recognise tautomeric relationships and to create a normalised form is nevertheless important at Beilstein for three reasons. First, by implementing the tautomer normalisation software as a post-registration step, i.e., after the BRN assignment, it is possible to recognise by a simple sort of the normalised BRCTs the tautomer families which exist in the Beilstein file. Each family represents a compound and has as its members all separately known tautomeric forms of that compound; in principle, although this has not been done in practice, each family can be assigned a Beilstein Tautomer Number (BTN) which can be related to the BRN of each tautomeric form.

The second reason why this is important for Beilstein is that the normalised form represents not only the individual registered members of the family but also the space of all possible tautomeric forms, within the given definition of tautomerism. The normalised BRCT which defines this space is used as input to the Beilstein TAUFE program which, as mentioned earlier, generates the preferred tautomeric form on the basis of rules previously employed in the preparation of the Beilstein Handbook. In the case where maintaining the distinction between tautomeric forms is not considered important for publication in the Handbook, the preferred tautomeric form is used for naming and illustrating the compound in the Handbook.

Third, the tautomer information from the normalised BRCT is utilised in the tautomer search option of the Softron S4 search system. Although this is not yet fully implemented, S4 uses the tautomer information to control the mapping of the query structure onto all valid tautomeric forms of candidates in the Beilstein file.

CHARGE MESOMERISM

The situation regarding charge mesomerism is somewhat different. For several reasons, we decided that charged structures should be standardised to a preferred form at registration. Primarily this is to minimise the variation of input forms, which can be high in the case of charges situated in conjugated systems, particularly those involving one or more heteroatoms. Several situations are encompassed by the registry software, admittedly at the expense of losing the distinction between possibly separately existing charge mesomers.

The registry software here provides two principal functions: charge neutralisation and charge localisation. The emphasis of charge neutralisation is to deal with zwitterionic-type structures in which opposite charges are located either on adjacent atoms or within the same pi-system. The software attempts wherever possible to bring the structure into a non-charged form, and is accompanied by a redistribution of p-electrons and, where appropriate, a revaluation of formal valency (see Figure 1).

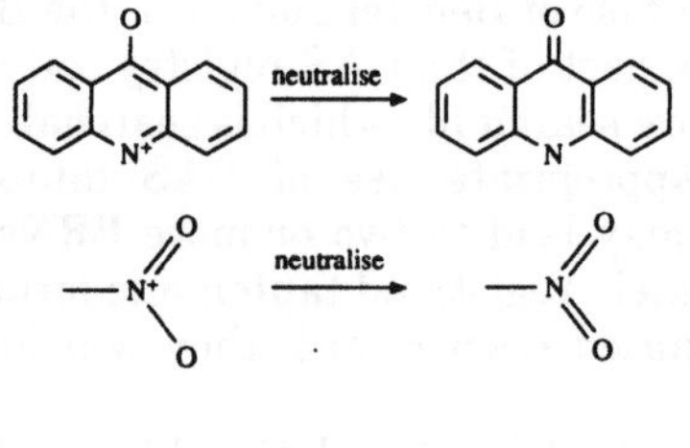

Figure 1. Charge neutralisation

Betaine-type structures and others in which opposite charges are not linked through a pi-system are left unchanged by the charge neutralisation procedure. Remaining charges, if any, are then subjected to charge localisation, which deals with delocalised charges and structures input with charges localised on arbitrary atoms within a pi-system. This procedure attempts to relocate each charge onto a preferred atom within the pi-system, with an associated redistribution of p-electrons (see Figure 2).

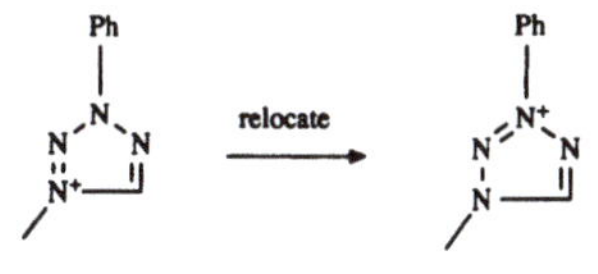

Figure 2. Charge localisation

If the charge is already situated on the preferred atom, or is not situated in a pi-system, then no charge relocation takes place. The selection of the preferred atom is purely algorithmic, making use of standard electronegativity values and ultimately, where a choice between atoms still remains, on the basis of their Morgan numbers. This guarantees that an atom is uniquely selected, and that the same atom is always selected.

STEREOCHEMISTRY AND UNIQUE NUMBERING

The treatment of stereochemistry and the generation of a unique numbering are described here together in order to emphasise the close interaction between these procedures. In short, the Wipke and Dyott SEMA algorithm[14,15] is used, with some modifications in its implementation, in order to generate a unique numbering which describes unambiguously both the constitution and the configuration of each structure. The overall purpose is, of course, to ensure that stereoisomers of a molecule are recognised as different and are registered differently. The close interaction arises from the fact that the configuration of each stereocentre in the molecule is coded in terms of a parity[16], whose value depends on the numbering assigned to the ligand atoms of the centre, and that the numberings are so assigned as to result in the optimal parity. In general terms, where the structure contains several stereocentres, the algorithm ensures that the combination of parities, or *parity record*, for the entire molecule is optimal. Note, the unique numbering generated here is not necessarily the same as that generated by the SEMA algorithm itself nor any similar algorithm. However, it is nevertheless canonical according to the criteria used, and is consistently reproducible; these are the two requirements of a canonicalisation program.

The algorithm is best described in three parts. Part I partitions the atoms of the structure into classes on the basis of their different structural attributes. The emphasis here is to maximise the number of classes into which the atoms are partitioned, in order to reduce the number of cycles required in Part II of the algorithm. The numbering assigned in Part II depends on the partitioning criteria, specifically the attributes used and the order in which they are applied.

Readers will be familiar with the concept of partitioning based on extended connectivity. We incorporated some of the modifications described by Moreau[17] in order to take into account not only extended connectivity but also all other discriminating attributes in the initial partitioning. The atom attributes are as follows, applied in the order given: extended connectivity, p-electron count, atomic number, localised charge, isotopic mass, valence, localised radical, and attached hydrogen isotopes. Note that stereochemistry is not considered in the partitioning process. The reason for this is to ensure that the partitioning is the same for all stereoisomers of a compound, with the result that all equivalent spanning trees based on the constitution of the molecule are considered in Part II of the algorithm.

If the final number of partitions equals the number of atoms in the structure then only one spanning tree can be grown, and this will provide the unique numbering for the structure, and will also determine uniquely the stereochemical parity record, if any.

Part II of the algorithm starts by ranking the partitioned atoms and identifying the *start atom* (or start atoms), and proceeds to generate a spanning tree. A spanning tree is grown from the start atom in a recursive, breadth-first fashion, in which neighbours of the start atom are first assigned in ascending order of their ranks; the neighbour with the highest rank is assigned to the spanning tree first, followed by the neighbour with the next highest rank, and so on. For each neighbour so assigned, each of its neighbours is similarly assigned, excluding any previously assigned neighbours, and so on level-by-level until all atoms in the structure graph have been assigned. The assignments result in the construction of a *from list*, a *ring closure* list and, if appropriate, a stereochemical parity record (see Figure 3).

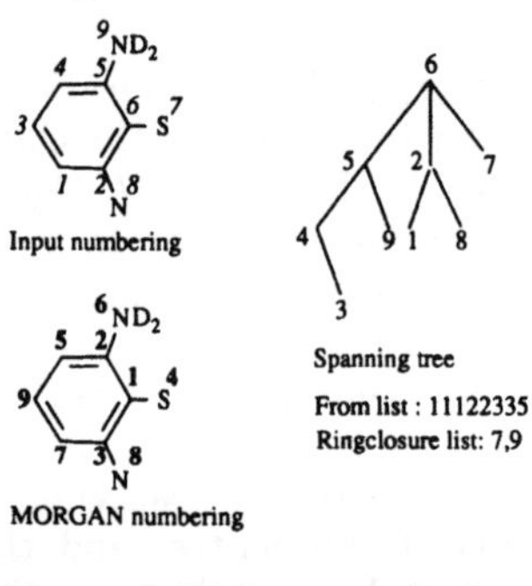

Figure 3. Unique numbering

Where an atom has two or more neighbours of identical rank (*choice points*) then all possible permutations of neighbour assignments are made, each leading to the generation of a further spanning tree. Some additional tricks are incorporated into Part II of the algorithm, for example the recognition of equivalent terminal ligands and ligand groups, and special provision for highly symmetric ring systems. These are introduced in order to control the number of spanning trees which must be grown, and therefore the time and space requirements of the program.

If a neighbour is assigned which completes the ligand assignments around an asymmetric atom then a stereochemical *atom parity* is computed and appended to the parity record of the current numbering. If the neighbour is assigned to a steric double bond then a *bond parity* is computed only if the other atom of the double bond and all of its respective ligand atoms have already been assigned, i.e., are already present in the current spanning tree. Similarly in the case of an allene-type chirality axis, all relevant atoms must be numbered before the *axis parity* is computed.

The parity assignment is illustrated with a simple example. Figure 4 illustrates the assignment of a Cahn-Ingold-Prelog (CIP) descriptor to a typical asymmetric atom; chemists will be familiar with the principle of ordering the ligand atoms according to the CIP Sequence Rule, and determining the *handedness* of the tetrahedron when viewed with the least significant ligand oriented to the rear. The computation of a parity proceeds in an analogous manner, albeit based on the principle of ligand numbering and not CIP priority.

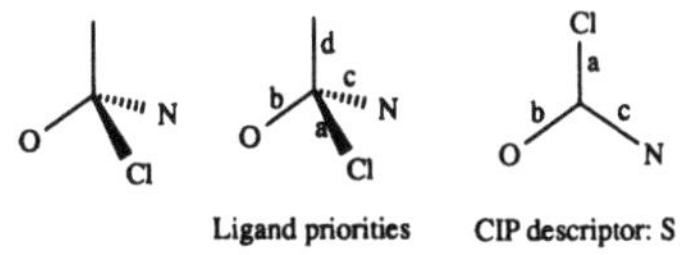

Figure 4. Assignment of a Cahn-Ingold-Prelog descriptor

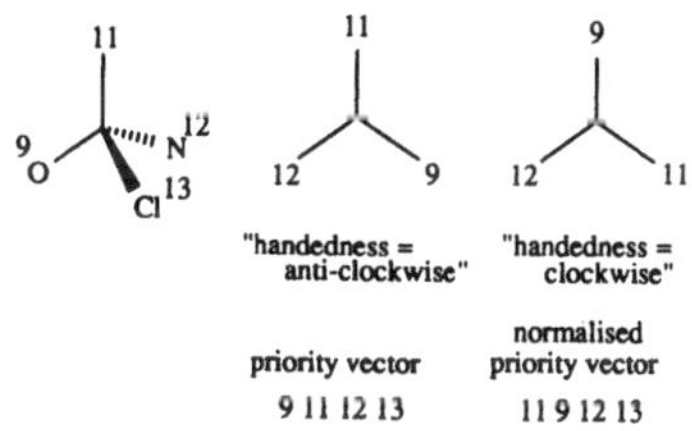

Figure 5. Computation of handedness

Figure 5 shows the same asymmetric atom with arbitrary input numbering of the ligand atoms. The tetrahedron is viewed with the atom of highest input number oriented to the rear, and the handedness is computed on the basis of the input numbering of the remaining ligands. If necessary, the handedness is normalised into a clockwise sense by interchanging the first two entries in the priority vector. At the point of parity assignment, the (normalised) priority vector is substituted by the Morgan vector of the current spanning tree, and the number of *transpositions* required to sort the Morgan vector into ascending order is computed (see Figure 6).

In the case of a tetrahedron this number is either odd or even (or zero) and this leads directly to a parity of 1 or 0 respectively (see Figure 7).

The situation is analogous for steric double bonds and allene-type chirality axes since, although the transposition rules are somewhat different, the number of transpositions required to bring the Morgan vector into its optimal form is always odd or even.

On completion, each spanning tree will have associated with it a certain parity record. Part III of the algorithm takes the final remaining spanning trees and

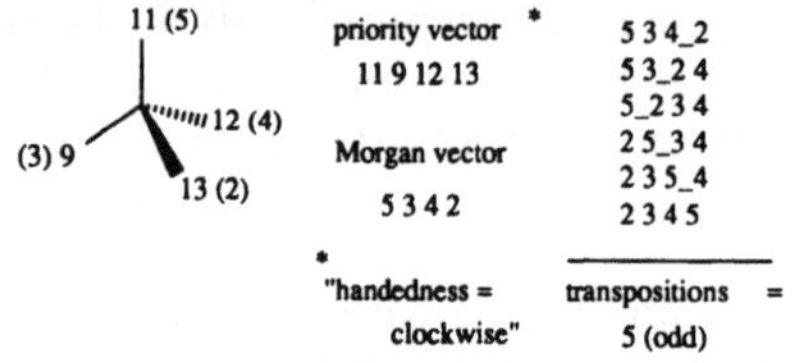

Figure 6. Computation of transpositions

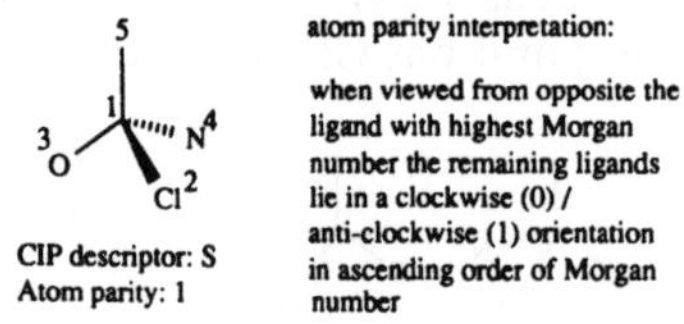

Figure 7. Interpretation of atom parity

compares the parity records both to identify unambiguously the true stereocentres in the structure and to determine the optimal parity record. Of course, either by accident or oversight, structures may be drawn and input with graphic stereo-bonds at atoms which are not stereocentres; it is therefore essential for correct registration that the software identifies and discards any *phantom* stereocentres. The most elaborate part of Wipke and Dyott's algorithm does just this, and is implemented similarly in the Beilstein registry system.

Once the optimal parity record has been identified, the respective *stereo lists* are generated and the BRCT is assembled on the basis of the numbering of the final spanning tree.

The registry software also computes wherever possible a CIP descriptor for each asymmetric atom, steric double bond and chirality axis. These descriptors are stored in the BRCT and are available for display, but are not used for the purpose of registration. They are utilised however to assist in the interpretation of CIP descriptors used during structure input. A wide variety of stereochemical descriptors may be used at input, and the registry software attempts to interpret these in terms of a graphical representation from which parity codes can be computed for the purpose of registration. The simple example in Figure 8 illustrates this situation.

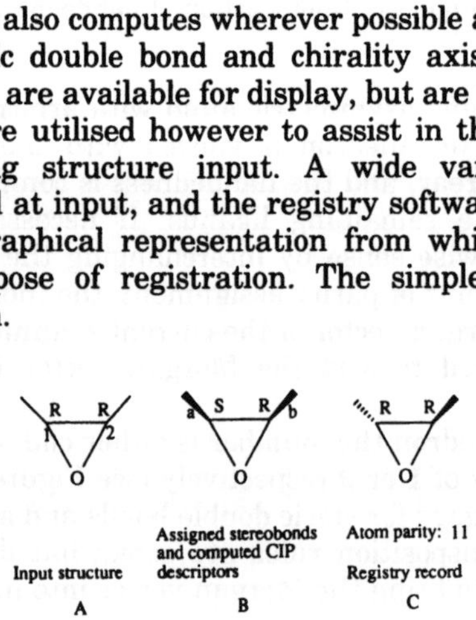

Figure 8. Graphical representation of stereochemistry

The structure **8A** is input with both centres marked R. For each centre, the registry system selects a ligand atom, assigns by convention an above-plane stereobond, and computes a CIP descriptor(**8B**). The computed descriptor is compared with the input descriptor and where these are different the assigned stereobond is inverted, thereby inverting the configuration of the stereocentre. A parity code is then computed for the centre in the normal way on the basis of the assigned stereobond, and is used for registration(**8C**).

Finally, in a joint project with the Gmelin Institute and Chemplex GmbH, the registry software and SDF format have been extended significantly during the past 18 months to accommodate organometallic and inorganic substances, with particular emphasis on stereochemistry and the representation of non-tetrahedral geometries, for example pentavalent phosphorus compounds and octahedral metal complexes. Again, algorithms similar to those described by Choplin and Wipke[18] have been used for this purpose, which are closely related to those described here.

REFERENCES

1. *Beilstein Online Database*; Heller, S., Ed.; ACS Symposium Series 436; American Chemical Society: Washington, 1990.
2. Jochum, C. 'Building a Structure-Oriented Numerical Factual Database'. In *Chemical Structures: The International Language of Chemistry*; Warr, W.A., Ed.; Springer-Verlag: Berlin, 1988, pp. 187–193.
3. Details of individual programs may be obtained from the Computing Division, Beilstein Institute, Varrentrappstr. 40 42, W-6000 Frankfurt, Germany.
4. Dittmar, P.G.; Stobaugh, R.E.; Watson, C.E. 'The Chemical Abstracts Registry System. I. General Design'. *J. Chem. Inf. Comput. Sci.* **1976,** *16,* 111–121, and subsequent papers.
5. Barnard, J.M.; Jochum, C.; Welford, S.M. In *Chemical Structure Information Systems*; Warr, W.A., Ed.; ACS Symposium Series No. 400; American Chemical Society: Washington, 1989, pp. 76–81.
6. *Stereochemical Descriptors in the Beilstein Handbook of Organic Chemistry*, printed in the preface to each volume of the Beilstein Handbook, or available separately from Springer-Verlag.
7. Morgan, H.L. 'The Generation of a Unique Machine Description for Chemical Structures – a Technique Developed at Chemical Abstracts Service'. *J. Chem. Doc.* **1965,** *5,* 107–113.
8. Welford, S.M. 'Die Beilstein Datenstruktur für organische Verbindungen'. In *Software Entwicklung in der Chemie 1*; Proceedings des Workshops 'Computer in der Chemie', Hochfilzen/Tirol, 19–21 November 1986; Gasteiger, J., Ed.; Springer-Verlag: Berlin, Heidelberg, 1987, pp. 5–11.
9. A complete description of the SDF/BRCT format can be obtained from the Computing Division, Beilstein Institute, Varrentrappstr. 40–42, W-6000 Frankfurt, Germany.
10. Hicks, M.G. 'Substructure Search Systems for Large Chemical Databases'. In *Software Entwicklung in der Chemie 3*; Proceedings des Workshops 'Computer in der Chemie', Tübingen, November 1988; Gauglitz, G., Ed.; Springer-Verlag: Berlin, 1989, pp. 9–23.
11. Loening, K.L. 'Conventions, Practices and Pitfalls in Drawing Chemical Structures'. In *Chemical Structures: The International Language of Chemistry*; Warr, W.A., Ed.; Springer-Verlag: Berlin, 1988, pp. 413–423.
12. Mockus, J.; Stobaugh, R.E. 'The Chemical Abstracts Service Chemical Registry System. VII. Tautomerism and Alternating Bonds'. *J. Chem. Inf. Comput. Sci.* **1980,** *20,* 18–22.
13. Welford, S.M. 'Tautomer Processing in the Beilstein Registry System'. In *Software Entwicklung in der Chemie 2*; Proceedings des Workshops 'Computer in der Chemie', Hochfilzen/Tirol, 18–20 November 1987; Gasteiger, J., Ed.; Springer-Verlag: Berlin, 1988, pp. 35–43.
14. Wipke, W.T.; Dyott, T.M. 'Simulation and Evaluation of Chemical Synthesis. Computer

Representation and Manipulation of Stereochemistry'. *J.Am.Chem.Soc.* **1974,** *96,* 4825–4834.
15. Wipke, W.T.; Dyott, T.M. 'Stereochemically Unique Naming Algorithm'. *J. Am. Chem. Soc.* **1974,** *96,* 4834–4842.
16. Petrarca, A.E.; Lynch, M.F.; Rush, J.E. 'A Method for Generating Unique Computer Structural Representations of Stereoisomers'. *J. Chem. Doc.* **1967,** *7,* 154–165.
17. Moreau, G. 'A Topological Code for Molecular Structures. A Modified Morgan Algorithm'. *Nouv. J. Chim.* **1980,** *4,* 17–22.
18. Choplin, F.; Marc, R.; Kaufmann, G.; Wipke, W.T. 'Computer Design of Synthesis in Phosphorus Chemistry: Automatic Treatment of Stereochemistry'. *J. Chem. Inf. Comput. Sci.* **1978,** *18,* 110–118.

STRUCTURE AND SUBSTRUCTURE SEARCHING ON DIALOG

Peter F. Rusch
Dialog Information Services Inc, 3460 Hillview Avenue, Palo Alto, California 94304, U.S.A.

ABSTRACT

Development of the computer-readable version of the Beilstein database prompted an extended review of factual and substructure search techniques. In order to provide appropriate access to the three-dimensional structure information in combination with the extensive range of physical property information, new methods of online access had to be developed. The Dialog Information Retrieval software is a long-established text searching system that had undergone some modifications to provide more extensive full-text and limited numeric range search. Access to the Beilstein database required that substructure and full structure search techniques be fully integrated into the existing capabilities to provide maximum user convenience and accessibility of the combination of factual and structure data. What has evolved is a mix of components that provides an entirely new system for structure query formulation and transmissions, a new structure search process, and enhancements to numeric searching. This paper describes the components, their integration, and the resulting service. Evaluation of the service and possible extension to it are reviewed.

INTRODUCTION

The Dialog Information Retrieval Service has successfully brought online a variety of chemical information databases since 1972. Among these databases are those containing text, numerical information and images of chemical structures. With the availability of the Beilstein database for online searching, it became important to turn attention to chemical databases containing graphical representations of chemical structures. Both graphical formulation of queries and graphical output of structure representations are required for a complete access to chemical structure information in the form of connection tables. An investigation was carried out concerning the most effective means of providing for structure and substructure search with subsequent display. It became apparent that there were many systems currently in the commercial sector that would be unavailable to Dialog. In addition, it was clear that the chemical information community was becoming better equipped with local computational power at the desktop. Although there was a predominance of IBM and IBM-compatible PCs, there was a significant penetration of Macintosh and other computational systems and workstations. What resulted from this investigation was the decision to implement

W. A. Warr (Ed.)
Chemical Structures 2

a substructure search system that was highly modular, allowing for multiple forms of access and output while integrating the retrieval system as closely as possible into the Dialog search process.

This paper will describe the systems analysis that led to these decisions; the integration of the various components with Dialog; the implementation of the system for availability in the commercial sector, and finally a review of enhancements that will be undertaken in the future.

SYSTEMS ANALYSIS

For more than 15 years, the Dialog Information Retrieval Service has been used successfully to do some forms of chemical substructure searching. Among the first of these was a text search technique known as substructure search *via* nomenclature (SSVN)[1]. Later forms of text search added ring information. A systematic nomenclature, such as the Chemical Abstracts (CA) Index Nomenclature is well suited to storage and retrieval as text. With proximity search operations, a long time standard of the Dialog Information Retrieval Service, the precision of substructure searching *via* nomenclature is easily established. Indeed, it is possible to provide for highly precise searches of millions of chemical substances names with very high recall. There are, however, classes of chemical substances the preferred names for which are trivial. Examples of such classes are natural products and some antibiotics. In these instances, the more complicated the substitution of these products, the more likely it is they will adopt a trivial name to represent a large, underlying structure, including in some cases, stereochemical descriptors. While such a diversion from systematic nomenclature is completely understandable and useful, even for the printed indexes of chemical substance names, it fails to serve the purposes of substructure search *via* nomenclature. In most of these cases, the classes and chemical substances so named have well-defined structures that are represented by connection tables.

There are also, of course, many chemical substances which have trivial or trade name nomenclature which is totally unamenable to substructure search *via* nomenclature although they may be well characterised. Unfortunately, some of these are not characterised and even structure searching is not an adequate way to access them in an online information environment.

For many years it has been recognised that the presence of rings can be used to characterise chemical substances. Indeed, among the first indexes[2] to chemical substances were those for molecular formulae and for ring systems. Beginning in the early 1980s, Dialog undertook a research and development project to generate ring information using the connection tables of the CA Registry System database. The result was a series of ring descriptors in text form that were used for both search and display. In addition, Dialog carried forward from the CAS Registry Structure File, a data element known as the ring identifier. Such ring information is quite compact and can be used as search terms that are predictable and are discriminating either individually or in combination.

With most chemical substance searching, it is useful to have as many descriptors as possible in order to discriminate the desired chemical substance or class of substances from an otherwise enormous set of information. Typically the CAS Registry, as an example, has approximately ten million chemical substances, the vast majority of which are not the answer to any given query. Accordingly, it is useful to perform online chemical substance searches using combinations of nomenclature, molecular formula, ring information and other descriptors in order to obtain the highest recall and precision possible commensurate with appropriate and low cost.

The Dialog Information Retrieval Service runs on some of the largest world-class mainframes that are available. Indeed the entire process is contained on a series of mainframes, each of which is self-sufficient. The technology is a standard inverted file and inverted index technology that has been described earlier[3]. In general, such a large information retrieval process is used by multiple customers simultaneously and is in control of all of the mainframe computing resources. The typical customer interaction is through a series of communication computers operating as front-ends to the mainframe and handling the communications protocol required. The mainframe system itself is responsible for all syntax checking, information retrieval, intermediate set formation, retrieval functionality, formatting, and delivery of output. Output can consist of either immediate output into the online environment, output to the electronic mail system, or output to a printer for subsequent despatch to the customer.

One of the characteristics of this system is that it allows multiple access to databases, indeed, currently more than 350 databases with a single logon. There is a multi-file environment, and the customer is benefited by the presence of only one means of accessing all of the different databases and their content. This imposes the additional burden on any new system that it be fully compatible with this mainframe environment and the appropriate input/output processors.

The systems analysis phase of this project focused first on query formulation and output of the information. Throughout the decade of the 1980s a *de facto* standard developed in other online chemical structure information systems. This was essentially a Tektronix 4010 standard which permitted both graphical formulation of queries that were transferred to the retrieval system and output of graphical structure information. In some cases, stand-alone retrieval systems were built for local processors, typically IBM or IBM-compatible processors. The standards could be as idiosyncratic as the author wished since their compatibility with other systems was not of primary importance. It was always possible to create files of information for interchange among systems. There are currently some standards now being proposed for such interchange of information[4].

The significant portion of queries and output transmitted to online information retrieval systems was using a Tektronix standard, and the input was either by graphic formulation, using this standard, or by some simplified keyboard input. There are several simplified keyboard inputs that have been widely adopted, one of which is that developed by Richard Feldmann[5] and the other, of which there are many variations, of a simple line description of a chemical substance[6] using an arbitrary numbering of the nodes present. Both of these forms of keyboard input were of greater interest when local computing power was restricted or eliminated by the use of a teletype or its equivalent. As the computer power increased at the user site, other options became available and needed to be included in the system that would be used for structure searching on Dialog.

The Dialog Information Retrieval Service is currently operating on multiple NAS 909010 and 9010X processors containing up to as many as 32 megabytes of directly accessible memory. Part of the system analysis was whether or not this processing system, with its ability to use inverted files and inverted index search technology would be an appropriate means for the basis of a structure searching system.

Another characteristic of the Dialog Information Retrieval Service is its large number of online direct access storage devices. Currently the configuration consists of approximately 900 gigabytes of such direct access storage. Part of the analysis was concerned with how this storage would be used to implement the structure and substructure searching system.

Finally, the Dialog Information Retrieval Service is not a database management service *per se*. The information loading and updating portion is not a part of the online retrieval function, therefore, it was important to have a structure searching system that was independent of the online searching modules and that was capable of updating any database that might have structure information. Of immediate interest, of course, was the Beilstein database, but other databases, such as the structure portion of the CA Registry System database, need to be considered as possible candidates for the structure searching system to be implemented.

INTEGRATION

The systems analysis showed that it was important for Dialog to adopt a fully integrated subsystem that would effectively perform structure and substructure searches. Such a subsystem would permit the Dialog program to remain in control of all the communications functions to the user's terminal, but would use searching techniques that were different from those of Dialog and more appropriately suited to the handling of structure information.

Indeed, part of the processing, particularly that for the substructure searching, needed to be fully separate from the processing of a normal text query by the Dialog Information Service. Although the screening phase of a normal substructure searching system could be adequately handled by an inverted file, inverted index data structure, it was found to be appropriate to keep that as a part of the subsystem rather than to use the Dialog processes themselves. Furthermore, the Dialog storage processes and formats for inverted files were to be retained for the text portions of certain databases; thus the screening and iterative search processes for substructure and structure search should be carried out on their own file structures, different from and separate from the Dialog file processes and formats.

While parts of this process are more appropriately separated, other parts are more appropriately shared and handled directly by Dialog. In particular, the processes for display were most easily handled by using the Dialog file structures and processes currently used for display. This would prove to be a good choice for reasons that will be described shortly.

IMPLEMENTATION

Key to the implementation of the substructure searching system is a chemical structure representation known as the ROSDAL[6] string. This is an ASCII character string that conveys sufficient information to convey fully a structure representation, including stereochemistry. A subset of a structure, that is to say, a substructure query, can also be represented by the ROSDAL string. Use of the ROSDAL string provides two immediate advantages for any new substructure searching system. First, the string needed to represent a substructure can be generated easily by a person. It is a relatively straightforward process using an arbitrary numbering of chemical structure nodes. Second, short cuts are permitted to represent kinds of attachments or nodes with unlimited or open attachments. Even if it is not of interest to generate a structure query using the ROSDAL string, it is an easy process to build a computer program that will interpret graphic input into a ROSDAL string for transmission. Several examples of these kinds of computer programs[7] exist, and they permit the maintenance of an open architecture for the input of structure queries. The only requirement is that the ROSDAL string be sufficient and invariant so that the queries can be formulated and sent.

One advantage of the ROSDAL string is that it allows for incoming communication from the customer location as an ASCII set of characters. This is precisely the kind of communication that is handled best by telecommunications networks worldwide and is subject to very high speed transmission with low error rates.

Another advantage of choosing the ROSDAL string as the means of communicating structure information in the system is that it can be used for output. Once again, the output is ASCII character string, easily handled with high speed and low error rates by all telecommunications networks. In addition, the ROSDAL string is easily enhanced with co-ordinates so that local processing power, such as that found in the programs mentioned[7] above, can translate the appropriate ROSDAL string with co-ordinates into a chemical structure display with stereochemistry.

The subsystem that was chosen to do the substructure searching and provide a list of Dialog accession numbers is the S4 system from Softron GmbH in Munich[8]. The subsystem effectively interprets the ROSDAL string structure query and initiates both a screening and iterative search of its own formatted files. The results of the screening search lead to an intermediate result such that if fewer than 500 structures respond, the search process ends. Subsequent to the screening search, if required, a unique iterative search is executed leading to the final result. This result of Dialog accession numbers of structures is returned to the Dialog system and presented to the user as the next sequential set in his search. Thus it is possible to mix a structure query with a text query at any time. The components of the system in block diagram are shown in Figure 1 where the user site communicates with the front-end or communications computers of Dialog using a ROSDAL string. The user site, of course, can contain sufficient computational power to generate ROSDAL strings from graphical input and produce graphical output from ROSDAL strings, or it may be a terminal handling only the ASCII characters of the ROSDAL string. Independent of the query formulation, search and output, is a separate file updating system that updates files for use exclusively by the S4 subsystem or by Dialog for display purposes.

In order for the Dialog customer to initiate a substructure or full-structure query, it became necessary to introduce a new command. This is the QS or query structure command which invokes the subsystem for structure searching. The QS command assumes that a ROSDAL string will be entered to represent the query structure. The ROSDAL string is terminated with a unique character string, and upon receiving the terminating character string, the subsystem automatically begins the search process.

Output from the search process is a fully compatible Dialog set, thus the output of a query search (QS) may be combined with the output from a text search or

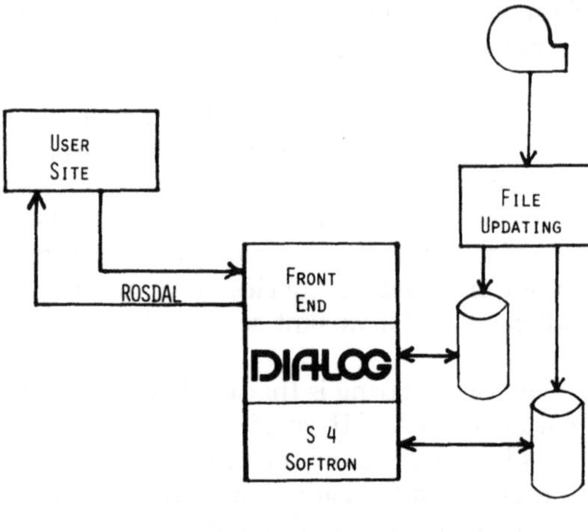

Figure 1. Block diagram of the Dialog Structure Searching System

another query search. It may be subjected to any of the appropriate Dialog processes for sets such as Sort, Type, Print, Combine or Select. The output of structure information and text information can be handled in the normal Dialog way as either pre-formatted records containing specified fields, or through user-defined formats displaying fields selected by the customer.

The QS command is unique to Dialog and offers significant flexibility in the execution of a structure or substructure search. There is a set of arguments that may be given to the QS command to invoke different structure searches. The complete range of arguments is shown in Figure 2. With any of these arguments, the customer may enter the keyword ID followed by an equals sign and an appropriate identification of the structure to be searched. This may be any text string up to 200 characters. If it is not present, the system substitutes the ROSDAL character string of the query in place of any characters from the customer.

Figure 3 shows a relatively straightforward example of the use of the QS command to invoke a substructure search in the Beilstein database online from Dialog. Those familiar with the Dialog Information Retrieval Service will note that there is the characteristic file banner shown for file 390, Beilstein Online. Every Dialog interaction uses the question mark as the command prompt. Following that command prompt, Figure 3 shows the QS command followed by an identification of 'primary amines'. Upon receiving the carriage return, the subsystem is invoked and sends forth a message requesting the ROSDAL connection table be entered. Shown in Figure 3 is an extremely simple ROSDAL string of 1N-2G.@ representing a nitrogen atom bonded to anything. Two hydrogens on nitrogen are assumed in the ROSDAL representation shown here, as the nitrogen has no free valances indicated. Upon receiving the termination character, the subsystem process is initiated, ending with the result of the first Dialog set with over 300,000 answers that respond to the substructure query representing primary amines. Figure 3 shows that set 1 is fully compatible for use with other sets in the same database. The example illustrates the Dialog command

QS = Query Structure*

Command	Example	Notes
QS	QS	The basic structure/substructure search command. It is entered before the ROSDAL string (linear representation of the structure) is entered or uploaded from MOLKICK. The system uses the ROSDAL string to identify the created set.
QS TEST **QS T**	QS TEST QS T	Searches a small (approximately 5%),representative portion of the database. (Free)
QS FULL **QS F**	QS FULL QS F	Use to extend a TEST search to the entire database without having to upload or reenter the ROSDAL string. Must be used before other QS commands.
QS RANGE *n* **QS R** *n*	QS RANGE 20000 QS R 20000	Searches a range of 100,000 consecutive Beilstein Registry numbers beginning with *n* . In the example, the records in the range of BN=20000 through BN=119999 are searched.
QS *n*	QS 1733531	Searches for the stereoisomers of the substance identified by BN=*n*. The system uses the entered Beilstein Registry Number and the corresponding Chemical Name to identify the created set. In the example, BN=1733531, CN=L-LYSINE, the created set includes four records: one for the entered isomer and others for BN=1722530, CN=D-LYSINE, BN=1722532, CN=DL-LYSINE, and BN=1616991, CN=LYSINE (stereoisomers of unknown configuration).
QS < > **ID=**< >	QS ID=amines QS T ID=search1 QS F ID=search1 QS R 1500 ID=Smith2	Any QS command can be identified in the system response by adding a name after the ID=. The QS command with its options and set identifier can have a maximum of 240 characters. If the ID= feature is not used, the system uses the ROSDAL string as the set identifier.

* When the QS command is incorporated in a SearchSave or ALERT strategy, DIALOG changes it to QC (Query Command). QC is displayed when you RECALL or EXECUTE a saved strategy that contains a structure or substructure search.

Figure 2. Structure search comments

```
File 390:  Beilstein Online-3,415,097 substances Apr 1990
           Copr. Springer-Verlag
 *** Now search CAS Registry Numbers (RN=) for substance of record ***

      Set  Items  Description
      ---  -----  -----------

? qs id=primary amines

QS/QC Query Structure Version 1.02
Enter ROSDAL connection table
Line No.  1 (enter '.@' to end or 'ABORT' to quit)

1n-2g.@

DASD 2FF  DETACHED
Processing - Atom-by-atom match started

      S1  300871  QS  ID =PRIMARY AMINES
? s s1 and ammonia (s) nickel/pr
          300871  S1
           13020  AMMONIA/PR  (see also ammoniak, NH3)
           15076  NICKEL/PR  (see also Ni)
             367  AMMONIA/PR(S)NICKEL/PR
      S2     271  S1 AND AMMONIA (S) NICKEL/PR
```

Figure 3. Dialog structure search for primary amines

to combine set 1 with the text term 'ammonia' in proximity to 'nickel' in the preparation field of the Beilstein database. The result is the set number 2, with 271 records from the Beilstein file containing records for primary amines showing preparations based on ammonia and nickel.

OUTPUT

Figure 4 shows an example of output of a simple structure that can be displayed with or without stereochemistry. There are two ROSDAL strings available for such display. Each of the possibilities, either GS (the graphic structure string with stereochemistry) or GR (without stereochemistry) contains co-ordinates sufficient to draw the structures as shown. The structures produced in Figure 4 were displayed using the MOLKICK software. The nature of the ROSDAL string with co-ordinates permits other local computational programs to produce structure displays.

The ROSDAL string is maintained by the Beilstein Institute and is essentially in the public domain, thus it is available for use by software developers and information specialists to formulate queries for the Dialog Information Retrieval Service. The advantage of the ROSDAL string is that it contains an ASCII set of characters and can be easily communicated using standard telecommunications networks. In addition, the string is easily generated manually and can be interpreted manually to provide structural display.

CONCLUSION

The Dialog Information Retrieval System is a large text retrieval system using mainframe hardware and inverted file, inverted index search technologies. It is in complete control of a variety of resources, including a massive array of direct access storage devices. In order to implement substructure searching in the Dialog

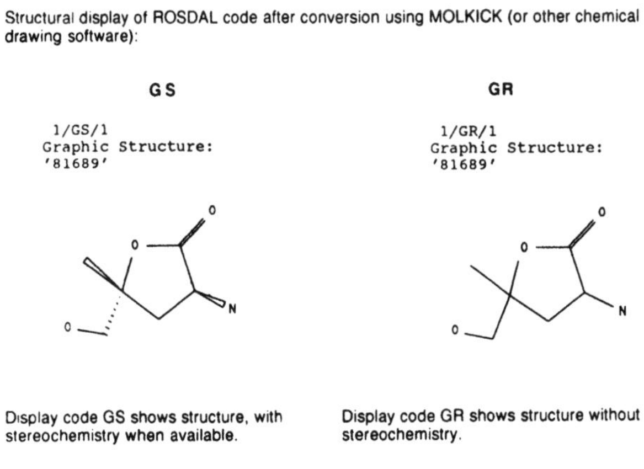

Figure 4. Structural display of ROSDAL code after conversion using MOLKICK (or other chemical structure drawing software)

system, a subsystem was integrated using the ROSDAL string for input of structure queries and output of results. The QS command has been integrated into the Dialog system to invoke the subsystem and accept the ROSDAL string query. Using this open architecture permits a variety of input and output mechanisms at the user site. They may be manual, or machine-based, with ROSDAL as the common communication language.

ACKNOWLEDGMENTS

The author gratefully acknowledges the assistance of Softron GmbH and its staff in the implementation of the S4 system, the Beilstein Institute for its assistance and encouragement in developing the system and providing the database in a usable format, and the author especially acknowledges the work of Shelley Giles, who was responsible for the integration of both the software and the database into the Dialog system.

REFERENCES

1. Fisanick, W.; Mitchell, L.D.; Scott, J.A.; Vander Stouw, G.G.; 'Substructure searching of Computer-Readable Chemical Abstracts Service Ninth Collective Index Chemical Nomenclature Files'. *J. Chem. Inf. Comput. Sci.* **1978,** *15(2),* 73–84.
2. *The Ring Index*; Patterson, A.M.; Capell, L.T., Eds; 1st ed.; American Chemical Society: Washington, 1940.
3. Summit, R.K., 'Lockheed Experience in Processing Large Data Bases for its Commercial Information Retrieval Service'. *J. Chem. Inf. Comput. Sci.* 1975, *15(1),* 40–42.
4. Barnard, J.M., 'Draft Specification for Revised Version of the Standard Molecular Data (SMD) Format'. *J. Chem. Inf. Comput. Sci.* **1989,** *30(1),* 81–96.
5. Feldmann, R.J.; Bacon, C.L.T.; Cohen, J.S., 'Versatile Interactive Graphics Display System for Molecular Modelling by Computer'. *Nature* (*London*), **1973,** *244(5411),* 113–115.
6. Barnard, J.M.; Jochum, C.J.; Welford, S.M., 'A Universal Structure/Substructure Representation for PC-Host Communications'. *Chemical Structure Information Systems*; In ACS Symposium Series 400, American Chemical Society: Washington, 1989; pp. 76–81.
7. MOLKICK, a product from the Beilstein Institute. Available from Springer Verlag New York, Inc. 175 Fifth Avenue, New York, NY 10010, U.S.A. or Springer Verlag Heidelberg, New Media Department, Tiergartenstrasse 17, W-6900 Heidelberg, Germany. ChemConnection, a product of SoftShell International, Ltd. Available from SoftShell International, Ltd. 2754 Compass Drive, Suite 375, Grand Junction, CO 81506, U.S.A. ChemTalk, part of the Chemist's Personal Software Series. Available from Molecular Design Limited, 2132 Farallon Drive, San Leandro, CA 94577, U.S.A.
8. Hicks, M.G.; Jochum, C.J., 'Substructure Search Systems. 1. Performance Comparison of the MACCS, DARC, HTSS, CAS Registry MVSSS, and S4 Substructure Search Systems'. *J. Chem. Inf. Comput. Sci.* **1990,** *30(2),* 191–199.

THE DARC IN-HOUSE PACKAGES AS A LIBRARY OF STAND-ALONE FUNCTIONS FOR BUILDING APPLICATIONS IN HANDLING CHEMICAL INFORMATION

Pascal Huguet and Osman Sultan
Questel, Le Capitole, 55 avenue des Champs Pierreux, 92012 Nanterre, France.

ABSTRACT

The variety of applications aimed at handling chemical information require the use of common tools corresponding to specific functions, for instance, structure editing, molecular formula or substructure searching, displaying, printing, and communicating with other software. This paper refers to the building of such stand-alone tools on the one hand, and how these tools can be assembled to form specific applications on the other hand. The implementation of such capabilities is obtained through an optimal use of the operating system, especially VMS on VAX.

THE TOOLKIT CONCEPT

No one supplier can hope to offer a monolithic solution to the computing needs of an enterprise such as a modern pharmaceutical or chemical company and, as a result, strategic alliances are evolving. Increasingly, developers of chemical information systems are building toolkits which enable system integrators to construct the 'ideal' system for the particular corporate computing environment found in a given company. In this paper we shall examine the components of one such toolkit, the DARC Toolkit from Questel (see Figure 1).

In the 'toolkit' concept, a 'tool' fulfils a specific function, can be used independently, may easily be replaced by another tool as necessary and is fully documented with respect to its function, implementation and use. A simple example of a tool is a molecular formula search module, which may be implemented for use within a DBMS, is easily replaceable by another molecular formula search module and whose molecular formula search capabilities and calls from any application are fully documented. DARC tools are classified as simple tools and advanced tools. Examples of simple tools are:

- structure editor
- non-duplicate checker
- structure registration
- molecular formula search
- substructure search
- full structure search
- display
- print
- import/export of connection tables

W. A. Warr (Ed.)
Chemical Structures 2

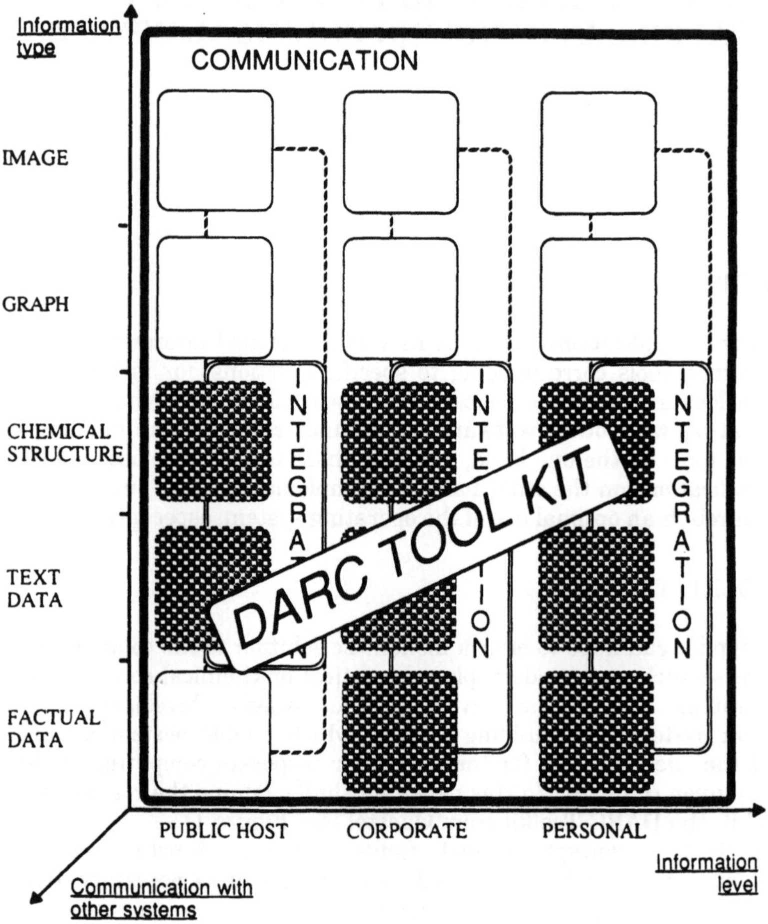

Figure 1. The DARC toolkit

whilst advanced tools may include:

- communication with external hosts
- pull-down menu generator
- screen-oriented application designer

In addition there are primitive tools which may either be used by the simple or advanced tools used above or which may be used independently. Examples of primitive tools include tools for:

- atomic co-ordinate calculation
- bond normalisation/relocalisation
- detection of rings and ring systems
- query answer set handling
- sorting
- paging
- list handling

DARC TOOLS

DARC tools exist in one of two forms: either a library of subroutines which can be linked to an application; or as shareable images in the DEC VAX/VMS environment. They may be called from any programming language, from DBMS application builders, from the DARC pull-down menu system and from Digital's DECwindows application environment. Tools must be easy to set up, efficient with respect to response time and the use of computer resources, and easy to maintain leading to maximal use of the capabilities of the VAX/VMS operating system. Where possible re-entrant functions are used for optimisation of oft-repeated tasks and shareable images give independence between the tool and the application, ease of maintenance and optimal performance. Some tools use interprocess communication techniques. The DARC toolkit is aimed at *de facto* or established standards such as:

- SMD format for structures
- X25, for networks
- PostScript for printing
- CDA/DDIF for handling of compound documents
- Microsoft Windows, Macintosh and DECwindows application environments

EXAMPLES

One combination is the linking of the structure editing capability and the VAX Packetnet System Interface (PSI) allowing X25 connection for query transfer to public hosts. Another example is the use of VMS interprocess communication and shared memory for the mixed DARC/DBMS display or print. A third example is the use of events flags for synchronisation when cross-registration of DARC structures and of DBMS data is undertaken.

ACKNOWLEDGEMENT

The summarising of the contents of this paper is largely the work of Dr William G Town of William Town Associates.

THE STANDARD MOLECULAR DATA (SMD) FORMAT

John M. Barnard
Technical Secretary, SMD Subgroup of the Chemical Structure Association
46 Uppergate Road, Stannington, Sheffield S6 6BX, England

ABSTRACT

During the past few years interest has grown in the development of standard formats for the machine-readable presentation of chemical structures, and a few proposals have been published. One of these is the Standard Molecular Data (SMD) format, developed by a group of European chemical companies. Under the auspices of the Chemical Structure Association, a series of technical working groups have examined the original version of SMD format and proposed a number of revisions and extensions. This poster paper describes the revised version of the format using annotated examples, and discusses the areas where further extension is required.

WHAT IS SMD FORMAT?

SMD format[1–7] is a computer file format which is being proposed as a standard for the exchange of chemical structure information between different organisations and computer programs.

SMD Format has been developed by a series of technical working groups which have included participants from a wide range of chemical companies and chemical information providers in the United States and Europe. The format is structured as a series of independent blocks, different types of information being shown in different blocks; only those blocks which are actually needed in a particular case need be included in the file.

WHY IS A STANDARD FORMAT NEEDED?

It is needed to allow computer-readable chemical structure data to be exchanged between different programs (perhaps from different suppliers) with the minimum of special conversion software.

Connection Tables

Computer representations of chemical structures have been used since the 1950s. With the dramatic reduction in the cost of computer storage over the past two decades, storage forms such as line notations have given way to the less compact atom-bond connection table, in which the atoms and bonds contained in the structure are explicitly listed, often in tabular form. Line notations still retain some applications, mainly as rapid input formats.

W. A. Warr (Ed.)
Chemical Structures 2

Internal and External Connection Tables

Some connection tables form the basic internal data structure on which algorithms operate in order to derive some information from the structure. Others are used by particular systems to store chemical structure data on disk files. It is this second class of connection table with which we are concerned.

Proprietary Connection Table Formats

Most computer systems have their own connection table formats for disk storage, and such connection tables can only be written or read by one particular system. Some manufacturers who produce a range of programs for handling chemical structures use a single connection table format which can be written or read by all programs in their range. Such formats do, however, remain proprietary products of individual companies, and their exact descriptions are not always publicly available.

An Exchange Format

The main purpose of a standard format is to reduce the number of conversion programs which are required to interconvert the variety of internal connection table formats in use by different application programs. A standard format is not envisaged as a replacement for these formats; nor is it envisaged simply as a standard database distribution format, though it may have some application (perhaps in a compacted form) in this respect. Rather, it is intended as a standard interchange format, which will allow different programs (possibly from different suppliers) to exchange structure descriptions.

The principal advantage of a standard format is that it will allow the transfer of chemical structure data between different programs without the need for special conversion software. Such a format will only be successful, however, if it meets the requirements of those likely to use it across a range of user communities. Considerable care therefore needs to be taken in its design, with input being taken from wide variety of individuals and organisations.

WHAT CAN SMD FORMAT BE USED FOR?

- Downloading of structures from online databases to microcomputer-based structure-handling programs.
- Exchange of data between chemical structure database systems and molecular modelling programs.
- Exchange of structure data between automatic structure elucidation programs (perhaps based on spectroscopic analysis) and chemical structure databases.
- Import of chemical structure and property data by systems for quantitative structure-activity relationship (QSAR) analysis and property prediction.
- Distribution/exchange of structure-based databases between different organisations.

- Exchange of structure data between structure databases and chemical reaction indexing and synthesis analysis programs.
- Uploading of query structures to a variety of different substructure search systems (both in-house and online).
- Representation of chemical structure diagrams within word-processed text, perhaps in association with the Standard Generalised Markup Language (SGML).

The use of independent blocks in SMD format allows a program to skip over blocks which it does not require, or is unable to process. This should facilitate exchange of structures between totally different programs, such as systems for chemical structure database management, molecular modelling, chemical synthesis planning, QSAR and spectroscopy-based structure elucidation.

A core of common information (for example, atom types and bonds) is likely to be needed by virtually all programs.

WHAT OTHER STANDARD FORMATS ARE THERE?

Several proposals have been made for 'standard' connection table formats in recent years. These have generally been put forward by particular user communities, and are geared to the requirements of that user community.

The Brookhaven Protein Databank Format

This format was developed for the representation of three-dimensional protein structures in the databank of such structures built by the Brookhaven National Laboratory. It has, however, been widely adopted for showing protein structures.

The Standard Crystallographic File Structure[8]

This was developed by a joint working party of the Data and Computing Divisions of the International Union of Crystallography, which reported to the Union's 1981 Congress in Ottawa. It contains features for showing crystallographic data such as Space Groups etc., as well as the basic connection table. Further proposals (the Crystallographic Information File, CIF) are to be brought to the Union's conference in Bordeaux, France, in July 1990.

The CODATA Format for Amino Acid Sequence Data[9]

This specialised format was developed under the auspices of the CODATA Task Group on Co-ordination of Protein Sequence Databanks.

Guidelines for Publication in Molecular Modelling[10]

This seems to have been proposed by a relatively informal group of medicinal chemists from several countries. It contains features for representing much data of interest to molecular modellers and medicinal chemists.

The JCAMP-CX Format[11]

This format has been developed by the Joint Committee on Atomic and Molecular Physics (JCAMP) to complement the JCAMP-DX format for spectroscopic data[12]. There is some overlap in the membership of the committee drafting this format, and the Working Groups for SMD Format, and the possibility of integrating the two formats is being considered. In their present forms, JCAMP-CS is quite close to being a subset of SMD Format (indeed, it corresponds very closely to the 'core' of the format, identified at the Technical Working Group meeting in May 1990 – see the section on 'What does SMD Format Consist Of', below), and at the very least facile interconversion should be possible.

RELATIONSHIP BETWEEN THE DIFFERENT FORMATS

There is little point in arbitrarily extending SMD Format to compete with existing standard formats in fields where these are already established. However, there is an urgent need to unite the efforts behind standard formats currently being developed for particular purposes, in order to prevent the emergence of a series of incompatible and competing standards, used by different groups of users. It is possible that SMD format may be able to become a sort of 'umbrella' for more specialised formats.

The American Society for Testing and Materials (ASTM) has recently begun an activity in the area of standard interchange formats for computerised chemical information, and meetings took place in Philadelphia in November 1989 and in Boston in April 1990, and were attended by a number of individuals who have been concerned with the development of SMD format. It is hoped that it may be possible to integrate these parallel activities.

WHAT DOES SMD FORMAT CONSIST OF?

A draft formal specification for the revised Version of SMD Format has recently been published for comment in the *Journal of Chemical Information and Computer Sciences*[1]. The format is defined by means of a set of syntax diagrams, similar to those used in the definition of programming languages such as Pascal.

An SMD file is an ordinary line-structured ASCII or EBCDIC file. The lines used for the description of a structure are divided into a series of self-contained blocks, each block containing a different sort of information. If the program reading the file does not need to use the information about, for example, stereochemistry, it can simply skip over the stereochemistry blocks. Similarly, if stereochemical information is not available to a program writing an SMD file, it need not write any stereochemistry blocks. Certain defaults are assumed for missing information.

The blocks are grouped together into Sections, which describe complete chemical entities like molecules or reactions; a reaction section is able to refer to the molecule sections for the molecules participating in the reaction.

The blocks available for description of molecules in SMD format, as defined in the draft specification[1], are as follows:

>NODE Block	Lists the atoms present
>CONVENTION Block	Divided into several sub-blocks, listing bonds present, hydrogen counts for each atom, charges and radical states. These are given together because they may depend on conventions being used for aromaticity, delocalisation etc.
>VALENCY Block	Specifies unusual valencies
>ISOTOPE Block	Specifies isotopes
>STEREO Block	Describes stereoconfiguration
>CIP—ATOM Block	Cahn-Ingold-Prelog descriptors
>CIP—BOND Block	Cahn-Ingold-Prelog descriptors
>ATTACHMENT Block	Specifies 'free-site' attachment in substructures
>RING—CONN Block	Specifies exact/embed match for ring systems in query substructures
>BOND—ENV Block	Specifies ring/chain environment in query substructures
>COORD Block	Actual 3-D co-ordinates for the atoms
>DISPLAY—COORD Block	2-D co-ordinates for graphical display
>NODE—PIC Block	Display information for atoms
>BOND—PIC Block	Display information for bonds
>NODE—DATA—PIC Block	Display information
>BOND—DATA—PIC Block	Display information

Special records are also used to specify defaults and 'audit data' (date/time written etc). Each Section, Block, Sub-block etc. starts with a header line with a distinctive tag character, and ends with an explicit terminator line. Some examples of SMD Format structure descriptions are given in the published draft specification[1].

A meeting of the Technical Working Groups in May 1990 recommended some minor revisions to the published draft[1], and also identified a 'core' subset of the format, which would allow only the description of simple molecules; extensions to this core can be used to describe generic structures, reactions etc. It is hoped that a formal document describing the revised draft specification will be issued during the summer of 1990.

WHAT INFORMATION CAN BE INCLUDED IN SMD FORMAT?

The draft specification as published covers the following classes of structure-related information:

- Simple organic molecules
- Shortcuts for commonly-occurring groups and residues
- Query structures for substructure search systems
- Generic structures
  where a list of alternative atoms or groups is given for a variable part of the molecule
- chemical reaction descriptions
  including identification of reaction sites and atom correspondences between reactant(s) and product(s)

- Stereochemical features
  including R and S designators, wedge bonds and a 'rotational list' description able in principle to handle any type of stereocentre
- Isotope labels
- 2-D co-ordinates for graphical display
- Actual 3-D atom co-ordinates

The specification also allows a variety of structure conventions (bond orders for aromaticity, tautomerism, alternating single and double bonds etc.) to be shown.

WHAT EXTENSIONS ARE NEEDED?

The revised version of the format has been designed in such a way as to permit further extension whilst retaining backwards compatibility. Amongst areas which the format as presently drafted is unable to handle (or cannot handle fully) are:

- Inorganic and organometallic compounds
- Pi-complexes
- Salts
- Bio-macromolecules
- Polymers
- Polypeptides
- Ancillary data (including text descriptors, property data, fragment-, atom- and bond-centred values)

Preliminary ideas have already been developed by the working groups for some of these, though they are not included in the present draft specification. It is planned to convene meetings of additional technical working groups to examine these problem areas, and to propose further extensions to the format. Particular attention will need to be given to the relationship between SMD format and other standard formats which have been proposed, or are in use, for some of these areas. In addition, it may be desirable to develop a compressed version of the format for the representation of large amounts of data.

WHAT IS THE HISTORY OF SMD FORMAT?

Mid-1980s	SMD Format developed by the 'CASP' group of European chemical and pharmaceutical companies, as an aid to their co-operation in the development of systems for computer-assisted chemical synthesis planning.
February 1987	Technical description of Version 4.3 produced by Bayer AG, Leverkusen, Germany.
May 1988	Open meeting at Frankfurt, Germany, attended by about fifty persons from chemical and pharmaceutical companies, chemical database providers, software houses and interested academics from the United States and Europe. The meeting identified several areas in which the format could usefully be extended or improved, and set up a series of technical working groups to propose modifications and extensions.

June 1988	Chemical Structure Association sets up 'SMD Subgroup' to control development pending formation of a new independent organisation.
July 1988	First meetings of technical working groups at Paris (Télésystèmes Questel) and Cambridge (Cambridge Crystallographic Data Centre).
September 1988	Open workshop to discuss SMD Format at ACS National Meeting, Los Angeles, CA.
November 1988	Second meetings of technical working groups at Frankfurt (Beilstein Institute).
December 1988	Open meeting in London to report on progress of working groups.
February 1989	Description of Version 4.3 of SMD format published in *Journal of Chemical Information and Computer Sciences.*
October 1989	Third meeting of technical working groups at Basel (Molecular Design MDL AG).
February 1990	Draft specification for Version 5.0 of SMD format published in the *Journal of Chemical Information and Computer Sciences.*
May 1990	Fourth meeting of technical working groups to consider comments received on Draft Specification (Bayer AG, Leverkusen, Germany).
June 1990	Formal report of working group recommendations to Open Meeting at Noordwijkerhout, The Netherlands.

WHERE CAN I GET FURTHER INFORMATION?

A mailing list is being maintained of those interested in the development of SMD Format. Anyone wishing to be included on it should contact the Secretary of the Chemical Structure Association.

Technical questions concerning the format should be addressed to the Technical Secretary (the author of this paper) who should also be contacted by anyone interested in participating in the activity of the working groups.

ACKNOWLEDGMENTS

Thanks are due to Dr Wendy Warr (ICI Pharmaceuticals) and Mr Peter Nichols (Hampden Data Services) for comments on the draft of this poster, and to Mr Roger Schenck (Chemical Abstracts Service) who presented the poster at the American Chemical Society Meeting in Boston in April 1990.

The following individuals participated in all or some of the technical working groups, which proposed the draft specification for Version 5.0 of SMD format. Affiliations shown are those which were current at the time the person concerned attended working group meetings.

J.M. Barnard (Barnard Chemical Information Ltd, Sheffield); H. Braun (F. Hoffmann-La Roche AG, Basel); J. Buckingham (Chapman and Hall, London); C. Buse (Sandoz AG, Basel); B. Carrabin (Polygen (Europe) Ltd, Paris); A.P.F. Cook (ORAC Ltd, Leeds); J. Dill (Molecular Design MDL AG, Basel); U. Hegi (Sandoz AG, Basel); U. Heigl (Molecular Design MDL AG, Basel); M. Hicks (Beilstein Institute, Frankfurt); K. Higgins (ORAC Ltd, Leeds); P. Hoever (Bayer AG, Leverkusen); P. Huguet (Télésystèmes Questel, Paris); S. Hull (Fraser Williams (Scientific Systems) Ltd, Macclesfield); H. Jacob (Ciba-Geigy AG, Basel); A. Kos (Molecular Design MDL AG, Basel); Herr Krause (Information und Kommunikation, Freiburg); J. Mockus (Chemical Abstracts Service, Columbus); M. Ott (CAOS/CAMM Centre, University of Nijmegen); Herr Posselt (Information und Kommunikation, Freiburg); B. Rhode (Ciba-Geigy AG, Basel); J. Römelt (Bayer AG, Leverkusen); G. Russell (Hampden Data Services Ltd, Nottingham); H. Saller (Chemodata Computer-Chemie GmbH, Gröbenzell); R. Schenck (Chemical Abstracts Service, Columbus); W. Sieber (Sandoz AG, Basel); R. Steppuhn (Schering AG, Berlin); J. Theodosiou (Molecular Design MDL AG, Basel); W.G. Town (Hampden Data Services Ltd, Abingdon); D. Watson (Cambridge Crystallographic Data Centre, Cambridge); S. Welford (Beilstein Institute, Frankfurt); P. Youkaribache (Polygen (Europe) Ltd, Paris); C. Zirz (Bayer AG, Leverkusen);

Financial support for the development of SMD Format during 1988 and 1989 has been provided by the following organisations:

Bayer AG (Germany); CAOS/CAMM Centre, University of Nijmegen (Netherlands); Chemical Design Ltd (U.K.); Chemical Abstracts Service (U.S.A.); Chemical Structure Association (U.K.); Chemodata Computer-Chemie GmbH (Germany); Ciba-Geigy AG (Switzerland); Dialog Information Services Inc (U.S.A.); European Communities Joint Research Centre (Italy); Finnigan Corporation (U.S.A.); Fisons plc Pharmaceutical Division (U.K.); Fraser Williams (Scientific Systems) Ltd (U.K.); Glaxo Group Research (U.K.); Hampden Data Services Ltd (U.K.); Imperial Chemical Industries PLC (U.K.); Institute for Scientific Information (U.K.); Molecular Design MDL AG (Switzerland); ORAC Ltd (U.K.); Pfizer Central Research (U.K.); Polygen (Europe) Ltd (France); F Hoffmann-La Roche AG (Switzerland); Sadtler Research Laboratories (U.S.A.); Sandoz AG (Switzerland); Schering AG (Germany); Smith Kline and French Research Ltd (U.K.); Télésystèmes Questel (France).

BIBLIOGRAPHY

Draft Formal Specification of SMD Format

1. Barnard, J.M. 'Draft Specification for Revised Version of the Standard Molecular Data (SMD) Format'. *J. Chem. Inf. Comput. Sci.* **1990**, *30*, 81–97.

Previous Version of SMD Format

2. Bebak, H. 'The SMD File Format. Version 4.3'. Copies will be provided on request to Dr W.T. Donner, Bayer AG, ZF-DID, Geb. Q18, W-5090 Leverkusen 1, Germany.
3. Bebak, H.; Buse, C.; Donner, W.T.; Hoever, P.; Jacob, H.; Klaus, H.; Pesch, J.; Römelt, J.; Schilling, P.; Woost, B.; Zirz, C. 'The Standard Molecular Data Format (SMD

Format) as an Integration Tool in Computer Chemistry'. *J. Chem. Inf. Comput. Sci.* **1989,** *29,* 1–5.

Development of Revised SMD Format

4. Barnard, J.M. 'Towards a Standard Interchange Format for Chemical Structure Data'. In *Proceedings of the 12th International Online Information Meeting*; Learned Information: Oxford, 1988; pp. 605–609.
5. Barnard, J.M. 'Standard Representations for Chemical Information'. In *Proceedings of the Montreux 1989 International Chemical Information Conference, Montreux, Switzerland, 26–28 September 1989;* Collier, H., Ed.; Springer-Verlag: Heidelberg, 1989; pp. 209–219.
6. Barnard, J.M.; Cook, A.P.F.; Rohde, B. 'Storage and Searching of Stereochemistry in Substructure Search Systems'. In *Chemical Information Systems: Beyond the Structure Diagram. Proceedings of a Conference held at the College of St Hild and St Bede, University of Durham, UK, 17–20 July 1989*; Bawden, D.; Mitchell, E., Eds; Ellis Horwood: Chichester; pp. 29–41.

Symposium on Standard Formats

7. *Chemical Structure Information Systems. Interfaces, Communication and Standards*; Warr, W.A., Ed.; ACS Symposium Series 40; American Chemical Society: Washington, 1989.

Other Standard Formats

8. Brown, I.D. 'Standard Crystallographic File Structure'. *Acta Crystallogr.* **1983,** *A39,* 216–224.
9. George, D.W.; Mewes, H.W.; Kihara, H. 'A Standardised Format for Sequence Data Exchange'. *Protein Seq. Data Anal.* **1987,** *1,* 27–39.
10. Gund, P.; Barry, C.D.; Blaney, J.M.; Cohen, N.C. 'Guidelines for Publications in Molecular Modelling Related to Medicinal Chemistry'. *J. Med. Chem.* **1988,** *31,* 2230–2234.
11. Gasteiger, J.; Hendriks, B.M.P.; Hoever, P.; Jochum, C.; Somberg, H. 'JCAMP-CS. A Standard Exchange Format for Chemical Structure Information in Computer-Readable Form'. Submitted for publication in *Appl. Spectrosc.*
12. McDonald, R.S., Wilks, P.A. 'JCAMP-DX: A Standard Form for Exchange of Infrared Spectra in Computer-Readable Form'. *Appl. Spectrosc.* **1988,** *42,* 151–162.

This paper describes the state of development of standard formats at the time of the conference (June 1990). Several significant developments have taken place since then, and further developments are expected in the early part of 1992. This paper is therefore included in the Proceedings purely as an historical record.

GEMINI: A GENERALISED CONNECTION TABLE LANGUAGE AND INTERPRETER

David Weininger and Arthur Weininger
Daylight Chemical Information Systems Inc., 111 Rue Iberville, Suite 610, New Orleans, Louisiana 70130, U.S.A.

ABSTRACT

The design of a generalised program for interpreting connection tables is presented, including a language for describing the external encoding of structural information. Such a program is useful for format conversion or as a front-end to chemical application software. This program has been implemented (GEMINI), and is discussed with examples.

There are only three fundamental ways in which connection table formats differ: what information is stored, how that information is represented, and how those representations are encoded. GEMINI achieves a high degree of generality by dividing its task into these three parts.

Internally, GEMINI uses a rich set of data types with well-defined relationships which can include redundancy or ambiguity. Such information includes atomic number, connectivity, bond order, chirality, charge, co-ordinates, etc. Fortunately, there are very few different representations that are used for each kind of information, and algorithms are readily available for their interconversion.

A string-processing language is used to specify a connection table format. This language allows succinct specification of both the content and encoding of most external formats that we have encountered, but is currently limited to character-oriented files.

BACKGROUND

Chemicals are usually stored in computer-readable format as data which represent a valence structure. For historical reasons, the external format for such data is commonly referred to as a connection table. A majority of connection table formats are simply formatted versions of an internal data structure written out in tabular format. However, a significant number of non-tabular formats are used for external computer representation of chemicals.

There is a large number of different connection table formats in active use today (probably in excess of 100). Several factors contribute to the proliferation of different formats for the same fundamental purpose. The most important of these are:

1. Connection table formats are often invented to serve a specific application. Different applications require storage of different kinds of information.
2. Different formats are designed for different trade-offs between conciseness, generality, and interpretation speed.

W. A. Warr (Ed.)
Chemical Structures 2

3. There is no standard connection table format that has been uniformly accepted by the chemical information community.
4. Some connection table formats are proprietary.

There are both good and bad effects of having such a variety of connection table formats. For any given application program, highest efficiency is achieved by using a special-purpose format. On the other hand, using multiple external formats introduces serious compatibility problems that limit the utility of a library of application programs. Such compatibility problems are growing more serious as computer programs become the medium of exchange for chemical intelligence.

NATURE OF THE PROBLEM

A connection table format can be considered a special-purpose external database format which is used to store a limited amount of chemical information. Our goal is to produce a method which allows accurate description and interpretation of connection tables in the general case.

Although there appears to be a wide variety of connection table formats, there are only three fundamental ways in which such formats can differ: what information is stored, which representations are used, and how those representations are encoded.

Information content

Compared to databases, connection tables contain very few different types of information. Typically, connection table formats implicitly limit the kinds of information that they can contain, i.e., their format defines the kind of information that is contained, rather than implementing semantics for general data storage. Kinds of information used to represent a valence structure include: atomic properties (element type, formal charge, isotopic mass), bond properties (connectivity, order), configurational properties (chirality, configuration about double bonds) and molecular properties (name or other identification). Provision is often made for storage of 2-D and 3-D co-ordinates. For the purposes of connection table interpretation, information which is not used for structural specification is irrelevant.

Not all connection table formats contain all these types of information. Formats can therefore be *information deficient* with respect to one another.

Representation of information

Each connection table format specifies one or more *representations* for each kind of information that it can contain. A representation is a rigorous system for characterising a particular kind of information. An adequate representation must not be information-deficient with respect to the information that it represents. For example, internal co-ordinates, local chirality, R/S, and bond parity are four different adequate representations of absolute chirality information. Adequate representations can be *converted* between each other algorithmically. Fortunately, there are very few different representations used for the types of information that are stored in connection tables.

Encoding representations

Connection table formats specify an *encoding* for each kind of informational representation which can be used. An encoding is a convention used to indicate the possible values that representations can take. The number of possible encodings for any given representation is very high. For example, the R/S representation of absolute chirality can be encoded as the characters 'R' and 'S', the digits '1' and '2', or many others.

Changing between encoding conventions for a given representation is done by a process known as *translation*. Algorithms used for translation are much simpler than those used for conversion (changing between representations). For instance, standard atomic symbols can be translated to atomic numbers becuse they are both encodings of element type. Translation cannot be used to change representations, e.g., chirality can be derived from 3-D co-ordinates only by a conversion algorithm.

A complete encoding convention for a connection table format also specifies how the representational encodings are organised (e.g., in a file or character stream).

APPROACH

The approach used here for connection table interpretation assumes that there are limited numbers of kinds of information which exist in the desired formats, and that it is known how to convert the information from each of these to a suitable internal form. The interpretation of any instance of a connection table can then be divided into two completely independent parts: extracting the information from the table and converting it to internal form. This allows automated interpretation, given the meaning (representation and encoding) of each field of a connection table format.

The following tasks must be accomplished to make this approach work:

1. Establish a set of information types which is adequate to describe the information contained in the desired connection table formats. Select a single data representation and encoding for each type of information; this will become GEMINI's *internal encoding*.
2. Collect the set of data representations used in the desired connection table formats, and select a single encoding convention for each one. This will be known as the set of *transfer variables*.
3. Implement algorithms to convert the transfer variable representations to those used in the internal encoding.
4. Establish a language for specification of connection table data encoding. This language needs to be general enough to specify all desired transfer variable encodings. The specification of a format in this language will be referred to as a 'GEMINI connection table format'.
5. Develop a program which integrates the above elements to interpret a connection table given its format specification.

IMPLEMENTATION

Internal encoding

The choice of the set of variables used internally to represent transfer information is not critical as long as it accurately describes all information to be derived from the desired connection table format. A slight benefit is associated with choosing a set of variables which store information independently. Any encoding convention may be used which is adequate to specify all desired values (including missing, as appropriate).

The set of variables used for GEMINI's internal encoding is: molecule label, atom count, atomic number, atomic charge, atomic weight, atom label, bond count, bond ends, bond type, bond label, bond direction, local chirality, 2-D and 3-D co-ordinates, and co-ordinate set labels.

Transfer variables

There are few enough different informational representations used in connection tables that they can be explicitly enumerated: this is the set of transfer variables. An encoding convention is selected for each one. The GEMINI transfer variables are shown in Table 1.

Conversion

Since target connection tables use representations other than those used in GEMINI's internal encoding, conversions will sometimes be needed to produce an internal encoding. The methods used for this are known as conversion algorithms.

Note that the set of transfer variables is redundant with respect to the variables used in the internal encoding (i.e., an internal variable can be specified by more than one transfer variable). This is a natural consequence of the multiplicity of representations available for connection table information. Furthermore, a connection table format may be inherently redundant (i.e., the same information may be specified more than once, or in more than one way). Therefore, another algorithm is needed to specify the order that conversion algorithms are applied to the transfer variables to produce an internal encoding. This algorithm is listed in Table 2.

Specification of connection table encoding

Specification of encoding of connection tables must be done in a very flexible manner because the number of ways to encode information is essentially limitless. GEMINI uses a string processing language designed especially for the purpose of specifying connection table formats (CTF). The object of a CTF specification is to direct GEMINI's processing of the input file in such a way that appropriate character strings (translated to the correct encoding, if needed) are assigned to transfer variables. The CTF language provides both format and translation

Table 1. Description of GEMINI transfer variables

Variable name	*Context*	*Class*	*Notes*	*Expect*
NATOMS	control	integer	(1)	1 to MXATM
NBONDS	control	integer	(1)	0 to MXATM +MXCY
NCHIRAL	control	integer		0 to MXATM
NAME	none	string-1		any non-null string
DESCRIPTION	none	string-1		any non-null string
COMPATABILITY	none	string-1	(2)	special string
FILETYPES	none	string-1		special string
CT SOURCE	none	string-1		any non-null string
2D SOURCE	none	string-1	(3)	any non-null string
3D SOURCE	none	string-1	(3)	any non-null string
NOTE	none	string-1		any non-null string
OTHERID	none	string-1		any non-null string
ATOM.NUMBER	atom	integer		0 to 100
ATOM.SYMBOL	atom	string-4	(1)	atomic symbol in 1st 2 chars
ATOM.CHARGE	atom	integer	(1)	-9 to 9
ATOM.HCOUNT	atom	integer	(1)	-9 to 9
ATOM.LABEL	atom	string	(1)	atomic symbol
ATOM.CHIRAL	atom	integer	(1)	0 or 1 or 2
ATOM.RS	atom	string-3	(1)	R or S
ATOM.PARITY	atom	string-2	(1)	"+" or "-" or " "
ATOM.MASS	atom	integer	(1)	0 to 400
ATOM.DEFECT	atom	integer	(1)	-50 to 50
ATOM.X2	atom	integer	(1,3)	anything, rounded to 0.01
ATOM.Y2	atom	integer	(1,3)	anything, rounded to 0.01
ATOM.X3	atom	real	(1,3)	anything, rounded to 0.01
ATOM.Y3	atom	real	(1,3)	anything, rounded to 0.01
ATOM.Z3	atom	real	(1,3)	anything, rounded to 0.01
BOND.FROM	bond	integer	(1)	1 to NATOMS
BOND.TO	bond	integer	(1)	1 to NATOMS
BOND.ORDER	bond	integer	(1)	1, 2, 3, or 4
BOND.LABEL	bond	integer	(1)	first 8 bytes significant
BOND.DIR	bond	string-2	(1)	"+" or "-" or " "
BOND.EZ	bond	string-3	(1)	"E" or "Z"
BOND.INOUT	bond	string-2	(1)	"+" or "-" or " "

Transfer variable classes

integer	accept [+-]?[0-9][0-9]*.*, integer, warn if delimiter other than space, tab, comma, or semicolon found.
real	accept [+-]?[0-9.][0-9.]*.*, real, warn if delimiter other than space, tab, comma, or semicolon found.
string-1	accept any string, delete non-printable characters, trim leading and trailing blanks, squeeze multiple blanks to single blanks, truncate at maximum string length.
string-2	accept any string, delete all non-printable characters & blanks, truncate to maximum string length.
string-3	accept any string, delete all but letters and digits, convert to upper case, truncate to maximum string length.
string-4	accept any string, delete everything but letters and digits, convert first letter to upper case, all others to lower case, truncate to maximum string length.

(1) Variables with .'s in their names (ATOM.XXX or BOND.XXX) have an implicit context when read. If they do not occur in a loop (ATOM_LOOP or BOND_LOOP), NATOMS is implicity incremented by ATOM.NUMBER or ATOM.SYMBOL, and NBONDS by BOND.TO (connections are always to the last-encountered BOND.FROM).

(2) COMPATIBILITY strings are expected to be in the form "3.61 3.62" with all previous compatible releases listed.

(3) Two sets of ATOM coordinate variables are provided. (.X2) and . (Y2) are intended for 2-D coordinates, (.X3), (.Y3), and (.Z3) for 3-D coordinates. Either or both sets may be specified.

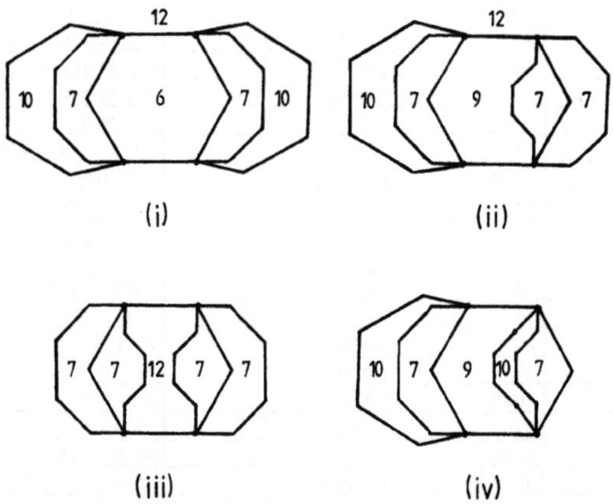

Figure 7. Structure with four embedments

Instead they all contain some intermediate simple-faces (alternative embedment regions).

The regions side of Figure 3 is defined further in Figure 8 to reflect the classification of simple-faces given above.

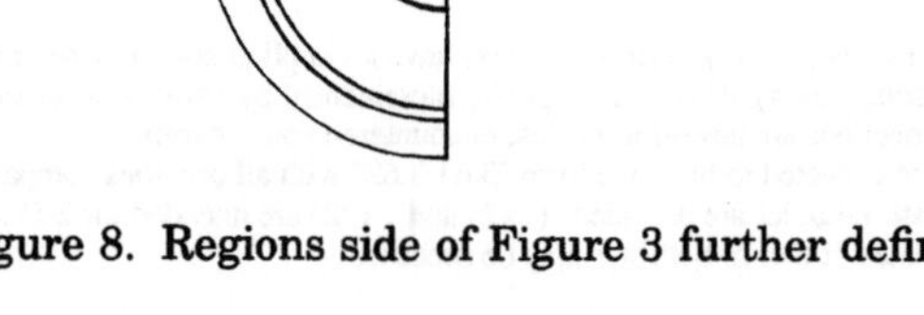

Figure 8. Regions side of Figure 3 further defined

Non-regions and Cut-faces

Let us now consider those cycles that cannot be drawn as regions of any embedment. As with regions, there exist non-regions that are not simple-cycles due to the presence of Doppelpunkte or Nachbarpunkte. Those non-regions that are simple-cycles will be referred to as '*cut-faces*'. The set of cut-faces can be differentiated in terms of their size in relation to adjacent simple-cycles.

Primary cut-faces are those in which at least one edge has the cut-face as its smallest simple-cycle.

facilities. This language is based on a general string processing model and is designed to be easily extensible. The language is simplified by including only those capabilities which are needed for CTF interpretation as demonstrated by an extant CTF format. A complete description of the current CTF specification language follows.

A CONNECTION TABLE FORMAT (CTF) DESCRIPTION LANGUAGE

General

A CTF specification is a single file which describes the format of a single type of connection table. The specification is composed of printable characters only, and is divided into the following components:

Comments: Matching braces ({}), and all intervening characters, are ignored.
Commands: Specific words are reserved as commands: see below and Table 3.
Transfer variable names: Reserved words enclosed in parentheses (Table 1).
Numbers: Some commands expect positive integers (a series of digits).
Strings: A string is a series of characters enclosed in double quotes ("). Strings may be concatenated with a plus sign (+). Within a string, non-printable characters are represented by a backslash followed by one or more characters:

newline	NL (LF)	\n	
horizontal tab	HT	\t	
backspace	BS	\b	
carriage return	CR	\r	
form feed	FF	\f	
backslash	\	\\	
double quote	"	\"	
bit pattern	octal digits	\ddd	(1, 2, or 3 trailing digits are OK)
nul	ASCII(0)	\0	(an example of above bit pattern)

Whitespace: spaces, tabs, and newlines not in a string; ignored except as a delimiter.

Utility commands

Four commands are provided to make the usage of the resultant file more convenient and robust. These commands are not required, and have no direct effect on connection table interpretation. Each command takes one string as an argument. The DESCRIPTION command allows identification of the type of connection table to which a particular specification applies. Notes about the usage of a particular CTF can be delivered to the user at runtime via the NOTE command.

The FILETYPES command is used to assert the normal file extensions which are used for such files of the intended type; e.g., if the argument is '.NCI .NBS', a warning will be written if the file is used for a file with a different file extension from those two. Since it is expected that GEMINI software will be updated in the future, known compatibility should be asserted with the COMPATIBILITY command; e.g., a compatibility string in the form '3.61 3.62' will cause a warning to be written if the file is used with an unlisted version of GEMINI.

Table 3. GEMINI CTF specification commands

Command name & arguments	*Description*
DESCRIPTION "*string*"	Description of connection table. No-op.
NOTE "*string*"	A note applying to this specification. No-op.
FILETYPES "*string*"	Assertion of normal file types for this format. No-op.
COMPATIBILITY "*string*"	Assertion of GEMINI compatibility level. No-op.
BEGIN_CT	Indicate beginning of connection table input block.
END_CT	Indicate end of connection table input block.
REWIND_CT	Flow control: go to BEGIN_CT.
BEGIN_LOOP	Indicate beginning of repetitive input block.
END_LOOP	Indicate end of repetitive input block.
EXIT_LOOP	Exit repetitive input block.
BEGIN_ATOM_LOOP	Indicate beginning of atom-oriented input block.
END_ATOM_LOOP	Indicate end of atom-oriented input block.
BEGIN_BOND_LOOP	Indicate beginning of bond-oriented input block.
END_BOND_LOOP	Indicate end of bond-oriented input block.
IF .. ELSEIF .. ELSE .. ENDIF	Conditional execution control.
EXACTLY, EQUALS, CONTAINS	IF / ELSEIF / ELSE test; followed by string.
GETRECORD [*pint*]	Advances input record; optionally limit to *pint* bytes.
GETRECORD_STARTING "*string*" [*pint*]	Like GETRECORD, but keeps reading until match.
GETRECORD_ENDING "*string*" [*pint*]	Like GETRECORD, but keeps reading until match.
GETRECORD_INCLUDING "*string*" [*pint*]	Like GETRECORD, but keeps reading until match.
UNGETRECORD	Push current record back onto input buffer.
READ *pint*	Read next *pint* bytes from record into the input buffer.
READ "*char*"	Read record to next delimiter into input buffer.
READ *	Read record to next ',', ';', or whitespace into buffer.
READ EOR	Read rest of record into input buffer.
MAP_ONLY "*old*" "*new*"	Replace strings of characters in the input buffer which match those in "*old*" with corresponding ones in "*new*" ("*new*" may be null strings).
MAP_DEFAULT "*def*" "*old*" "*new*"	Like MAP_ONLY except unmatched strings are replaced with "*def*" ("*def*" and "*new*" may be null strings).
LEFT [*pint*]	Left justify the input buffer and truncate to *pint* bytes.
RIGHT [*pint*]	Right justify the input buffer and truncate to *pint* bytes.
RESET	Restore buffer to state after last READ.
FILL (*variable*)	Assigns the current buffer to given transfer variable.

Program scope commands

The section of a GEMINI specification which specifies the actual connection table information must be delimited by **BEGIN—CT** and **END CT**. This delimits the repetitive structure specification block in formats which allow multiple connection tables in one file.

Buffer control commands

The GEMINI CTF specification language assumes that there are two logical buffers: a 'current record' which is read in from the connection table, and an 'active buffer' which is read from the current record and contains a string of characters to be assigned to a transfer variable or operated on in some other way.

The GETRECORD command gets the next record from the input file. An optional integer argument specifies the logical record length in bytes (longer records will be truncated, and shorter records will be blank padded to this length). Three commands, GETRECORD—STARTING, GETRECORD—ENDING, and GETRECORD INCLUDING, are similar but continue obtaining records from the input file until one is found which contains a specified string in the appropriate place, which becomes the current record. The UNGETRECORD command pushes the last record read back into the input buffer.

All string operations are done on a portion of the current record (the 'active buffer') which is controlled by the READ command. There are four forms of the read command:

1. If followed by a number, that number of bytes will be transferred from the current record to the active buffer. For example, the command READ **3** will cause the active buffer to be set to the next three characters in the current record.
2. The command READ * will cause the next item to be read into the active buffer using 'free format' conventions. Specifically, this means that leading whitespace will be skipped, then all characters up to the next comma, semicolon, or more whitespace will be transferred to the active buffer.
3. If another set of delimiters is required, they may be listed in a string argument following the read command, i.e., READ * is equivalent to READ " \T,;".
4. Finally, a special form allows the rest of the record to be read into the active buffer: READ EOR (read to end-of-record).

Transfer variable assignment commands

The FILL command is used to assign the active buffer to a given transfer variable. The exact name of the transfer variable in parentheses must follow the command, e.g. FILL (NAME) indicates that the active buffer contains the name of the molecule.

Conditional execution

A section of CTF code can be conditionally executed by using the IF, ELSEIF, ELSE, and ENDIF commands. Every IF must be terminated with an ENDIF; ELSEIF and ELSE are optional. Intervening commands are conditionally executed based on a lexical test of the current read buffer. IF and ELSEIF must be followed by a binary Boolean operator and a string. Allowed operators are CONTAINS (read buffer contains the string), EQUALS (the read buffer is the same as the string discounting trailing blanks) and EXACTLY (read buffer is exactly equal to the string). Null

Table 4. An example GEMINI connection table specification

```
DESCRIPTION    "TriPos' MOL-1 connection table format (used by SYBYL)"
FILETYPES      ".SYB .MOL .MOL1"
COMPATIBILITY "3.62"

BEGIN_CT

   GETRECORD { header record }
      READ 4      FILL (NATOMS)
      READ 5
      READ 40     FILL (NAME)

   BEGIN_ATOM_LOOP
      GETRECORD
         READ 4
         READ 4   MAP_ONLY

                 { /N / S/ S/ N/Si/ D/Al/Li/Ca/ K/Na/LP/ N/ S/ I/ F/Cl }
                 { /Br/ H/ P/ N/ S/O/O/N/N/C/C/C/C/C/ }

{ Sybyl codes }   "/31/30/29/28/27/26/25/24/23/22/21/20/19/18/17/16/15" +
                  "/14/13/12/11/10/9/8/7/6/5/4/3/2/1/"

{ atomic wts }    "/7 /32/32/ 7/14/ 1/13/ 3/20/19/11/ 0/ 7/16/53/ 9/17" +
                  "/35/ 1/15/ 7/32/8/8/7/7/7/6/6/6/6/"

                  FILL (ATOM.NUMBER)

         RESET    MAP_DEFAULT " " "/31/" "/1/"
                  IF EQUALS " " { do nothing } ELSE FILL (ATOM.CHARGE) ENDIF

         READ  9  FILL (ATOM.X3)
         READ  9  FILL (ATOM.Y3)
         READ  9  FILL (ATOM.Z3)
         READ  4  FILL (ATOM.LABEL)
   END_ATOM_LOOP

   GETRECORD
   READ 4     FILL (NBONDS)
   BEGIN_BOND_LOOP
      GETRECORD
         READ  4
         READ  4  FILL (BOND.FROM)
         READ  4  FILL (BOND.TO)
         READ  9
         READ  4  MAP_ONLY "/1/2/3/4/5/6/7/8/9/10/"
                           "/1/2/3/1/4//"
                  FILL (BOND.ORDER)
   END_BOND_LOOP

   GETRECORD { trailing record }
END_CT
```

scopes are allowed, e.g., IF EQUALS "N/A" ELSE FILL (NAME) ENDIF would specify the name whenever the read buffer was *not* "N/A".

Looping commands

Two different methods are used for repetitive specifications in connection tables. The most common method precedes the repetitive records by a specification of how many items are to follow; this is often the way atoms and bonds are specified. Special CTF commands are provided to accommodate this method. If this method is used, the specification for the number of atoms should be read and filled with FILL (NATOMS), then the repetitive atom specification block should be delimited with BEGIN—ATOM—LOOP and END—ATOM—LOOP. Atomic properties (i.e., transfer variables such as atom.number) can be assigned within this block. The commands

Buffer control commands

The GEMINI CTF specification language assumes that there are two logical buffers: a 'current record' which is read in from the connection table, and an 'active buffer' which is read from the current record and contains a string of characters to be assigned to a transfer variable or operated on in some other way.

The GETRECORD command gets the next record from the input file. An optional integer argument specifies the logical record length in bytes (longer records will be truncated, and shorter records will be blank padded to this length). Three commands, GETRECORD—STARTING, GETRECORD—ENDING, and GETRECORD INCLUDING, are similar but continue obtaining records from the input file until one is found which contains a specified string in the appropriate place, which becomes the current record. The UNGETRECORD command pushes the last record read back into the input buffer.

All string operations are done on a portion of the current record (the 'active buffer') which is controlled by the READ command. There are four forms of the read command:

1. If followed by a number, that number of bytes will be transferred from the current record to the active buffer. For example, the command READ **3** will cause the active buffer to be set to the next three characters in the current record.
2. The command READ * will cause the next item to be read into the active buffer using 'free format' conventions. Specifically, this means that leading whitespace will be skipped, then all characters up to the next comma, semicolon, or more whitespace will be transferred to the active buffer.
3. If another set of delimiters is required, they may be listed in a string argument following the read command, i.e., READ * is equivalent to READ **" \T,;"**.
4. Finally, a special form allows the rest of the record to be read into the active buffer: READ EOR (read to end-of-record).

Transfer variable assignment commands

The FILL command is used to assign the active buffer to a given transfer variable. The exact name of the transfer variable in parentheses must follow the command, e.g. FILL (NAME) indicates that the active buffer contains the name of the molecule.

Conditional execution

A section of CTF code can be conditionally executed by using the IF, ELSEIF, ELSE, and ENDIF commands. Every IF must be terminated with an ENDIF; ELSEIF and ELSE are optional. Intervening commands are conditionally executed based on a lexical test of the current read buffer. IF and ELSEIF must be followed by a binary Boolean operator and a string. Allowed operators are CONTAINS (read buffer contains the string), EQUALS (the read buffer is the same as the string discounting trailing blanks) and EXACTLY (read buffer is exactly equal to the string). Null

Conversion

Once interpreted, the conversion between informational representations can be done by widely-used methods. Novel conversion algorithms were needed in only two instances so far.

The first is in conversion of CTs which represent 2-D pictures with chirality indicated by hash/wedge bonds (BOND.INOUT). In these cases, a Z co-ordinate is artifically added, and this representation is turned into a 3-D representation for conversion to local chirality.

The second is in conversion of CT formats which do not store connection information, e.g., output from MO programs. In these cases, connectivity is deduced by Van der Waals' radii overlap and valence assumptions are used to deduce bond order for conversion to a valence representation. This method is robust for common elements, but is inadequate where valence assumptions are uncertain, e.g., transition metal complexes.

Program output

For convenience, the developmental GEMINI implementation writes a THOR data-tree as its output which is a language for general chemical information transfer. In practice, GEMINI would typically serve as a front-end to an application program which accessed the GEMINI's standard representations and no output file would be written.

CONCLUSION

The GEMINI approach to connection table interpretation provides a useful front-end for programs which process chemical information. CTF descriptions are compact enough that they can be included along with conection tables when transmitting data between sites with incompatible formats.

The CTF description language appears to be flexible enough to describe the vast majority of connection table formats in use today, and extensible enough to be adapted to those likely to arise in the future. A practical limitation of the language as described here is the inability to handle binary data conveniently. If interpretation of binary connection table formats becomes important, binary-oriented translation operators could be added to the language.

An intriguing possibility is creating a language to describe the converse operation, i.e., one which describes how to write a connection table, rather than read it. This would, in principle, allow any connection table format describable in the language to be converted to any other.

RINGS – THE IMPORTANCE OF BEING PERCEIVED

Dr.Geoff Downs*,
Barnard Chemical Information Ltd, 46 Uppergate Road, Stannington, Sheffield S6 6BX, England.

ABSTRACT

In areas such as retrieval, QSAR, synthesis design, reaction indexing, and structure display, ring analysis is required as a descriptive utility and to complement other structural analyses. Finding those rings necessary and sufficient for unambiguous representation, in an efficient manner and for the worst cases is not trivial. Issues include the 2-D representation of 3-D structures, the definition and perception of a ring set, and whether vertex and cut-vertex graphs simplify the analysis. Generic and partial/sub structures cause particular problems. Substructure queries can be the worst defined, introducing problems for structural conventions and ring/chain differentiation. After ring perception, it is necessary to select and represent the information relevant to a particular application. This can be used as a condensation, to enable more efficient matching, and as an expression, to give more detail to reduced graphs.

INTRODUCTION

The aim of this paper is to change your perception of rings! As basic justification for such an attempt I commend to you Mike Lynch's keynote address to the last conference[1] in which he looked forward to the further development of graph theory to enable a 'sufficiently abstract way [to talk] about characteristics of ring systems', and drew attention to Lavoisier's observation that 'to change a science, one needs to change the language of the science'.

The connection table and related adjacency/incidence matrices are widespread two-dimensional representations used for chemical structures, and can be used as input to a variety of perception routines to detect a wide range of structural characteristics. Rings are just one of many such characteristics, but their perception is not quite as simple as it may seem at first.

Ring perception is one of those background tasks mainly of interest to programmers and system designers. Users rarely give the matter a second thought, until unexpected results or annoying limitations are revealed.

Part of the problem is that ring systems are becoming increasingly more complex. The theoretical possibilities of a few years ago are now synthesised realities. A recent issue of *Chemical Reviews*[2], devoted to strained organic compounds, says much about trends in this direction, and yields some very

* It was with great sadness that the author received the news that Cheng Qian of Chemical Abstracts Service died on 1 August 1990 after a long fight against cancer. His charming personality, dedication and deep mathematical understanding of areas such as ring perception make his death a tremendous loss to us all, and the author therefore wishes to dedicate this paper to his memory.

W. A. Warr (Ed.)
Chemical Structures 2

interesting ring systems from the perception as well as the chemical point of vi Interest in cage structures in particular seems to have increased over the I decade, an essential feature about them being that they are seen to be th dimensional objects. It is this three-dimensionality that is a large part of tl fascination for chemists, and is also the root of many problems for ring percept when the representation used is a two-dimensional connection table or mat These problems are the main subject to be addressed by this paper since t challenge current ring perception theory.

IMPORTANCE

Rings and ring systems are acknowledged to be an important feature of majority of organic structures. As Gray[3] states:

> 'The overall importance of the ring systems in chemistry is reflected conventional chemical nomenclature and in the various ring indices detail known structures.'

Most ring systems are simply isolated rings or paired fusions; the task of find these rings is trivially simple. What is important is to be able to treat even most complex cases in the same way. Complex cases will take considerably lon in machine-processing terms, but the results must be consistent with analyse simpler ring systems.

Ring perception is a routine task, a utility to provide a description of cyclic pa of a structure in such a way as to complement other topological analyses, such fragmentation, functional group detection, and topological indices.

The nature of the perception and the set of rings to be found are somew application-dependent. For structure retrieval and reaction indexing the empha is on providing total recall with as high a precision as possible. For synthe design the construction of 'good' synthetic routes to cyclic compounds depends equally 'good' analysis of their cyclic structure. The requirement is for detectio those rings that are deemed of synthetic significance or importance: a suita fuzzy concept! For structure-activity relationship studies the important rings those which either singly or as an aggregate contribute to certain obser activities or reveal structural similarity between ring systems with simi properties. From this, it can be concluded that the rings perceived and included a representative ring set can be a critical factor for performance in a gi application.

Ring perception is thus an important component in a wide range of applicati requiring structural analysis. The task is to perceive the most appropriate set rings for the particular application in an efficient and consistent manner. For t it is necessary first to classify rings in some way so that comparisons a appropriate selections can be made.

CLASSIFICATION

This section concentrates on the nature of rings rather than on methods to fi them. It is necessary to delve into some theoretical aspects and define concepts tl will explain certain complex situations found in real structures. Much of tl material is covered more fully by Downs *et al.*[4], but some of the concepts and terms used here are an extension to that original paper.

Simple-cycles

When is a cycle a ring? This is a matter of definition and convention. The term 'ring' is rather vague, and preferably should be used only as a general term. A cycle is a path through the structure that starts and ends at the same vertex. If any other vertex occurs more than once in the sequence then this is a '*Doppelpunkt*'[5], such as the *cut-vertex* (a vertex whose removal separates the system into more than one component) in Figure 1(i) which separates the 12-edged cycle into two 6-edged cycles. If any non-consecutive vertices in the sequence are adjacent in the structure then they are '*Nachbarpunkte*'[5], such as those forming the *cut-edge* (an edge whose removal separates the system into more than one component) in Figure 1(ii) which separates the 10-edged cycle into two 6-edged cycles. A cycle without Doppelpunkte or Nachbarpunkte is a '*simple-cycle*'.

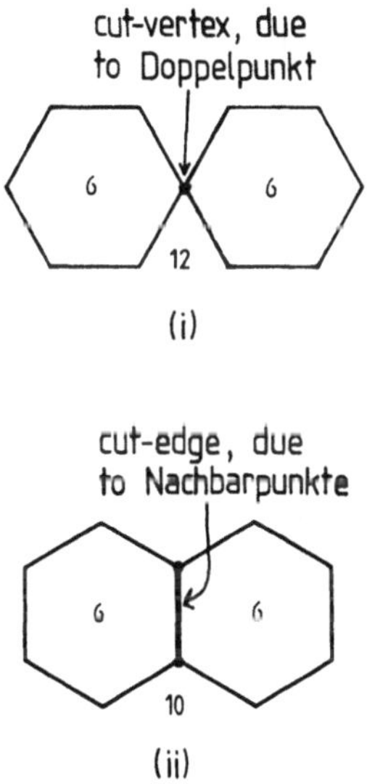

Figure 1. Doppelpunkt and Nachbarpunkte

Regions and Simple-faces

Let us now move on to the concepts involved in projecting a three-dimensional structure in two dimensions. For this we shall only consider structures for which this is possible (planar structures), although the concepts do have parallels in the non-planar environment. A planar structure is one which can be drawn (embedded) on a flat surface (plane) or sphere so that edges do not cross, they only meet at vertices. Planar, or spherical, embedment gives a series of regions on the plane or sphere. The unbounded part is known as the infinite region, and t bounded part(s) as the finite region(s), as shown in Figure 2. Any finite region c be reorientated to become the infinite region, and *vice versa* by changing t projection (viewing or drawing the embedment from a different viewpoint). This easier to visualise by projection onto a sphere; finite and infinite regions can interchanged by rotating the sphere to reveal the different possible projections.

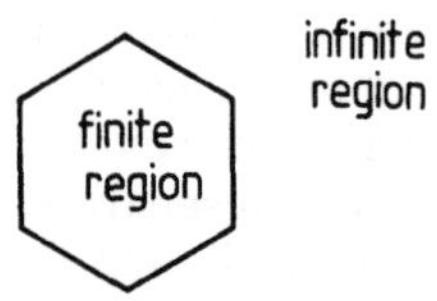

Figure 2. Regions

From the set of all cycles, some can be drawn as regions of a planar embedme and some cannot. The set of all simple-cycles is a subset of all cycles, which can divided similarly into regions and non-regions. This four-way interrelationship regions/non-regions and cycles/simple-cycles is represented in the Venn-li diagram in Figure 3.

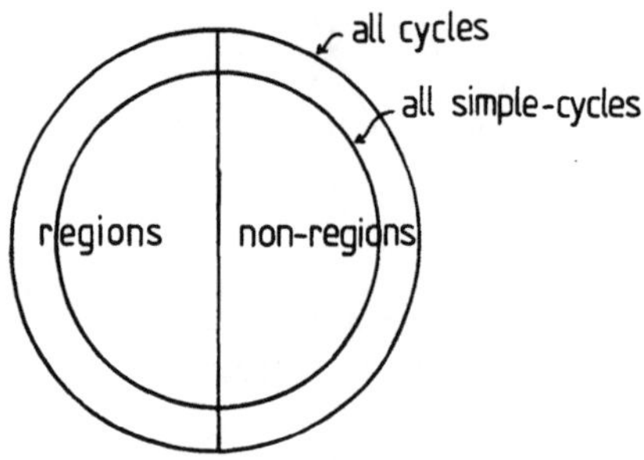

Figure 3. Interrelationship of regions/non-regions and cycles/simple-cycles

The planar embedment of norbornane is given in Figure 4; as can be seen, the are two distinct projections, one has a 6-edged infinite region, the other a 5-edg infinite region. The set of all regions, (5,5,6), is identical for both. So, should tl ring set for norbornane be (5,5), (5,6), or (5,5,6)?

For the structure in Figure 5, there are two distinct embedments possible. Tl first is shown in three distinct projections, 5(i)——5(iii), and the second in two distinct projections, 5(iv) and 5(v). The two embedments defi

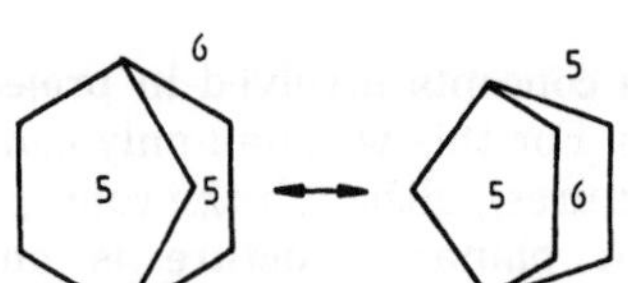

Figure 4. Planar embedment of norbornane

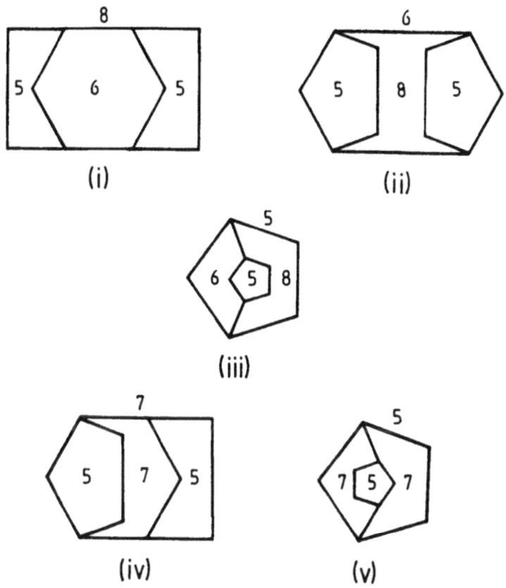

Figure 5. Structure with two distinct embedments

different regions. So, should the ring set for this structure be (5,5,6), (5,5,6,8), (5,5,7,7), (5,5,6,7,7), or (5,5,6,7,7,8)?

Those regions that are also simple-cycles will be referred to as '*simple-faces*'. In most cases, there exists a single embedment with a projection in which the smallest simple-faces, the *minimal* simple-faces, are enclosed by a single largest simple-face, the *maximal* simple-face. For isolated rings, the infinite and finite regions define the same simple-cycle, and so the maximal and minimal simple-faces are identical. If Doppelpunkte or Nachbarpunkte are involved, as in Figure 1, then a region exists that is not a simple-cycle and hence not a simple-face.

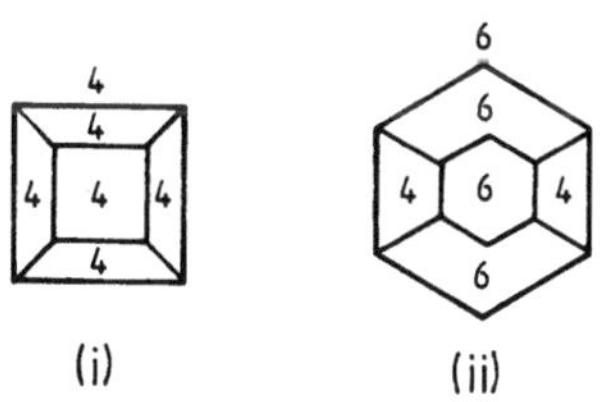

Figure 6. Simple-faces and symmetry

For certain structures, such as cubane, Figure 6(i), symmetry means that there is no distinction between maximal and minimal simple-faces since they are all the same size and interchangeable. In other cases, such as the structure in Figure 6(ii), there are minimal simple-faces, but there is no single maximal simple-face: there are several 6-edged simple-faces to choose from, referred to as intermediate simple-faces. There are also cases in which no single projection of any embedment contains all the minimal simple-cycles. The structure in Figure 7 has the four embedments shown, but none contains the five minimal simple-cycles (6,7,7,7,7).

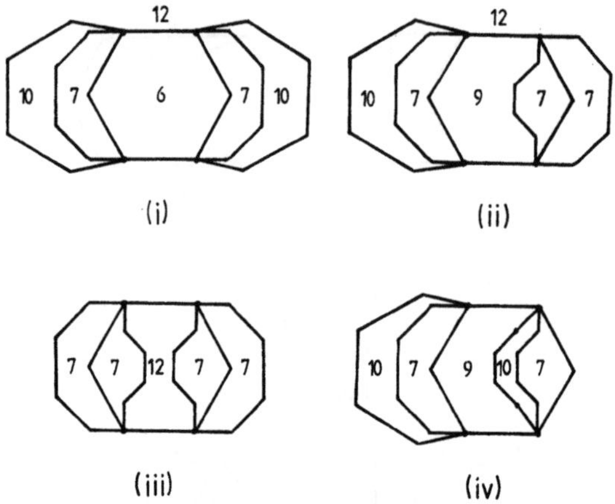

Figure 7. Structure with four embedments

Instead they all contain some intermediate simple-faces (alternative embedment regions).

The regions side of Figure 3 is defined further in Figure 8 to reflect the classification of simple-faces given above.

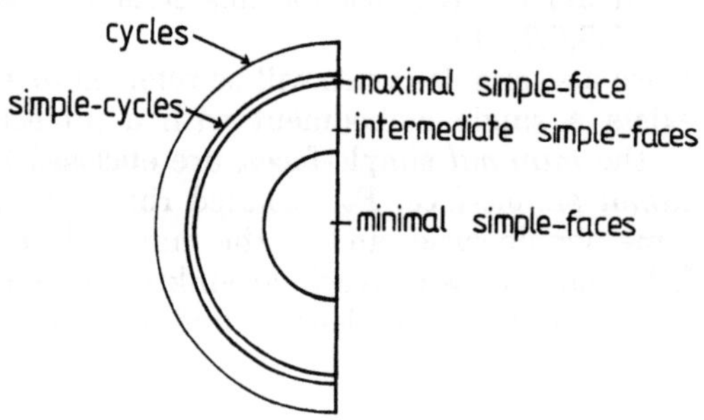

Figure 8. Regions side of Figure 3 further defined

Non-regions and Cut-faces

Let us now consider those cycles that cannot be drawn as regions of any embedment. As with regions, there exist non-regions that are not simple-cycles due to the presence of Doppelpunkte or Nachbarpunkte. Those non-regions that are simple-cycles will be referred to as *'cut-faces'*. The set of cut-faces can be differentiated in terms of their size in relation to adjacent simple-cycles.

Primary cut-faces are those in which at least one edge has the cut-face as its smallest simple-cycle.

Secondary cut-faces are those in which at least one edge has the cut-face the same size as the smallest simple-cycle associated with that edge.

Tertiary cut-faces are those in which each edge is associated with smaller simple-cycles.

The majority of cut-faces are tertiary and are of little interest. In contrast, the presence of primary and secondary cut-faces does have important consequences for ring perception and ring system representation, as can be seen by considering the examples in Figures 9, 10 and 11. In pagodane, Figure 9, the 4-edged cut-face happens to be the smallest simple-cycle in the structure. Each of its edges is associated with 5-edged simple-faces, none smaller, hence it is a primary cut-face. In trishomocubane ([1.1.1] triblattane), Figure 10, the next smallest simple-cycles after the six 5-edged simple-faces are the 6-edged cut-faces marked in bold. Each of their edges is associated with a smaller simple-face, hence they are tertiary cut-faces. However, the situation changes if each of the bridge vertices (1, 2 and 3) is replaced by two vertices to give the [2.2.2] triblattane in Figure 11. The six simple-faces are now 6-edged, and so the three 6-edged cut-faces are now secondary cut-faces. If another vertex is added, to give a [3.3.3] triblattane, then these become primary cut-faces since they are smaller than all of the adjacent simple-faces!

The non-regions side of Figure 3 is defined further in Figure 12 to reflect the classification of cut-faces given above.

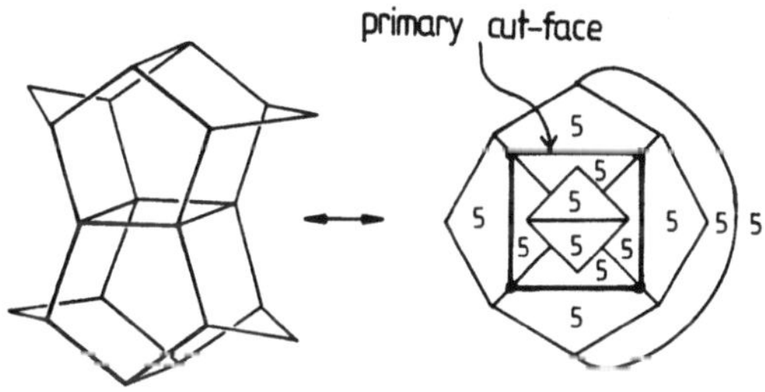

Figure 9. First cut-face example (pagodane)

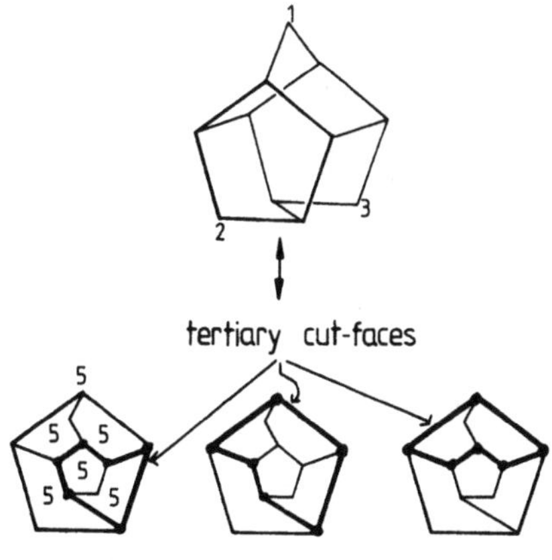

Figure 10. Second cut-face example (trishomocubane)

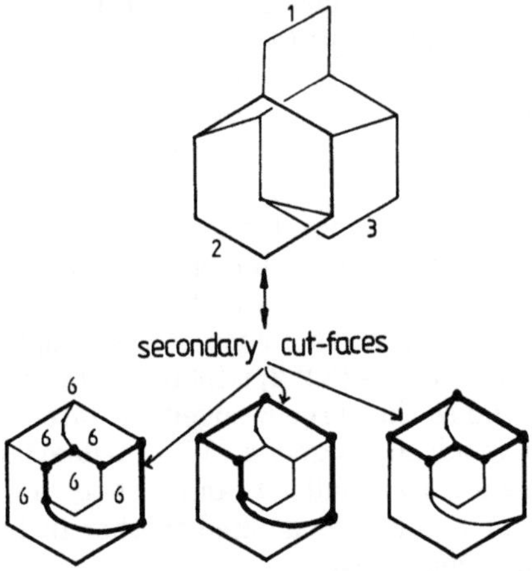

Figure 11. Third cut-face example (triblattane)

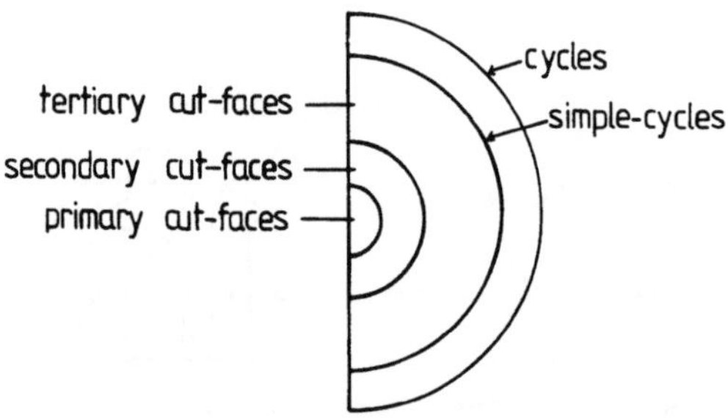

Figure 12. Non-regions side of Figure 3 futher defined

SELECTION

The above classification of ring types is very useful when one considers the ring sets that are used in the majority of applications, since it reveals those areas likely to be treated inconsistently for complex ring systems.

The number of finite regions, μ (the nullity, Frerejacque number, cyclomatic number, cyclicity, number of spanning tree chords etc.), in any projection of any embedment is a constant for a given ring system, and is calculated by Cauchy's formula:

$$\mu = \epsilon - \nu + 1$$

where ϵ is the number of edges and ν the number of vertices in the ring system.

The Smallest Set of Smallest Rings (SSSR) is a ring set containing μ of the smallest rings in the ring system. For trivial ring systems this seems quite sensible and can be seen to correspond to inclusion of the minimal simple-faces. However, a problem occurs where there is no maximal simple-face and a selection has to be made from several equivalent minimal or intermediate simple-faces. As a result, a given SSSR may not be an adequate description of the ring system, leading to problems of comparison. This problem has been recognised many times and so most ring sets add these equivalent rings to a basic SSSR.

A second problem occurs where cut-faces are present. A primary cut-face will always be the smallest ring for some edge and hence will occur in any SSSR, arbitrarily displacing another simple-cycle from the ring set. Similarly, secondary cut-faces may displace equivalent sized simple-cycles from the ring set. For instance, [2.2.2] triblattane in Figure 11 has three secondary cut-faces and six simple-faces, all the same size. The main criterion for an SSSR is that the set of μ rings should include all edges and vertices of the ring system; the cut-face/simple-face distinction is not recognised. There are also cases in which a tertiary cut-face, due to its size, will always displace a simple-cycle from the SSSR, as shown in Figure 13 in which the 8-edged tertiary cut-face is smaller than the two 9-edged simple-faces[6]. This second problem has yet to be addressed by any SSSR-finding algorithm.

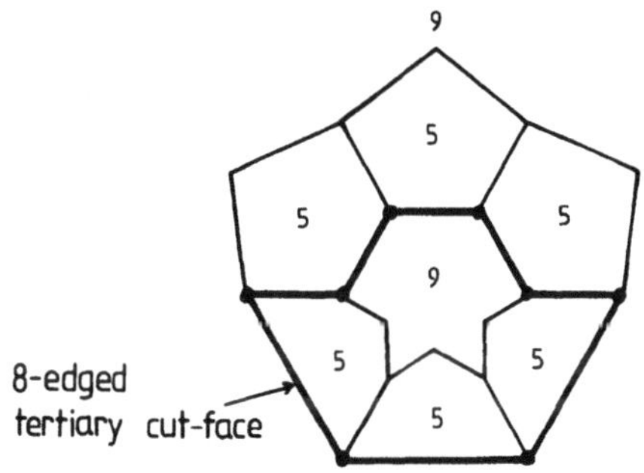

Figure 13. Tertiary cut-face displaces simple-cycle in SSSR

To overcome the first problem, Plotkin[7] proposed the set of κ-rings, which is the union of all possible SSSRs. This set will thus include all six simple-faces of cubane rather than the five in each SSSR. It will also include all primary and secondary cut-faces, and the tertiary cut-face of Figure 13. However, it will still not include all intermediate (alternative embedment) simple-faces, or any maximal simple-face. For instance, in norbornane, Figure 4, the 6-edged maximal simple-face is not included, and in Figure 5, the 7-edged intermediate simple-faces and the 8-edged maximal simple-face are not included. The situation in Figure 7 is particularly interesting since there is only one SSSR (6,7,7,7,7), yet no one projection of the structure contains all of these rings. Careful consideration is necessary to determine whether reliance on just the smallest rings adequately represents such ring systems for a given application.

Rings such as the six-edged maximal simple-face of norbornane are often regarded as important[8] and many attempts have been made to include them in the ring set, but without acknowledging their nature or treating them consistently. Hence, the SSSR or set of κ-rings has been supplemented by the addition of all rings with up to seven[9] or eight[10] edges. This, however, can lead to the inclusion of large numbers of tertiary cut-faces. In cubane, for instance, all 6-edged tertiary cut-faces will be included, such as the one shown in Figure 14.

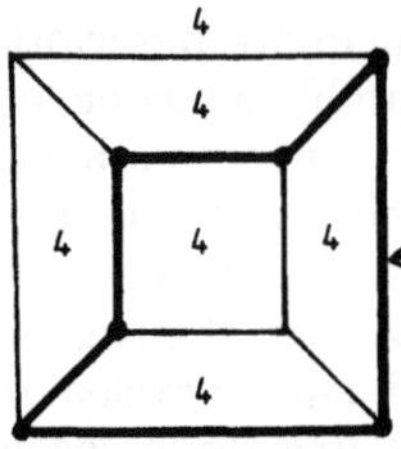

Figure 14. A tertiary cut-face in cubane

An alternative approach is that of the β-ring concept[5]. A β-ring must be a simple-cycle, which is where the concept originated. From the set of all simple-cycles, any with three or four edges are β-rings, and then any remaining that are not the linear combination of three or more smaller β-rings are also β-rings. Since there are only two smaller β-rings in norbornane, the 6-edged maximal simple-face is classified as a β-ring. Where there are three or more minimal simple-faces, any adjacent larger maximal simple-face or intermediate simple-faces will not be β-rings. It is also possible to classify tertiary cut-faces as β-rings, as in adamantane (Figure 15), in which the 8-edged tertiary cut-faces are β-rings since each is only the linear combination of two of the 6-edged simple-faces.

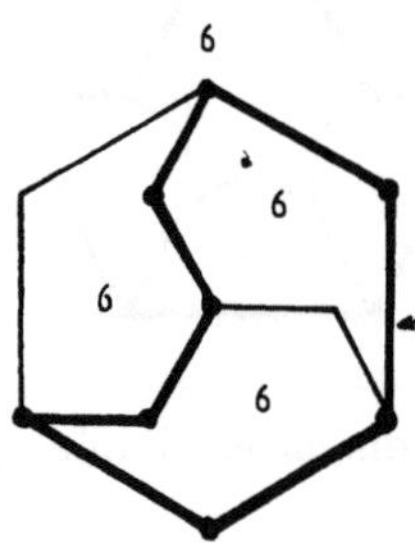

Figure 15. A tertiary cut-face in adamantane

To overcome these problems, the Extended Set of Smallest Rings (ESSR) was developed as part of the research on the Generic Structures Project at Sheffield University. The ESSR includes all simple-faces plus all primary and secondary cut-faces, as defined here; it does *not* include regions or non-regions that are not simple-cycles, or non-regions that are classified as tertiary cut-faces. An algorithm to find the ESSR has been described[11].

The ESSR is used at Sheffield as the basis for the ring representation as part of the screening stage. It is also utilised in other applications discussed in these proceedings: in the reduced graphs outlined by Gillet *et al*[12]; in the GENSAL-GREMAS conversion project outlined by Stiegler *et al*[13]; and in the clustering for structure-activity analysis outlined in the poster session by Downs *et al*[14].

PERCEPTION

An extensive review of ring perception algorithms has been published[15]. This section summarises a few points that make life easier when implementing an algorithm to find a particular ring set, and outlines some of the more general issues.

Time can be saved by pre-processing the connection table to differentiate ring and chain bonds, if they are not already differentiated. This can be done by growing a spanning-tree and back-tracking to set ring bonds. The ring-closures (chords) found can also be used to find an initial fundamental basis for certain ring perception methods.

When processing the connection table or matrix, it is best to consider each ring component separately, and preferably differentiate components further into blocks separated by cut-vertices (spiro-fusions) and even cut-edges.

There are several matters to consider when deciding upon the perception method. The major decision is whether the initial set is to be a subset or a superset of the required set of rings. Matrix manipulation, such as by the Welch-assembly-Gibbs algorithm[10,16], quickly finds all cycles in a ring system, but then time and effort are required to select the required subset. For complex ring systems, vast numbers of cycles can be present. Due to the additional complexities encountered in generic structures, I am biased in favour of starting with a simple subset and gradually adding any extra ones as required! A simple initial set can be produced by path-tracing, growing a spanning-tree, or by matrix manipulation. If path-tracing is chosen, then this can be done in a depth-first or breadth-first manner. Depth-first is easier to implement, but breadth-first is more efficient for complex ring systems. If matrix manipulation is used, linear combinations of the resultant basis set can be used to generate any other cycle. However, for complex ring systems this requires a lot of combinations, most of which are unproductive.

SUBSTRUCTURES AND CONVENTIONS

Substructures present many problems for ring system representation and interpretation which must be answered when selecting a particular ring set; a few of the issues are outlined briefly in this section.

The most immediate problem is that of free sites. It is necessary to adopt one of two conventions: atoms either are or are not regarded as potential free sites unless otherwise stipulated. The preferable default for ring analysis is to assume not. This forces the user to indicate actively those sites to which other chains or rings may be attached. For instance, in Figure 16, should (i) be regarded as a substructure of (ii), or should it be represented as (iii) or (iv)? In the strictest sense, (i) has no possible fusion-points and hence should not match against the 6-edged simple-faces of (ii). In (iv) each vertex can fuse to further rings, so should it be regarded as a substructure of norbornane or even cubane? This is where the importance and effect of the choice of ring set to represent structures becomes apparent. As shown in Figure 14, cubane contains 6-edged tertiary cut-faces in which each vertex is a fusion-point, just as in Figure 16(iv). Similarly, in Figure 17, is (i) a substructure of (ii), i.e., does the convention used allow potential fusion-points to join across a single edge with no intervening vertices? If the answer to

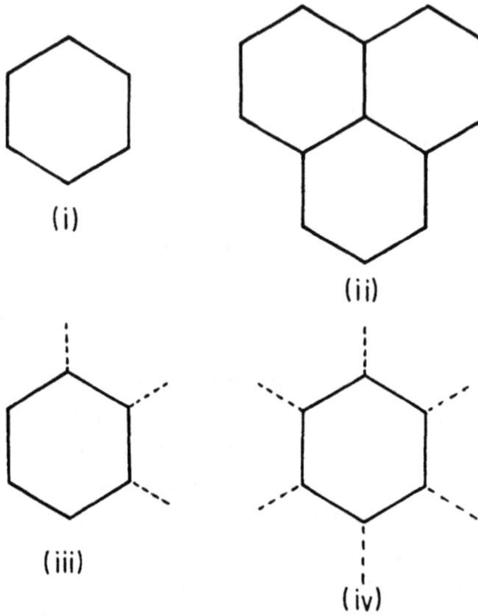

Figure 16. The problem of free sites: first example

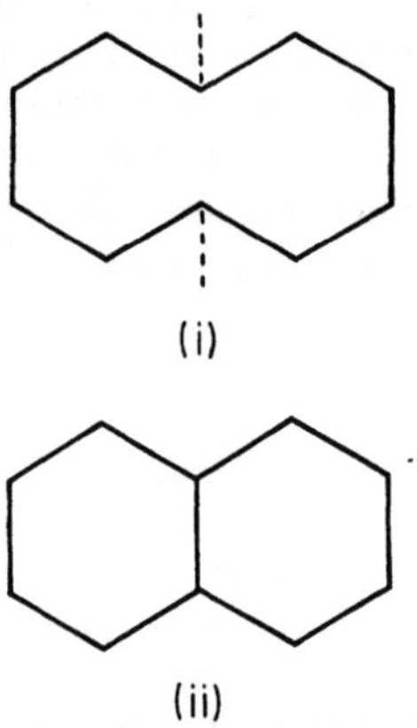

Figure 17. The problem of free sites: second example

these questions is yes, then the ring set to use is the set of all simple-cycles, and simple substructures will match against the majority of ring systems. If the answer is no, then a subset of the set of all simple-cycles has to be chosen and the consequences have to be explained to users.

Figure 16, (iii) and (iv) also show that ring/chain differentiation needs some thought. If the convention is that a free site will allow undifferentiated substitution, then a facility should be provided to allow restriction to either ring or chain. A slightly different matter involves very large rings with long sequences with no fusion-points to other rings, or which form rings of rings. Such structures have caused problems for linear notations, necessitating conventions to decide such contentious issues as whether these sequences have more in common with chains or rings, and whether it is worth perceiving such rings.

Conventions, and awareness of them by the user, are also required for metal bonds, tautomerism and aromaticity. These issues are too large to discuss here.

Overall, it can be seen that a number of conventions and user overrides are necessary. The user needs to be aware of the defaults, to be able to define substructures as loosely or tightly as required, and to be conscious of the consequences of the conventions and the ring set used.

CONCLUSION

In conclusion it can be stated simply that no ring set solves all the issues and problems mentioned in this paper; for some it is just the start of further problems! However, the ability to differentiate ring types based on the classification outlined here will enable implementors of ring perception routines to categorise and choose a particular desired set of rings, and assess the consequences and potential inconsistencies associated with that choice. With reference to the introduction: a start has been made to the necessary developments in language and graph theory.

REFERENCES

1. Lynch M.F. 'R & D in chemical information science: retrospect and prospect'. In *Chemical Structures, the International Language of Chemistry*: Warr W.A., Ed.; Springer-Verlag: Berlin, 1988; pp. 1–10.
2. 'Strained Organic Compounds'. *Chem. Rev.* **1989**, *89* (5).
3. *'Computer-assisted Structure Elucidation*; Gray, N.A.B.; Wiley Interscience: New York, 1986.
4. Downs G.M.; Gillet V.J.; Holliday J.D.; Lynch M.F. 'Theoretical aspects of ring perception, and development of the Extended Set of Smallest Rings (ESSR) concept'. *J. Chem. Inf. Comput. Sci.* **1989,** *29,* 187–206.
5. Nickelsen, H. 'Ringbegriffe in der Chemie-Dokumentation'. *Nachr. Dok.* **1971**, *3*, 121–123 (and associated microfiche).
6. Grateful thanks to Cheng Qian of Chemical Abstracts Service for provision of this example.
7. Plotkin, M. 'Mathematical Basis of Ring-finding Algorithms in CIDS'. *J. Chem. Doc.* **1971,** *11,* 60–63.
8. Corey E.J.; Wipke W.T.; Cramer R.D.; Howe W.J. 'Techniques for Perception by a Computer of Synthetically Significant Structural Features in Complex Molecules'. *J. Am. Chem. Soc.* **1972,** *94 (2),* 431–439.
9. Corey E.J.; Petersson G.A. 'An Algorithin for Machine Perception of Synthetically Significant Rings in Complex Cyclic Organic Structures'. *J. Am. Chem. Soc.* **1972,** *94 (2),* 460–465.
10. Wipke W.T.; Dyott T. 'Use of Ring Assemblies in a Ring Perception Algorithm'. *J. Chem. Inf. Comput. Sci.* **1975,** *15,* 140–144.
11. Downs G.M.; Gillet V.J.; Holliday J.D.; Lynch M.F. 'Computer Storage and Retrieval of Generic Chemical Structures in Patents. 9. An Algorithm to Find the Extended Set of Smallest Rings (ESSR) in Structurally Explicit Generics'. *J. Chem. Inf. Comput. Sci.* **1989,** *29,* 207–214.
12. Gillet V.J.; Downs G.M.; Holliday J.D.; Lynch M.F.; Dethlefsen W. 'Searching a Full Generics Database'. In these Proceedings.
13. Stiegler G.; Maier B.; Lenz H. 'Automatic Translation of GENSAL Representations of Markush Structures into GREMAS Fragment Codes at IDC'. In these Proceedings.
14. Downs G.M.; Poirette A.R.; Willett P.; Walsh P. 'Evaluation of Similarity Searching Methods Using Activity and Toxicity Data'. In these Proceedings.
15. Downs G.M.; Gillet V.J.; Holliday J.D.; Lynch M.F. 'A Review of Ring Perception Algorithms for Chemical Graphs'. *J. Chem. Inf. Comput. Sci.* **1989,** *29,* 172–187.
16. Matyska L. 'Fast Algorithm for Ring Perception'. *J. Comp. Chem.* **1988,** *9(5),* 455–459.

COMPUTER REPRESENTATION AND SEARCHING OF CHEMICAL SUBSTANCES

James G. Nourse, W. Douglas Hounshell, Burton A. Leland, Alan J. Gushurst, and David G. Raich.
Molecular Design Limited, 2132 Farallon Drive, San Leandro, CA 94577, U.S.A.

ABSTRACT

Chemical structures are typically represented in computer programs as simple graphs, where the atoms are represented by a list of nodes and the bonds by a list of non-directional edges. We will describe an extension to such a representation which allows properties to be identified with a defined subgraph in a structure; these properties are fully searchable at a substructure level. A description of the representation will be given and examples of searching capabilities will be illustrated. An implementation of these techniques has been applied to homopolymers, copolymers, non-stoichiometric mixtures, formulations, and 'superatoms'. Other potential uses will also be discussed. These extensions to chemical representation allow us to represent and search a much broader class of chemical substances than the set of discrete chemical structures which have previously been handled.

BACKGROUND

Chemical structures are typically[1] represented in computer programs as bi-directional graphs used to depict a valence-bond theory model of the chemical structure. Atoms are represented by a list of nodes and bonds by a list of edges. Typically there are some constraints imposed for chemical sense (e.g., valence control), and only predefined properties can be associated with each (e.g., formal charge, bond order). Other data can be associated with the full structure (e.g., chirality, molecular formula, or any other data stored in database fields) (see Figure 1).

THE SGROUP: AN EXTENSION TO CHEMICAL REPRESENTATION

There are many examples of other types of data which one might wish to associate with parts of a chemical structure. For example, one might associate: (a) with atoms: stereochemical purity, isotopic distribution, text label; (b) with bonds: stereochemical purity; (c) with fragments or components: percentage presence in a mixture; (d) with connected pieces of the chemical structure: polymer repeat; or (e) with arbitrary pieces of the chemical structure: atoms that are part of a pharmacophore, (i.e., active group), electron withdrawing groups.

We have therefore introduced the notion of an 'SGroup' as an extension to chemical structure representation, and have implemented SGroups in a MACCS-II

W. A. Warr (Ed.)
Chemical Structures 2

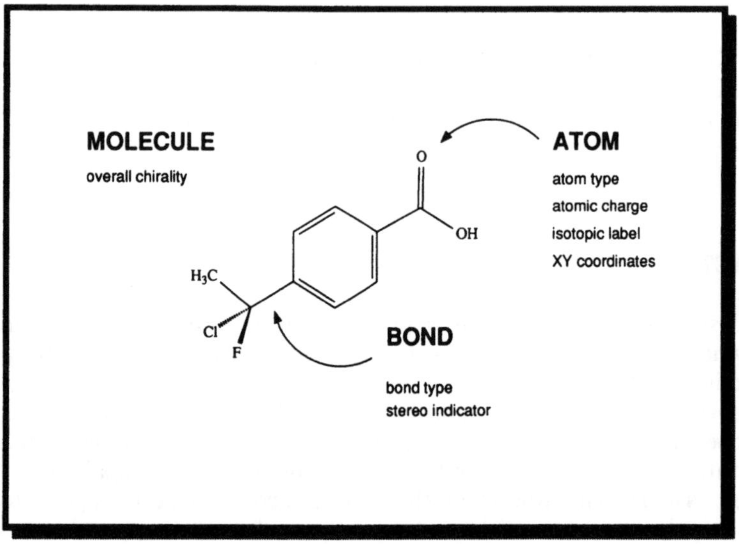

Figure 1. Typical chemical representation

environment[2]. We define an SGroup as a persistent collection of atoms and bonds, i.e., as an identified substructure which is maintained as a part of the representation of a chemical structure when stored in a database or in a transfer file. We have further delineated two types of SGroups, 'chemical SGroups' and 'data SGroups'.

Since there are certain chemical uses of these SGroups which are general, we have implemented three broad classes of chemical SGroups: (a) polymers, (b) components/mixtures/formulations, (c) superatoms/multGroups. We place two restrictions on these. Firstly, all chemical SGroups are composed of intact fragments and substructures delimited by a limited set of crossing bonds. A crossing bond has one atom contained in the SGroup and one outside the SGroup; the bond itself is not considered to be in the SGroup. Secondly, SGroups must be hierarchical, that is, if any two chemical SGroups include any atoms or bonds in common, then one must be a subset of (or equal to) the other.

Data SGroups can be defined by any arbitrary substructure. A data value is associated with this substructural piece only, and not with the entire structure. While data SGroups can also be defined in terms of crossing bonds, it is convenient to think of them as arbitrary substructures carrying their own data. There is no hierarchy limitation on data SGroups; arbitrary overlap is permitted.

CHEMICAL SGROUPS

The chemical SGroups outlined above are now considered in turn.

Polymer Representation

With a polymer SGroup, we associate the following internally supported attributes:

1. Polymer type: monomer, SRU (structural repeating unit), copolymer, modification, graft, crosslink. There is also a special query polymer type (AnyP).
2. Polymer subtype: random, block, alternating, unspecified. These designations of the type of repeat pattern of the polymer can only be associated with a polymer of type=copolymer.
3. Connectivity: head-to-tail, head-to-head, unspecified.

In addition there are restrictions on the number of crossing bonds that the various polymer types may have. A monomer may not have any crossing bonds. SRUs and all other polymer types (except for crosslinks) can have two to four crossing bonds when registered. Crosslinks may have four to eight. Queries can have zero to eight crossing bonds.

The monomer polymer type allows the registration of monomers for which the structure of the resulting polymer is not known or cared about. The SRU polymer type is used to register known structural repeating units with the resulting crossing bonds implicitly connected based on the connectivity information (see Figure 2).

End groups can be specified if known but there is no requirement that they be known. If unknown they are represented by a special 'null' atom type, represented by the symbol '*'. This atom serves as a sort of placeholder and has many attributes of a generic 'any' atom (see Figure 3).

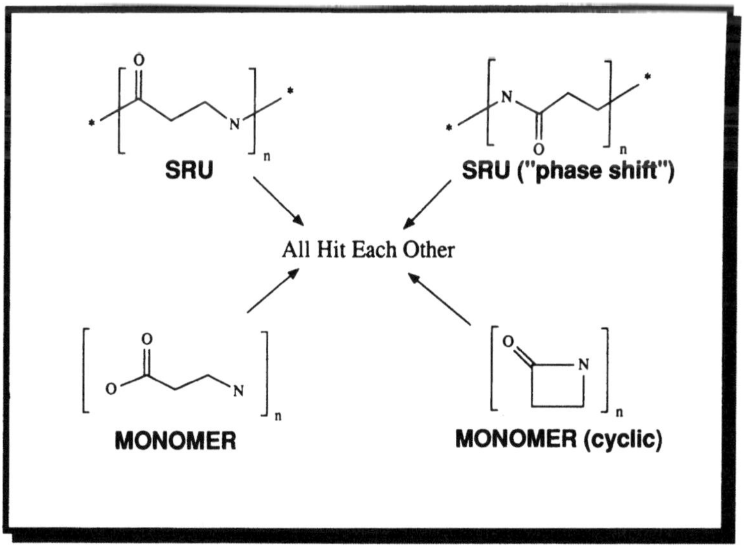

Figure 2. Examples of homopolymers: monomer and SRU

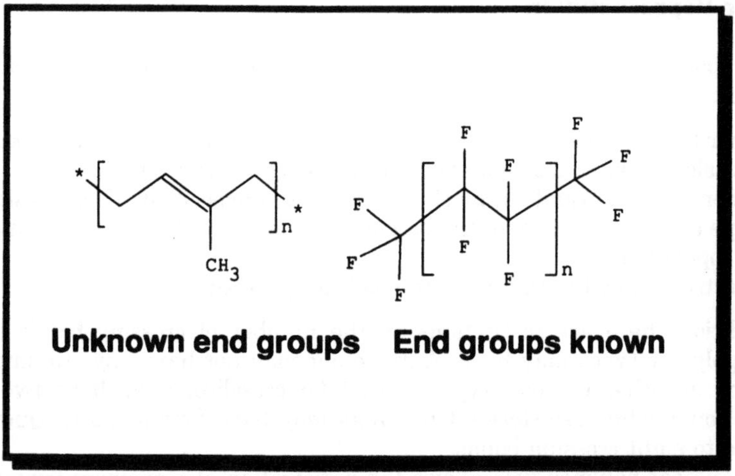

Figure 3. End groups can be specified if known, or as '*' atoms if not

Copolymers are polymers containing more than one SRU or monomer. A wide variety of such combinations are permitted to accommodate the likely knowledge of copolymer structure (see Figure 4). Modifications are used to represent changes that have been done in a post-processing reaction on a polymer. Graft and

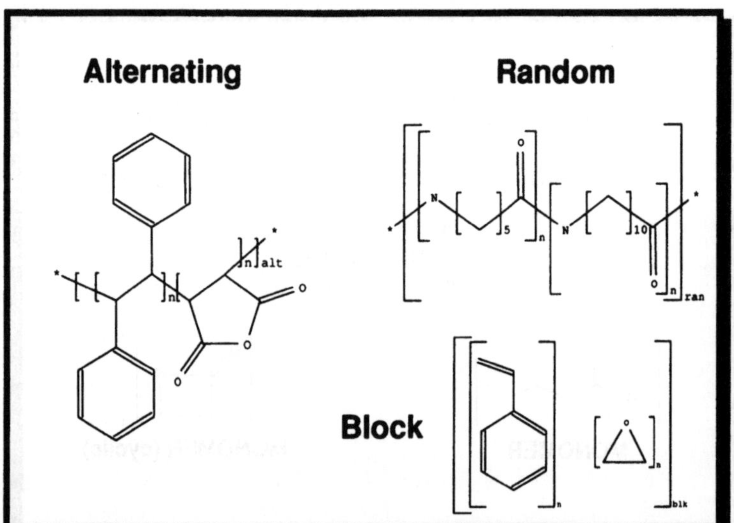

Figure 4. Examples of copolymers. Subtypes can be random, alternating, block, or unknown

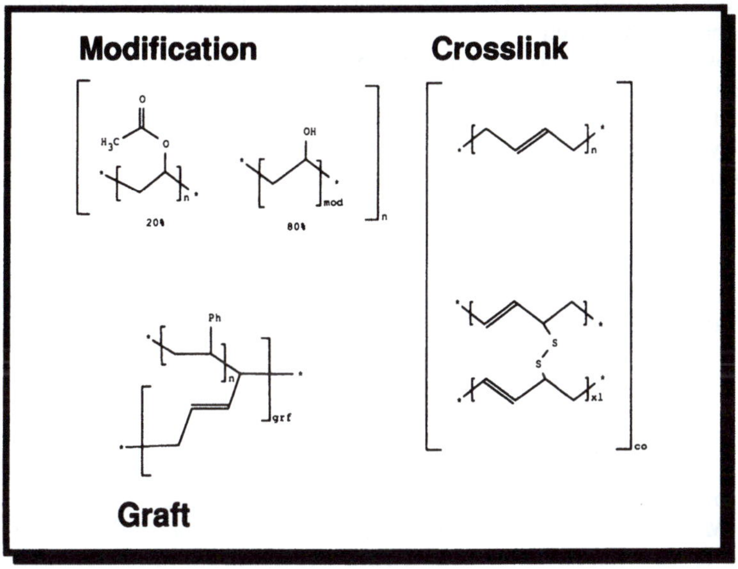

Figure 5. Examples of modification, graft, and crosslink co-polymers

crosslink are used to indicate these important structural features of connecting polymers (see Figure 5).

Specialised Polymer Structure Searching

A user can search for an exact match of a query by considering or ignoring the polymer SGroups. If the SGroup definitions are considered, the user can additionally choose whether to consider or ignore: (a) the polymer type, subtype, connectivity, and (b) the endgroups.

It is also possible for monomers and SRUs of the same polymer to find each other. This includes both addition polymers (e.g., polystyrene) and condensation polymers (e.g., polyamides). This feature was included to permit the registration of polymers with less well defined structures (e.g., only monomer known) with the hope of finding the equivalent polymers with defined structures. An algorithm was developed which can rapidly assess and map the structures involved in most addition and condensation polymerisations.

Exact match searching can also take into account the 'phase shift' possibility of SRUs. The choice of the crossing bonds for any SRU is arbitrary. While the bonds are usually chosen to correspond to those formed in the polymerisation reaction, such a choice is not required. The searching algorithm used has been developed in a way that does not require the same choice of crossing bonds in the query and database structure.

With these features, all of the structures depicted in Figure 2 will find each other in an exact match search.

In a substructure search, the query property 'any polymer' permits a query to hit any of the various polymer types. Polymer sub-types and connectivity can also be searched for as desired. If these properties are not indicated on a substructure query, any will be hit.

Substructure search generally works intuitively, in that substructures which match must have corresponding polymer SGroup properties as well. For substructure search we recognise 'phase shift', and we traverse crossing bonds in all possible combinations. However, we do not currently allow for constructive substructure search; for example, a query with two instances of a repeating polymer unit will not hit the structure in the database which includes just one instance.

Mixtures and Formulations

For mixtures and formulations, we define an underlying 'component' chemical SGroup. Mixtures are composed of a set of components, and formulations are composed of an ordered set of components (see Figure 6). Each component must have no crossing bonds, and all pieces of structure within a mixture or formulation SGroup must belong to a component SGroup. Nested (hierarchical) mixtures and formulations are allowed, but there must be intervening component SGroups.

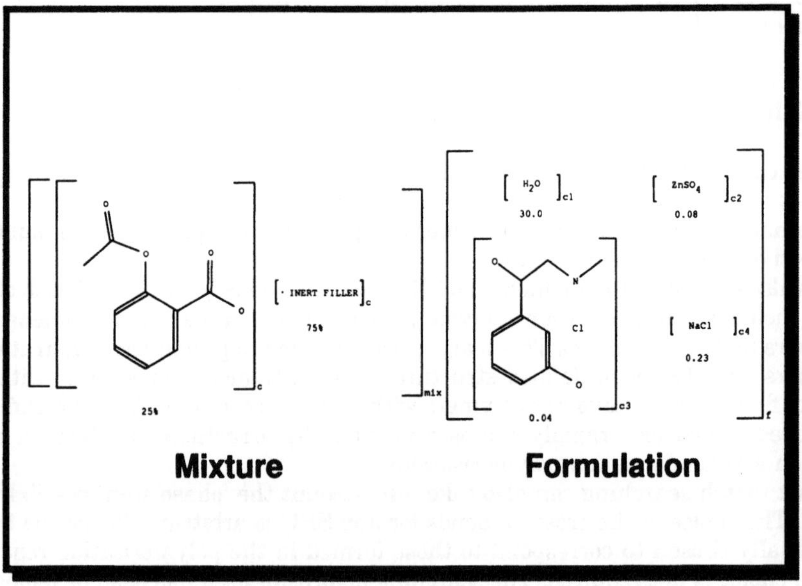

Figure 6. Examples of mixture and formulation

Exact match searching allows the user to specify whether, in addition to matching the chemical graphs, the corresponding component/mixture/formulation SGroups are also required to match. For substructure searching, the component ordering in formulations is considered 'relative', i.e. the query hits a database structure if the components appear in the same order in both the query and the database structure (see Figure 7).

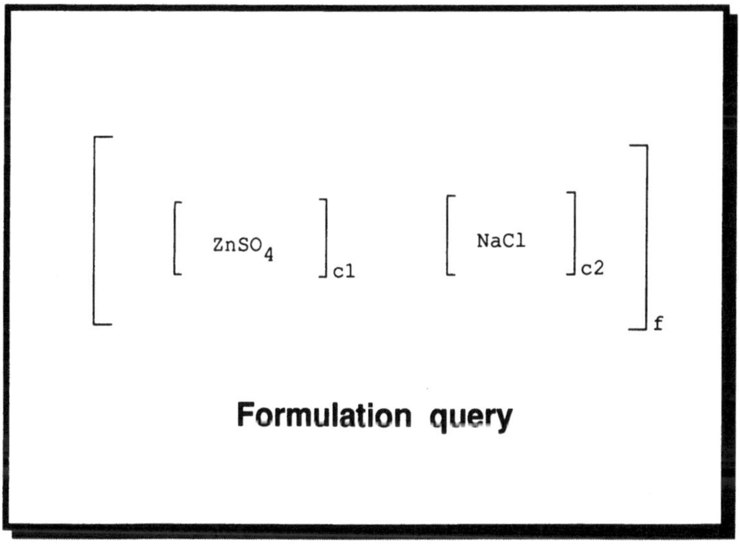

Figure 7. Substructure searching for formulations requires that the 'relativc' order of components match

Superatoms and MultGroups

We have implemented two additional types of chemical SGroups to simplify structure display and input.

Superatoms are chemical SGroups with zero to eight crossing bonds that can be 'contracted' (i.e., removed from the structure), and replaced in display by a user-specified label. The user can invoke a replicate function to find all groups identical to an existing superatom and have the program construct superatom definitions for those groups. Superatoms are useful for simplifying the display of structures, e.g. in the case of peptide chains or standard protecting groups (see Figure 8).

MultGroups are chemical SGroups which are to be repeated a specified integral number of times. These SGroups may have zero or two crossing bonds. These can be used both as drawing short cuts and for display simplification (see Figure 9).

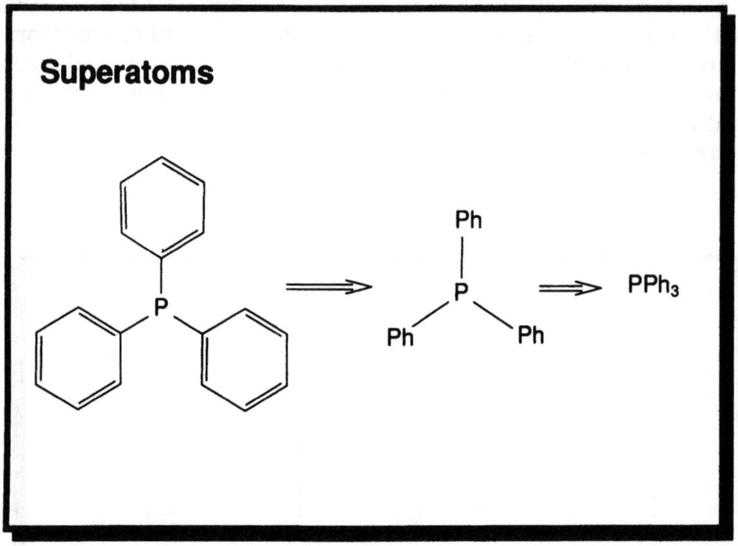

Figure 8. Examples of superatoms

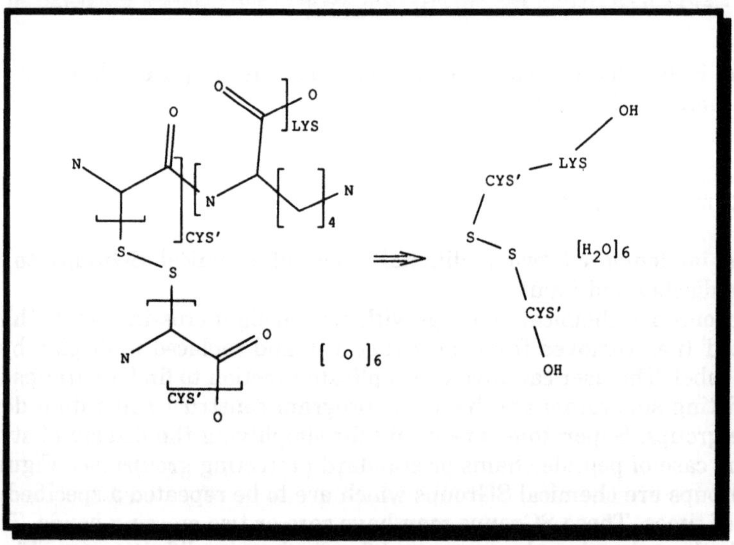

Figure 9. Examples of multGroups and superatoms

DATA SGROUPS

The definition of a data SGroup (set of atoms and bonds) can be either tied directly to a chemical SGroup, or to an arbitrary set of atoms and bonds (except that once a bond is included, the two atoms of which it is composed are automatically included). Each data SGroup has associated with it a field and a value of data. The data fields to be allowed are defined by a database administrator as a part of the data dictionary. In the MACCS-II implementation, the program allows the same data field control as for data attached to the structure (text, numeric, and formatted).

These data are available and searchable for both exact match and substructure searching. In a search query, the user defines each required SGroup; each of these has a set of atoms and bonds, a field, and a value. In substructure searching, a candidate is considered a hit if it contains a data SGroup which is defined by the same set (or superset) of the atoms and bonds in the query data SGroup for the particular field specified by the query, and with a value matching the query's value. In the MACCS-II implementation, the language for matching data values is identical to that used in searching other data fields.

Examples of data that one might wish to tie directly to a chemical SGroup include percentage composition (e.g., of a component in a mixture or formulation, or of an SRU in a copolymer), additional information about how a polymerisation was conducted, or average molecular weight of a polymer (see Figures 4, 5, and 6).

Data attached to arbitrary atoms (and bonds) can be used to: (a) annotate a particular set of atoms, e.g., as part of a pharmacophore or as 'reactive' sites; (b) annotate fragments, e.g., with stoichiometric multipliers of a salt or solvate, as major or minor, or as active or inert; (c) describe an unknown portion of a structure by attaching descriptive data to a 'null' or '*' atom; (d) explain more fully the nature of a particular site (atom), e.g., stereochemical purity, isotopic purity (see Figure 10). In short, the use of SGroup data permits a user-extensible chemical structure representation. The user can define new data fields and attach values to atoms. These values are fully a part of the connection table and are searchable both with exact match and with substructure searching.

SGROUPS MUST BE HANDLED IN ALL AREAS OF THE PROGRAM

Within the operation of a chemical information computer program, the representation of a chemical structure will take various forms (see Figure 11). Any structural feature must be representable in each of these forms, and must have appropriate manipulation operations available with each. Central to the processing is the main connection table (generally data structures defined in a computer programming language). However, there are other major embodiments of the structure. We have implemented SGroups in all of these environments.

A Database

The user must be able to store structures in a database and retrieve them. The user must be able to do various structural searching operations, such as exact match, substructure search, similarity search, and formula search.

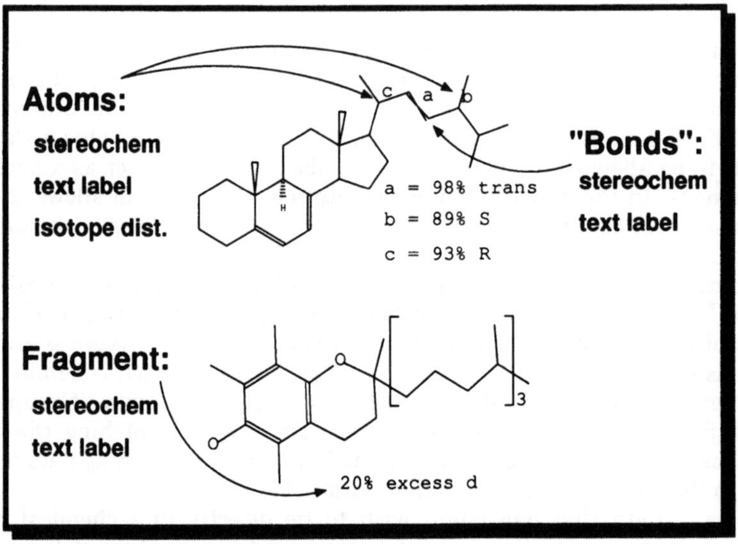

Figure 10. Examples of data SGroup display

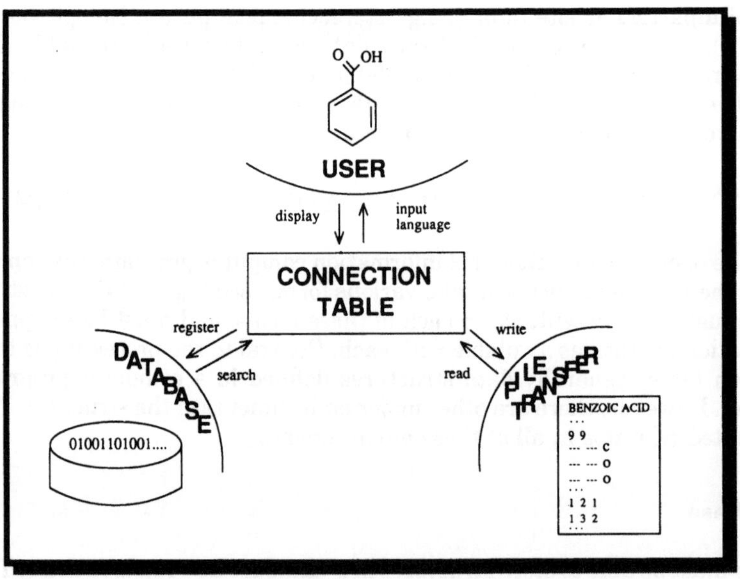

Figure 11. Various forms in which a structure must be represented

In MACCS-II the database representation is in the form of a SEMA name[3], which is a compact bit string representing the connection table in a spanning tree format and including fields with the properties of the nodes (atoms) and edges (bonds). We have implemented extensions to the SEMA name for storing the chemical and data SGroup information (atom and bond set; and property or data values). The fields for data SGroups are defined by the database administrator and are incorporated into the data dictionary for the database. We discussed searching implications above.

A Transfer File

Structures must be transferable to other programs that understand the same structural representation rules *via* an editable file format.

In MACCS-II this representation is in the form of a Molfile. Extensions have been added to the Molfile description that allow SGroup information to be stored and transferred to other programs.

A User Interface

The representation is a graphical depiction of the structure, including an indication of all of the structural features present. The user must be able to define ('draw') and manipulate these graphical images and have the program interpret the structure into the underlying connection table.

Besides having a graphical user interface, MACCS-II also provides access to structures through MEDIT[4], a structure editor which is available from the command line or from a MACCS-II program. The MEDIT language has been extended to allow for the definition, identification, modification, and deletion of SGroups.

SGROUP USER INTERFACE

A new side menu for drawing and manipulating structures with SGroups has been added to the MACCS-II structure drawing editor (see Figure 12).

User definition of the chemical SGroups is through a mechanism where the user identifies the crossing bonds and 'seed' atoms, which are 'grown out' until hitting a crossing bond (or encompassing the entire fragment). It is convenient to think of chemical SGroups in this sense as connected (as opposed to arbitrary) substructures. A data SGroup is defined by specifying the chemical SGroup to which it is tied, or by specifying each atom and bond to be included.

The display shows traditional brackets with identifiers in the lower right corner. The brackets' positions are determined automatically when a chemical SGroup is defined, but may be altered by the user. The position of SGroup data is also determined at time of definition using defaults, but the user has a variety of methods which affords a great deal of flexibility in modifying how the data is to be displayed.

The user interface has operations to enforce rules and to simplify certain operations. For example, in order to define a simple modification polymer, the user

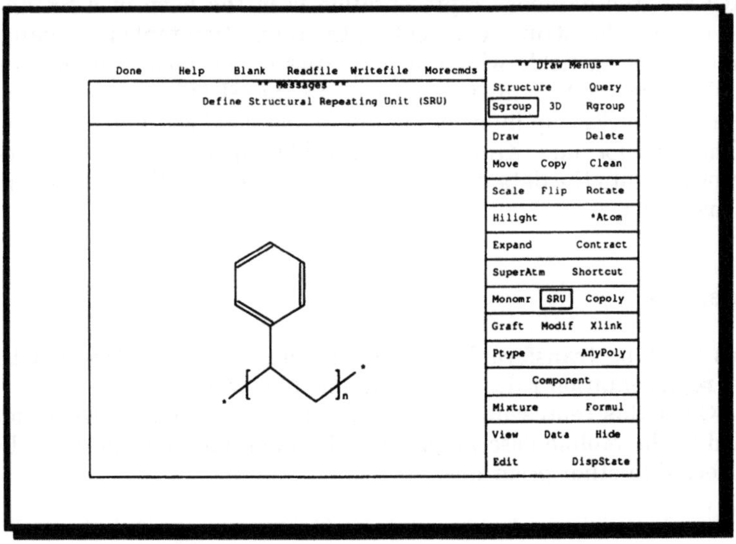

Figure 12. The draw menu for SGroups in MACCS-II

first draws the unmodified SRU, then picks the 'Modification' button and indicates the SRU which was modified. The program then makes a copy of that SRU, changes the type of the copy to 'Modification' and encloses both in another polymer SGroup. The user can then make the chemical modification on the copy of the SRU (see Figure 13).

The hierarchy of components and mixture or formulation SGroups is enforced by the user interface. The user interface also includes simplifications such that component SGroups are automatically defined when a mixture or formulation SGroup is defined; but the user has the option of defining components, e.g., for the purpose of collecting multiple fragments as one component (as in the case of a salt).

CONCLUSIONS

In this paper we have shown how the representation of chemical structures in computer programs can be enhanced by the addition of SGroups to the representation. We have implemented specialised handling for classes of polymers, mixtures, and superatoms. The implementation of data SGroup allows users to add properties to the connection table and have them searchable as an integral part of the structure.

ACKNOWLEDGMENTS

The authors gratefully acknowledge helpful discussions with many of our colleagues in the chemical industry, and the contributions of many other

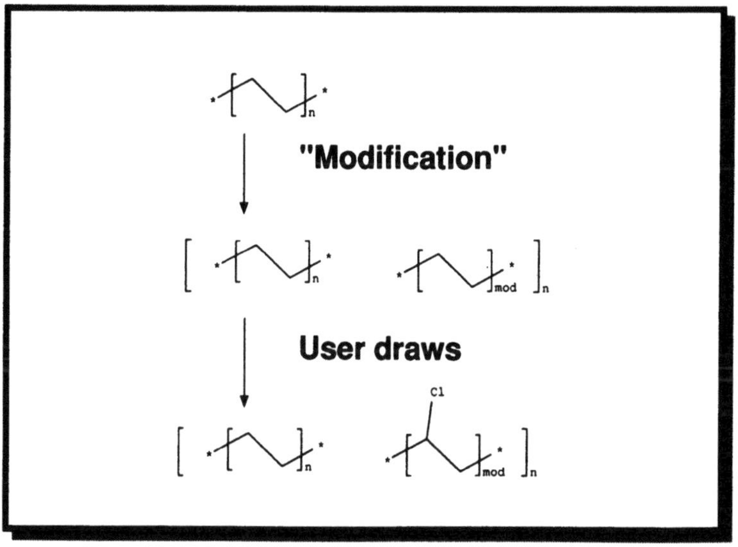

Figure 13. Operation of the 'Modification' button in structure drawing

employees of Molecular Design Limited, particularly those of James Dill and David Hughes.

REFERENCES

1. *Chemical Structures: The International Language of Chemistry,*; Warr, W.A., Ed.; Springer-Verlag: Heidelberg, 1988.
2. MACCS-II is a proprietary product of Molecular Design Limited, 2132 Farallon Drive, San Leandro, CA 94577, U.S.A. See: Ahrens, E.K.F. 'Customisation for Chemical Database Applications'. *In Chemical Structures: The International Language of Chemistry;* Warr, W.A., Ed.; Springer-Verlag: Heidelberg, **1988**; pp. 97–111.
3. Wipke, W.T.; Dyott, T.M. 'Stereochemically Unique Naming Algorithm'. *J. Am. Chem. Soc.* **1974**, *96*, 4834–4842.
4. MEDIT (the Molecule EDITor) in MACCS-II is a set-oriented structure manipulation language. This language will be the subject of a future paper.

INFORMATION INTEGRATION: DISTRIBUTED CHEMICAL INFORMATION MANAGEMENT SYSTEMS

Dennis H. Smith, James Barstow, Raymond E. Carhart and John Laufer.
Molecular Design Ltd, 2132 Farallon Drive, San Leandro, CA 94577, U.S.A.

ABSTRACT

The new generation of distributed computing systems offers exciting new ways to bring together chemical information, including chemical structures and reactions, data, text and graphics. The connection of high performance workstations to networks of geographically distributed computers rather than to single hosts provides the connectivity to support information integration. The next challenge is to construct software systems that actually deliver the potential functionality. We will discuss software architectures designed to achieve integration at both the workstation and host. End-user control over how information is accessed and presented, independent of its geographical location, will be shown to be an essential part of such systems. Emerging standards for transport of information among diverse applications will be discussed as the 'glue' that makes distribution the enabling technology for integration.

INTRODUCTION

Information systems in the decade of the 1990s will be based on radically different computer hardware and software architectures compared to those of the 1980s. We are already in the midst of a revolution in computer hardware, in operating environments, and in application software.[1] The ultimate effects of this revolution on the utility and use of information systems can today be predicted in general terms. This paper examines several aspects of the revolution, using as examples information problems we face today. We show how those problems will be solved in the future by systems that are just now becoming available.

Distributed Systems

The new computer architecture of distributed systems is the enabling technology for the revolution. In the 1980s, information systems were built based on central mainframe computers accessed by users through terminals, as illustrated in Figure 1. These 'undistributed' systems met many needs of both corporations and end-users. The central facility provided a focal point for organising and maintaining information resources (files, databases, and so forth). The performance of the terminals was perceived by most users as adequate. But as the end of the decade neared, dramatic improvements in workstation hardware and software technology served to highlight the limitations of this information environment and unmasked its inherently poor interactivity, graphics, and general performance.

W. A. Warr (Ed.)
Chemical Structures 2

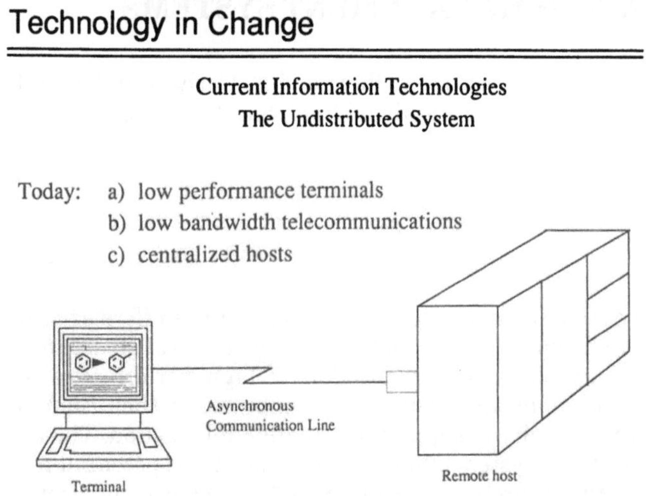

Figure 1. Computer technology of the decade of the 1980s. Until late in the decade, all significant chemical information systems possessed a centralised, 'undistributed' architecture

At the end of the decade, the new architecture of distributed systems, Figure 2, began to emerge as a solution to many of the limitations of the older architectures. The advantages of this architecture have been discussed thoroughly by many people. For the purposes of this problem/solution oriented presentation, we mention three advantages that are particularly important for scientific information. First, the architecture divides the computing power among several machines, and allocates the power appropriately to these machines. Thus, a workstation manages the graphical user interface and all its manipulations, leaving the tasks of complex computations and access to large corporate databases to those machines that handle them best. Second, a graphical user interface of high, and predictable, performance is provided to the user who will require such interfaces in his or her desktop environment. Third, the inter-machine connectivity provided by computer networks solves many of the problems of access to globally distributed information resources.[2]

Information Integration and Standards

People are requiring access to many different information resources, supported by many different hardware and software systems. Distributed systems, emerging standards for information exchange, and desktop integration of results will support these requirements.[1,2] A simple example is shown in Figure 3. Access to combined corporate chemical and relational databases is supported by the windowing system of the graphical user interface (GUI). Other applications may be running in other windows, for example, access to a local database, or a document processor. Other applications may appear as icons on the screen, to be activated at the touch of a button. Integration of results among applications that

Technology in Change

Future Information Technologies
The Distributed System

Tomorrow: a) high performance workstations
b) high bandwidth networks
c) centralized compute and database servers

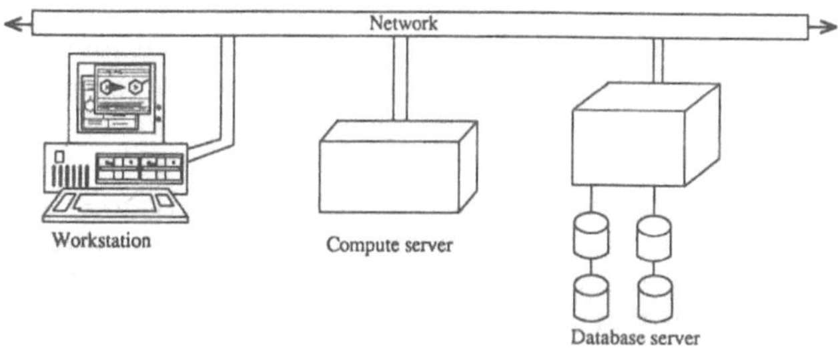

Figure 2. Computer technology of the decade of the 1990s. By the end of the decade we can anticipate that all significant chemical information systems will possess a distributed architecture. Desktop workstations will supplement the central hosts of the 1980s, freeing these host computers for tasks for which they are best suited

Workstations
Gateway to Several Information Resources

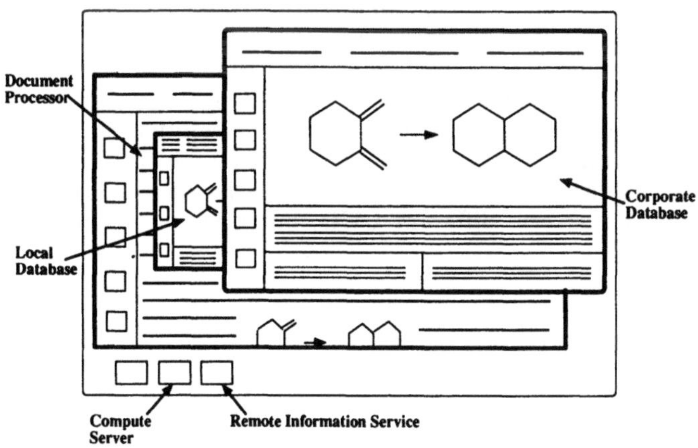

Figure 3. A schematic representation of information integration at the desktop workstation. The workstation's graphical user interface (GUI) provides a windowing system that is used as a gateway, with multiple viewports, to diverse sources of information

comply to standards for data interchange is achieved by simple point and click operations.

Information integration also occurs at the database servers in a distributed system. Here too, adherence to standards, for example, standards for data models and query processing, is critical for providing integrated applications. Problems that are solved at the server are discussed in the later section on information integration at distributed hosts.

The Emerging Role of End-users

The revolution in computing is accompanied by an equally important revolution in the community of end-users. This community is demanding an increased rôle in making corporate decisions on information systems. Most users of chemical information systems are not paid to sit in front of workstations. Thus, users are demanding consistent, easy-to-learn, easy-to-use graphical user interfaces, coherent access to varied information sources, and interfaces and applications that can be tailored, or customised, to suit individual preferences. The popularity of the Macintosh user interface has driven all vendors to providing similar graphical user interfaces for workstation software. We expect the user community to continue to apply pressure on vendors to provide coherent, logical views of varied chemical, relational, textual and graphical information. And the vendors will have to provide solutions that are highly customisable.

INFORMATION INTEGRATION AT THE DESKTOP

In this section we discuss how information may be shared among applications on the desktop workstation. Although our discussion in this section is in terms of local access to information, we emphasise from the beginning that the same application user interface can be used to access remotely distributed information. This is discussed in a later section. Once the information is available at the workstation from whatever source, it can be shared with other applications as discussed below.

A critical problem that is addressed only indirectly in sections that follow is that the workstation environment in most organisations is highly heterogeneous. When a software application is investigated as a solution to a problem, it is essential to understand the range of workstations on which this application will run. Emerging standards for GUIs will make it easier for vendors to provide the same application on several different GUIs, but this is not generally the case today.

Windowed Operating Environments

The operating environment on the desktop workstation will be a windowed, graphical user interface as illustrated in Figure 3, for access to both local workstation and remote host application software. Users will require such interfaces for many reasons, including responsiveness, high quality graphics and colour, WYSIWYG presentation, ease of use, and so forth. People who have used such interfaces will not return either to terminals or to any user interface that displays a system prompt. The advantages of GUIs extend much further, however. In this operating environment, information can be moved among applications very quickly and easily. It will be trivial to justify the cost differential between terminals and workstations based on the workstation's capabilities for integration and just plain getting work done.

Methods for Integration

There are two methods for moving information between applications on the desktop, a system clipboard and interapplication data exchange. Cut and paste using the system clipboard is one standard method. All GUIs have the concept of a 'clipboard' to which information is moved ('cut') from one application, and from which the information is inserted ('pasted') into another application. Interapplication data exchange is becoming more common and offers more transparent integration. There are four levels of information exchange made possible by these two methods.

Bitmaps and Text. In its simplest form, cut and paste moves a designated portion of the contents of a window to the clipboard in the form of a bitmap of graphical images, or, directly, text itself. A bitmap is simply a graphical image consisting of all of the pixels in the image. This image can usually be moved and resized after pasting into another application, but original graphics and text are no longer editable. This loss of information results in a loss of fidelity of the graphical image as the image is rendered on other devices such as printers.

Cut and Paste of a Higher Level Object. Some GUIs allow cut and paste of a higher level representation. In such applications, the underlying data describing the graphical object are moved; the GUI provides facilities for rendering the image. A good example of such data is the collection of vectors and text comprising a chemical structure diagram. The recipient application can display the graphical object, and may allow editing of the object. However, if the object is cut and pasted back to the original application all chemical significance in this example would be lost.

Cut and Paste with 'Private' Data. Some applications perform cut and paste as described above, but also include additional, application-specific, 'private' data in the data structure passed to the recipient application. Private data preserve a higher level of information about the graphical object related to how the object was created. An example of such private data is the connection table for a chemical structure diagram. The recipient application may retain or discard the private data. If the recipient retains and understands the private data, it may allow editing with complete retention of the fidelity of the image and the private data. Even if the recipient application is capable only of preserving the data, cut and paste of the image back to the original application preserves the data and yields an intact and editable object.

Preserving the integrity of private data requires a level of co-operation among donor and recipient applications. There are no general standards that applications are required to follow. The net result is that cut and paste should work between products from the same vendor. It may work among products from different vendors if they follow the GUI-specific standards and preserve private data.

Interapplication Data Exchange. A fourth method for achieving information integration at the workstation is to adopt a platform-specific method, if available, for interapplication communication (IAC) to move graphics, data, and/or text between applications. This method is a direct exchange of data and does not require use of cut and paste. However, in order for anything to work, both donor and recipient applications must add functionality supporting the GUI-specific standards for data exchange. Successful data exchange *via* IAC between applications supplied by different vendors requires a close working relationship between the vendors, or published, detailed specifications of the data structures themselves.

Each GUI has its own standards for interapplication data exchange. Popular examples are today Dynamic Data Exchange (DDE) in Microsoft Windows, Inter-Application Communication (IAC) in the upcoming version 7 of the Apple Macintosh operating system, and Application Communication Services (ACS) on the Digital platforms under DECWindows.

Customisation

Every application within a given GUI is expected by the user to conform to the style guidelines for the GUI. This reduces learning and re-learning time and training costs, because all compliant applications have a base set of fundamental operations (file, edit options, window sizing, scroll bars, and so forth) that work the same way in every application. Given an application that meets these 'work-alike' criteria, users have additional requirements for modifying aspects of the look and feel of an application. Modifications are driven by several requirements, for example:

1. standard, but easily modifiable, report formats including chemical graphics;
2. modified user interfaces tailored to specific applications or segments of the user community;
3. rapid creation of form and table output for *ad hoc* queries.

The users' basic requirements are for systems that will accomplish much of the work he or she needs to do without having to find a programmer to develop a new or modified application. Integrated software systems allowing this level of power and flexibility are becoming available.

Specific Problems and Solutions

Here are several examples of integration of information among applications that can be achieved today in the new workstation environments. We have selected those with direct relevance to chemical information, but we stress that the same techniques are equally applicable in many other scientific and technical areas.

Integrating Graphical Information. One important requirement is to bring together graphical objects, which may be in the form of sketches (vector-based) or bitmaps. Many software packages provide these capabilities. We present a simple example that combines a sketch and a bitmap in Figure 4.

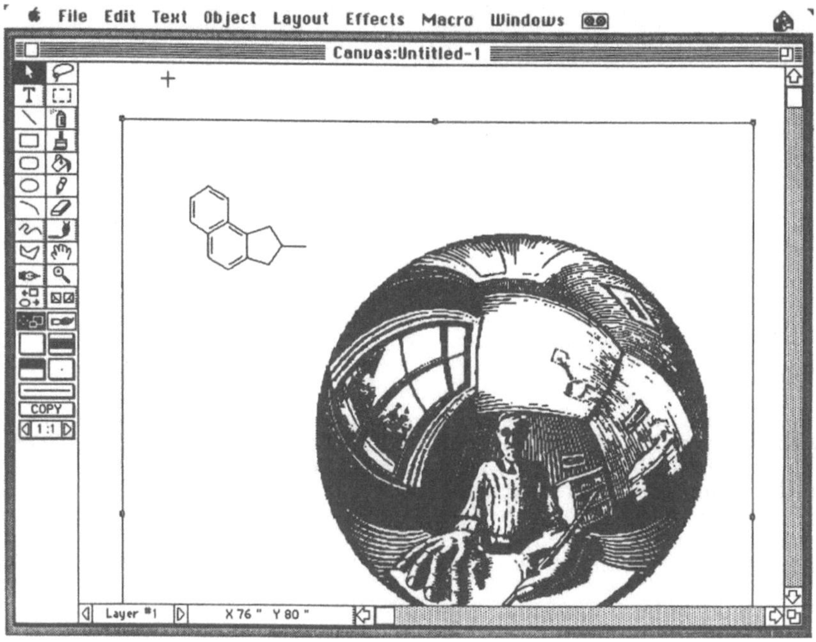

Figure 4. A screen dump from the Apple Macintosh program Canvas, illustrating the integration of graphical sketches and bitmaps. The structure diagram from a chemical drawing program (another application) was cut and pasted to the Canvas window; the bitmap was imported to Canvas from a scanned image

Integrating Graphics and Chemistry. Chemists need chemical graphics, often combined with other graphical objects, for a wide range of applications including searching chemical databases, reports, presentation graphics, and so forth. The example in Figure 5 illustrates one approach to integrating and manipulating sketch and chemical graphics in a single editor. The problem solved by this application is to maintain for the user a clear distinction among different types of graphical objects, so that sketches and bitmaps remain sketches and bitmaps, and chemical graphics retain their underlying chemistry. These graphical objects can be edited at will without compromising their identities.

The use of standards for GUIs greatly reduces the development efforts required to 'localise' software for other languages. The chemical drawing application illustrated in Figure 5 ports easily to the kanji version of Microsoft Windows running on a personal computer that is not 100% IBM compatible. The kanji version allows input of mixed English and Japanese characters in the text editor, as illustrated in Figure 6.

Integrating Chemistry and Local Data. Workstations are becoming very powerful computers. Many organisations adopting a distributed systems strategy will take advantage of this fact by using workstations to store local information. Given a

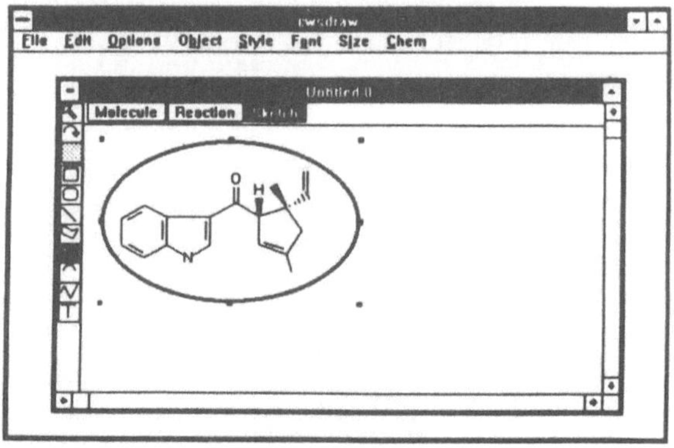

Figure 5. A screen dump of a chemical drawing application in the Microsoft Windows GUI, illustrating integration of chemical graphics and sketches. The structure diagram (a chemical object) is highlighted by a surrounding ellipse (a sketch object). Either object can be edited with retention of its object type

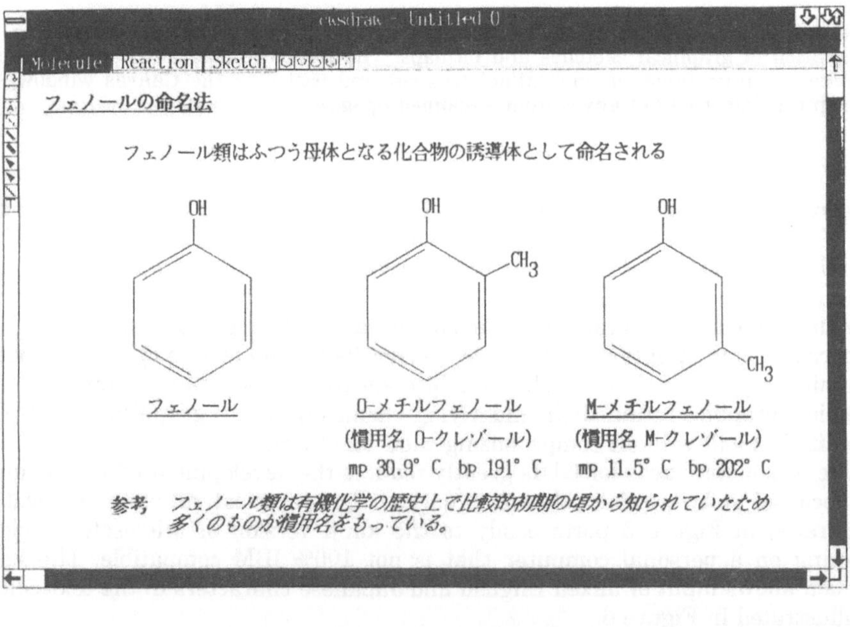

Figure 6. A screen dump from a chemical drawing application running on a Fujitsu FMR-70 under the kanji version of Microsoft Windows specially adapted for the Fujitsu machine and the Japanese language. The chemical structure diagrams have been annotated with kanji, katakana and hiragana characters, together with roman text and arabic numerals

database management system with capabilities for storing chemical and other numerical or textual data, it is possible to search and display local chemical information in a variety of ways. With a customisable interface, one can quickly construct form- and table-based presentations for this information, as illustrated in Figure 7.

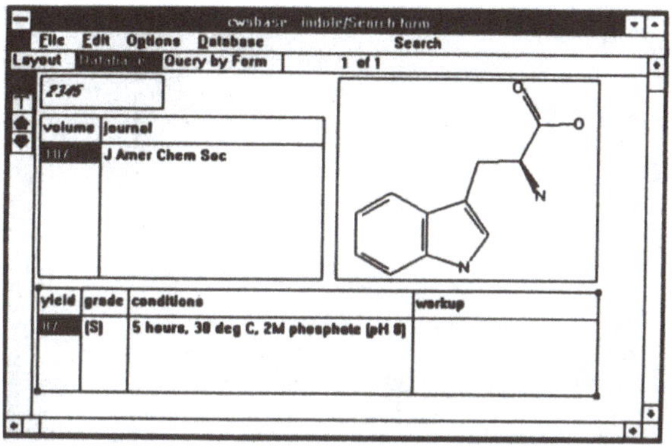

Figure 7. A screen dump from a chemical database application, illustrating integration of chemical structures and data from a local workstation database. This application views tables as a special kind of form box, and allows graphical objects to be presented in either single form boxes or in tables (see Figure 8)

Integration with other Applications. There are many applications where sharing information is required. The most common is to bring the results of specialised applications, for example, search results from chemical databases, portions of spreadsheets, statistical analyses, and word processing text, together in a word or document processing program. We illustrate the integration of graphical table output resulting from a local chemical database search into a word processing program in Figure 8.

INFORMATION INTEGRATION AT DISTRIBUTED HOSTS

The computer architectures available in the 1980s resulted in the creation of many large databases on centrally located machines. These computer facilities and their databases will retain their value far into the future, as these machines are ideally suited for central control of and access to information, rapid search of very large databases, and central maintenance and backups of systems, software, and databases. Users' needs for access to information in the 1990s will require systems that are truly distributed, systems capable of bringing together information from many different hosts. Distributed system architectures seek to take maximum advantage of the power and information resources of the hosts that exist today.

The discussion in this section is focused on software systems running on the host, accessible through the desktop workstation. The combined systems operate completely transparently for the user, 'behind' the workstation GUIs discussed in

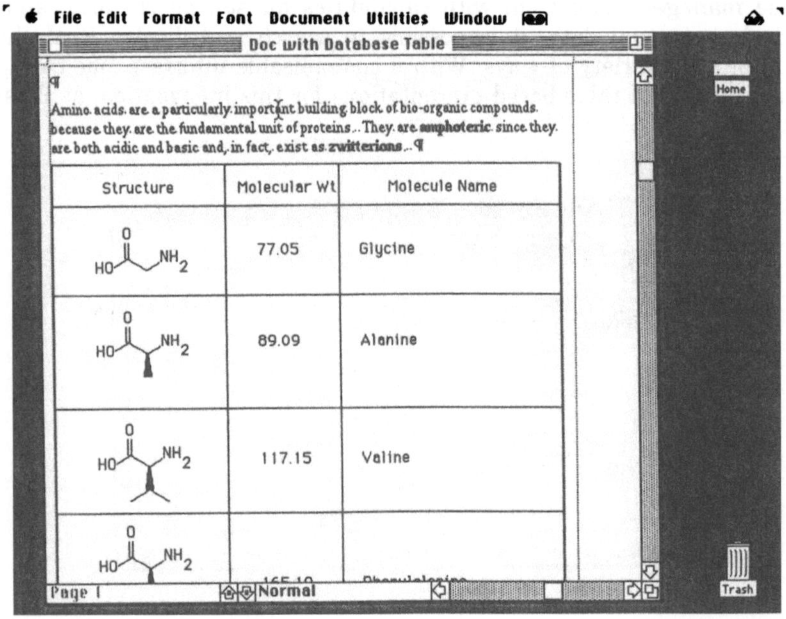

Figure 8. A screen dump from Microsoft Word on the Apple Macintosh, illustrating the integration of output from a chemical database application, as a graphical table, with a word processing application. The table was moved into Word *via* the clipboard (cut and paste)

the previous section. This section introduces several technical issues that may seem unimportant to users who just want to get their job done. But these issues are of critical importance to designers and application developers of distributed systems, and critical to the transparent operation of the systems for the end-users.

Global Distribution of Information

There are many technical and financial reasons that make distributed systems attractive for today's organisations. There is now, however, growing realisation of another important and very practical driving force toward such systems. Chemical information systems, both public and private, reside on many different machines in many different locations around the world. Many firms in the chemical and pharmaceutical industry have installations in at least the US, Europe and Japan; recent mergers serve to emphasise the growing global nature of the industry.

The rapidly improving quality and speed of computer networks makes it possible to link efficiently multiple hosts together with workstations. In principle, correctly designed and well built distributed systems can bring individuals together with information independent of its location. In practice, several technical issues remain to be solved in order to provide users with simple, transparent access to this information. Two important issues, consistent data models, and coherent search and update techniques, are discussed in the next two sections. Obviously, the quality of the computer networks linking machines is an important issue as well, but we do not address this issue directly.

In the discussions below, the reader should assume that the examples of database access that are described can be carried out over distributed databases on distributed machines.

Consistent Data Models over Diverse Data

Users want to bring together chemistry, data, text, graphics, and now voice, as required to solve particular problems. From a system designer's point of view, this simple-sounding requirement presents a real challenge. This information is currently stored in many systems from many vendors, in a wide variety of proprietary and non-standard formats. The data 'models' are fundamentally different among these systems. For example, many database management systems (DBMSs) use the relational model for storage of and access to data (RDBMSs). A database of chemical structures ('CDBMS') may be a flat file, or a hierarchy of arbitrary complexity. A database of textual information ('TDBMS') may use a segmented document as a data model. Integration of such systems so that information in one system can be related to information in another is extremely difficult.

One approach to creating a consistent data model is to look for a standard. The best standard available currently is the relational data model. This standard does not specify the actual physical storage of data. Therefore, if a collection, or database, of for example, chemical objects, or documents, can be viewed as a collection of relations, then it becomes possible to design and build systems that view disparate data in similar ways. Recent results obtained by adopting a consistent, logical view of chemical and relational databases are described in a later section. A recent report discusses both the storage of chemical structures in a relational system, and the methodology for and performance of various types of searches carried out over these structures.[3]

Other database vendors are moving rapidly to adopt or adapt the relational model. For example, traditional vendors of TDBMSs are adopting a more relational view of and access to text, and traditional RDBMS vendors are adding more capabilities for handling text. Some systems based in part on graphics, or other data objects, such as object-oriented database management systems, may have an underlying RDBMS. However, further discussion of these systems is beyond the scope of this paper.

Coherent Search and Update of Diverse Data

A consistent data model is a necessary but not sufficient condition for a distributed system that provides data access methods acceptable to users. It is also essential to provide a coherent way to search for and update information in multiple, diverse databases. This has implications for applications running in part at the workstation GUI, in that users want a single interface to diverse information.

There are, however, important implications for host-based software as well. With a consistent data model in hand, it is possible to define a more coherent search system. Such a system makes it possible to mask many of the idiosyncrasies of the component DBMSs by orchestrating complex searches over multiple, distributed databases 'behind the scenes', so the user does not need to understand what data are being searched where. In the next section, we briefly discuss some recent results obtained by adopting a unified approach for searching and updating data in combined chemical and relational database management systems.

Specific Problems and Solutions

We have designed and implemented an architecture for host-based access to diverse databases that meets some of the requirements and objectives described above. This architecture defines a consistent data model for chemical databases (structures and reactions) and relational databases. The data model is hierarchical, and is formally related to the 'non-first normal form' (NF2). This model has a well-developed theory[4], a long history, and is now being discussed as one approach to meeting modern requirements for object-oriented databases.

We have implemented software, called Distributed Systems Host[5], or 'DSHost', based on the architecture and data model. DSHost provides:

1. methods for host to workstation and host to host communication of information in a distributed computing environment;
2. methods for co-ordinating access to different numbers and types of DBMSs; and
3. methods for co-ordinating combined database search/retrieval and update operations over the different DBMSs.

With NF2 as the data model, hierarchies of arbitrary complexity can be defined that relate chemical and relational tables to one another. These hierarchical views, which we call 'HViews' in this paper, refer to the, possibly multiple, DBMSs involved by type (relational, chemical) and name. An HView, which is described in an external file, also contains information on the network addresses of the various databases, and the fields used to link the tables together in the hierarchy, as illustrated in Figure 9.

An HView defines a 'logical' view of the contents of a set of (possibly) disparate databases. This logical view provides a data dictionary that is used by both the workstation and DSHost software to access the hierarchy. Thus, the simple HView definition summarised at the bottom of Figure 9 is all that is required to give access to all fields in each of the defined segments.

DSHost uses the logical view to control access to databases, and to control the search and update procedures themselves. HViews may contain entries for multiple chemical databases and multiple relational databases from multiple vendors. At one limit, an HView can define access to a single table in a single relational database if that is all that is required for an application. At the other limit, a complex HView can define multiple, distributed chemical and relational databases, which are to be accessed at the same time.

The architecture also defines a coherent method for orchestrating combined searches over relational and chemical data. Queries combining search terms about chemical and/or relational data are received by DSHost from the workstation, and are passed to the appropriate DBMS (automatically using results from one

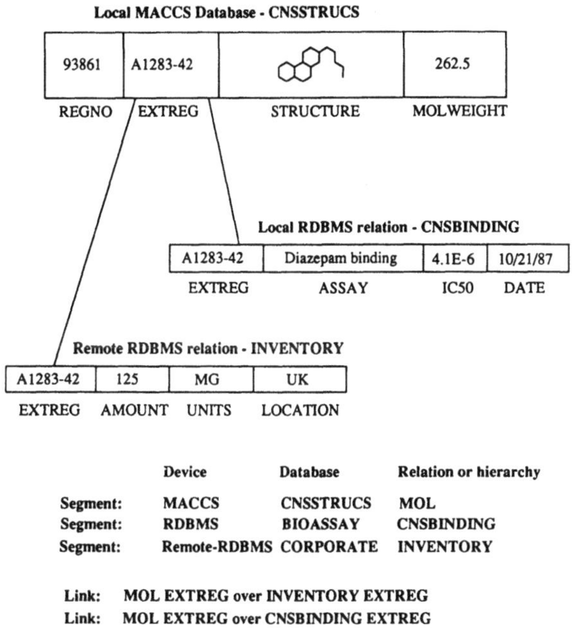

Figure 9. A symbolic description of a hierarchy consisting of a MACCS-II database named CNSSTRUCS linked to two tables, one of each in two RDBMSs. One RDBMS is local to the same machine, the other resides on a remote, but networked, machine. The data field that links the databases together is a registry number, EXTREG, a field common to the linked databases. The HView definition is summarised at the bottom of the figure. It specifies three 'segments' of three database 'devices', one a MACCS database, and a local (no prefix) and a remote (prefix 'Remote-') RDBMS. The databases' physical names and the constituent relations or hierarchies are defined in the next two columns. Finally, the linking information is defined

database search to control the search over the other database(s)). Results are passed back over the network to the workstation.

The implementation is distributed in that individual databases may reside on any network-accessible host. Apart from the added complexities of networks and communication software, co-ordination of combined searches over distributed databases works the same, independent of the physical location of the database. This effectively isolates the end-user from any requirement to know where data reside, and in what DBMS.

Integrating Chemical Databases. The simplest chemical DBMS (CDBMS) application is, of course, access to a single molecule or reaction database. This is base-level DSHost functionality which we do not explicitly illustrate here. Many interesting applications of distributed systems require access to multiple CDBMSs, perhaps on multiple machines. We have demonstrated several applications where searches in multiple CDBMSs are controlled by DSHost. One example is illustrated in Figure 10.

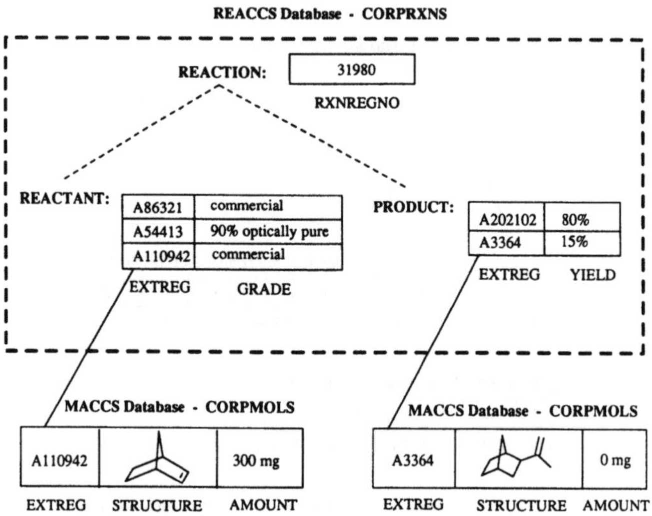

	Device	Database	Relation or hierarchy
Segment:	REACCS	CORPRXNS	REACTION
Segment:	MACCS	CORPMOLS	MOL (a)
Segment:	MACCS	CORPMOLS	MOL (b)

Link: REACTION.REACTANT.EXTREG over MOL(a).EXTREG
Link: REACTION.REACTANT.EXTREG over MOL(b).EXTREG

Figure 10. A pictorial view of the hierarchy and the HView definition of one way to link molecule and reaction databases. A REACCS database named CORPRXNS has the same MACCS-II database, named CORPMOLS, linked by EXTREG to both reactants and products. The linkage is a logical one, not a physical one; no data are duplicated. For additional description, see caption to Figure 9

There are several interesting applications which can be developed with the HView of Figure 10. For example, one can determine what reactions are available, at high yield, to synthesise molecules located in a corporate database by identity or substructure search. Or, one can identify novel transformations of a given molecule in the corporate database, qualified by the amount available in the sample inventory.

Another example demonstrates how chemical databases of the same type can be searched as if they were a single database. DSHost supports HViews that express the idea of 'concatenation'. Thus, multiple reaction or molecule databases can be searched in sequence as though they were one. A sample HView and pictorial description of this application of DSHost are given in Figure 11. DSHost carries out a search over two or more concatenated databases as if they were one database. Concatenating multiple molecule or reaction databases has obvious applications where chemical databases are distributed for whatever reason. Considering the example molecule databases of Figure 11, such distribution may result from separate molecule registration processes, subsetting a database to two or more different locations, or different physical databases kept for special purposes but with requirements in some applications that the databases be searched together.

A final example illustrates a solution to a common problem in managing chemical information. The previous examples, Figures 9, 10 and 11, each had a linking field available that could relate information in one database to that stored in another. Often, however, chemical databases must be correlated, or linked, and no linking field such as EXTREG in the previous examples is available. To solve this problem, DSHost allows linking on chemical structure, as illustrated in Figure 12. In the example of Figure 12, one can pose searches over information in one molecule database and relate it to information in another database in the absence of any common field beyond that of the molecular structure itself. For example, one can perform a substructure search for a set of molecules in a copy of the Fine Chemicals Directory on a machine in the U.S.A, to determine which of those structures was registered in the corporate database in Europe and assigned to which department for testing. This linkage is relatively slow because it requires a molecule identity search at each step, but it is useful when no other method of correlating databases is available.

Integrating Chemical and Relational Databases. From the point of view of DSHost, databases are databases. Whether they contain chemical or relational data, the conceptual model is the same. Therefore, it is easy to extend the examples

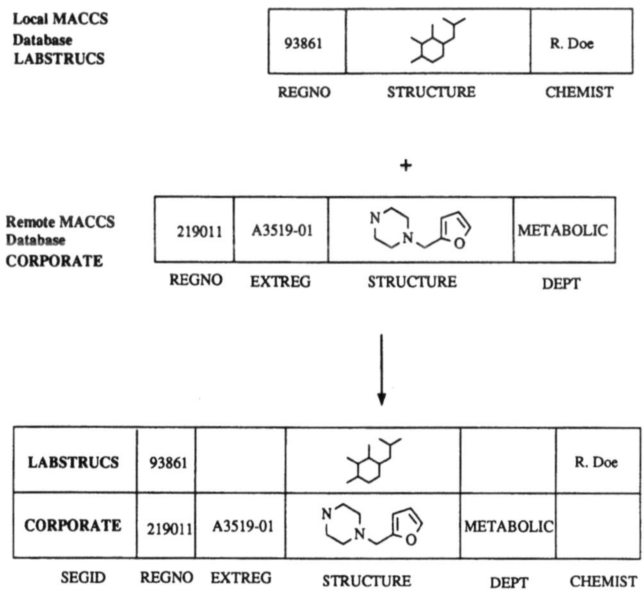

Figure 11. A pictorial view of two concatenated molecule databases, one called LABSTRUCS, the other called CORPORATE. In this example, the databases physically reside on different machines on the network. The process of concatenation yields a logical view which represents the union of the fields in the individual databases

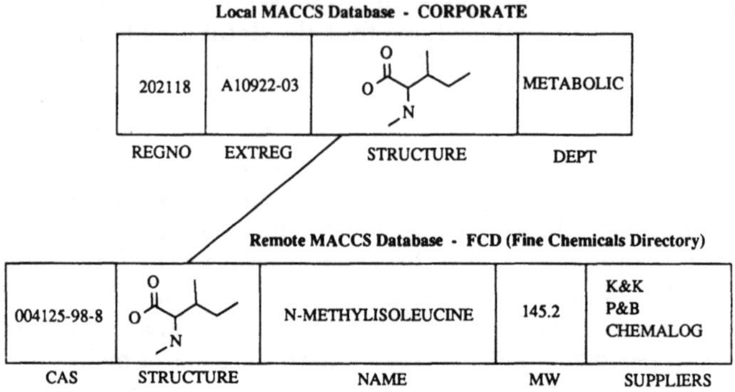

	Device	Database	Relation or hierarchy
Segment:	MACCS	CORPORATE	MOL (a)
Segment:	MACCS	FCD	MOL (b)
Link:	MOL(a).STRUCTURE over MOL(b).STRUCTURE		

Figure 12. A pictorial view, and the HView definition, of two distributed molecule databases, named CORPORATE, and FCD, respectively. In the absence of a linking field such as a registry number common to both databases, the linking field is specified as chemical structure. Identity of chemical structures is used to satisfy the 'equality' required by the link

of CDBMS access in the previous section to include access to RDBMSs. It is also easy, although we do not provide an example here, to forget chemistry altogether, and use DSHost solely for access to RDBMSs.

There are many scenarios for combined CDBMS and RDBMS access that can be conceived and demonstrated today. But perhaps the best simple example to illustrate the power of the methods is shown in Figure 13. The HView of Figure 13 allows one to issue single queries over different fields in multiple different databases supplied by multiple vendors. The implications of this capability for organisations that have globally distributed information in several different DBMSs should be obvious.

Integration with 'Non-Conforming' Database Management Systems. The NF2 data model in DSHost is derived from a relational model. We have made several changes in our CDBMS access to make our systems appear sufficiently 'relational' that they fit adequately into the NF2 model. We are taking a strongly relational view of information. The DSHost technology is just now maturing to the point where we understand its strengths and limitations. DSHost's greatest strength, its ability to provide a consistent logical view of different databases, is also, in a sense, its greatest weakness, That is, another DBMS, to conform with the model, must support a data model that is in some sense 'relational'.

There are many other types of DBMSs in the world. Often we are asked the question, 'Can I integrate DBMS XYZ with DSHost so that it will work together as an integrated system?'. The answer that is emerging is, to the extent that system XYZ is based on a strong relational model, probably 'yes'. If the system is based on another, unrelated model, the answer might still be 'yes', but it will be much more difficult. We have not yet demonstrated integration with such a 'non-conforming' DBMS application. This is currently an area of research for us.

Integration with Other Host Applications. DSHost has no definition of a user interface. It is strictly a database 'server' that responds to the right commands from the outside world. Therefore, integration with other applications will occur at a fairly low level. There are, in principle, two ways to integrate its capabilities with other host-based applications. The first would be to define an application programming interface to allow other applications to make use of DSHost's database integration, distribution, and search and update capabilities. This would allow construction of systems that require DSHost's functionality as part of a larger application. The second would be to add ancillary functions to DSHost to allow it to: a) control other applications, by spawning processes, for example; or b) provide results for other applications in the form of files of results.

These specific techniques are now under investigation. Whatever capabilities may be developed, however, from the user's point of view, integration will still be occurring at the workstation GUI. This is the subject of the next section.

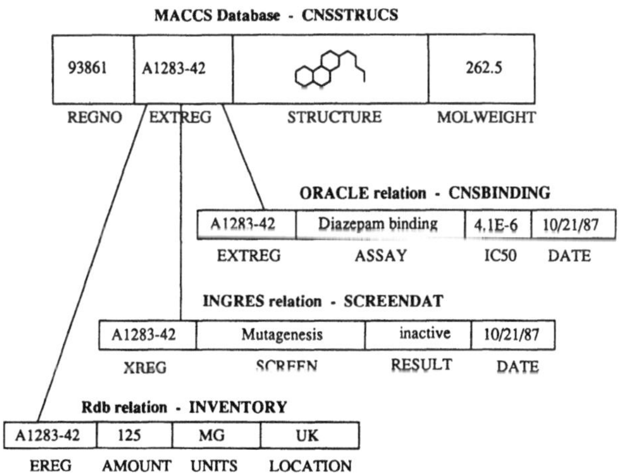

	Device	Database	Relation or hierarchy
Segment:	MACCS	CNSSTRUCS	MOL
Segment:	ORACLE	BIOASSAY	CNSBINDING
Segment:	INGRES	SUMMARY	SCREENDAT
Segment:	Rdb	CORPORATE	INVENTORY

Link: MOL.EXTREG over CNSBINDING.EXTREG
Link: MOL.EXTREG over SCREENDAT.XREG
Link: MOL.EXTREG over INVENTORY.EREG

Figure 13. A pictorial view, and the HView definition, of a molecule database linked with three different RDBMSs, Oracle, Ingres and Rdb/VMS (from Digital Equipment Corporation). Any of the databases could reside on a remote machine on the network

THE INTEGRATED, DISTRIBUTED SYSTEM

The previous section discussed the two major components of a distributed system, the desktop workstation and the host-based database servers, largely in isolation. In this section we bring them together, briefly discussing what by this point should be obvious: consistent, coherent workstation access to distributed host information vastly expands the power of desktop computing. The workstation moves from being a useful tool for managing local information, to a full participating member of a global information environment.

In the following sections we describe some general goals and some specific problems and solutions in the context of our application software.

Workstation Access to Host Information

The challenge in developing application software in this area is to make remote host information as easy to access as is workstation information. In practice, this means that whatever tools are provided to the user for accessing workstation information work exactly the same way for accessing a networked host. A workstation program used to draw graphical representations of chemical objects, such as illustrated in Figures 5 and 6, must produce query structures and/or substructures in exactly the same way independent of the location of the database that is the target of the query. Similarly, queries over chemical or relational data must be expressed in the same way by the user, whether posed over workstation or remote information. If graphical forms and tables are built to control access to data, then the tools for building them, and their behaviour when used to review or update information, must be independent of the location of the data.

These objectives are not always easy to realise completely, given the very broad spectrum of capabilities of workstations and host DBMS's that users expect to use and access. But on some configurations of workstations and hosts, these objectives can be largely realised today.

Integration of Workstation and Host Information

If a distributed system were comprised entirely of homogeneous machines and networks, total integration of workstation and host information would essentially come 'for free'. To the extent that machines and operating environments differ, this goal is more difficult to realise. We have established an interim goal that is realisable. There must be sufficient overlap in functionality between the workstation and host systems to preserve essential information; beyond that, the systems may differ based on their respective capabilities. For example, a workstation chemical database can be made upward compatible with corresponding databases on the host. This enables use of workstation databases to update information on the host with no loss of fidelity. But workstation graphical manipulations will generally exceed those available in systems composed of terminals and a host. And host database manipulations in heterogeneous environments will generally exceed those available on workstations.

Once information has been obtained from databases either on a workstation or a host, then it can be moved to other workstation applications in exactly the same ways as described in an earlier section, resulting in presentations similar to that of Figure 8. These capabilities for integration are the same independent of the source of the information, as long as the workstation application has access to all the underlying data, for example, connection tables for molecules and reactions.

Specific Problems and Solutions

We briefly discuss some examples that illustrate some of our approaches to a common user interface for information access.

Consistent User Interface. The user interfaces for the chemical drawing and database access applications are the same, independent of the location of the database which is being accessed or to which information is being moved. Drawing of chemical structures and queries is the same. Defining a graphical form or table for viewing search results is the same. Most of the differences between workstation and remote access are masked from the user. As a simple example, consider the database access application window of Figure 14. In this example, the user does not need to know or care whether the data dictionary represents a workstation or a remote HView. And if it is a remote HView, the user does not need to know how the databases are distributed.

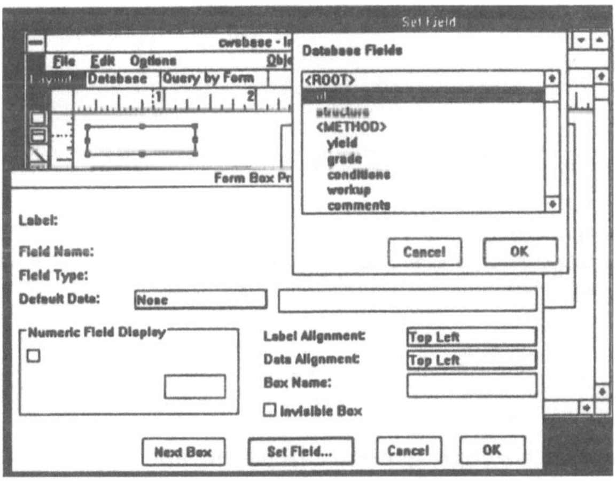

Figure 14. An example of window and dialogue box layout for a database access application program. The upper right dialogue box displays the data dictionary representing the logical view of the database(s) created from a defined HView. Point and click operations select a specific field name for insertion into the 'Field Name:' of the second dialogue box occupying most of the lower portion of the window. This latter dialogue box also controls the layout of information in the selected box in the graphical forms or tables used to view information. The form itself is largely obscured by the dialogue boxes, but the selected box is visible in the upper right corner. The specific selection of 'id' will result in the identification number of a chemical structure to be displayed in the selected box whenever structures are retrieved. Alternatively, it becomes the assigned identification number for update

Consistent Chemical Structures and Queries. A complex substructure search query created in a chemical drawing application is presented in Figure 15. This query is completely compatible with workstation or host substructure search systems. Since both workstation and host chemical representations are equivalent for these query properties, the query can be submitted to a either a workstation or remote chemical database, or to both in consecutive searches.

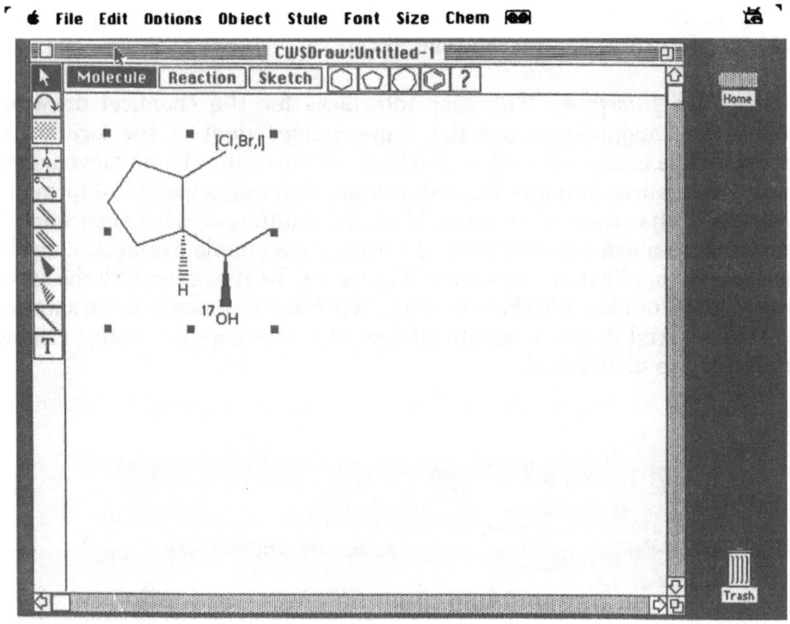

Figure 15. A screen dump of a chemical drawing application in the Macintosh GUI illustrating a substructure that can be submitted as a query to either or both a workstation or host chemical database

CONCLUSIONS

We have attempted, through general discussion and specific examples, to meet three objectives in this paper. Our first objective was to demonstrate that access to distributed information sources leads to information integration. This should be patently obvious apart from any of our presentation, but people are often confused in the juxtaposition of the terms 'distribution' and 'integration'. We hope we have provided sufficient examples of bringing together information at the desktop, and sharing it with other applications, to demonstrate the principles.

Our second objective was to emphasise the importance of end-users in designing and implementing information systems. We have involved users in our own developments, and will continue to do so. We presented much of this paper from the point of view of advantages for users. These people have the most to gain from using information systems in the 1990s, and are therefore asking for greater participation in making decisions. Yet the systems are merely tools to aid them in the jobs they are really paid to do. The tools must be intuitively obvious to learn and re-learn.

Our third objective was to stress the importance of standards in enabling new technologies in the 1990s. We mentioned some specific examples in discussions of the kanji version of Microsoft Windows, of interapplication data exchange, and of the relational data model. There are many other advantages to standards that were not stated, such as the extensive printer support provided by GUIs, ANSI standard SQL, and so forth. Some of the 'standards' are not yet universally accepted, but progress is being made. Their existence is already paying off handsomely.

As a final conclusion, management of chemical information is about to undergo a revolution, and it will be very exciting to be a part of it.

REFERENCES

1. Smith, D.H. 'Information Integration in an Incompatible World'. In *Chemical Structure Information Systems*, W.A. Warr, Ed.; ACS Symposium Series 400, American Chemical Society: Washington, 1989, pp. 18–40.
2. Smith, D.H., 'Consumer-Driven Access to Scientific and Technical Information, 1990s and Beyond'. Paper presented at The Division of Chemical Information Herman Skolnik Award Symposium, 197th Meeting of the American Chemical Society, Dallas, Texas, April 11, 1989.
3. Hagadone, T.R.; Lajiness, M.S. 'Capturing Chemical Information in an Extended Relational Database System'. *Tetrahedron Comput. Methodol.* **1988,** *1,* 219–230.
4. Pistor, P.; Traunmüller, R. 'A Data Base Language for Sets, Lists, and Tables'. *Technical Report TR 85.10.004*; Heidelberg Scientific Centre (IBM), 1985.
5. This name has been changed to ISIS/Host. ISIS is a trademark of Molecular Design Limited.

INTEGRATING CHEMICAL STRUCTURES INTO AN EXTENDED RELATIONAL DATABASE SYSTEM

Thomas R. Hagadone and Michael S. Lajiness
The Upjohn Company, Kalamazoo, MI 49001, U.S.A.

ABSTRACT

This paper proposes extensions to the relational database model that allow chemical structure and other complex data types to be included in a relational database. It is argued that this approach provides benefits over the common practice of storing chemical structures in a chemical database system and associated research data in a relational or other general database system. The design, implementation, and usage patterns of an extended relational system are discussed in the context of the Upjohn Cousin compound information system. Emerging extensibility features that support the proposed approach within commercially available database systems are reviewed.

INTRODUCTION

Over the last decade, chemical database management systems (CDBMS) and relational database management systems (RDBMS) have evolved independently. CDBMSs have developed to meet the special needs of storage and retrieval of chemical structures. RDBMSs have developed to meet efficiently the requirements of a broad range of business, engineering, and scientific applications. Because of the strengths of each type of system, both are commonly used to manage scientific research data in the chemical and pharmaceutical industries.[1] A CDBMS is employed to manage collections of proprietary and published chemical structures and a RDBMS, or other general DBMS, is used to manage biological response data, physico-chemical data, and other compound-related data.

Because of the intimate relationship between data in the CDBMS and the RDBMS, there is frequently a need simultaneously to access and correlate data from the two systems. This requirement has led to a number of approaches to integration of the two databases which have been previously reviewed.[2] Most approaches result in a duplication of software functionality and user training requirements, as well as restricted retrieval and reporting capabilities. Since CDBMSs and RDBMSs normally support different data models and data definition and manipulation languages, any approach to integration based on multiple DBMSs will inevitably have to address many thorny compatibility issues.

In this paper, we advocate an approach to the integration problem in which an RDBMS based on the SQL language[3] is extended to encompass chemical structure and other complex data types. This approach integrates all data normally stored separately in a CDBMS and RDBMS into a single extended RDBMS. The resulting system provides a single uniform data model and data manipulation language that can be applied to chemical structures and associated data simultaneously. The

W. A. Warr (Ed.)
Chemical Structures 2

benefits over the traditional approach are reduced duplication of software functionality and training requirements and the ability to retrieve easily a broad range of data in unanticipated ways.

To demonstrate the feasibility of the extended RDBMS approach, a prototype extended system was designed and implemented as part of the Upjohn Cousin compound information system.[4,5] The prototype design included support for chemical structure, molecular formula, chemical name, and registry number data types. After testing and refinement, the extended system was made available for general use at Upjohn in 1984. Experience gained since that time has provided insight into the strengths and weaknesses of our design and the extended RDBMS approach in general.

In this paper the issues involved in extending an RDBMS will be discussed and the particular design that was adopted at Upjohn will be described. The utility of the approach will be demonstrated through a series of Cousin query examples in an extended SQL language. Although complete familiarity with SQL is not required to appreciate the benefits of the approach, a basic understanding of SQL, such as is found in reference[6], will be helpful. Following the discussion of the extensions, the performance of the system and the level of user acceptance will be described. Finally, some of the limitations of the present design will be explored, followed by a discussion of current DBMS research that promises to improve DBMS extensibility in the future.

ABSTRACT DATA TYPES

An RDBMS based on the SQL language consists of a number of components as shown in Figure 1. The system kernel supports the data definition and manipulation functions of SQL as well as providing database recovery management, automatic concurrency control, distributed database support, and an authorisation mechanism. Surrounding the kernel are application programs for entering and editing data, performing *ad hoc* queries, creating reports, developing new applications, etc. While this general approach serves well, a major limitation of existing RDBMSs is the small number of data types allowed. Most systems

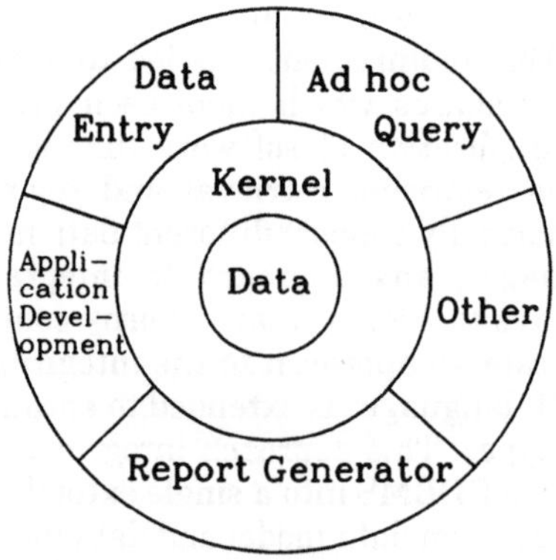

Figure 1. Standard RDBMS architecture

provide support for only integer and floating point numbers, character string, and date/time data types. There is no provision for common data types encountered in chemical research such as chemical structure, molecular formula, chemical name, and registry number.

As a solution to this problem, the concept of abstract data type (ADT) has been proposed.[7,8] The concept provides a mechanism for adding new data types, such as chemical structure, to a DBMS by supplying a framework that allows the following within the DBMS environment:

1. The creation of new user-defined data types.
2. The definition of new operators for data types.
3. The implementation of new access methods for data types.
4. Optimised query processing for commands containing new data types and operators.

Support software must be developed in each of these areas and integrated into the DBMS for every new abstract data type. For example, to create a chemical structure ADT, a storage format, such as a connection table format, must first be defined. Then operators for full structure, substructure, and other types of structure searches must be defined. Access methods need to be implemented to provide fast structure searches and finally, the characteristics of the various structure searches must be made known to the query optimiser to allow efficient overall query execution.

Since extensible DBMSs are still largely in the research stage, the Cousin prototype was designed to implement ADTs through a software shell inserted between the relational kernel and application programs as shown in Figure 2. The shell provides each of the required components within the ADT framework for the ADTs we implemented. The *ad hoc* query interface, report generator, and database update programs operate through the ADT shell which provides transparent access to the ADTs.

The most interesting and challenging ADT to implement was that for chemical structure. In the following sections we illustrate the Cousin ADT approach by first giving examples of structure-containing SQL statements and then describing the underlying support required for their execution.

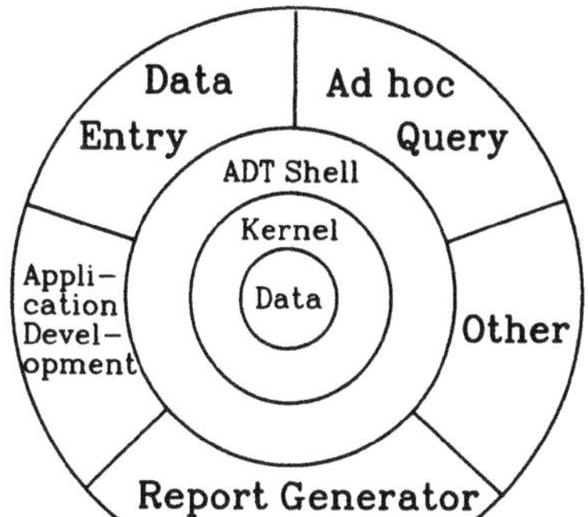

Figure 2. RDBMS with Cousin ADT support

CHEMICAL STRUCTURE AS AN ADT

For purposes of discussion, suppose a simple database table (relation) named COMPOUND exists, as shown in Table 1, containing columns for registry number (RN), chemical structure (STRUCTURE), and inventory amount (INVENTORY).

Table 1. Compound table

COMPOUND		
RN	STRUCTURE	INVENTORY
123	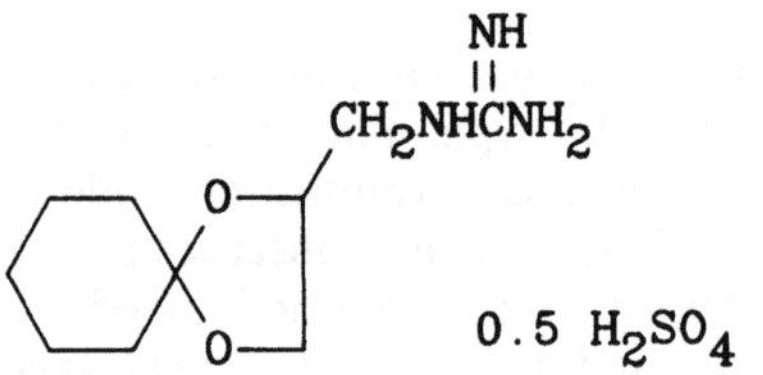	10.5

In a multiple DBMS approach, this table might be split into two parts with registry number and structure stored in the CDBMS, and registry number and inventory stored separately in the general DBMS. Queries on this data would require two subqueries, one for each DBMS, followed by a data merging step.

In the following examples, extended SQL commands are used to perform single-step queries on the COMPOUND table. The commands can be typed directly into Cousin. Results are automatically formatted for output as part of the *ad hoc* query interface. The first example retrieves all data for a compound of known structure:

SELECT * FROM COMPOUND WHERE STRUCTURE=?

This is a normal SQL statement except for the fact that STRUCTURE is not a standard SQL data type. When the ADT shell command processor discovers from the data dictionary that the STRUCTURE column is a chemical structure it invokes the molecule editor and prompts the user to enter the desired structure. Following structure entry, the search is performed and any COMPOUND table row retrieved from the database is displayed, including the structure.

The next query finds all compounds with a particular substructure that have more than 10 grams of inventory:

SELECT * FROM COMPOUND WHERE STRUCTURE>? AND INVENTORY>10

In this case the > operator associated with the structure ADT causes the substructure query editor to be invoked. The user draws the desired substructure query and then the substructure and inventory searches are performed to find any qualifying compounds, which are then displayed.

In the third example, a search is done for all compounds with registry numbers in a predetermined list that are similar to a specified compound and that have more than 10 grams of inventory:

```
SELECT * FROM COMPOUND WHERE RN IN FILE LIST1 AND
  STRUCTURE LIKE ? AND INVENTORY>10
```

The LIKE operator is used here to request a similar-structure search and the IN FILE predicate is an extension to the standard SQL IN predicate that allows files of registry numbers to be manipulated. In this case LIST1 is a list of registry numbers saved from a previous search.

The IN FILE predicate provides a convenient way to manipulate registry number lists. The following example finds all registry numbers in LIST1 or LIST2 that are also in LIST3 or LIST4:

```
SELECT RN FROM COMPOUND WHERE (RN IN FILE LIST1 OR
  RN IN FILE LIST2) AND (RN IN FILE LIST3 OR RN IN FILE LIST4)
```

Definition of the structure data type

The first step in the creation of a new data type is the definition of the data structure for the type. In Cousin, structures are stored in a typical connection table format with atom and bond counts, and lists of atom and bond attributes. Atom attributes include atom type, co-ordinates, charge, isotope, and attached hydrogen count. Bond attributes are bond type and connecting atom numbers.

The Create statement is used to create a new table. The table in the examples can be created with the statement:

```
CREATE TABLE COMPOUND
  (RN CHAR(6),
  STRUCTURE STRUCT,
  INVENTORY FLOAT)
```

where RN, STRUCTURE, and INVENTORY are column names and CHAR, STRUCT, and FLOAT are their respective data types. The STRUCT data type alerts the ADT shell that this column is to hold chemical structures. A varying character string column is created to hold the connection table. To support fast structure searching, the shell also creates two additional tables.

The first table supports full structure and component structure searching and has the form shown in Table 2. The RN column contains the registry number for the compound and the HASH_VALUE column is a character string containing a

Table 2. Compound_structure_hash table

COMPOUND_STRUCTURE_HASH	
RN	HASH_VALUE
123	01101...0
123	10100...1
123	11001...1

hash value for each connected component of the molecule and for the overall molecule. The hash value is a 16 byte value calculated from a connection table in which tautomer and aromatic bond types have been normalised. It has the properties of being independent of the ordering of atoms and bonds in the connection table and of having a low probability of matching the hash value of a different molecule. The 'low order' bit of a hash value is set to zero for complete structures and to one for structure components to allow for both full structure and component structure searches as described below.

The second table supports substructure and similar-structure searching and has the form shown in Table 3. Again, the first column contains the registry number. The SCREEN_BITS column, which is effectively 48 bytes long, contains a bit string indicating the presence or absence of the members of a set of molecular fragments. In practice, it was found that performance suffered too much if fragment screening data were kept in a database table; therefore, they are currently stored in a file outside of the RDBMS environment.

Table 3. Compound_structure_screen table

COMPOUND_ STRUCTURE_ SCREEN	
RN	SCREEN_ BITS
123	100100110...

When a structure is inserted, updated or deleted, the ADT shell performs the corresponding operation on the hash and screen tables to keep them synchronised with their associated structure-containing table.

Implementing structure search operations

The ADT shell is assigned the task of translating the original query into a new form that efficiently carries out the request. This section and the following section describe some of the details involved in the translation process.

Four structure search operations are supported. Full structure and component search take advantage of the hash table for their implementation and substructure and similar-structure search use the screen table. For a full structure or component match the ADT shell creates a new SQL statement of the following form:

```
SELECT RN FROM COMPOUND_STRUCTURE_HASH
  WHERE_HASH VALUE = 'hvalue'
```

where 'hvalue' is the hash value calculated for the query structure. The 'low order' bit of the hash value is used to specify a full structure search or a component search. For a full structure search, the low order bit set to zero. For a component search, the low order bit is set to one. The ADT shell creates an index on the HASH—VALUE column to provide good search performance. The RDBMS simply looks up the query hash value in the index, then goes directly to the associated row(s) in the hash table and retrieves the corresponding registry number(s). This approach results in search response times of the order of one second. Since different structures may occasionally be assigned the same hash value, candidate matches should be passed through an atom-by-atom match to achieve complete accuracy. The atom-by-atom match will be fast and will not significantly affect response time since only a few structures will need to be checked and rejections will be rare.

Substructure and similar-structure searching use the screen file for their implementation. Substructure searching is performed using a traditional two-step process in which an initial fast screening operation is performed over the complete file, followed by a detailed atom-by-atom match.[9] The similar-structure search is performed in a single step by calculating the Tanimoto coefficient[10] value between the query structure fragment screen and each of the database structure fragment screens. Coefficient values are ranked from high to low value and cut-off values limit the results to a maximum number of structures and a minimum coefficient value. The registry numbers of molecules selected in the search are passed back into the RDBMS environment by the method described in the next section. Substructure and similar-structure searches require of the order of twenty seconds elapsed time to complete for a 100,000 molecule database.

Query optimization

In query optimisation, the DBMS selects a strategy for implementing a query that makes the best use of the available underlying database access paths. For some types of searches, it is possible for the ADT shell to translate the original query into a new form that the DBMS can execute directly. For example, the query in the first example:

```
SELECT * FROM COMPOUND WHERE STRUCTURE=?
```

can be translated to the nested subquery form:

```
SELECT * FROM COMPOUND WHERE RN IN
  (SELECT RN FROM COMPOUND STRUCTURE_HASH
  WHERE HASH_VALUE = 'hvalue')
```

if the follow-up atom-by-atom match is not needed.

For cases that require processing outside of the DBMS facilities, as in substructure searching, a different mechanism must be employed. The substructure search in the second example:

```
SELECT * FROM COMPOUND WHERE STRUCTURE>? AND INVENTORY>10
```

is translated into a substructure search in combination with the following statement:

```
SELECT * FROM COMPOUND WHERE RN=? AND INVENTORY>10
```

This statement is then executed repeatedly with the ? successively replaced with each registry number returned from the substructure search. Although this may seem inefficient, the capability of most relational systems to pre-compile a statement and then execute it many times with differing parameter values (registry numbers) results in fast searches.

A more significant problem is that this technique gives top priority to the substructure search part of the query which may, in fact, be the wrong strategy. For example, if only one per cent of the compounds in the database have more than 10 grams of inventory and there is a fast access path available on inventory, then it may be more efficient to retrieve the compounds with more than 10 grams of inventory first and then perform the substructure search on the resulting structures. An efficient optimiser considers all available access paths and then chooses a search strategy based on the best available path.

OTHER ADTS

Although chemical structure is the most interesting and challenging case, other useful ADTs have been implemented in Cousin, including registry number, molecular formula, and chemical name.

Registry number

The registry number ADT is implemented in a simple fashion by providing an input routine that converts all registry numbers appearing in a query into a standard form. For example, the Upjohn registry numbers U-1,234, U-1234, U1234, and 1234 are all used and all refer to the same molecule. Each is automatically converted to U001234 on input to the ADT shell to provide accurate retrieval without requiring the user to employ the standard form of registry number.

Molecular formula

Molecular formula is stored in a variable length character string and in an additional support table. For example the formula $C_{20}H_{40}N_6O_8S$ is stored as the character string 'C20 H40 N6 O8 S1 and also in the support table in the form shown in Table 4.

Table 4. Compound_mf_search table

COMPOUND_MF_SEARCH						
RN	CCNT	HCNT	NCNT	OCNT	SCNT	OTHER
123	20	40	6	8	1	NULL

The support table contains atom counts for carbon, hydrogen, nitrogen, oxygen, sulphur and the other elements. Counts for the other elements are encoded and packed into a single database column. Database indexes are created on carbon count (CCNT) and hydrogen count (HCNT) and are used to implement fast searches for exact formulae as well as for formula queries with variable acceptable range counts for elements (e.g., SELECT * FROM COMPOUND WHERE MF= 'C20-22 N4-6 O8 S CL0'). As with structure searching, the ADT shell translates the user's original query into SQL statements that take advantage of the support table to provide fast searches.

Chemical name

A chemical name can contain several systematic and/or trivial synonyms for a molecule. A name is stored as a character string and also in a supporting table. For example the following name:

Guanidine, (1,4-dioxaspiro[4.5]dec-2-ylmethyl)-, sulphate or Guanadrel sulphate or Anarel or Hylorel

is stored as it appears and also in a support table in the form shown in Table 5.

Table 5. Compound_name_search table

COMPOUND_ NAME_ SEARCH	
RN	NORMALIZED_ NAME
123	GUANIDINE14DIOXASPIRO45DEC2YL...
123	GUANADRELSULFATE
123	ANAREL
123	HYLOREL

For each synonyn, a normalised form is stored in which punctuation and blanks are removed, and lower case letters are converted to upper case. When a name search is requested, the query name is normalised and then looked up in the supporting table which is indexed on normalised name. Fast searches are thus provided when a complete name or name prefix is entered as the query. Name substring searching is handled by a slow linear search technique using the standard SQL LIKE operator. The original unnormalised name is used for display and hardcopy output.

JOIN OPERATIONS

A research environment requires the management of data of greater complexity and variety than the simple table used in the above examples. In particular, biological response data in a variety of forms must be accommodated. As an example of how additional data tables can be manipulated in Cousin, imagine that we have two simple tables of test result data with multiple test results per compound as shown in Table 6.

An actual table would contain test result data for many compounds. For our examples however, the data for a single compound will suffice.

Suppose we desire a report containing data from the COMPOUND table and corresponding test result data for the two tests. A common way to approach this problem would be to combine the three tables with a natural join[11] operation over registry number. However, since a natural join forms the Cartesian product of the

Table 6. Test_a,test_b tables

TEST_A	
RN	ED_50
123 123 123	10 MG/KG 12 MG/KG 12 MG/KG

TEST_B	
RN	LD_50
123 123	40 MG/KG 50 MG/KG

rows that match on the join column, the result would be a table containing all possible combinations of rows in the three tables. For our sample test result tables, a natural join would produce three copies of the TEST_B data and two copies of the TEST_A data. Since this is not what is usually wanted in a research data report, a special join operation was implemented as part of the Cousin ADT shell. The following query uses this join to combine data from the three tables:

SELECT * FROM COMPOUND ALSO TEST_A ALSO TEST_B
WHERE COMPOUND.RN=TEST_A.RN AND COMPOUND.RN=TEST_B.RN

Table 7 is produced as a result.

Table 7. Compound_test_a_test_b table

COMPOUND_TEST_A_TEST_B				
RN	STRUCTURE	INVENTORY	ED_50	LD_50
123	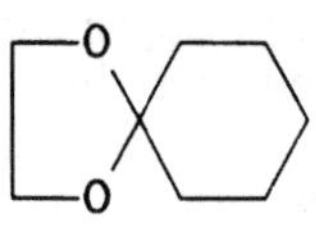	10.5	10 MG/KG 12 MG/KG 12 MG/KG	40 MG/KG 50 MG/KG

Like the natural join, the Cousin join combines the rows of the joined tables by equal registry number and produces a single registry number column in the result table. The duplicate data problem is eliminated by allowing each row of the joined tables to appear once, at most, in the result table. Additionally, the Cousin join provides the capability of the outer join[12] to supply null values for missing data. For example, if the query above was performed and there was no data in the TEST_B table, then the data for the COMPOUND and TEST_A tables would still appear, and null values would be supplied for the TEST_B table.

The Cousin join operation can be used to define complex queries within a single statement. For example, the following query finds all compounds that contain a specified substructure, have more than 10 grams of inventory, and have an ED50 less then 15 mg/kg and a LD50 greater than 30 mg/kg:

SELECT COMPOUND.RN FROM COMPOUND ALSO TEST_A ALSO TEST_B
WHERE COMPOUND.RN=TEST_A.RN AND COMPOUND.RN=TEST_B.RN
AND STRUCTURE>? AND INVENTORY>10 AND ED—50<15
AND LD—50>30

THE REAL USER INTERFACE

In a pharmaceutical research environment, one learns quickly that users are not willing to contruct SQL queries such as those shown in the previous examples. A simpler user interface is required. To meet this need a menu-based interface was built on top of SQL. This interface is data dictionary driven and consists of three levels that move from broad to successively narrower categories:

1. Broad categories (e.g., chemical data, cardiovascular data, CNS data)
2. Table level menus (e.g., basic compound data, acute hypotensive screen, CNS receptor binding assay)
3. Table content menus (e.g., registry number, structure, ED_{50})

The user defines a query beginning at the top level and moves to lower levels selecting desired tables and table columns along the way. Search conditions can be applied to any column in a selected table to limit the data retrieved. Up to 128 tables and any number of search conditions can be specified in a single query. A particularly useful menu command, the ALL command, can be used to select all data in the current menu and in all lower level menus for retrieval. ALL is a useful retrieval shortcut in cases where the user wants to see all data available on a set of compounds at the project, therapeutic area, or global level.

When menu selection is complete, the user's query is translated into extended SQL statements which are then passed to the ADT shell for execution. In translating the query the assumption is made that tables will be joined *via* the special join operation described above and that a compound must pass all specified search conditions to qualify for retrieval. This simple interface allows users to create most reports for themselves without assistance. A consultant's help is available for producing special reports that require complex SQL commands and/or command procedures.

USER EXPERIENCE

The success of an information system can be measured in terms of the level of end user acceptance the system enjoys. Statistics gathered on each Cousin session and summarised in monthly reports allow Cousin usage to be tracked. Figure 3 charts Cousin usage statistics since 1981. Both the number of users and the number of substructure searches have increased steadily to a level of about 500 per month. The total number of sessions has increased more quickly however to about 4,000 per month. This increase is partly due to a rise in the availability of IBM PC based graphics workstations beginning in 1985. Much of the increase however, results from the large scale addition of biological test result data to the database and the ability to report this data easily using the facilities described in this paper. The use of the menu system for user-defined queries parallels overall growth and is charted in the non-routine query curve in Figure 3.

CONCLUSIONS

This paper has demonstrated the feasibility of extending a RDBMS to encompass data types commonly found in chemical and pharmaceutical research, particularly

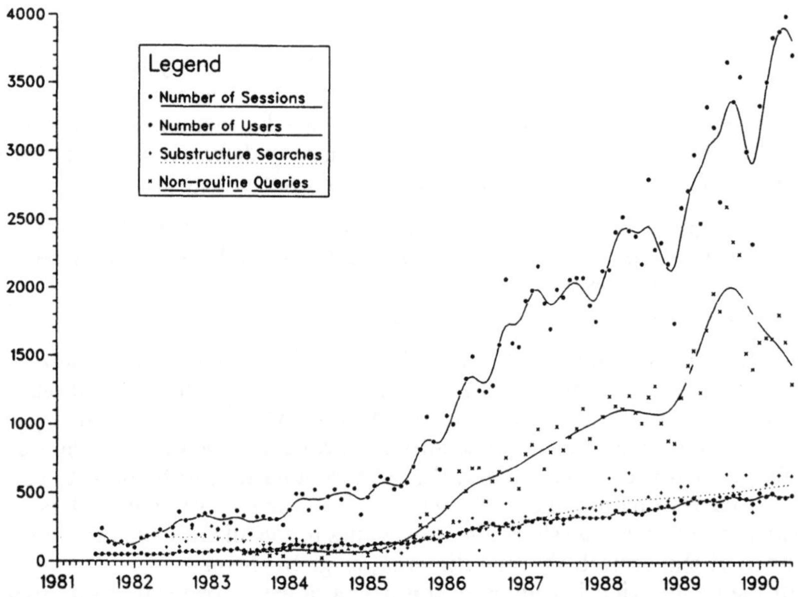

Figure 3. Upjohn Cousin system monthly usage levels

chemical structure. The main benefit of this approach is the ability to store a wide range of compound-related data within the framework of the relational data model and to manipulate the data with the SQL language, thus eliminating software duplication and interface conflicts associated with multiple DBMSs. The approach allows chemical software developers to concentrate on chemical issues by minimising the need to duplicate functionality already available in an RDBMS.

The Cousin approach to ADT support has a number of limitations. The implementation was achieved by enveloping the relational kernel with a shell to provide the ADT capability. This technique makes it difficult to optimise many queries and requires the creation of new application programs (*ad hoc* query, report generator, etc.) to take advantage of the ADTs defined. A more satisfactory solution would be achieved by the full integration of the ADT concept at all levels of the DBMS. This would result in the opportunity to provide better query optimisation and to use standard vendor-supplied application programs.

Database systems researchers have become aware that the standard data types available in DBMSs today are inadequate for many scientific and engineering applications[13]. New DBMS prototypes are addressing this problem by implementing mechanisms to allow ADTs to be defined. Some systems attempt relatively straightforward extensions to existing RDBMSs. The Ingres Object Management Extension and the Oracle Object database project are examples of this approach. Others attempt to provide ADTs through 'object-oriented' data models that follow the themes of object-oriented programming languages.

GemStone from Servio-Logic, VBASE from Ontologic, and ORION from Artemis Technology are examples of the object-oriented approach.

Although none of the commercially available extensible DBMSs currently provide all of the features necessary to support satisfactorily a chemical structure ADT, such support is the goal of these companies. Whichever approach eventually prevails, it appears clear that next-generation DBMSs will include extensibility features and graphical user interfaces. Inclusion of chemical research data, including structures, in such systems is an opportunity that will provide clear benefits.

REFERENCES

1. Ahrens, E. K. F. 'Customisation for Chemical Database Applications'. In *Chemical Structures: The International Language of Chemistry*; Warr, W.A., Ed.; Springer-Verlag: Berlin, 1988; pp. 97–111.
2. Hagadone, T. R. 'Current Approaches and New Directions in the Management of In-House Chemical Structure Databases'. In *Chemical Structures: The International Language of Chemistry*; Warr, W.A., Ed.; Springer-Verlag: Berlin, 1988; pp. 23–41.
3. *Database Language – SQL*; Technical Committee on Database, X3H2; American National Standards Institute: New York, 1986.
4. Howe, W. J.; Hagadone, T. R. 'Molecular Structure Searching: Computer Graphics and Query Entry Methodology'. *J. Chem. Inf. Comput. Sci.* **1982,** *22,* 8–15.
5. Hagadone, T. R.; Howe, W. J. 'Molecular Substructure Searching: Minicomputer-based Query Execution'. *J. Chem. Inf. Comput. Sci.* **1982,** *22,* 182–186.
6. *An Introduction to Database Systems*; Date, C.J.; 3rd Ed.; Addison-Wesley: Reading, Massachusetts, 1982.
7. Stonebraker, M. 'Adding Semantic Knowledge to a Relational Database System'. In *On Conceptual Modelling*; Brodie, M. L.; Mylopoulos, J.; Schmidt, J. W., Eds; Springer-Verlag: New York, 1984; pp. 333–343.
8. Stonebraker, M. 'Inclusion of New Types in Relational Data Base Systems'. *In Proceedings of IEEE/Data Engineering* **1986,** 262–269.
9. Barnard, J. M. 'Problems of Substructure Search and their Solution'. In *Chemical Structures: The International Language of Chemistry*; Warr, W.A., Ed.; Springer-Verlag: Berlin, 1988; pp. 113–126.
10. *Similarity and Clustering in Chemical Information Systems*; Willett, P.; Research Studies Press: Letchworth, 1987.
11. Codd, E. 'A Relational Model of Data for Large Shared Data Banks'. *Commun. ACM* **1970,** *13,* 377–387.
12. Codd, E. 'Extending the Database Relational Model to Capture More Meaning'. *ACM Transactions on Database Systems* **1979,** *4,* 397–434.
13. Carey, M., Ed. 'Special Issue on Extensible Database Systems'. *Database Engineering*, **1987.**

MULTIPLATFORM CHEMICAL STRUCTURE MANAGEMENT

Marc Peeters and Danny Verbinnen
Janssen Research Foundation, Turnhoutseweg 30, 2340 Beerse, Belgium.

ABSTRACT

In search of excellence, the Chemical Metaform Concept offers an open, independent medium of exchange, with high quality, which integrates chemistry in compound documents.

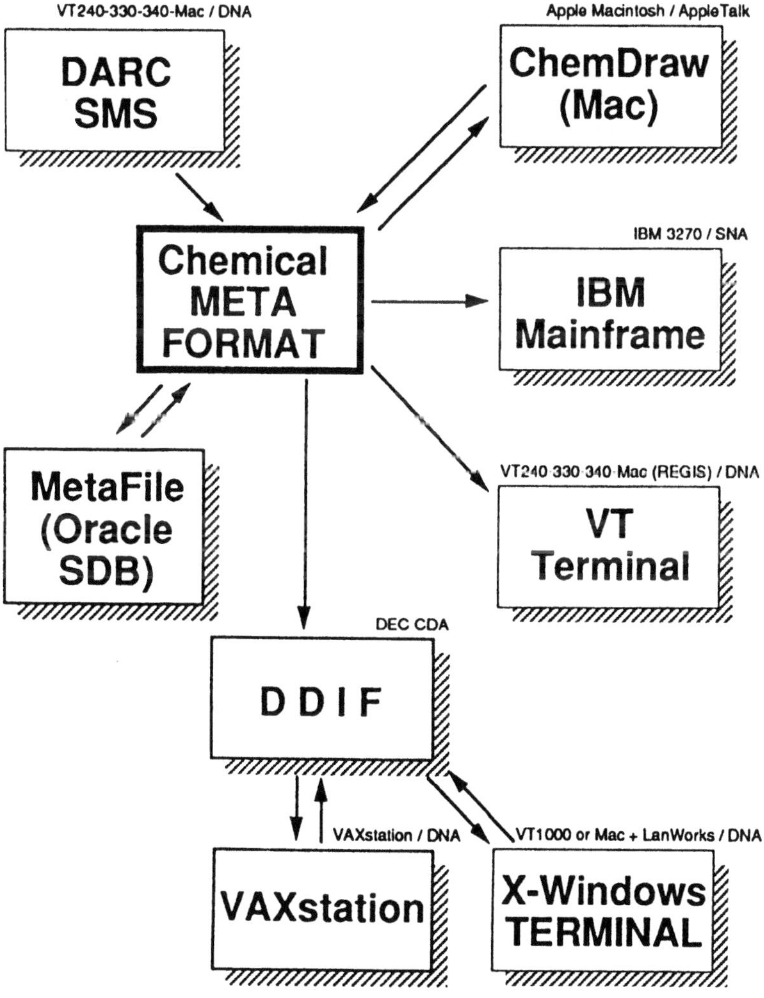

W. A. Warr (Ed.)
Chemical Structures 2

POSTER OUTLINE

In chemical information systems we lack the availability of multiplatform chemical structure representation. We felt that this representation should not be limited to one platform, e.g., the Macintosh. Therefore we defined the Chemical Structure Metaformat.

The Metaformat is a universal, hardware/software independent description of chemical structures. Any structure available to us from either a DARC-SMS or a ChemDraw source should be usable in revisable form on Macintosh, VAX VT-terminal, VAXstation, X-windows terminal, IBM 3270 terminal, etc. The Metaformat converters take care of this task. Every structure is either stored in its Metaformat form, or passes through the Metaformat filter before display or printing.

The Metaformat is a corner stone of our systems: it is being used as intermediate between ChemDraw, Disspla, DARC, PostScript, Regis, DDIF, etc.; it makes it possible e.g., to obtain editable full function documents in ChemDraw on Macintosh with structures from a central VAX SMS DARC database, and this completely transparent to the user.

Together with open-ended standard chemical software packages it provides the complete toolbox for multiplatform chemical computing.

REPRESENTATION AND SEARCHING OF 3-D PROTEIN STRUCTURES

Peter J. Artymiuk and David W. Rice
Krebs Institute of Biomolecular Research and Department of Molecular Biology and Biotechnology, University of Sheffield, Western Bank, Sheffield S10 2TN, England

Helen M. Grindley, Eleanor M. Mitchell, Elizabeth C. Ujah and Peter Willett
Krebs Institute of Biomolecular Research and Department of Information Studies, University of Sheffield, Western Bank, Sheffield S10 2TN, England

ABSTRACT

This paper summarises recent work on the representation and searching of the 3-D macromolecules in the Protein Data Bank. The molecules are stored as labelled graphs in which the nodes denote α-helix and β-strand secondary structure elements and in which the edges denote the distances and angles between these elements. A program, POSSUM, is described that allows the automatic identification of all occurrences of user-defined secondary structure motifs in the Protein Data Bank using Ullmann's subgraph isomorphism algorithm. Examples of the use of POSSUM demonstrate its effectiveness for exploratory studies of protein structure. A procedure is then described for the ranking of POSSUM output in order of decreasing structural overlap with a query motif. Current studies involve the use of a maximal common subgraph isomorphism algorithm to investigate protein homology at the secondary structure level.

INTRODUCTION

The increasing availability of 3-D structural data from X-ray crystallography, NMR and quantum mechanical studies has led to the development of sophisticated molecular graphics systems that can be used for the interactive analysis and display of the 3-D structures of chemical molecules. However, these systems have generally not possessed facilities that allow the co-ordinate data to be used as the basis for 3-D searching operations analogous to those available for the searching of databases of 2-D chemical molecules. This limitation provided the initial impetus for a programme of research at the University of Sheffield into algorithmic techniques for substructure searching in databases of 3-D chemical structures (specifically, the small molecules in the Cambridge Structural Database). This work, which commenced in 1983, has resulted to date in the development of efficient techniques for the searching of user-defined 3-D patterns of atoms and for the identification of patterns of atoms common to two or more 3-D structures. An overview of these studies is provided by Brint *et al*[1].

Over the last few years, rapid developments in molecular biology and biotechnology, in particular, the emerging discipline of protein engineering, have

W. A. Warr (Ed.)
Chemical Structures 2

led to the widespread adoption of database systems for the handling of information about macromolecular species. As with small 3-D molecules, the primary focus of interest has been in the implementation of highly sophisticated systems for the graphical analysis and display of 3-D protein structures[2,3] and there have been few published descriptions of retrieval systems for 3-D macromolecular species. This paper summarises some of the work that has been carried out in our two departments to develop retrieval techniques for use with the Protein Data Bank, the primary source of 3-D data for macromolecules[4,5]. Specifically, we discuss the design and implementation of a program, called POSSUM (Protein Online Substructure Searching - Ullmann Method), that allows searches to be carried out for secondary structure motifs in the Protein Data Bank using graph-theoretic techniques that are based on those used for 2-D and 3-D substructure searching in databases of small molecules[6–8]. In the paper, we describe the machine-readable structure representation that has been developed for the storage and retrieval of protein secondary structures. This is followed by several examples of the use of POSSUM to identify common secondary structure motifs in the Protein Data Bank and by an evaluation of a ranking procedure that can be applied to the output of POSSUM searches. The paper concludes with some preliminary results from a new program that permits the automatic detection of structural homology at the secondary structure level.

REPRESENTATION OF PROTEIN SECONDARY STRUCTURES

Developments in protein crystallography have led to the availability of several hundreds of proteins for which an accurate 3-D structure is available. Analysis of this data has led to the identification of secondary structure motifs, i.e., patterns of α-helices, β-strands and loops that occur repeatedly within protein structures, and this, in turn, has led to a requirement for systems that can be used to search for motifs in files of 3-D protein co-ordinates. When only a few structures were known this could be carried out by eye, but as the number of structures has increased, several programs have been devised for the automatic identification of motifs[9–12]. These programs have used a range of techniques to allow searches for motifs to be carried out but they have all adopted an analogous type of machine-readable structure representation. Specifically, they (and we) have chosen to represent the common (predominantly α-) helix and β-strand secondary structure elements by lines in 3-D space, so that the overall topography of a protein can be represented by the geometric relationships that exist between these linear secondary structure elements. [It should be noted that we have also considered the representation and searching of protein structures using coordinate data for the C^{α} atoms[13] but this represents too detailed a level of description for a retrieval system that is based on the secondary structure of proteins.] The remainder of this section describes the methods used to process the structures in the Protein Data Bank to yield such linear representations. The version of the Protein Data Bank used in the experiments reported here contained the co-ordinate set released in April 1988, from which all non-protein files and a number of outdated co-ordinate sets had been removed. This data-set is being updated with each new release of the databank as it becomes available.

The database creation programs contain three main stages, these being the identification of the secondary structure elements in a protein, the calculation of the major axes for the helix and β-strand elements, and then the creation of the matrix representation that is used for motif searching. The reader is referred to the paper by Mitchell *et al.*[14] for fuller details of these stages.

The first stage, the identification of the secondary structure elements, arises from the fact that while the depositors of a set of protein co-ordinates will usually provide secondary structure assignments, these assignments are occasionally controversial and there are also many proteins for which such data are not available. Accordingly, an automated procedure is used to ensure that a consistent set of assignments is used. The main source of assignments is the algorithm developed by Kabsch and Sander[15], which was distributed with the Protein Data Bank at one time and which identifies the secondary structure elements by calculating the hydrogen bonding pattern of the atoms from the X-ray co-ordinates of the carbon, oxygen and nitrogen atoms. The Kabsch and Sander criteria cannot be used on 33 files which consist of only C^α atoms; however, 19 of these have secondary structure assignments made by the original depositor, and these were used instead. The remaining 14 files do not have secondary structure assignments and so a new program was developed to find the secondary structures in these files. This program takes a probe structure of four C^α atoms with ideal angles for the helix and β-strand, and then fits the probe to each four-residue long portion of the protein. If the fit produces a root mean square error of fit (RMSE) of less than a user-defined tolerance figure then the program flags the starting C^α of this segment as being in the secondary structure and moves on to look at the following four-residue segment, commencing with the next C^α in the chain. It was decided to delete symmetry related sub-units within a Protein Data Bank file to leave a single entity to represent each structure.

The residue numbers for the starting and ending residues for each of the secondary structure assignments were then used as the input data to a program, substantially adapted from one supplied by Barlow and Thornton[16], which calculates the axes of the helix and β-strand secondary structure elements in a protein. The modified program superimposes an 'ideal' model secondary structure onto the actual secondary structures defined in the input file, using a least squares procedure, and calculates an RMSE between them. If this RMSE is below a user-defined threshold level then the axes are considered to be an acceptable fit. RMSE values greater than these thresholds are usually associated with kinked or curved structures that are far better approximated by more than one straight line, the subdivision being effected by visual inspection on an Evans and Sutherland PS300 graphics system using the FRODO molecular fitting program[3].

Finally, once all of the major axes for a protein have been identified, the program calculates the angles between every pair of axes, the distances between their mid-points and the distances of closest approach: these data are then stored in a matrix characterising that protein. The April 1988 release of the Protein Data Bank that was used in these experiments gave a total of 326 such matrices and this constituted the searchable database that was used for the experiments described in the remainder of this paper. Entirely analogous matrix representations are used for the query motifs that are searched against this database[14].

IDENTIFICATION OF SECONDARY STRUCTURE MOTIFS

Use of Subgraph Isomorphism Algorithm

We have noted above that several other groups have used a linear representation of protein structure as the basis for retrieval systems of various kinds. However, none of them has taken any account of the techniques that have been devised for 2-D and 3-D substructure searching in databases of small molecules and that make use of isomorphism algorithms from graph theory to encompass the inherently combinatorial nature of structure matching problems[14]. In this section, we describe a program called POSSUM (Protein Online Substructure Searching – Ullmann Method) that uses such graph-theoretic techniques to provide a highly efficient means of identifying the occurrence of secondary structure motifs in the Protein Data Bank.

It is possible to use graph-theoretic methods for the searching of protein and motif structures since they may be regarded as labelled graphs, with the nodes of the graph corresponding to the linear representation of the helices and strands, and the edges to the inter-line angles and distances. A search of a protein structure for a motif can then be regarded as a search for a subgraph isomorphism, i.e., for the presence of a subgraph consisting of a set of nodes and edges within another graph[17]. Many such algorithms have been used for chemical substructure searching; work in Sheffield over the last few years has highlighted the general utility of the subgraph isomorphism algorithm due to Ullmann[18–20] and this has been used for the implementation of the structure matching procedure that lies at the heart of POSSUM. A brief description of the Ullmann algorithm is given in the paper by Rasmussen *et al.* in these proceedings[21] while Mitchell *et al.* describe the modifications that were required to the basic algorithm to enable it to be used for motif searching[14]. POSSUM has been implemented in Fortran 77 and runs on a MicroVAX-II. Matches to the query pattern are output by the program in a format compatible with the FRODO graphics package, thus allowing the immediate inspection on an Evans and Sutherland PS300 terminal of the highlighted secondary structures that match the query patterns.

Query Specification

The data-set matrices consist of inter-axial angles and mid-point and closest approach distances. The availability of these three types of data provides a range of searching options depending upon the requirements of a particular query. At present the working version of the algorithm allows the user to take one of three options when searching for a pattern:

1. A search based on both angular and distance criteria.
2. A search based on purely angular criteria; this is achieved by setting the distance error to a very large figure, e.g., 500 Å, although this option cannot provide a very specific search.
3. A search based on purely distance criteria; this is achieved by setting the angular error to a very large figure, e.g. 180°.

Options 2 or 3 may be used as an early screening stage to remove the large majority of the structures in the database, before the more detailed search allowed by the first option. This allows the user a certain amount of freedom in the searches that are carried out, and enables the search to be tailored to particular patterns.

Two types of distance criterion can be used in options 1 and 3. The first of these is used when the pattern is known to contain secondary structural elements that are separated by large distances, i.e., greater than 10 Å; in this case, the user can specify a distance error for the distance between the mid-points of the axes lines. Alternatively, if the distances between the elements are small, i.e., less than 10 Å, the distance error can be specified for the closest approach distance. The tolerances can be expressed as absolute quantities, e.g., a certain number of Å, or as a proportion of the the distance concerned, e.g., 10%. In practice, the closest approach distance has been found to be more effective for retrieving secondary structures that interact with each other in some sense, e.g., neighbouring pairs of β-strands in a sheet, or helices in van der Waals contact with each other. However, the mid-point distance appears more appropriate for the case where the secondary structure elements are more remote from one another in the structure. The user can also specify whether or not the order that the secondary structure elements occur along the protein chain is important in the search for a particular query motif.

Use of the Program

The utility of POSSUM for searching within the Protein Data Bank is illustrated here with reference to the three following secondary structure motifs:

1. Eight-stranded β-barrel: this is a pattern of considerable structural importance[22], the query being derived from the motif's occurrence in subunit A of triose phosphate isomerase (Protein Data Bank code 1TIM).
2. NAD-binding fold: this is a pattern which represents part of the nucleotide-binding domain found within the protein lactate dehydrogenase (Protein Data Bank code 4LDH) and has two α-helices and five β-strands. This structure is a modification of the classic Rossmann-fold[23] suggested for NAD-binding proteins by the inclusion of an α-helix from a non-contiguous part of the polypeptide chain (as discussed further below)[24].
3. Calcium binding fold: this is a pattern of two β-strands and three α-helices that represents part of a calcium-binding domain found within the muscle protein Troponin-C (Protein Data Bank code 2TNC).

Eight-stranded Parallel β-barrel

There are 14 known occurrences of this motif[22], but only six of these were present in the release of the Protein Data Bank used in this work: enolase (1ENL), 2-keto-3-deoxy-6-phosphogluconate aldolase (1KGA), pyruvate kinase (1PYK), triose phosphate isomerase (1TIM), D-xylose isomerase (1XIA) and taka-amylase (2TAA). The β-barrel from sub-unit 1 of 1TIM was used to define the motif, and all distances and angles were employed in the search, together with the specification of the order in which the secondary structure elements were to occur. A closest approach search with angular and distance tolerances of 55° and 30%, respectively, found four of the six expected occurrences of the motif and found no unexpected matches. The proteins it failed to find were 1KGA and 1XIA. The first of these contains only an incomplete set of C^{α} co-ordinates and the β-barrel is not represented in the data. 1XIA also is comprised of C^{α} co-ordinates, and contains strands that are very twisted. The error values used in these runs are not large enough to ensure their retrieval; when the distance error value is increased to 40%, the motif in 1XIA is found, with no additional hits. The search was also carried out without specifying the order of the strands in the motif; this led to a large increase in the number of hits owing to the regularity of the motif.

NAD-binding Fold

Schulz and Schirmer[25] discuss the NAD-binding fold and note that the structures of the four known NAD-dependent dehydrogenases, i.e., lactate dehydrogenase (1LDX, 3LDH, 4LDH and 5LDH), s-malate dehydrogenase (2MDH), liver alcohol dehydrogenase (4ADH, 5ADH, 6ADH and 7ADH) and D-glyceraldehyde-3-phosphate dehydrogenase (1GPD, 2GPD and 3GPD), all contain the same sheet topology. Birktoft and Banaszak[26] have noted that there are considerable differences in the orientation of the helices in the NAD-binding fold, with the exception of the helix marked B in Figure 1 that immediately follows the the first strand. This helix and the connection to the strand form important contacts to the NAD, thus explaining its spatial conservation. In addition, Baker *et al*[24]. have remarked on the spatial conservation of a second α-helix (X in Figure 1) which, whilst not normally considered part of the nucleotide-binding fold, provides residues that interact with the nicotinamide moiety. Though spatially conserved, the portion of this helix is not conserved in the sequence, in some cases occurring before and in some cases after the nucleotide-binding domain.

The query pattern chosen in Figure 1 contains five parallel β-strands labelled a to e in a sheet, and two α-helices labelled B and X. The labelling of the β-strands a to e indicates the folding pattern of the strands within the sheet, with strand a at the start of the sequence going through to strand e finally. Helix B is situated in sequence between strands a and b and helix X is from a non-contiguous part of the sequence. The two α-helices lie in a plane above the plane of the paper. As discussed in the paragraph above, the four known NAD-dependent dehydrogenases all contain the same sheet topology.

A non-sequential search for this pattern was conducted using POSSUM with tolerance parameters of 35° angular error and 50% midpoint distance error. Of the 24 hits found when this pattern was used to search the proteins in the databank, 16 occurred in known dehydrogenases. The other hits were in glutathione reductase (2GRS and 3GRS), subtilisin novo (2SBT), ATCase (4ATC, 5ATC and 7ATC) and p-hydroxybenzoate hydroxylase (1PHH). In the case of glutathione reductase and p-hydroxybenzoate hydroxylase, the hits occur in the FAD-binding domain of the protein which consists of a parallel β-sheet flanked by helices on one side, but by an anti-parallel sheet on the other. It is therefore quite different in some respects from the Rossman fold, yet still retains the additional X helix outlined in the query pattern. The motif detected in ATCase is a nucleotide binding type fold in the catalytic domain. This forms part of the 'CTP-binding'

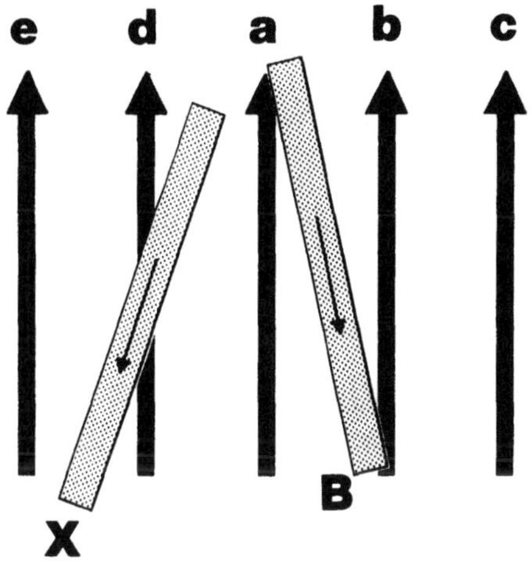

Figure 1. Diagrammatic representation of the NAD binding fold motif. The five parallel betα-strands are labelled a to e and the two alpha-helices B and X.

crevice in the enzyme, as distinct from the regulatory CTP binding site[27]. The presence of a nucleotide binding fold in subtilisin is mentioned by Cantor and Schimmel[28], although they also list it as being present in flavodoxin, adenylate kinase, rhodanese and phosphoglycerate kinase. In these latter examples, when the structure is examined on the graphics, it is found that the additional helix used in this pattern is missing, which explains why the pattern is not found.

Calcium Binding Fold

The search was run using tolerance parameters of 35° and 50% mid-point distance. Hits were found in nine proteins, with one of these being in the source protein turkey troponin (2TNC) and one in the related chicken troponin (3TNC); six of the other structures – four hits in calmodulin (1CLN), two in each of three calcium-binding parvalbumins (1CPV, 2CPV and 3CPV) and three in each of two vitamin D-dependent calcium-binding proteins (2ICB and 3ICB) - contain regions that bind calcium, similar to that found in troponin. In addition, a hit was obtained for 7ADH. When this protein is checked on the graphics, it is found that the correct secondary structural elements are present, but in a completely different orientation to that of the other hits. This example shows that the representation of the 3-D structure that is used here can provide an ambiguous definition of the relative orientation of the secondary structural elements for small motifs. For example, if it is assumed that there are n secondary structures in a protein, represented by n lines, then there are $n(n\text{-}1)/2$ distinct pairs of lines. Since each pair of lines is represented by two geometric criteria, a 3-D structure is represented by $n(n\text{-}1)$ parameters, whereas Richards and Kundrot[12] note that six parameters are needed for a complete specification of the relationship between each pair of lines. Accordingly, when n is small, the description used here is not sufficient to provide an exact structure for the query motif.

Conclusions

Despite this potential problem with small patterns, we believe that the results that have been described here justify the use of this definition of the structural elements, as it provides a valuable and compact representation for rapid searching procedures. POSSUM is now in routine use in the Department of Molecular Biology and Biotechnology in the University of Sheffield for studies of protein structure. In 1977, for example, Richardson reviewed all of the 37 β-strand patterns then known[29]. Searches were carried out in the Protein Data Bank for 34 of these motifs (the other three not being searched since they are very simple motifs that occur in nearly all strand-containing proteins). Given the extensive studies that have been carried out using the Protein Data Bank and given the relatively small number of structures in it, it is remarkable to find that no less than 15 of these searches resulted in the identification of at least one occurrence of the query motif that had not been observed by Richardson[30]. As another example, we have recently established that there is a striking resemblance in the tertiary fold of the *Salmonella typhimurium* CheY chemotaxis protein and that of of the GDP-binding domain of *E. Coli* Elongation Factor TU (EF Tu). These two protein structures are representatives of two major macromolecular classes: CheY is a signal transduction protein with sequence homologies to a wide range of bacterial proteins involved in regulation of chemotaxis, membrane synthesis and sporulation; whilst EF Tu is one of the family of guanosine nucleotide binding proteins which includes the *ras* oncogene proteins and signal transducing G proteins. The similarity revealed by POSSUM extends far beyond previously recognised resemblances of each protein's fold to that of a generic nucleotide binding domain, thus demonstrating the utility of the program for exploratory studies of protein structure[31].

As well as being highly effective in terms of the structures that are retrieved, the program is also efficient in execution, with a search of the entire Protein Data Bank requiring about 5 CPU minutes (on average) on a MicroVAX-II.

RANKING OF POSSUM OUTPUT

The Matching Algorithm

POSSUM is an example of a *partial* match retrieval system, i.e., one where the retrieved structures contain the query pattern in addition to other substructural features. *Best match*, or *nearest neighbour*, searching systems allow the identification of structures that are most similar to an input target structure. Such systems have been extensively studied in the context of databases of 2-D chemical molecules for applications such as structure-property correlation and the selection of compounds for biological screening[32]. In this section, we present a best match search algorithm that can be used to rank the output from a POSSUM search so as to allow the user's attention to be focused on those database motifs that are most similar to the query motif. It should be noted that the techniques described here could, in principle, be used to rank the output from alternative types of retrieval system that are based on 3-D searching, e.g., the topological system described by Rawlings[11].

When POSSUM was first designed, the intention was that searching should be based solely upon the inter-line angles. However, initial results demonstrated clearly that both angular and distance information need to be specified in a query if an effective search is to be carried out, and a range of searching facilities is now available in POSSUM as described above. There are, however, two characteristics of the system that can cause problems in some circumstances:

1. The precise form of the distance constraints needs to be selected by the user, and is dependent on the type of query that is to be searched for in the Protein Data Bank.
2. No explicit account is taken of the size of the secondary structure elements, i.e., the numbers of residues contained within them, as manifested by the length of the lines in 3-D space. It is thus possible to obtain hits which, while satisfying the information in the query statement, differ considerably from the actual motif represented by this statement. In many searches this is, in fact, advantageous, but there are other circumstances when such information is of importance, e.g., searching for topography around an active site.

The best match algorithm described in this section seeks to overcome these two limitations of the basic algorithm by means of a post-processing step. This results in a ranking of the output from an initial, broadly-defined POSSUM search so that the structures at the top of the ranking are those which contain motifs that are structurally most similar to the query motif.

The basic idea of the algorithm is to approximate the overall topography of a motif, either in a query or in a database structure, by the distribution of inter-line distances between the component secondary structure elements. The basic substructure searching algorithm in POSSUM utilises only a single distance to characterise the structural relationship between each pair of lines; that described here utilises large numbers of distances between each such pair. Specifically, each linear secondary structure element, L, is automatically assigned a series of NL points P_1, P_2, ... P_{NL}, located at equal distances along its major axis. Distances are then calculated between each distinct pair of points for each distinct pair of lines and the frequency distribution of these distances calculated. The degree of similarity between a query motif and a database motif resulting from the initial search is then measured by the extent of the agreement between the two distance distributions. If the overall shape of the secondary structure elements of a database motif is very similar to that of the query motif then the two distributions should be broadly comparable. However, the distributions will differ if one or more of the secondary structure elements in the database motif is not of comparable size to the corresponding secondary structure element in the query motif (these correspondences having been identified in the initial POSSUM search so that it is known, e.g., which particular β-strand in the matched protein structure matches the corresponding β-strand specified in the query motif).

Once an initial search has been carried out, and potential query motif-to-database motif matches identified, the user of the system is asked to specify two parameters:

1. L: a real value that defines the length of the segments along the lines representing the secondary structure elements.
2. R: a real value that defines the width of the range categories in the frequency distributions that are used for the subsequent ranking calculation.

The algorithm involves the following steps:

1. Each of the lines defining the longitudinal axes between the start and end points of the secondary structure elements specified in the query motif is split up into segments so as to obtain a number of points, NL, along the lines for which the atomic X,Y,Z co-ordinates can be calculated. The distance between all the points from the start point along the line up to the point before the end point are all equal in length. This length is the value L which is specified by the searcher at the start of a run; thus, the smaller the value of L, the greater the number of points there is along the lines.
2. Once all of the points have been identified, a frequency distribution is constructed containing a number of inter-point distance range categories. The R parameter specified at the start of a run defines the range of distances to be included within a category. Thus, if R is specified as 1.0 (all distances are assumed to be in Å) then the distance range categories will be 0.0→0.9, 1.0→1.9, 2.0→2.9 etc., the categories starting at 0.0 and increasing in equal-sized steps.
3. The inter-point distance is calculated between each point on a line and every other point on every other line within the query motif.
4. As each distance is calculated, it is used to increment the appropriate range category of the frequency distribution, denoting the occurrence of an inter-point distance falling between the minimum and maximum values of the category. When all of the inter-point distances for the query motif have been calculated, the resulting frequency distribution can be regarded as a summary of the overall shape of the motif.
5. The four previous steps are repeated for each of the database motif hits resulting from the initial search, so that there is a frequency distribution representing each such hit.
6. The similarity between the query motif and each of the database motifs is then estimated by the similarity between the corresponding frequency distributions. This similarity can be measured by means of the χ^2 statistic or by means of the sum of the squared differences; other similarity measures are discussed by Grindley[33].

Evaluation of the Algorithm

Once the similarities for the output from the initial search have been calculated, the structures can be displayed *via* FRODO for visual inspection and evaluation of the rankings. The effectiveness of the algorithm was tested using the three motifs described above i.e., the eight-stranded β-barrel, the NAD-binding fold and the calcium-binding fold.

The ability to specify values for the parameters L and R means that a wide range of experimental conditions is available for investigation. An increase in L results in a decrease in the number of points in each linear secondary structure element and hence in the number of inter-point distances that must be evaluated. An increase in R results in a reduction in the number of categories included in the statistical calculation. In addition, the similarity measure that is used will also affect the rankings. In each case, the following procedure was adopted:

1. The initial substructure search was executed with tolerance values deliberately set to ensure that a very large number of hits was produced to evaluate the ranking procedure. Higher angle and distance tolerances than usual were used in the search, and no account was taken of the order of the secondary structure elements in the primary sequence.
2. Rankings of the hits resulting from this search were generated using each of the different similarity measures and using a range of L and R values: $0.5 \leqslant L \leqslant 10.0$ and $0.5 \leqslant R \leqslant 5.0$. The discussions below correspond to the rankings obtained when the χ^2 statistic was used for the similarities and when both L and R were set at 0.5; the results obtained with other parameter settings are discussed by Grindley[33].
3. For comparison, rankings were also obtained from the outputs of POSSUM searches resulting from the use of angular and mid-point distance tolerances of 35° and 50%, respectively.

Eight-stranded Parallel β-barrel

The search was run with tolerances of 30° and 20 Å and gave 382 hits from eight different proteins of which 60 hits were distinct, the remainder being variations of the same set of eight β-strands. Nearly all combinations of parameters resulted in rankings in which the first three hits were 1TIM, which is the protein from which the query pattern is taken, and then 1PYK (pyruvate kinase) and 2TAA (Taka-amylase) both of which are known to have eight-stranded parallel β-barrels.

Apart from two 1AAT hits, the remainder comprise sets of β-strands in immunoglobulins. Immunoglobulins are large protein structures containing many β-strands, usually in the form of several anti-parallel β-sheets. An average distance apart for adjacent β-strands in sheets is approximately 4.5 Å, and thus the use of a midpoint tolerance value set at 20 Å, leads to the retrieval of many hits. The protein structure ranked fourth (from 1MCP) was viewed on the graphics system and the pattern located in this molecule was of eight compact parallel strands. However, the cylindrical barrel shape was not so apparent (although two semi-circular motifs could be seen on top of each other). A 3FAB hit was ranked fifth: this did not have a recognisable β-barrel structure, the parallel strands being in fact two strands apart in the immunoglobulin anti-parallel β-sheets and the whole set of eight strands being neither particularly parallel or cylindrical.

For this query motif, the ranking search is redundant since the normal POSSUM search using tolerances of 35° and 50% mid-point distances retrieves just the three β-barrels that were identified at the top of the ranked search. This is because the query pattern is very distinctive.

NAD-binding Fold

The initial search was run with angular and mid-point distance tolerances of 35° and 10 A. The latter value is small compared with the previous run; even so, no less than 181 separate hits were obtained from 34 different proteins. The ranking obtained was inspected to discover the whereabouts of the matched patterns corresponding to strands a to e and helices B and X in the hits for the dehydrogenases mentioned in the previous section on the NAD-binding fold. In some cases these hits were also viewed on the PS300 graphics system to determine the overall relative orientations and distances involved compared to the query pattern.

The first two structures at the top of the ranking are hits from 4LDH. The 4LDH hit placed first in the ranking corresponds exactly to the query pattern, containing as it does the set of helices and strands from which the query pattern is derived. The 4LDH hit ranked second is the same as the first hit, except that an extension of helix X is found instead of X itself.

The correct folds of the other lactate dehydrogenases 1LDX, 3LDH and 5LDH occur further down the ranking at positions 12, 54 and 82 respectively. Although this does not appear very encouraging, in fact in all three cases the correct fold hit is in the second position with respect to the total number of hits for each particular protein, i.e., in 1LDX the correct fold hit is in the second position out of 10 1LDX hits (and similarly for 3LDH and 5LDH with totals of seven and eleven hits, respectively). In each case the highest ranked hit for each of these three proteins is a match that contains the two correct α-helices and four correct β-strands a to d but the fifth β-strand is a strand f that invariably occurs in the sheet next to β-strand e.

The method performs better with the liver alcohol dehydrogenases; the correct fold hits for 4ADH, 5ADH, 6ADH and 7ADH are at positions 3, 6, 8 and 33, respectively. In the case of 5ADH, 6ADH and 7ADH these correct fold hits are in second position with respect to the total numbers of hits in these proteins (17 in 5ADH, 25 in 6ADH and 16 in 7ADH). The hits at the fourth and fifth positions in the ranking are from 5ADH and 6ADH, respectively, and involve the correct β-sheet strands but include two α-helices on the opposite side of the sheet.

The hits for the correct fold in s-malate dehydrogenase (2MDH) sub-units A and B occur at ranking positions 124 and 142 respectively and are in third and fifth position with respect to the total of 10 hits in 2MDH. This is a disappointing result. The D-glyceraldehyde-3-phosphate dehydrogenase hits also provide similar results. In 1GPD, correct fold hits occur at ranking positions 91 and 113 (there are two correct fold hits for similar reasons as those described for 4LDH). These two hits are in first and second positions with respect to the total of four hits in this protein. 2GPD only has one hit, the correct fold, at position 131 and 3GPD has two hits, the correct fold being listed second at position 150.

It will thus be realised that the hits corresponding exactly to the query motif folding pattern occur throughout the whole ranking and are not clustered together near the top of the ranking. However, when each individual dehydrogenase protein is considered, the correct fold hits do, in general, appear in the top or second position of the total number of hits for that particular protein. There are several factors to be considered here. The frequency distributions corresponding to each of the database motifs are not noticeably dissimilar from each other and hence this motif represents a much more stringent test of the effectiveness of the ranking procedure than the previous two patterns. Secondly, the query motif is very similar to a Rossmann-fold, a ubiquitous motif within protein structures for which many hits will be found in non-dehydrogenase type proteins as well. Out of the 181 hits from the initial search, 123 were in dehydrogenase proteins and the remaining 58 were in non-dehydrogenase type proteins. In the top two thirds of the ranking, i.e., in the first 120 hits, 97 were from dehydrogenase proteins. Many of these hits contained a certain proportion of the strands and helices in the correct folding pattern but also contained some strands and/or helices in close proximity to the correct folding pattern set. The poor performance with 2MDH, where the correct fold appeared very low down the ranking, is probably due to the fact that although the strands and helices are in the correct fold, the lengths of the strands and helices differ greatly to those of the query pattern. So, although the overall shape of the frequency distribution will be similar to that of the query pattern, the peaks of the frequency distribution will be higher as more inter-line distances will be involved and thus the statistical ranking calculation will assign a low ranking.

Consideration of the non-dehydrogenase type hits reveals that the highest ranked hit is one from subtilisin (1SBT) at position 9. This does not contain quite the same folding pattern as the query motif. In 1SBT, strands b and c are swapped around and the equivalent helix B occurs in sequence between strands b and c. However, the strands and helices are in relatively good orientations to each other, although one helix is quite short. The corresponding hit in 2SBT occurs at position 55. ATCase, which is known to contain a nucleotide-binding domain[14], has hits in 4ATC and 5ATC at positions 123 and 93, respectively, that contain the correct fold of five strands and two helices. Most of the other proteins which are known not to contain a NAD-binding fold occur towards the bottom of the ranking, e.g., 1AAT, 1PYK, 1PGI, etc. On inspection, these hits have sheets that are a lot more twisted and curved than the query pattern sheet, their helices and strands are of very differing lengths, and the helices are not in favourable orientations with respect to each other or to the sheet.

When a standard, non-ranked POSSUM search is carried out with angular and mid-point tolerances of 35° and 50%, a total of 23 hits is achieved, these including all examples of the four types of dehydrogenase protein. Of the 23 hits, 21 are dehydrogenases and 13 of these are in the correct fold.

Calcium-binding Fold

The primary search selection procedure was run with tolerances set at 30° angular tolerance and 500 A mid-point distance tolerance (this latter figure meaning that the search is effectively based solely upon angular constraints). In all, 73 hits were obtained from 13 different proteins (42 of them coming from 2MDH).

The top ranking hit is one of those from 2TNC; this is hardly surprising since it comprises the set of strands/helices from which the query pattern was derived. The hits ranked second, third and fourth are from 1CPV and 2CPV, which both represent the protein calcium-binding parvalbumin B. The second and fourth ranked hits (1CPV and 2CPV) in fact contain the same set of strands/helices and are very similar to the query pattern. Calcium-binding parvalbumin B is known to have a calcium-binding site very similar to that of Troponin-C (2TNC)[34]. The 1CPV hit ranked third contains the same set of strands/helices as the 1CPV hit ranked above it, apart from one of the helices which is different. This does not make it such a good match although this different helix is in close proximity. The next two hits ranked fifth and sixth are both from calmodulin, 1CLN. Calmodulin in fact has two very similar domains linked by a long α-helix. The 1CLN hit ranked fifth has its set of strands/helices all taken from one domain, whilst that ranked sixth has its set of strands/helices all taken from the other domain. Both hits are very similar in structure to the query pattern and represent the two known calcium-binding sites in calmodulin. The two hits ranked seventh and eighth again come from 1CLN. Both of these hits contain four of the same strands/

helices in the two domains described above but in both cases one of the helices is different. In fact this different helix in the hit ranked eighth is the long linking helix which, although too long compared to that of the query pattern, is in a reasonably correct orientation. The hit ranked ninth is 2MDH. Malate dehydrogenase is not known to possess a calcium-binding site and the set of strands/ helices here do not resemble the compact query motif since they are all a fair distance apart on opposite sides of the molecule. The 2GRS hit is ranked tenth and the pattern found was not at all convincing in terms of similarity to the query pattern. The 1LDX hit was ranked next at position eleven - it had a reasonable resemblance apart from the fact that two of the helices appeared to point in the wrong direction with respect to the directions of the two crossed β-strands. Further 1CLN rankings in the list appear to take a selection of strands/helices from both domains and as such are not good matches at all, while other hits have strands/ helices that are too far apart to resemble the query pattern properly.

In summary, the ranking was successful in that the known calcium-binding folds of troponin-C, calcium-binding parvalbumin B and calmodulin were contained within the top six ranking positions. An inspection of the distance distributions for these six structures reveals a substantial degree of similarity with the distribution for the query motif, and similar comments apply to the 1CLN hits ranked seventh and eighth. After this, however, the distributions become markedly dissimilar.

When the normal POSSUM search was carried using the standard parameter settings without any subsequent ranking, there were just six hits from three different proteins (2TNC, 1CPV and 1CLN). This list does *not* contain the 2CPV and 1CLN hits which occurred at the fourth and sixth positions in the ranking. These are both very good structural matches to the query motif and thus the ranking procedure has enabled the identification of database motifs that are eliminated from the conventional, single stage search.

The operation of the ranking procedure is illustrated diagrammatically by the five frequency distributions shown in Figure 2. These have been calculated with both L and R set to 0.5 Å and show the dotted line that represents the distribution for the query pattern (referred to here as TNC2) and the continuous line that represents (a) the 1CPV hit that was ranked second, (b) the 1CLN hit that was ranked sixth, (c) the 1LDX hit that was ranked eleventh, (d) the 1CTS hit that was ranked thirty fifth and (e) the 2MDH hit that was ranked sixty ninth. It will be seen that the overlap between the two distributions drops off very quickly as the ranking is descended, this corresponding to the increasing structural differences between the query motif and the database motifs.

Conclusions

The three motifs tested here give rather different results. In the case of the β-barrel, the distinctness of the motif means that the standard POSSUM searching facilities are quite sufficient to give a complete and precise search. The advantages of the ranking procedure are clearly seen with the calcium binding search, since the algorithm successfully identifies occurrences of the query pattern that are not obtained in the normal, unranked POSSUM search. However, the NAD-binding search is much less successful, with the correct hits identified in the standard POSSUM search being distributed throughout the ranking.

Ranking provides an obvious means of dealing with the large amounts of output that can result from a conventional POSSUM search. However, the precise algorithm studied here would seem to need considerable developmental work before it could be used as a standard retrieval option.

IDENTIFICATION OF STRUCTURAL HOMOLOGY

We have recently begun to use our graph-based structure representations for the identification of *homology*, i.e., areas of structural commonality. This lies at the heart of much research into the primary structure of proteins, i.e., the main chain sequences, but we are now developing algorithmic procedures for the detection of homology at the secondary structure level using a maximal common subgraph algorithm. Given a pair of graphs, the maximal common subgraph is the largest set of nodes and/or edges that is common to the pair of graphs that is being compared. Such algorithms have been used on sets of small 3-D molecules that share some common activity characteristics to identify the structural equivalences that exist between them (since these common features may correspond to the pharmacophoric pattern that is responsible for the observed activity). A comparison of techniques for the identification of maximal common subgraphs in small 3-D molecules has been reported by Brint and Willett who show that the most efficient results are obtained using an algorithm that is based on *clique detection* in chemical graphs, where a clique is a subgraph in which every node is connected to every other node and which is not contained in any larger subgraph with this property[35,36].

Given two 3-D protein structures that are characterised by the linear secondary structure elements described previously, the program uses the clique detection procedure to identify all patterns of α-helices and β-strands in 3-D space that are common to the two structures and that are greater than some user-defined minimal size. This restriction is imposed to ensure that the user is not overwhelmed with a very large output consisting primarily of small common substructures that are of little structural significance, since clique detection algorithms invariably identify huge numbers of small cliques (as is exemplified, by the studies of Brint and Willett using small 3-D molecules[35]).

The program has been tested using sets of structurally related proteins and has been shown to be capable of identifying correctly the areas of structural commonality. For example, the protein myoglobin (1MBD), which contains nine α-helices, was used as a query protein structure to search our coded version of the Protein Data Bank. A minimum clique size of seven secondary structure elements was specified, i.e., the maximal common subgraph should contain at least seven secondary structure elements in common with the target protein. The angular and distance tolerances were set at 50° and 40% (closest approach distance), respectively, together with the requirement that matching cliques should have the same sequence ordering of secondary structures as in the query protein, 1MBD.

In the first run, 39 cliques were found. Of the eight myoglobin entries in the database (1MB5, 1MBC, 1MBD, 1MBN, 1MBO, 1MBS, 2MBN and 3MBN), one clique of size nine, i.e., a match of the whole query protein, was found with 1MBC, 1MBD, 1MBO, 1MBS, 2MBN and 3MBN. In the other two myoglobin structures, 1MB5 and 1MBN, two cliques each of size eight secondary structure elements were found. Matches were obtained not only with the myoglobin structures in the Protein Data Bank but also with some of the 16 haemoglobin structures. No

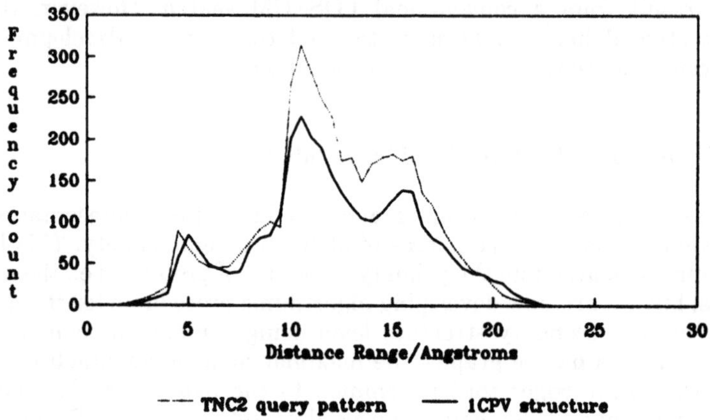

Frequency Count
350
300
250
200
150
100
50
0
0
5
10
15
20
25
30
Distance Range/Angstroms
TNC2 query pattern
1CPV structure

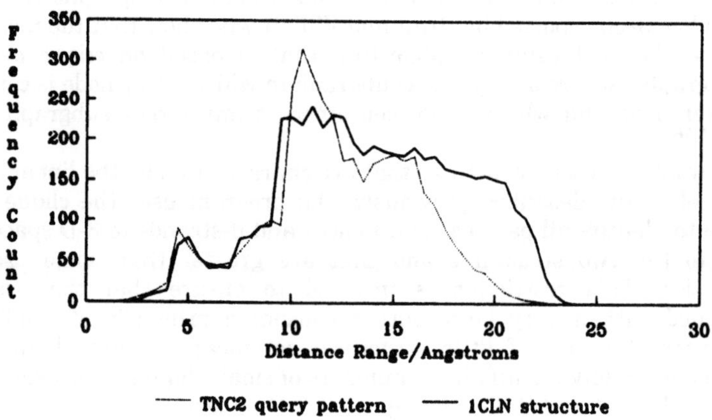

Frequency Count
350
300
250
200
150
100
50
0
0
5
10
15
20
25
30
Distance Range/Angstroms
TNC2 query pattern
1CLN structure

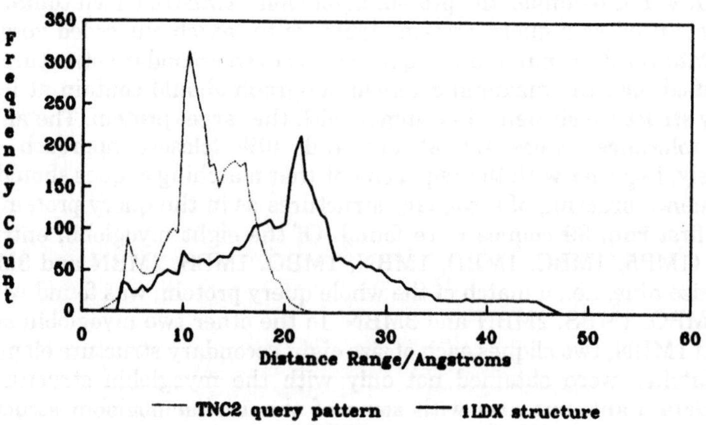

Frequency Count
350
300
250
200
150
100
50
0
0
10
20
30
40
50
60
Distance Range/Angstroms
TNC2 query pattern
1LDX structure

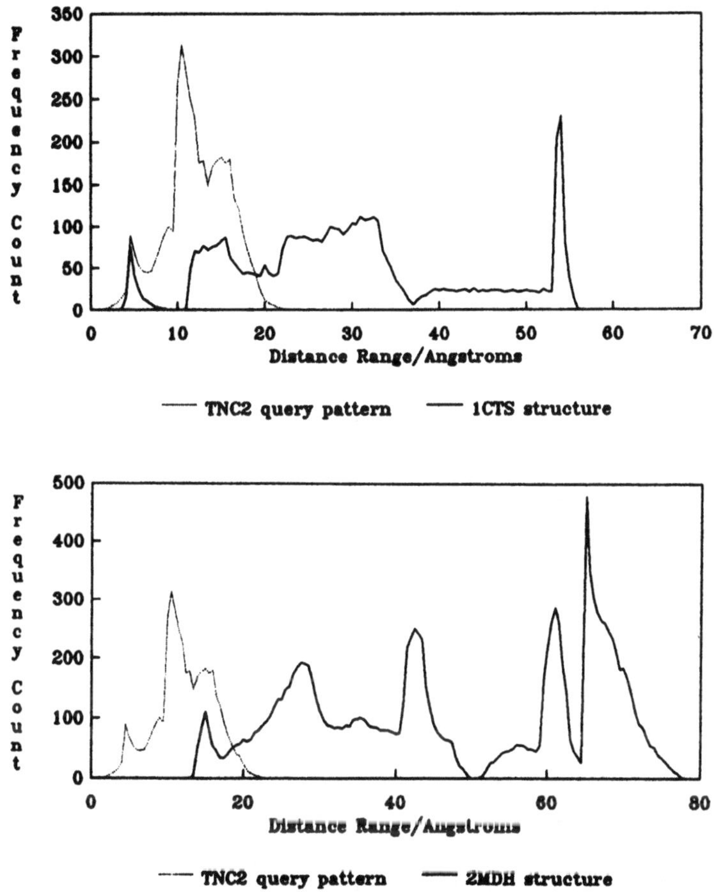

Figure 2. Frequency distributions for ranking of motifs in order of decreasing similarity with part of a calcium-binding domain found within the muscle protein Troponin-C (called here TNC2). The distributions represent the hits that came (a) second (b) sixth (c) eleventh (d) thirty-fifth and (e) sixty-ninth. In each case, the distribution for the query motif is shown by a dashed line and that for the database protein motif by a continuous line.

cliques of the requisite size were found in the haemoglobin structures 1HBS and 1HDS but in each of the other 14 structures, either one, two or three cliques of size seven secondary structure elements were identified. In the cases of 2HHB, 2MHB, 3HHB and 4HHB, cliques were identified in both the A and B domains. No hits were found with any non-globin proteins. A second run was carried out with angular and closest approach distance tolerances of 10° and 10%, respectively. In this case, the same number and size of cliques were matched in the myoglobins as described previously. However, no matches with any haemoglobin structures were found.

These results indicate that the clique detection program is effective in finding matches with other proteins of similar three-dimensional structure, e.g., structures belonging to the same family of proteins. The myoglobin query protein is a relatively small structure and only contains α-helices but larger proteins including β-strands can be used successfully as query structures. Generally, with larger query proteins, very large numbers of cliques can be identified, and we are currently implementing techniques that will allow efficient processing of the large amount of data that is produced from such a run.

CONCLUSIONS

When the protein data bank was first set up in the late seventies, it was intended to act as an archive for the deposition of sets of macromolecular co-ordinates. The burgeoning demands of research in molecular biology and biotechnology has led to increasing use being made of the information that is stored in databases of 3-D chemical substances; in particular, much attention has been devoted to the development of sophisticated molecular graphics systems. In this paper, we have demonstrated that this data can also be used for a range of searching operations, using relatively small modifications to well established techniques that were first developed for the storage and retrieval of small chemical structures. To date, we have considered only the linear α-helix and β-strand secondary structure elements but will shortly extend our methods of representation to the various loop and turn structures.

Our current work focuses on the homology procedure described in the previous section. The work with the globin structures was undertaken primarily to evaluate the program's ability to identify structural homology in an efficient and effective manner. While it is clearly of interest to be able to identify such regions automatically, rather than *via* the manual inspection of a graphics terminal, the program has been developed for the automatic identification of homologies that had not previously been detected. We are thus now taking individual protein structures in turn and then using each of them for a scan of the entire Protein Data Bank to identify the substructures that the query structure has in common with each of the other members of the database; in essence, this provides an effective mechanism for the measurement of the degree of resemblance between the query and all of the other proteins. In addition, if some particular common substructure is found to occur with several different structures from the database, this might suggest the presence of a novel type of previously unidentified secondary structure motif.

ACKNOWLEDGMENTS

We thank the British Library Research and Development Department, the Department of Education and Science, Pfizer Central Research and the Science and Engineering Research Council for funding this work. DWR is a Lister Institute Research Fellow. PJA is a Royal Society 1983 University Research Fellow and thanks the Medical Research Council for support.

REFERENCES

1. Brint, A.T.; Mitchell, E.M.; Willett, P. 'Substructure Searching in Files of Three dimensional Chemical Structures'. In *Chemical Structures: The International Languag of Chemistry*; Warr, W.A., Ed.; Springer-Verlag: Heidelberg, 1988.
2. Diamond, R. 'Applications of Computer Graphics in Molecular Biology'. *Compute Graphics Forum* **1984,** 3, 3–11.
3. Jones, T.A. 'Interactive Computer Graphics: FRODO'. *Methods Enzymol.* **1985,** *115* 157–171.
4. Abola, E.E.; Bernstein, F.C.; Bryant, S.H.; Koetzle, T.F.; Weng, J. 'Protein Data Bank' In *Crystallographic Databases: Information Content, Software Systems, Scientifi Applications*; Allen, F.H.; Bergeroff, G.; Sievers, R., Eds; Data Commission of th International Union of Crystallography: Cambridge, 1987.
5. Bernstein, F.C.; Koetzle, T.F.; Williams, G.J.B.; Meyer, E.F.; Brice, M.D.; Rodgers, J.R. Kennard, O.; Shimanouchi, M.; Tasumi, M. 'The Protein Data Bank: a Computer-base Archival File for Macromolecular Structures'. *J.Mol. Biol.* **1977,** *112,* 535–542.
6. *Communication, Storage and Retrieval of Chemical Information*; Ash, J.E.; Chubb, P.A. Ward, S.E.; Welford, S.M.; Willett, P., Eds; Ellis Horwood: Chichester, 1985.
7. Barnard, J.M. 'Recent Developments in Chemical Structure Handling'. *Perspectives i Information Management* **1989,** *1,* 133–168.
8. Lipscombe, K.J.; Lynch, M.F.; Willett, P. 'Chemical Structure Processing'. *Annu. Rev Inf. Sci. Technol.* **1989,** *24,* 189–238.
9. Abagyan, R.A.; Maiorov, V.M. 'A Simple Qualitative Representation of Polypeptid Chain Folds: Comparison of Protein Tertiary Structures'. *J. Biomol. Struct. Dyn.* **1988** *5,* 1267–1279.
10. Rawlings, C.J.; Taylor, W.R.; Nyakairu, J.; Fox, J.; Sternberg, M.J.E. 'Reasoning Abou Protein Topology Using the Logic Programming Language PROLOG'. *J. Mol. Graphic* **1985,** *3,* 151–157.
11. Rawlings, C.J. 'Artificial Intelligence and Protein Structure'. In *Biotechnology Infor mation '86*; Wakeford, R., Ed.; IRL Press: Oxford, 1987.
12. Richards, F.M.; Kundrot, C.E. 'Identification of Structural Motifs From Protein Co ordinate data: Secondary Structure and First-level Supersecondary Structure'. *Proteins Struct. Funct. Genet.* **1988,** *3,* 71–84.
13. Brint, A.T.; Davies, H.M.; Mitchell, E.M.; Willett, P.; 'Rapid Geometric Searching i Protein Structures'. *J. Mol. Graphics* **1989,** *7,* 48–53.
14. Mitchell, E.M.; Artymiuk, P.J.; Rice, D.W.; Willett, P. 'Use of Techniques Derived fron Graph Theory to Compare Secondary Structure Motifs in Proteins'. *J. Mol. Biol* **1990** *212,* 151–166.
15. Kabsch, W.; Sander, C. 'Dictionary of Protein Secondary Structure: Pattern Recognitio of Hydrogen-bonded and Geometrical Features'. *Biopolymers* **1983,** *22,* 2577–2637.
16. Barlow, D.J.; Thornton, J.M. 'Helix Geometry in Proteins'. *J. Mol. Biol.* **1988,** *201,* 601-619.
17. Tarjan, R.E. 'Graph Algorithms in Chemical Computation'. In *Algorithms for Chemica Computations*; Christoffersen, R.E., Ed.; ACS Symposium Series 46; American Chemica Society: Washington, 1977, pp.1–20.
18. Brint, A.T.; Willett, P. 'Pharmacophoric Pattern Matching in Files of 3-D Chemica Structures: Comparison of Geometric Searching Algorithms'. *J. Mol. Graphics* **1987,** *5* 49–56.
19. Downs, G.M.; Lynch, M.F.; Manson, G.A.; Willett, P.; Wilson, G.A. 'Transpute Implementations of Chemical Substructure Searching Algorithms'. *Tetrahedron Com put. Methodol.* **1988,** *1,* 207–217.
20. Ullmann, J.R. 'An Algorithm for Subgraph Isomorphism'. *J. Assoc. Comput. Mach* **1976,** *16,* 31–42.
21. Rasmussen, E.M.; Willett, P.; Wilson, T.; 'Chemical Structure Handling Using th Distributed Array Processor'. In these Proceedings.
22. Chothia, C. 'The 14th Barrel Rolls out'. *Nature* **1988,** *333,* 598–599.
23. Rossmann, M.G.; Moras, D.; Olsen, K.W. 'Chemical and Biological Evolution of a Nucleotide Binding Protein'. *Nature* **1974,** *250,* 194–199.

24. Baker, P.J.; Farrants, G.W.; Rice, D.W.; Stillman, T.J. 'Recent Progress on the Structure and Function of Glutamate Dehydrogenase'. *Biochem. Soc. Trans.* **1987,** *15,* 748–751.
25. *Principles of Protein Structure*; Schulz, G.E.; Schirmer, R.H.; Springer-Verlag: Heidelberg, 1979.
26. Birktoft, J.J.; Banaszak, L.J. 'Structure-function Relationships Among Nicotinamide-adenine Dinucleotide Dependent Oxidoreductases'. In *Peptide and Protein Reviews*; Hearn, M.T.W., Ed., Dekker: New York, 1984.
27. Honzatko, R.B.; Crawford, J.L.; Monaco, H.L.; Ladner, J.E.; Edwards, B.F.P.; Evans, D.R.; Warren, S.G.; Wiley, D.C.; Ladner, R.; Lipscomb, W.N. 'Crystal and Molecular Structures of Native and CTP-liganded Aspartase Carbamoyltransferase from *E.coli*'. *J. Mol. Biol.* **1982,** *160,* 219–263.
28. *Biophysical Chemistry. Part I: The Conformation of Biological Macromolecules*; Cantor, C.R.; Schimmel, P.R.; W.H. Freeman: San Francisco, 1980.
29. Richardson, J.S. 'β-sheet Topology and the Relatedness of Proteins'. *Nature* **1977,** *268,* 495–500.
30. Artymiuk, P.J.; Mitchell, E.M.; Rice, D.W.; Willett, P. 'Searching Techniques for Databases of Protein Secondary Structures'. *J. Inf. Sci.* **1989,** *15,* 287–298.
31. Artymiuk, P.J.; Rice, D.W.; Mitchell, E.M.; Willett, P. 'Structural Resemblance Between the Families of Bacterial Signal Transduction Proteins and of G proteins Revealed by Graph Theoretical Techniques'. *Protein Eng.* **1990,** *4,* 39–43.
32. *Similarity and Clustering in Chemical Information Systems*; Willett, P.; Research Studies Press: Letchworth, 1987.
33. Grindley, H.M. '3-D Distance Comparisons in Protein Secondary Structure Motifs For Substructure Search Refinement'. M.Sc. dissertation, University of Sheffield, 1988.
34. Kretsinger, R.H. 'Calcium-binding Proteins'. *Annu. Rev. Biochem.* **1976,** *45,* 239–266.
35. Brint, A.T.; Willett, P. 'Algorithms for the Identification of Three-dimensional Maximal Common Substructures'. *J. Chem. Inf. Comput. Sci.* **1987,** *27,* 152–158.
36. Brint, A.J.; Willett, P. 'Upperbound Procedures for the Identification of Similar Three-dimensional Chemical Structures'. *J. Comput.-Aided Mol. Des.* **1988,** *2,* 311–320.

USE OF ANGULAR INFORMATION IN THREE-DIMENSIONAL SUBSTRUCTURE SEARCHING SYSTEMS

Andrew R. Poirrette and Peter Willett
Department of Information Studies, University of Sheffield, Western Bank Sheffield S10 2TN, England.

ABSTRACT

Most published reports of 3-D substructure searching systems have been based on the use of interatomic distance information. This poster discusses work to date on a continuing project to investigate the utility of angular information for such systems. Specifically, a screen set selection algorithm is described that provides an extremely rapid way of identifying angular ranges that occur approximately equifrequently in a 3-D database. These ranges can then be used as screens to allow efficient 3-D substructure searches to be carried out. The algorithm has been developed for the selection of angle-based screens but is applicable to any situation requiring the selection of a set of approximately equifrequently occurring descriptors.

SCREEN SETS

Substructure searching involves the identification of all molecules in a chemical database that contain a user-defined partial structure[1]. It is usually implemented by means of a two-stage retrieval algorithm in which an initial *screening* search is used to eliminate from further consideration large numbers of molecules that cannot possibly contain the query substructure; only those molecules that match the query substructure at the screen level then undergo the second-stage *atom-by-atom* search, in which a detailed and time-consuming subgraph isomorphism algorithm is used to determine whether the query substructure is present. The last few years have seen an explosion of interest in the design and implementation of substructure searching systems for use with databases of 3-D chemical structures: this work is reviewed by Martin *et al*[2]. These systems have typically used interatomic distance information for both the screening and the subgraph isomorphism stages.

The atom-by-atom stage of a substructure search, whether in 2-D or in 3-D, is far more demanding of computational resources than is the screening search. The overall efficiency of a substructure searching system is thus crucially dependent on the *screenout*, i.e., the fraction of the database that is eliminated from the atom-by-atom search by the screening search. There has accordingly been considerable interest in the development of algorithmic techniques for the selection of fragment screens that will give high screenout. This work has demonstrated that screenout is maximised if the fragments which are chosen for inclusion in the screen set occur approximately equifrequently in the database that is to be searched. In this poster, we summarise the main features of an algorithm that can be used for the

W. A. Warr (Ed.)
Chemical Structures 2

selection of fragment screens for substructure searching. The algorithm has been developed for angle-based searching in the Cambridge Structural Database but can be used for the selection of any type of screens, given an input file of fragment occurrence data that has been sorted into alphanumeric order. Full details of the work are provided by Cringean *et al*[3].

THE ALGORITHM

The screen set selection algorithm described here derives from an algorithm developed by Cooper *et al.* for the creation of sorted dictionaries of words using a form of distribution sorting[4].

In essence, the algorithm involves the following three steps:

1. Assume that some class of fragment descriptor has been selected as the basis for the screening system that is to be implemented. For example, in a 2-D substructure searching system, this class might be the augmented atom, which consists of an atom together with the atoms and bonds that are immediately adjacent; in a 3-D substructure searching system, it might be pairs of atoms and the associated interatomic distances.
2. Generate all occurrences of this fragment class in a sample of chemical structures. Sort the fragment occurrences into increasing alphanumeric order and then cumulate them so that each particular type of the chosen fragment class is stored together with its frequency of occurrence. Previous studies have shown that even quite small sample files suffice to give reasonably accurate fragment occurrence data[5]. In what follows, we shall refer to this sorted file of fragment occurrences as a *fragment dictionary*.
3. Assume that a screen set is to be created that contains N screens. Partition the fragment dictionary into N partitions, each of which contains approximately the same number of fragment occurrences. The range of values encompassed within each such partition then corresponds to one of the screens that are available for assignment to database structures or to query substructures.

The algorithm described here provides an effective way of implementing the partitioning step, i.e., Step 3 above. The algorithm first roughly divides the dictionary into a number of *sections* and then more accurately divides these sections into the required number of *partitions*:

1. The ideal partition frequency, P, is calculated. This is done by summing the frequencies of occurrence for all of the fragments in the fragment dictionary and dividing this sum by the number of partitions required, N.
2. The dictionary is divided into sections. Each fragment with a frequency at least P becomes a section on its own while fragments with frequencies between $0.5P$ and P begin a new section.
3. Two adjacent sections are fused if their combined total frequency is less than $1.3P$.
4. A number of partitions is allocated to each section, depending on the total occurrence frequency of the fragments in the section. If a section contains just a single fragment then only one partition is allocated. The number of partitions allocated to other sections is calculated by an iterative procedure. The maximum level for the partition size in all partitions is first set to P, the ideal

partition size. The total number of partitions that would be created is then calculated. The maximum level is then gradually refined until the number of partitions is just less than or equal to N.

5. The mean partition size for each section is then calculated and the division points, i.e., the boundaries between one partition and the next, are placed as close as possible to the required positions in the complete fragment dictionary.
6. If the number of partitions created is less than N, then the partition containing more than a single fragment that has the largest frequency is split into two. This is repeated until precisely N partitions have been obtained.

The range of fragment values in each partition resulting from the use of the procedures above then defines one of the screens in the screen set. A database structure or query substructure is encoded by generating each fragment in turn, searching the dictionary to identify the partition in which it occurs, and then setting the corresponding location in the bit string that is used for the screening stage of the substructure search.

RESULTS

The algorithms have been used to produce sets of *valence angle*, or *bond angle*, screens for angle-based substructure searches of the crystallographic data in the Cambridge Structural Database. Specifically, a file of 5,000 structures was selected that contained only atoms from the following list of atomic types: B, Br, C, Cl, I, N, O, P and S. These were then conflated to the three classes: carbon; nitrogen or oxygen; everything else, with these three classes being denoted by C, X and Y respectively. Each of the 3-D structures in turn was then used to generate all possible fragments of the form $A_1A_2A_3V$, where A_1, A_2 and A_3 denote the atomic classes for the three atoms that define a specific bond angle in the chosen molecule and where V is the value of that angle in degrees (rounded to the nearest degree). A_2 is defined to be the vertex of the bond angle and the fragment is canonicalised by ensuring that $A_1 \leq A_3$. In all 81,622 fragments were generated from the 5,000 structures: after sorting and cumulation, these fragments gave rise to a total of 1,096 distinct fragment types in the fragment dictionary that provided the input to the screen set selection procedure. Pascal implementations of the algorithm on IBM 3083 and Prime 9950 equipment required less than one second of CPU time to produce screen sets containing 128, 256 and 512 screens. These are currently being used to investigate the efficiency of angle-based searches of the 3-D structures in the Cambridge Structural Database.

Cringean *et al.* note that the selection methodology is applicable to any application domain that requires the selection of a set of approximately equifrequently occurring attributes from an input dictionary that contains sample occurrence data[3]; they report further experiments in which the algorithm described above, and related algorithms, were also used with sets of interatomic distance, octuplet and word stem occurrence data. In all cases, the screen sets that were obtained exhibited a high degree of equifrequency.

The screen set procedures studied here are part of an ongoing project to investigate the use of angular information for 3-D substructure searching; work in the future will seek to extend the present bond angle screens to include torsion angle data and to compare the retrieval capabilities of angular and distance screens.

ACKNOWLEDGMENTS

We thank Janey K. Cringean for developing the original version of the screen set selection algorithm described here, the British Library Research and Development Department for funding and the Cambridge Crystallographic Data Centre for its active collaboration in this work.

REFERENCES

1. *Communication, Storage and Retrieval of Chemical Information*; Ash, J.E.; Chubb, P.; Ward, S.E.; Welford, S.M.; Willett, P., Eds; Ellis Horwood: Chichester, 1985.
2. Martin, Y.C.; Bures, M.G.; Willett, P. 'Searching Databases of Three-dimensional Structures'. In *Reviews in Computational Chemistry*; Lipkowitz, K. B.; Boyd, D. B., Eds; VCH: New York, **1990**, 213–263.
3. Cringean, J.K.; Pepperrell, C.A.; Poirrette, A.R.; Willett, P., 'Selection of Screens for Three-dimensional Substructure Searching'. *Tetrahedron Comput. Methodol.* **1990**, *3*, 37–46.
4. Cooper, D.; Dicker, M.E.; Lynch, M.F. 'Sorting of Textual Data Bases: a Variety Generation Approach to Distribution Sorting.' *Inf. Process. Manage.* **1980**, *16*, 49–56.
5. Gannon, M.T.; Willett, P. 'Sampling Considerations in the Selection of Fragment Screens for Chemical Substructure Search Systems'. *J. Chem. Inf. Comput. Sci.* **1979**, *19*, 251–253.

CONFORMATIONAL FREEDOM IN 3-D DATABASES

Nicholas W. Murrall and E. Keith Davies
Chemical Design Ltd, Unit 12, 7 West Way, Oxford OX2 0JB, England.

ABSTRACT

With the advent of databases capable of storing full 3-D information on chemic structures, it is now feasible to search such databases for particular arrangemen of atoms or functional groups. This is of particular importance in drug desig where known 3-D dispositions of atoms or groups (pharmacophores) are believed be responsible for the physiological action of the drug. When designing a 3- database system, it is necessary to include information on the conformation flexibility of a molecule, since a given molecule can exist in a variety of differe conformations. This paper presents a novel way of storing this information, i which disk storage requirements and search times are independent of the numbe of conformations stored. A system based on this approach thus enables a tru search to be performed, finding all possible matches to a given pharmacophore, n just those corresponding to the particular conformation stored. This pape describes the architecture of such a system and the processes that are carried ou during a search, with special emphasis on the 3-D keys used to store and retriev data.

INTRODUCTION

Recent years have seen a rapid growth in the use of 3-D database systems i chemical research. In particular, systems have been developed at Abbo Laboratories[1,2], Lederle Laboratories[3] and Sheffield University[4] as well b software manufacturers[5]. Typically these systems are used to identify molecul that match a particular pharmacophore that represents the key features necessar for an active drug of a particular class.

CHEMISTS' REQUIREMENTS

A typical drug molecule might be made up of several key features: a ring system, flexible side chain, H-bond donor or acceptor atoms, etc. Any system that is to b used to search for such molecules must, therefore, have the following capabilitie

1. Ability to handle rigorously conformational freedom since a large proportion the compounds of interest will be flexible.
2. Ability to identify key functional groups on which activity depends.
3. Ability to generate and store in a usable manner all compounds that match th query.
4. Ability to integrate both the query construction and the processing of th

W. A. Warr (Ed.)
Chemical Structures 2

results with modelling since this will be needed to analyse further the compounds suggested by the system.

DATABASE SYSTEM REQUIREMENTS

To implement these requirements the following features are prerequisite in the database system:

1. Three-dimensional screens relevant to the pharmacophore that enable a large proportion of the database to be eliminated very rapidly must be used.
2. Conformational data must be stored in a manner that allows rigorous searches of all available conformations whilst not impinging on either disk storage requirements or search times.
3. There must be a flexible substructure search facility to confirm the presence of functional groups.
4. Final checking that the candidate molecule matches the query must be performed by rapid pattern-matching techniques.
5. The results must be stored and presented in a manner suitable for further modelling.

All of these features, as incorporated in ChemDBS-3D, will be discussed in turn.

THREE-DIMENSIONAL KEYS

All systems offering rapid searching utilise 3-D keys, usually based on the distance between atoms. In ChemDBS-3D we base our keys on the generic properties of atoms that are important to pharmacophoric action; these are termed centres. We currently use four types of centre by default:

1. Heteroatoms which can act as H-bond acceptors.
2. Heteroatoms which can act as H-bond donors.
3. Heteroatoms with a formal charge.
4. Ring centroids.

Thus a histamine molecule would contain 2 H-donors, 1 H-acceptor, 1 ring centroid and 1 charge centre. Atoms can behave as more than one centre type.

The 3-D keys used indicate the presence or not of a particular interaction within a distance range and are similar to those of other systems[3,6]. Thus the N-H to N-H distance of 5.8Å found in histamine would be represented by setting of the donor-donor screen for the range 5.5-6.0Å. For implementational expedience each screen type is stored as a 32 bit bit-mask using one VAX word thus giving a total of 32 screens per interaction type.

STORAGE OF CONFORMATIONAL DATA

To store the conformational data a rule-based[7] conformational analysis is performed and the keys calculated for each low energy conformation are thus generated. When completed all screens of a particular type are combined to yield one screen set for that interaction type. Thus if one conformation had a H-donor to

H-donor separation of 5.8Å and another 5.3Å both the 5.0-5.5Å and the 5.5-6.0Å screens would be set for that compound:

0000 0000 0000 0110 0000 0000 0000 0000

SIDE EFFECTS

One side effect of using this method is the ability to identify key functional groups on the basis of the screens that are set, e.g., 1–3 substitution in a five-membered ring is typically 2.202Å whilst that in a six-membered ring is 2.390Å. To differentiate between these a small distance range (0.1Å) is used for screens between 1.7Å and 3.0Å.

QUERY SPECIFICATION

In order to specify a query all the chemist has to do is enter the essential features of the pharmacophore into Chem-X. This could be achieved by reading from a file, standard building technques or even removing non-essential features from a known active compound. Centres are automatically identified and distances, with tolerances to allow for non-precise matches are assigned. These tolerances can be changed by the user. The substructure entered in this way is also that subsequently searched for. The histamine H_1 example, therefore, requires a tautomeric nitrogen system (distance 2.2Å) with a nitrogen H-donor 5.1Å and 5.8Å from the two nitrogens.

SEARCH TECHNIQUES

A simple formula screen based on the numbers of each type of centre is scanned first. The presence of each interaction in the query is confirmed by setting all screens appropriate to the distance, ± tolerances, within a search mask and comparing this with the stored mask. This check removes all compounds that do not have any conformations with interactions within the required range. It does not indicate which conformation or even that all required interactions occur in a single conformation.

USE OF SUBSTRUCTURE SEARCHING

The substructure search, using standard subgraph isomorphism techniques[8], is used not only to confirm the presence of the required functional groups but also to identify all possible candidates for the key centres. This is used in the final stage of a search: conformation regeneration.

GENERATION OF CANDIDATE STRUCTURES

In order to determine which conformations actually match the input query, conformational analysis is performed on all candidates still remaining. This has been found to be quicker than reading from disk. For each conformation generated the following steps are carried out:

1. Ensure it is a low energy conformer (e.g., within 5 kcals of minimum).
2. Ensure the pattern of centres in the candidate matches that in the query. This utilises Ullman's subgraph isomorphism algorithm[9] as described by Willett[10].
3. If the match is satisfied, orientate the conformer so it is fitted onto the query, rename it according to the query numbering scheme and write it to a local results database.

INTEGRATION WITH MODELLING

Since the conformations resulting from a search are written to a local results database in the same orientation, they may be compared either visually or using standard modelling techniques such as spatial analysis. Here the conformations thus generated are displayed superimposed on one another or each can be scanned in rapid succession using standard animation facilities.

The further processing of the results is fully integrated into the Chem-X[11] modelling suite and in particular ChemStat[12]. This allows the automatic calculation of a wide variety of properties, either conformationally dependant or not, such as complex geometry variables which might involve vectors, spatial or even quantum mechanical properties and their subsequent analysis using statistics which can also be presented graphically in the form of scatter plots.

SUMMARY

To summarise, the ChemDBS-3D system consists of three main sections: the loading of the database and generation of the keys, searches on the database, and further processing of the results of a search. A total modelling/database system allowing the user effectively to search conformationally flexible molecules.

REFERENCES

1. Martin, Y.C.; Danaher, E.B.; May, C.S.; Weininger, D. 'MENTHOR, a Database System for the Storage and Retrieval of Three-dimensional Molecular Structures and Associated Data Searchable by Substructural, Biological, Physical, or Geometric Properties'. *J. Comput.-Aided Mol. Des.* **1988,** *2,* 15–29.
2. Van Drie, J.H.; Weininger, D.; Martin, Y.C. 'ALADDIN: an Integrated Tool For Computer-assisted Molecular Design and Pharmacophore Recognition from Geometric, Steric, and Substructural Searching of Three-dimensional Molecular Structures'. *J. Comput.-Aided Mol. Des.* **1989,** *3,* 225–251.
3. Sheridan, R.P.; Nilakantan, R.; Rusinko III, A.; Bauman, N.; Haraki, K.S.; Venkataraghavan, R. '3DSEARCH: a System for Three-dimensional Substructure Searching'. *J. Chem. Inf. Comput. Sci.* **1989,** *29,* 255–260.
4. Jakes, S.E.; Watts, N.; Willett, P.; Bawden, D.; Fischer, J.D. 'Pharmacophoric Pattern Matching in Files of 3-D Chemical Structures: Evaluation of Search Performance'. *J. Mol. Graphics,* **1987,** *5,* 41–48.
5. MACCS (Molecular ACCess System), 1989, developed and distributed by Molecular Design Ltd, San Leandro, California, U.S.A.

6. Jakes, S.E.; Willett, P. 'Pharmacophoric Pattern Matching in Files of 3-D Chemical Structures: Selection of Interatomic Distance Screens'. *J. Mol. Graphics,* **1986,** *4,* 12–20.
7. Dolata, P.D.; Leach, A.R.; Prout, K. 'WIZARD: AI in Conformational Analysis'. *J. Comput.-Aided Mol. Des.* **1987,** *1,* 73–85.
8. Tarjan, R.E. 'Graph Algorithms in Chemical Computation'. In *Algorithms for Chemical Computations*; Christoffersen, R.E., Ed.; ACS Symposium Series 46; American Chemical Society: Washington, 1977, pp.1–20.
9. Ullman, J.R. 'An Algorithm for Subgraph Isomorphism'. *J. Assoc. Comput. Mach.* **1976,** *16,* 31–42.
10. Brint, A.T.; Willett, P. 'Pharmacophoric Pattern Matching in Files of 3D Chemical Structures: Comparison of Geometric Searching Algorithms'. *J. Mol. Graphics,* **1987,** *5,* 49–56.
11. Chem-X molecular modelling software, 1990, developed and distributed by Chemical Design Ltd, Oxford, England.
12. Perry, N.C,; Davies, E.K. The Use of 3-D Modelling Databases for Identifying Structure Activity Relationships'. In *QSAR: Quantitative Structure-Activity Relationships in Drug Design*; Fauchere, J.L., Ed.; Alan R. Liss: New York, 1989, pp.189–193.

USING 3-D SIMILARITY SEARCHING TO DEVELOP SYNTHETIC TARGETS

William C. Ripka and Charles J. Eyermann
E.I. Du Pont de Nemours & Co, Inc., Wilmington, DE 19880, U.S.A.

ABSTRACT

One of the key challenges facing the medicinal chemist is converting data obtained from X-ray crystallography, biological screening, and mechanism studies into ideas for new synthetic targets. To develop these new targets, two problems must be solved. First, a pharmacophore model representing the 3-D arrangement of the functional groups required for biological activity must be developed. Then, a molecular framework must be found which can position these functional groups in the proper 3-D orientation. This paper reviews our approaches to developing pharmacophore models and how these models can be used as input for searching 3-D databases to find desirable molecular 'frameworks'. Our 3-D searching is done using the GEOSTAT software from the Cambridge Crystallographic Data Centre and Molecular Design Limited's MACCS-3D software. Results for several pharmacophores will be used to illustrate how the technique can be used to develop novel synthetic targets.

INTRODUCTION

In industry, the synthetic chemist is challenged to design and synthesise novel and patentable compounds which have a desired biological activity or physical property. Traditionally, chemists approach this challenge by using their intuition in combination with data from the literature, the existence of relevant natural product structures, or leads from screening programmes. While the success of these approaches is unequivocal, it is also well known that in the pharmaceutical or agrichemical industry these approaches often require synthesis and testing of tens of thousands of compounds before a commercial product is finally produced. The increasing cost of research and development has made it imperative that more efficient routes of discovery are found.

To reduce the total number of compounds to be synthesised, substantial efforts have been made to use computational chemistry, quantitative structure activity relationships (QSAR), and molecular graphics to discover, and then to optimise the activity/property of the new compounds[1–5]. This approach, often called 'rational' design, has been made possible by a rapidly expanding set of crystallographic data, reaction mechanisms, and activity/property data. The success of the rational design approach has, however, been largely limited to *optimising* the features of an existing class of structures. Until recently, very little progress had been made in developing techniques for *de novo discovery* ('design') of an entirely new class of structures which have a desired activity/property.

The recent development of techniques such as 2-D similarity searching[6], 2-D to

W. A. Warr (Ed.)
Chemical Structures 2

3-D structure conversion programs[7], 3-D database and search systems[2,8–10], probe interaction calculations[11], and distance geometry[12] provides the necessary tools to make *de novo* discovery possible. Several recent reviews on these techniques are recommended.[4,5]

During the past four years we have been using many of these techniques in conjunction with pharmacophore searching of 3-D databases as a tool for generating novel synthetic targets. In this paper we review our strategy, and also outline why we believe 3-D database searching will help integrate molecular modelling systems with 2-D and 3-D database/information systems.

SYNTHETIC TARGETS

Generating ideas for synthetic targets first requires suitable pharmacophore models, i.e., the specific 3-D arrangement of essential functional groups of a compound which is required for effective binding at a given enzyme/receptor site. From receptor mapping studies it is known, for example, that a simple α-adrenergic pharmacophore model requires a nitrogen atom, an oxygen atom, and phenyl group in the 3-D relationship shown in Figure 1.

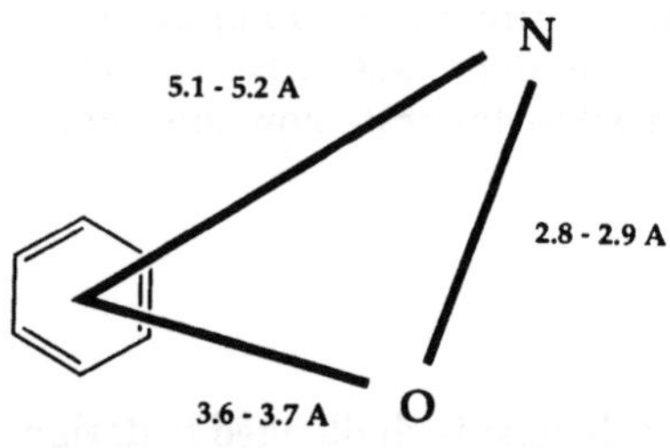

Figure 1. Alpha-adrenergic pharmacophore model

How are synthetic targets developed which have these functional groups in a proper 3-D orientation? Our strategy involves the following five steps:

1. Developing a pharmacophore model using methods such as systematic conformational search, QSAR, distance geometry, and probe interaction calculations.
2. Using the pharmacophore model to build a 3-D query.
3. Searching the 3-D database.
4. Removing 'undesired' atoms and groups from the hit structures.
5. Adding additional desired functionality to the molecular framework generated in 4.

To illustrate this strategy, consider the development of new synthetic targets for the nicotinic receptor. Dixon and coworkers used a distance geometry ensemble approach[13] with a series of nicotinic agonists and antagonists to develop a nicotinic pharmacophore (Figure 2). Using this pharmacophore arrangement, we searched the Cambridge crystal structure database[14] using the GEOSTAT[10] software to find structures which match the pharmacophore query. Example hits are shown in Figure 3. Many structures known to be active at the nicotinic receptor were found, for example acetylcholine. Using two of the rigid framework hits, croomin and stemonine, the synthetic target shown in Figure 4 can be proposed.

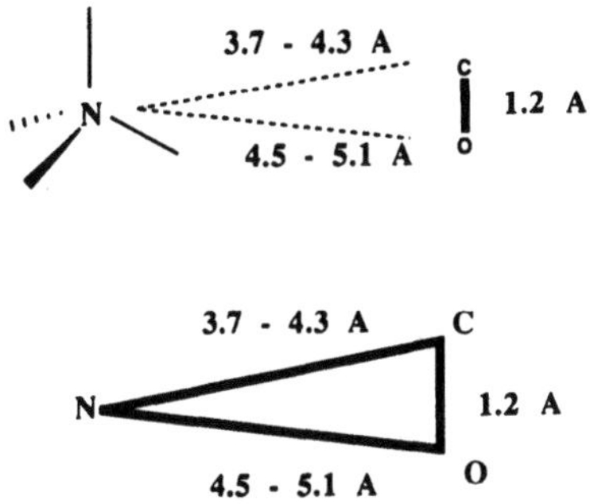

Figure 2. Nicotinic 3-D pharmacophore query based on the ensemble distance geometry derived model of Dixon *et al.*[13]

PHARMACOPHORE MODELS

Figure 5 schematically outlines the diverse range of structural information a synthesis chemist may have available for building a pharmacophore model. To illustrate this we will use the crystal structure[15] of the *l casei* dihydrofolate reductase (DHFR) methotrexate complex (Figure 6).

For the (rare) case where the 3-D structure of the receptor/inhibitor complex is known, the conformation of the inhibitor can directly be used to build a 3-D pharmacophore. For methotrexate, it is known that the pteridine ring system is

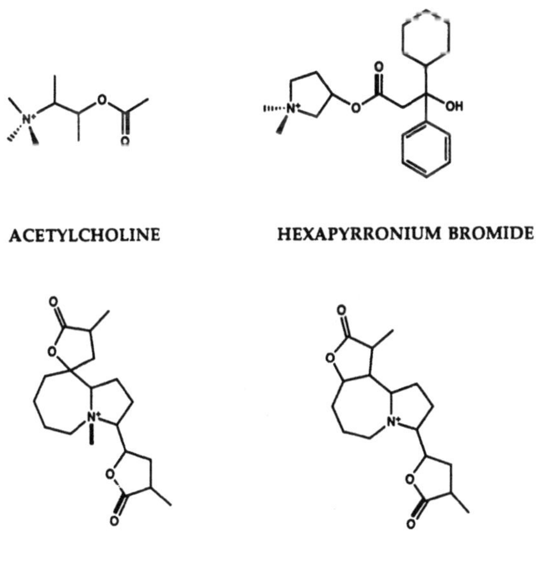

Figure 3. Example results for 3-D database search using nicotinic pharmacophore

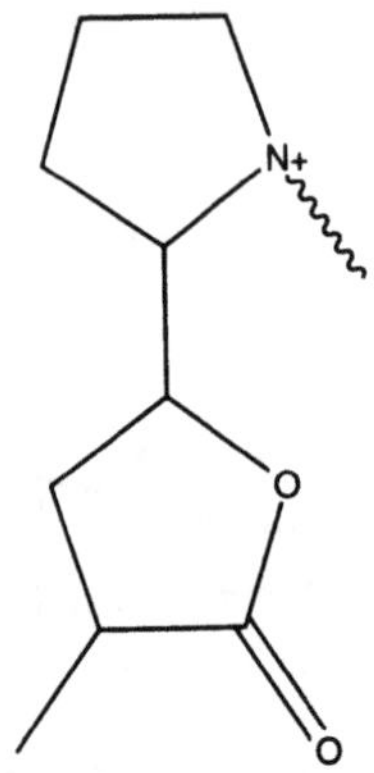

Figure 4. Synthetic target based on croomin and stemonine hits

essential for binding[16], therefore the pharmacophore model was built based on methotrexate's complexed conformation (Figure 7). Figure 8 shows four hits from a 3-D search of the Cambridge database. Figure 9 shows the hyperforin fit to the active site of *l casei* DHFR, and Figure 10 the obtusifolin fit to the site. These structures could be modified to develop a synthetic target specific for *l casei* DHFR.

If the 3-D structure of a single active compound is known, for example, a natural product, a hypothesis of the functional groups in the pharmacophore can be made, their geometric relationships measured, and a 3-D query generated.

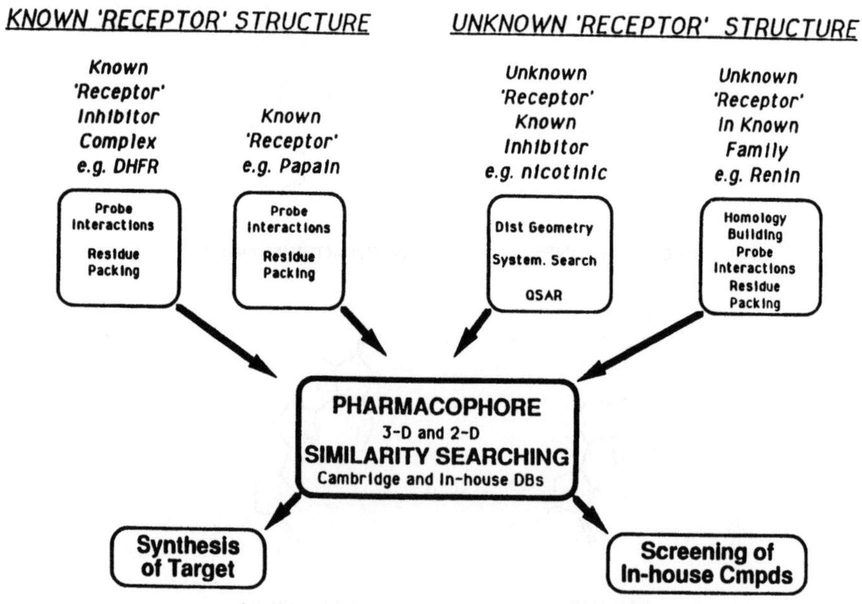

Figure 5. Synthetic target design strategy

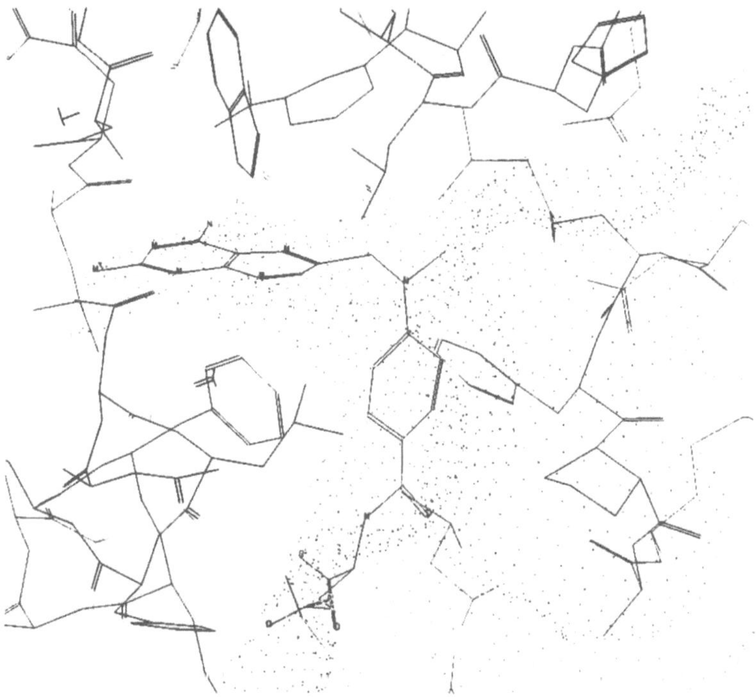

Figure 6. Methotrexate complexed with dihydrofolate reductase (DHFR). Dotted surface represents extended VDW surface for the DHFR active site residues. Methotrexate heteroatoms are labelled

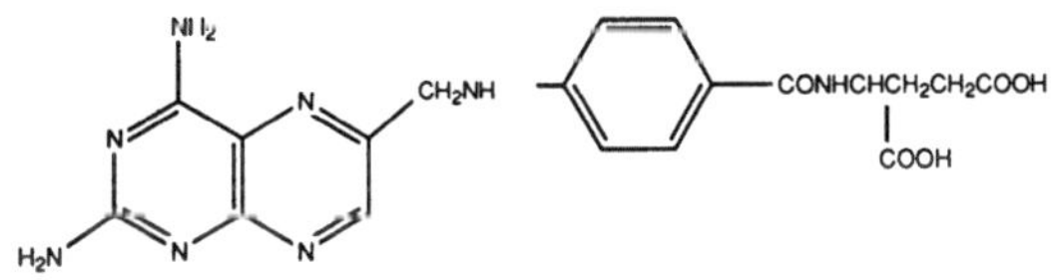

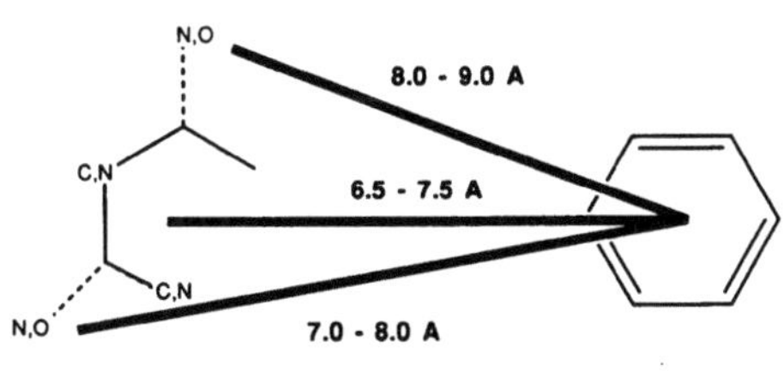

Figure 7. 3-D pharmacophore query based on methotrexate conformation when it is bound to DHFR

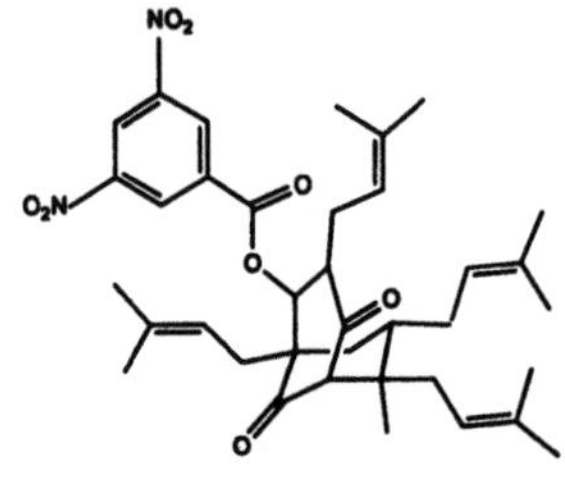

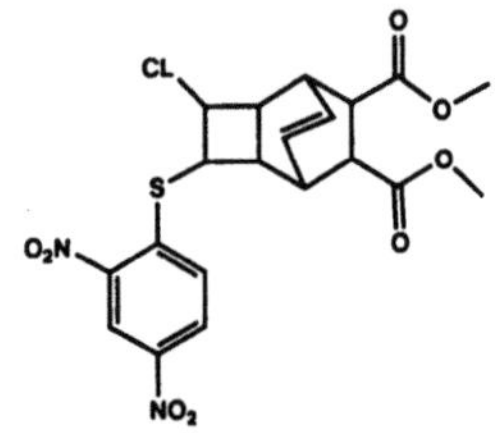

HYPERFORIN

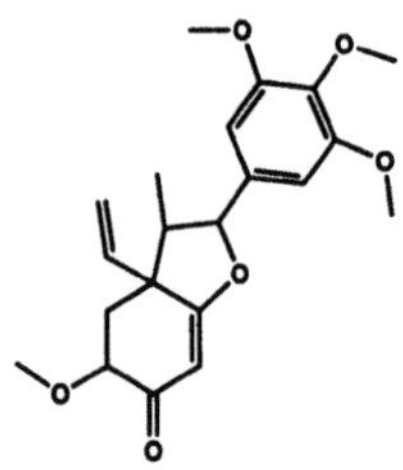

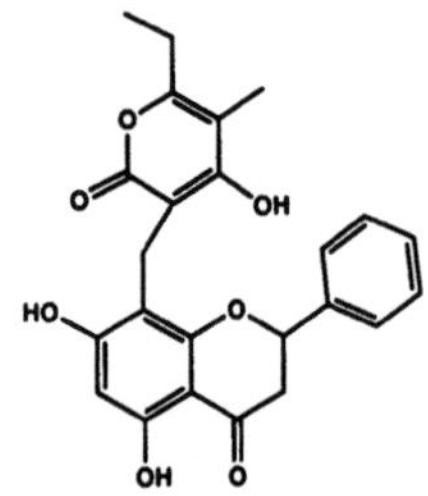

OBTUSIFOLIN

Figure 8. Example results for 3-D database search using methotrexate pharmacophore

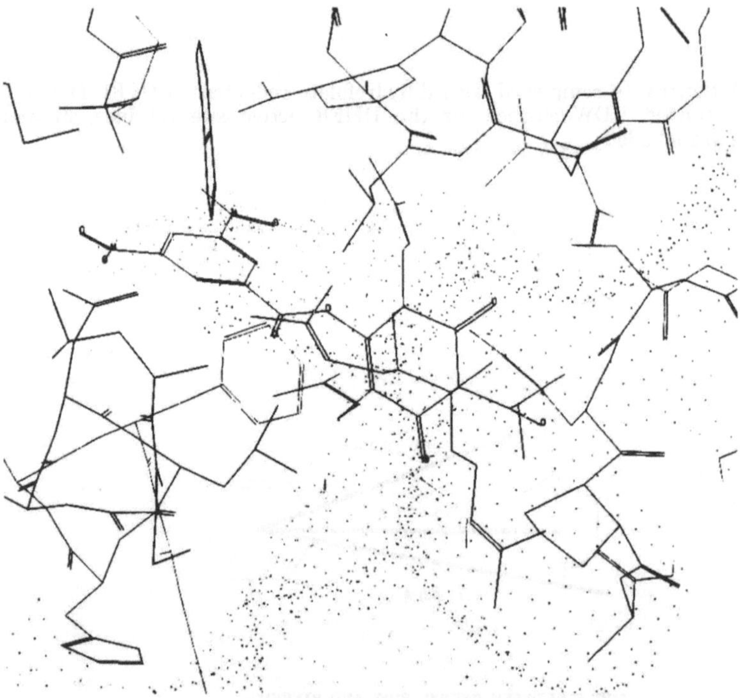

Figure 9. Hyperforin hit 'docked' in the active site of DHFR. Note that the bicyclo portion of hyperforin matches the phenyl portion of the pharmacophore query. The two nitro groups match the amino groups in the pteridine ring of methotrexate

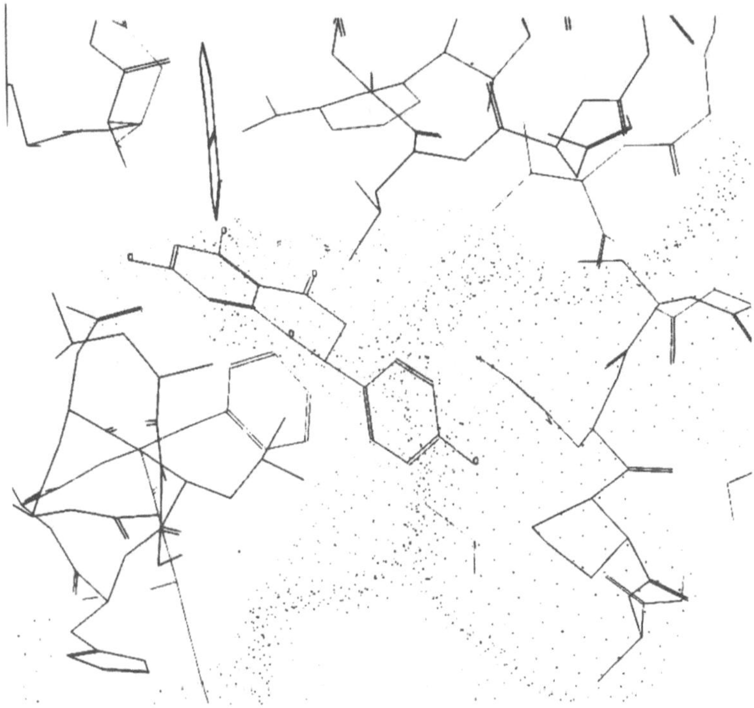

Figure 10.Obtusifolin hit 'docked' in the active site of DHFR. The two hydroxy groups match the amino groups in the pteridine ring of methotrexate

If only the structure of the enzyme/receptor is known, energy calculations between functional group probes[1] and the active site residues of the enzyme/receptor can be used to determine favourable positions ('hot spots') to place the functional group. These 'hot spots' can readily be converted into a pharmacophore model for use in a 3-D search.

Usually the 3-D structure of the enzyme/receptor is not known. In this case, receptor mapping techniques such as CoMFA[17] and conformational analysis techniques such as systematic search and distance geometry[12] are applied to a series of active and inactive structures to obtain a pharmacophore model for use in 3-D database searching.

3D SEARCHING SYSTEMS

In 1986, when we began to consider using 3-D database searching as a synthetic target design tool, we chose to use the Cambridge database system[14] because it gave us direct access to over 30,000 organic crystal structures, and a 3-D searching capability through the GEOSTAT[10] program. This saved us considerable software development time and immediately gave us 30,000 low-energy 3-D structures to search. We also were able to use GEOSTAT to search internal 3-D models which we converted to the Cambridge FDAT file format[14].

Since 1986, others have developed their own specialised 3-D searching systems. These systems include the ALADDIN system[2] developed at Abbott, the 3DSEARCH system[8] developed at Lederle, and the MACCS-3D system[9] developed by Molecular Design Limited in collaboration with an industrial consortium. Any one of these systems can be used to develop synthetic targets using the strategy outlined above.

INTEGRATING 2-D AND 3-D

As we continued to use 3-D database searching to develop synthetic targets, there was an increasing need to integrate 2-D chemical structure database facilities with molecular modelling facilities (Figure 11). This need, and a desire to provide the synthetic chemist with easily accessible lead discovery and optimisation tools (Figure 12), led us to choose MACCS-3D as our primary 3-D database tool.

2-D/3-D Data and Information

- 2-D Structure Registration
- Biological Data Storage
- Technical Reports
- [Physical Property Storage]

⇓

Data Analysis and Modeling

- 3-D Structure Storage and Retrieval
- Lead Discovery and Optimization
- Storing and Rapid Searching of 3-D Descriptors

Figure 11.Integrating 2-D and 3-D databases and tools

- **3-D Structure Storage and Retrieval**

  3-D structures generated from 2-D structures
  crystal structures (Cambridge and internal)
  structures generated from molecular modeling

- **Lead Discovery and Optimization**

  2-D similarity searching (random screening)
  3-D searching for pharmacophore (synthetic targets)
  Quantitative Structure Activity Relationships (QSAR)

- **Storing and Rapid Searching of 3-D Descriptors**

  Structure perception utilities
  Conformational Analysis tools (multi-conformations)
  Molecular Shapes, Volumes, Properties

Figure 12. Needs in chemical structure based data analysis and modelling tools

Our use of MACCS-3D has provided insight into the value of integrating 2-D data, such as biological activity, with 3-D structures. To illustrate the value of such integration, we will use the 2-D and 3-D versions of the MACCS Drug Data Report[18] (MDDR). A 2-D data search (Figure 13) of the MDDR database for all structures having adrenergic activity yields 124 hits (Figure 14). A 3-D search of these structures using an α-adrenergic pharmacophore yields 11 hits (Figure 15). Specific biological activity for each of these hits can then be viewed (Figure 16). This level of integration begins to provide powerful ways to hypothesise about 2-D and 3-D QSAR.

CONCLUSIONS

Three-dimensional database searching is a powerful new tool for developing *novel* synthetic targets. This technique can be used in cases where an enzyme/receptor 3-D structure is known, or alternatively in the case where the enzyme/receptor structure is unknown. In the latter case an active compound or series of compounds can be used to develop a pharmacophore model. Because 3-D database searching requires biological data stored in 2-D databases and molecular modelling techniques, we expect to see increased integration of 2-D and 3-D database/information systems.

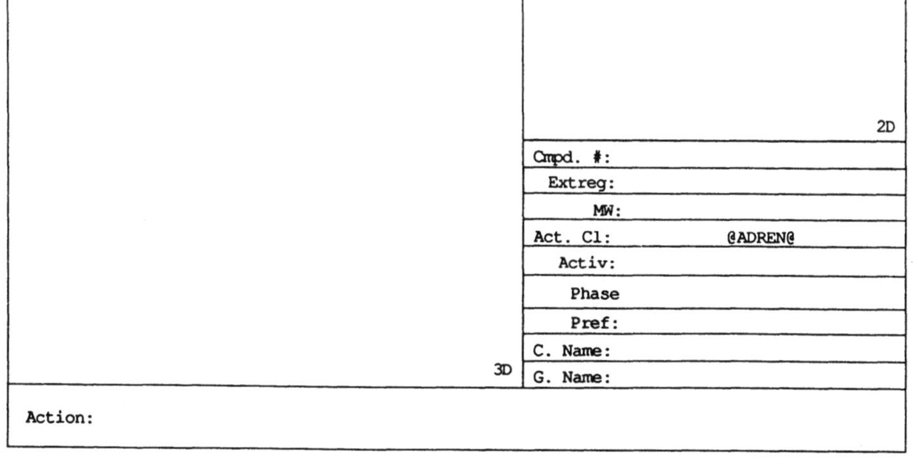

Figure 13. Search of MDDR for compounds which have 'adren' in the activity class data field

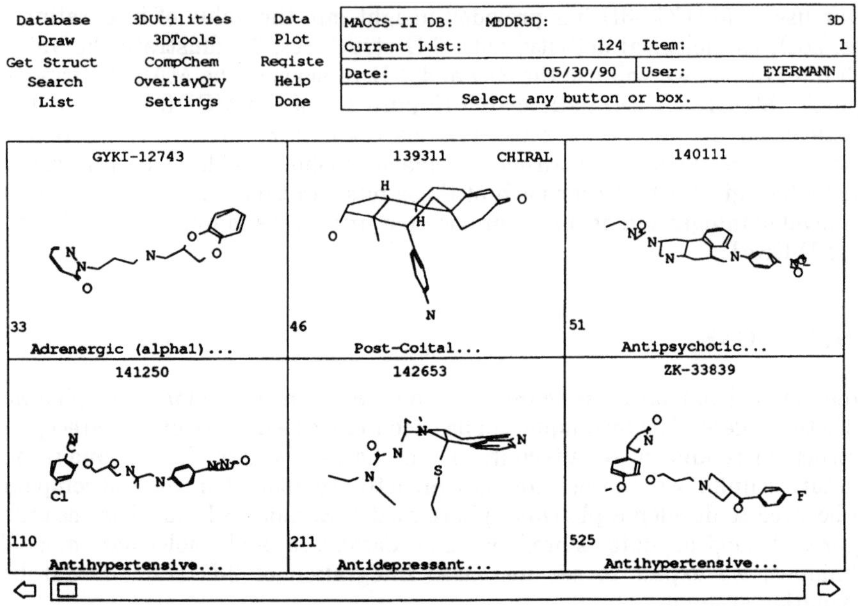

Figure 14. First six hits from data search of MDDR for compounds with 'adren' activity

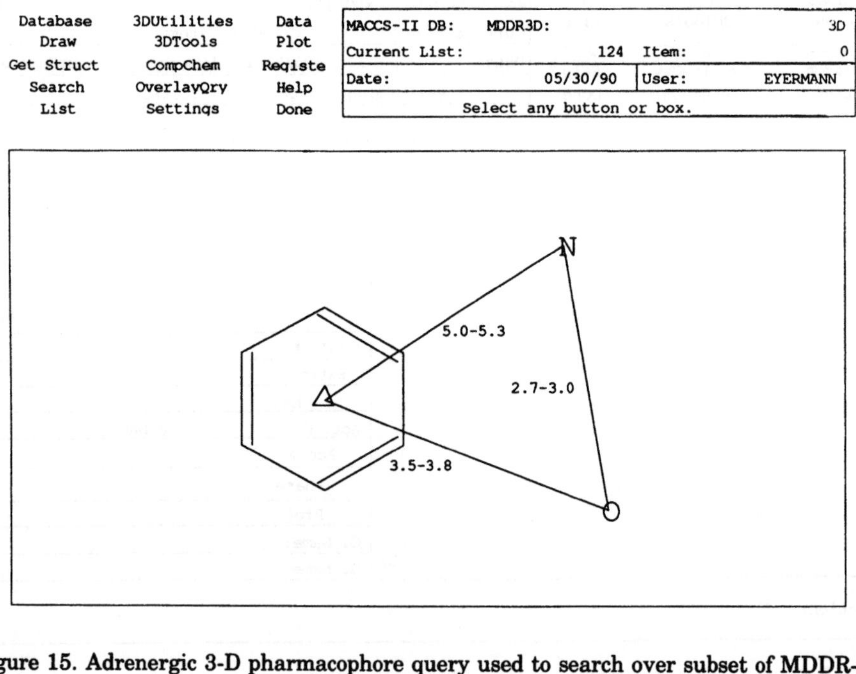

Figure 15. Adrenergic 3-D pharmacophore query used to search over subset of MDDR-3D database

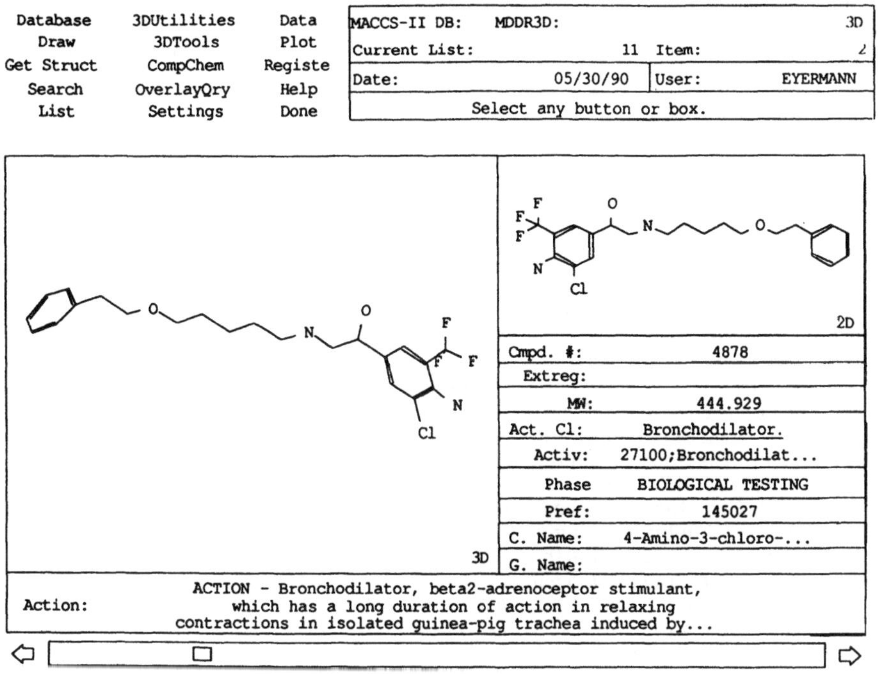

Figure 16. Special data form for MDDR-3D showing one of the 11 hits from the combined adrenergic data search and 3-D pharmacophore search

ACKNOWLEDGMENTS

We are grateful for the many helpful discussions with Jeff Blaney and Dan Kleier. We are also thankful for the assistance we received from the technical staff at the Cambridge Crystallographic Data Centre and Molecular Design Limited.

REFERENCES

1. Ripka, W.C.; Sipio, W.J.; Blaney, J.M. 'Molecular Modelling and Drug Design: Strategies in the Design and Synthesis of Phospholipase A2 Inhibitors'. *Lect. Heterocycl. Chem.* **1987,** *IX* 95–104.
2. Van Drie, J.H.; Weininger, D.; Martin, Y.C. 'ALADDIN: An Integrated Tool for Computer-assisted Molecular Design and Pharmacophore Recognition from Geometric, Steric and Substructure Searching of Three-dimensional Molecular Structures'. *J. Comput.-Aided Mol. Des.* **1989,** *3,* 225–251.
3. DesJarlais, R.L.; Sheridan, R.P.; Seibel, G.L.; Dixon, J.S.; Kuntz, I.D.; Venkataraghavan, R. 'Using Shape Complementarity as an Initial Screen in Designing Ligands for a Receptor Binding Site of Known Three-dimensional Structure'. *J. Med. Chem.* **1988,** *31,* 722–729.
4. Cohen, N.C.; Blaney, J.M.; Humblet, C.; Gund, P.; Barry, C.D. 'Molecular Modelling Software and Methods for Medicinal Chemistry'. *J. Med. Chem.* **1990,** *33,* 883–894.

5. *Molecular Foundations of Drug-Receptor Interaction*; Dean, P.M.; Cambridge University Press: Cambridge, 1987.
6. *Similarity and Clustering in Chemical Information Systems*; Willett, P. Wiley: New York, 1988.
7. Rusinko, A., III; Skell, J.M.; Balducci, R.; Pearlman, R.S. CONCORD, University of Texas at Austin, distributed by Tripos, Associates, St. Louis, MO.
8. Rusinko, A., III; Sheridan, R.P.; Nilakantan, R.; Haraki, K.S.; Bauman, N.; Venkataraghavan, R. 'Using CONCORD to Construct a Large Database of Three-dimensional Co-ordinates from Connection Tables'. *J. Chem..Inf. Comput. Sci.* **1989,** *29,* 251–255.
9. MACCS-3D, Molecular Design Limited, San Leandro, CA.
10. GEOSTAT software from the Cambridge Crystallographic Data Centre, Cambridge, England. See: Murray-Rust, P.; Raftery, J. 'Computer Analysis of Molecular Geometry, Part VI: Classification of Differences in Conformation'. *J. Mol. Graphics* **1985,** *3,* 50–59.
11. Goodford, P.J. 'A Computational Procedure for Determining Energetically Favourable Binding Sites on Biologically Important Macromolecules'. *J. Med. Chem.* **1985,** *28,* 849–857.
12. *Distance Geometry and Molecular Conformation*; Crippen, G.M.; Havel, T.F.; Research Studies Press: New York, 1988.
13. Sheridan, R.P.; Nilakantan, R.; Dixon, J.S.; Venkataraghavan, R. 'The Ensemble Approach to Distance Geometry: Application to the Nicotinic Pharmacophore'. *J. Med. Chem.* **1986,** *29,* 899–906.
14. Cambridge Structural Database and associated software, Cambridge Crystallographic Data Centre, Cambridge, England.
15. Bolin, J.T.; Filman, D.J.; Matthews, D.A.; Hanlin, R.C.; Kraut, J. 'Crystal Structures of *Escherichia coli* and *lactobacillus casei* Dihydrofolate Reductase Refined at 1.7Å Resolution'. *J. Biol. Chem.* **1982,** *257,* 13650–13762.
16. Blaney, J.M.; Hansch, C.; Silipo, C.; Vittoria, A. 'Structure-activity Relationships of Dihydrofolate Reductase Inhibitors'. *Chem. Rev.* **1984,** *84,* 333–407.
17. Cramer, R.D., III; Patterson, D.E.; Bunce, J.D. 'Comparative Molecular Field Analysis (CoMFA) .1. Effect of Shape on Binding of Steroids to Carrier Proteins'. *J. Am. Chem. Soc.* **1988,** *110,* 5959–5967.
18. Prous Drug Data Report, MACCS format distributed by Molecular Design Limited, San Leandro, CA.

IDENTIFYING UNIQUE CORE MOLECULES FROM THE OUTPUT OF A 3-D DATABASE SEARCH

Yvonne C. Martin
Computer Assisted Molecular Design Project, Abbott Laboratories, Abbott Park, IL 60064, U.S.A.

John H van Drie
Meridien Research Inc.
Present address: BioCAD Corporation, 1091, North Shoreline Blvd, Mountain View, CA 94043, U.S.A.

ABSTRACT

The computerised design of potentially bioactive molecules or the computer recognition of a new biological property of an existing molecule can lead to orders of magnitude more suggestions than are reasonable to explore experimentally. Thus we added to the ALADDIN Control Language new commands that describe the transformation of one chemical structure into another. Atoms or groups can be removed or replaced and bonds can be made, broken, or changed in bond order. We used this language to describe transformations that remove extraneous substituents from molecules suggested for synthesis as potential dopaminergic agonists[1] and demonstrated that such transformations reduce the number of compounds to be investigated by 66-85% without loss of geometric information. Similarily such transformations lead to the identification of the structural family to which an existing compound belongs.

BACKGROUND

The decade of the 1980s has given the medicinal chemist tools to examine structure-activity relationships not in traditional 2-D fashion, but rather in 3-D. Simultaneously, advances in molecular biology and protein crystallography have combined to produce 3-D structures of proteins at an ever increasing rate. These two types of information provide important 3-D structural information on which the scientist can base the design of more potent and selective bioactive molecules. To be efficient, the structure-based design of bioactive compounds requires specialised computer tools beyond traditional molecular graphics. As a result, recently several groups have developed computer programs to find in a database or to design molecules that match a particular 3-D requirement[2].

In our original report on the 3-D searching program ALADDIN we demonstrated that geometric and steric searching does indeed *recognise* molecules that had previously unsuspected D2 dopamine agonist activity[3]. A second example, forecast of the plant growth regulating activity of compounds in a commercial database, has recently been reported by Bures *et al*[4].

W. A. Warr (Ed.)
Chemical Structures 2

We have also investigated the computer *design* of potential D2 dopamine agonists by geometric searching of databases of 3-D structures with ALADDIN[1,3]. The idea of our method is to identify frameworks that can be used to hold the groups essential for biological activity (pharmacophore atoms)[5] in the required geometric relationships. We first search a number of databases for molecules in which atoms of any type match the key geometric constraints. We then transform these molecules into the 2-D structures of the molecules to be suggested for synthesis, generate the 3-D structures of the suggested compounds with CONCORD[6] and double-check that they meet the proposed geometric requirements for dopaminergic activity.

A difficulty common to all three of these studies is that too many hits are identified. Clearly it is impractical to examine thousands of compounds individually in a molecular graphics system or even as 2-D printed structures, to say nothing of in an experimental assay. The problem in compound design is thus to identify the unique core molecules that might be synthesised. The problem in using the 3-D properties of known compounds to suggest those for screening is to identify those molecules that provide essentially duplicate suggestions and so to test only some of each set of duplicates. For example, Figure 1 shows three database molecules, 1–3;

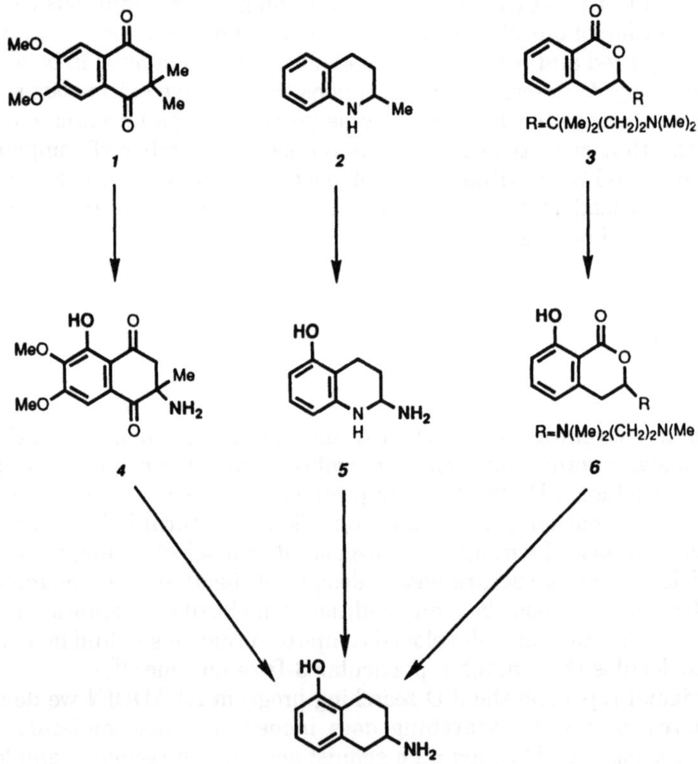

Figure 1. Three database molecules, 1–3; the molecules suggested from them, 4–6; and the structure that is the geometric essence of them, 7.

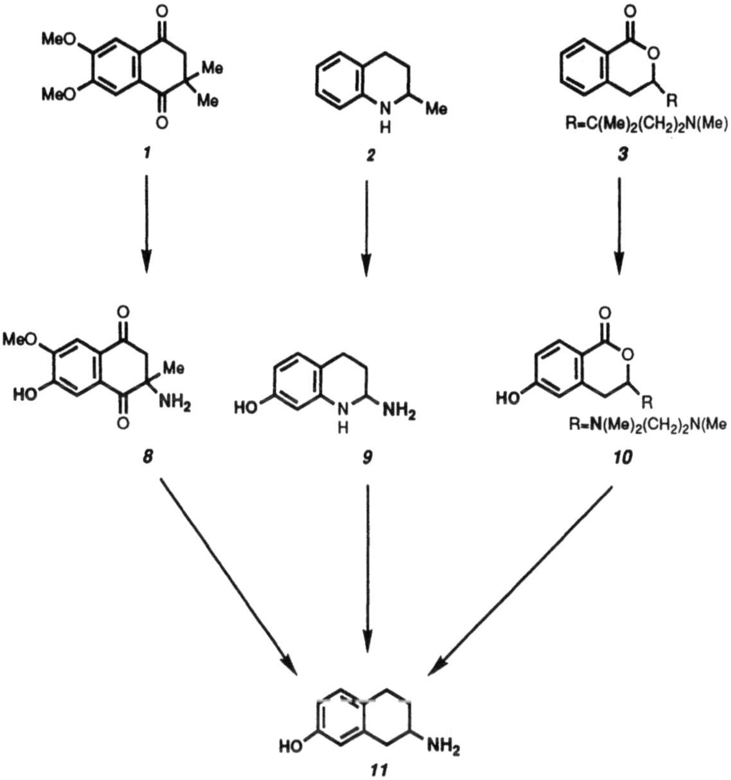

Figure 2. Three database molecules, 1–3; the different molecules suggested from them, 8–10; and the structure that is the geometric essence of these other structures, 11.

the molecules suggested for synthesis as dopaminergics, **4–6**; and the essential core of the three suggestions, **7**. A particular core molecule is thus identical to one molecule or a number of suggested molecules in the geometric relationships between the pharmacophore atoms and also the atoms necessary for establishing the geometric relationships between these pharmacophore atoms.

This report describes one of the solutions that we have devised for this problem. Although it is applicable to evaluating lists of known compounds for testing, we have used it more extensively in the design of new compounds. The essence of the strategy is that we prune the structures of substituents that are not part of the pharmacophore and transform geometrically similar atoms into one common atom. For example, database compounds **1–3** suggested the synthesis of compounds **4–6** respectively (Figure 1). This involves the addition of the pharmacophore OH and amino groups. However, if we remove extraneous substituents, convert all first row non-pharmacophore atoms into carbons, and change non-aromatic sp^2 atoms into sp^3 atoms, we see that the essential dopaminergic core in **4–6** is the known compound 5-hydroxy-2-aminotetralin **7**. By such a structure modification procedure, many formerly different molecules are transformed into the same molecule. On the other hand, as shown in Figure 2, these same four database compounds also lead to the suggestion of a second known dopaminergic compound, 7-hydroxy-2-aminotetralin **11**.

Additionally, since our design strategy involves the transformation of one molecule into another it was necessary to develop a means to do this automatically if it were to be efficient. Performing the chemical transformations with a text editor incurs the risk of generating nonsense chemical structures. Thus for this purpose we needed to describe to a computer program how to transform the database molecules to the compounds suggested for synthesis in such a way that only chemically valid structures would be produced.

ALADDIN CONTROL LANGUAGE COMMANDS TO PERFORM STRUCTURAL MODIFICATIONS

ALADDIN searches are controlled by a file of instructions written in ALADDIN Control Language, ACL. ACL is an extension of the Daylight sub structure specification language GCL that provides instructions for the program GENIE[7] For a geometric search the ACL file describes the substructure environment of the atoms of interest in standard GCL; the geometric objects (lines, points, and planes) to be constructed from these atoms in each XYZ dataset; and the distance, angle, plane angle, and torsion angle constraints to be applied to the relationships between these objects. For a subsequent steric test on a set of molecules with the 'hit' atoms already identified, the ACL file contains the name of the reference molecule, the superposition rule for orienting candidate molecules with it, and the name of the file that contains the surface points to test if the molecule fits within. For the work described in this report, we extended ACL to describe the transformations to be applied to the 2-D structures of the molecules.

In the Daylight software the 2-D chemical structure of a molecule is represented by a SMILES (Simplified Molecular Input Line Entry System) string derived from the hydrogen-suppressed graph of the molecule[8]. The MODSMI operations, the subject of this report, essentially involve:

1. identification of the atoms of interest,
2. substitution of one character or set of characters for another in the SMILES string of the compound, and
3. making the new SMILES unique.

Figure 3 shows examples of the four primitive MODSMI operations.

Since the atoms in a SMILES string can be labelled with isotopes, in one ALADDIN output file we tag the hit atoms with the 'isotope' that corresponds to the order of the geometric object in the ACL description of the search. For example, in the ALADDIN D2 design, geometric object 1 was an aliphatic non-benzylic atom and geometric object 2 was an aromatic atom. Thus structure 1 would be labelled in the corresponding SMILES strings as shown in Figure 4 for the two different hits. As will be shown below, this special labelling of atoms may be used to control the MODSMI operations.

The atoms on which the MODSMI transformations are to be applied are identified by a GENIE search that recognises atoms or bonds from their substructural environment[7]. They are highlighted for ALADDIN processing with the GCL command 'PRINT '/BOLDID/''. Because the 'hit' atoms from an ALADDIN geometric test are labelled with the geometric object to which they correspond, MODSMI transformations can be applied to all atoms that are in a particular substructural environment or to only such atoms with a particular

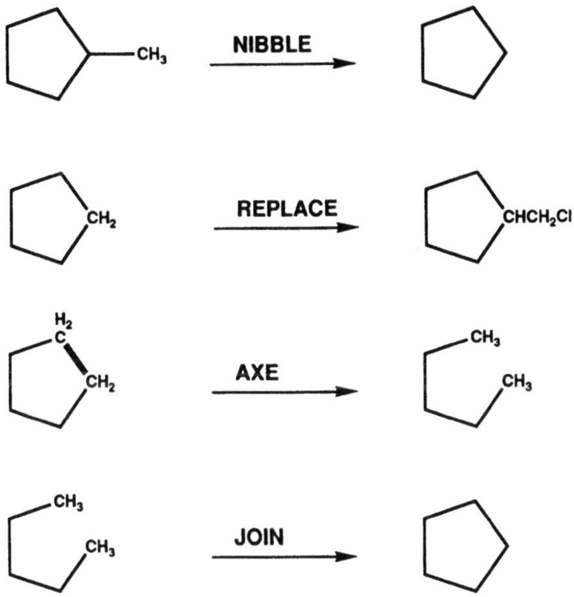

Figure 3. Examples of the four primitive MODSMI commands.

numeric identifier. The latter is done with the ACL command 'MARKHIT', for example, 'CALL 'MARKHIT' '1" specifies that the following commands apply only to atoms identified with 'isotope' 1. If all atoms regardless of 'isotope' are to be considered (normal GENIE processing), the command is: 'CALL 'MARKHIT' '-1". Finally: 'CALL 'MARKHIT' '0" causes GENIE to record hits only on non-labelled atoms.

The simplest MODSMI transformation made to the identified atom or group of atoms is NIBBLE. It removes the identified atom(s) from the SMILES string. Thus NIBBLE called iteratively will transform a pendant propyl into a pendant ethyl into a pendant methyl into no pendant group. For example NIBBLE was used to remove the OMe and Me groups of structure **1** as part of the transformation of it into structures **7** and **11**.

The MODSMI command REPLACE changes the identified atom or group of atoms into another. REPLACE allows one to change the atom type of the specified atom and/or to add a group onto that atom. Thus if a ring is used to hold certain groups in place, but it is immaterial whether the ring atoms are carbons, oxygens, or nitrogens, one can REPLACE on all non-carbon atoms in the ring with carbon atoms as in the transformation of structure **5** into structure **7**.

Importantly, with REPLACE one can add new groups onto an atom. For example, if a particular type of carbon is identified as an attachment point for a Br, one would identify the carbon atom with GCL and replace the identified carbon atom with a carbon atom with an attached bromine: REPLACE C(Br). The combination of MODHIT and REPLACE allows one to transform a 'hit' atom in a molecule into the particular atom type in a molecule to be suggested for synthesis. In the dopamine D2 agonist case, the identified aromatic atom would be transformed into into a carbon bearing an OH and the identified aliphatic atom an aliphatic nitrogen atom as shown in the transformation of 1 into 4 and 8 in Figure 4. The MODSMI commands are shown in Figure 5.

For bond transformations, GCL is used to identify the atoms to be present at each end of the bond. MODSMI provides two elementary transformations to be made to bonds. JOIN connects unconnected atoms. AXE *n* changes the bond order of an existing bond by n, either positive or negative. Thus AXE -1 would break a single bond.

One calls the MODSMI routines with the ACL statement HALT. For example, to NIBBLE away a non-hit Br atom the commands would be:

CALL MARKHIT '0'
IF 1- Br { PRINT '/BOLDID/' HALT 'MODSMI:NIBBLE' }.

The command HALT indicates that processing on that SMILES is to stop, the programmed MODSMI transformation is to be made, and re-execution of the entire ACL using the newly modified smiles is to be performed. The MODSMI transformations are performed iteratively until no more can be applied. By appropriate nesting very elaborate transformations can be made to a 2-D structure.

One can use standard GCL commands to print any current structure, intermediate or final in the MODSMI transformation, to an output file. However, an ACL command to stop modification of the current SMILES has also been added: 'HALT 'LOGSMI;QUIT''. When this statement is executed, the current SMILES is output into a file with a name derived from the name of the input ACL file plus an

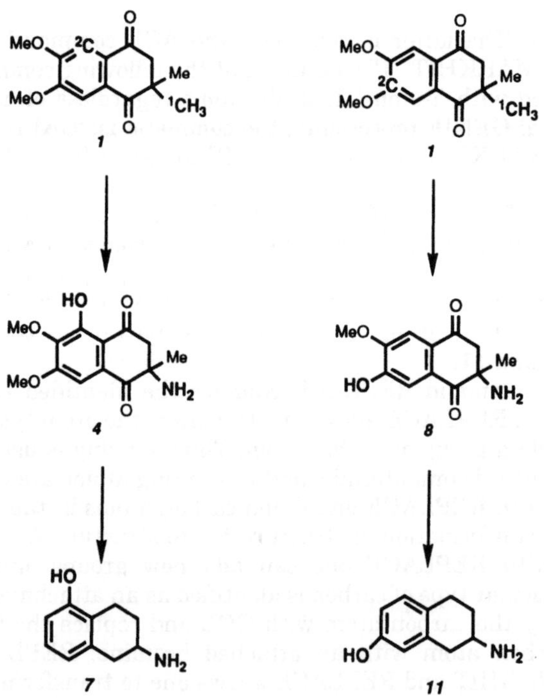

Figure 4. The 'isotope' labelling of the atoms in 1 that leads to 4 and 8 and thus to 7 and 11 respectively.

```
ALADDIN NO_THOR_LOOKUP
\
\ First we replace the hit aromatic atom with an aromatic carbon
\ bearing an OH group.  We keep the label on the carbon atom and
\ include one on the added O so that we don't remove the added OH
\ when we remove extraneous substituents.
\
CALL "MARKHIT" "2"
IF 1- [a&!D3]     { PRINT "/BOLDID/"
                    HALT "MODSMI:REPLACE [2c]([3OH])" }
\
\ Next we replace CHk with NHk-1 or other substituents to the N
\ that has the appropriate number of H's.  This atom retains its
\ label also.
\
CALL "MARKHIT" "1"
IF 1- [C&H3]      { PRINT "/BOLDID/"
                    HALT "MODSMI:REPLACE [1NH2]" }

IF 1- [C&H2]      { PRINT "/BOLDID/"
                    HALT "MODSMI:REPLACE [1NH]" }

IF 1- [C&H1]      { PRINT "/BOLDID/"
                    HALT "MODSMI:REPLACE [1N]" }

IF 1- [Cl,F,Br,I,S&H,O&H]     { PRINT "/BOLDID/"
                    HALT "MODSMI:REPLACE [1NH2]" }
```

Figure 5. The ACL commands that describe the change from structures **1–3** to **4–6** respectively.

incremented structure number. (To track database molecules from which a modified SMILES was produced, one can perform just the 'LOGSMI;QUIT' command in a separate run using an ACL with a different name. The original and the modified SMILES will have the same numeric identifier.)

Typically these MODSMI transformations are performed on SMILES previously identified by an ALADDIN search. However they may be applied to molecules identified by any procedure. When MODSMI is used the molecules are not in a database, but are presented as an input file.

EXAMPLES OF MODSMI TRANSFORMATIONS

Figure 6 shows the ACL that would fuse four saturated carbon atoms onto two adjacent unsubstituted aromatic carbon atoms. Notice that the operations are specified in reverse order from the order that one thinks of them. This is because of the contextsensitive execution of GCL. Hence, on the first pass of benzene through this ACL, all of the IF statements except $FIRSTATT would fail. The resulting ethylbenzene would pass the $SECONDATT IF statement. Finally, 1,2-diethylbenzene would pass the diethyl test. If one were to reverse the order of the $SECONDATT and $FIRSTATT statements above, MODSMI would continue adding ethyl units to all the unsubstituted carbon atoms in the input molecule.

The power of the context-sensitive execution of GCL becomes apparent when one considers the example in Figure 7, the ACL that was used to remove the unwanted substituents from database structures in the D2 design problem. This ACL automatically AXEs and NIBBLEs until no $FRGs (freely rotatable groups) or $TERMINALFRG's (terminal freely rotatable groups) are present. Only then are the hit atoms transformed into the target pharmacophore atoms.

```
\
ALADDIN NO_THOR_LOOKUP
\
\  The following statement says that if there are two or more
\  aromatic carbon atoms in the molecule that are in two rings, then
\  the SMILES of the current molecule should be logged to the output
\  file and processing of this SMILES stopped.
\
IF 2- [c;R2][C;R] { PRINT "/SMILES/" HALT "LOGSMI;QUIT" }
\
\  The following says that if there are two or more ethyl groups
\  on adjacent aromatic carbon atoms, then the methyl groups should
\  be joined by a single bond.
\
DEFINE $CH3Cc      [C&H3]CccC[C&H3]
IF 2- [$CH3Cc]     { PRINT "/boldid/" HALT "MODSMI: JOIN" }
\
\  The following statement says that an atom of the $SECONDATT type
\  is an aromatic carbon atom (with one hydrogen attached) that is
\  adjacent to an aromatic carbon atom with an unsubstituted ethyl
\  group attached.
\
DEFINE $SECONDATT [c&H1]c[C&H2][C&H3]
\
\  The following says that if there is an atom of the SECONDATT
\  type, then an ethyl group should be added to this atom.
\
IF 1- [$SECONDATT]  { PRINT "/BOLDID/"
                      HALT "MODSMI: REPLACE c(CC)" }
\
\  The following statement says that an atom of the $FIRSTATT type is
\  an unsubstituted aromatic carbon atom that is adjacent to another
\  unsubstituted aromatic carbon atom:
\
DEFINE $FIRSTATT  [c&H1][c&H]
\
\  The following says that if there is an atom of the FIRSTATT
\  type, then an ethyl group should be added to this atom:
\
IF 1- [$FIRSTATT] { PRINT "/BOLDID/"
                    HALT "MODSMI: REPLACE c(CC)" }
```

Figure 6. The ACL commands that add a fused ring to an aromatic ring with two adjacent sites of unsubstitution.

ADDITIONAL PROGRAMS FOR PROCESSING OF MODSMI STRUCTURES

The program ONESMILE removes duplicate structures from a sorted list of SMILES. Thus, after the MODSMI transformations, sorting and ONESMILE would be used to produce a file of the unique molecules. Notice that this is possible only because all MODSMI operations produce the unique SMILES with the result that each particular molecular structure is represented by the same SMILES string regardless of the order of the atoms in the structure from which it originated.

ATOMWT optionally computes the molecular weight for each of a set of SMILES and writes this as the first three characters in an output file. After sorting by atomic weight, ATOMWT can be run again, this time using the option to remove the molecular weight. The result is a SMILES file sorted by molecular weight.

By combining MODSMI, ONESMILE and ATOMWT operations, one can transform a list of ALADDIN hits into a much smaller list that preserves the essence of the original results with the smallest compounds at the beginning of the list.

RESULTS OF MODSMI TRANSFORMATIONS OF D2 HITS[1]

The ACL shown in Figures 5 and 7 was used to transform the database structures into molecules to be synthesised. The ALADDIN search of a database of carefully modelled 3-D structures used in studies of a variety of medicinal chemistry projects identified 13016 potential compounds for synthesis. The MODSMI transformations reduced this number to 2529, an 80% reduction. Two slightly different searches of CONCORD structures of compounds in the Fine Chemicals Directory produced 12996 and 6559 original hits that were transformed into 1874 and 2247 unique compounds for an 85% and 66% reduction, respectively. Finally, the ALADDIN search of CONCORD structures of compounds in POMONA89[8] resulted in 10319 hits which were transformed into 2314 unique compounds for a 78% reduction in the number of compounds to be considered. Clearly the MODSMI transformations reduce the number of compounds to be handled by later steps. Thus this is a very useful strategy for evaluation of hits from a 3-D search.

Of these, 259, 124, 76, and 160 compounds respectively match the total search target. The low percentage of designed compounds that meet the total criteria was a source of concern. However, it was shown not to be due to the MODSMI transformations but rather because firstly, CONCORD generates only one conformation that might be different from that of the database molecule and secondly, the original search used broad criteria in order not to miss any interesting molecules[1]. Each of these is responsible for approximately 50% of the 'failures'.

DISCUSSION

The MODSMI commands of ACL have proven to be very useful in reducing the duplication of information from a 3-D design while preserving the essential information.

Such transformations could have reduced the number of compounds to consider even further if other transformations had been added. For example, our ACL did not specify the removal of the C_4H_4 fragment of fused naphthalene rings or the C_3H_6 or C_4H_8 fragments of fused saturated rings, attached substituents that have a chain that ends in a ring other than benzene, or attached substituents with unsaturation.

The reason we did not program the above transformations is that we did not anticipate the need for them. Thus one limitation of our strategy is that an expert must recognise what transformations the program should perform. However, once this is programmed it may be used in other studies. We typically iteratively refine the MODSMI ACL with a subset of the compounds until the structures are pruned satisfactorily. Additionally, sometimes it is difficult to anticipate the exact 2-D structures from which one would want to remove certain substituents and those for which they should be left on the molecule; in other words it may be difficult to program, in 2-D structure terms, transformations that will have insignificant consequences for the 3-D arrangement of the pharmacophoric groups. At a certain

```
ALADDIN NO_THOR_LOOKUP
\
\  The first statements define one type of freely-rotatable bond
\  that has an SP3 atom that is not in a ring and that is connected
\  by a single bond to another SP3 atom.  The symbol D refers to
\  the number of attached heavy atoms.
\
DEFINE $CSP3       [C&H4&D0,C&H3&D1,C&H2&D2,C&H1&D3,C&H0*D4]
DEFINE $NQUAT      [N&H4&D0,N&H3&D1,N&H2&D2,N&H1&D3,N&H0*D4]
DEFINE $NSP3       [N&H3&D0,N&H2&D1,N&H1&D2,N&H0&D3]
DEFINE $OSP3       [O&H2&D0,O&H1&D1,O&H0&D2]
DEFINE $HALOGEN    [I,Br,Cl,F]
DEFINE $SP3        [$CSP3,$NQUAT,$NSP3,$OSP3,$HALOGEN]
DEFINE $NONRINGSP3   [$SP3$!R]
\
\  The end atom on a side chain is labeled $TERMINALFRB, and the
\  other atoms in the chain are labeled $FRB. If there is the
\  former in the molecule that is removed, if not, then if there
\  is an $FRB one bond to it (and not to a hit atom) is reduced by 1.
\  * designates "any atom".
\
DEFINE $TERMINALFRB [$NONRINGSP3&D1]-[!R]
DEFINE $FRB        [$NONRINGSP3$!D1]-[*]
CALL "MARKHIT" "0"
IF 1- [$TERMINALFRB]{ PRINT "/BOLDID/" HALT "MODSMI:NIBBLE" }
IF 1- *-[$FRB]     { PRINT "/BOLDID/" HALT "MODSMI:AXE" }
\
\  The following removes unsubstituted phenyl rings.  (In GENIE rings
\  are built by connecting atoms with the same numeric label.)
\
DEFINE $PENDANTPHE   [c&H0]2[c;H1][c;H1][c;H1][c;H1][c;H1]2
IF 1- *-$PENDANTPHE  { PRINT "/BOLDID/" HALT "MODSMI:AXE" }
\
\  The following removes pendant aldehydes and -SO2's.
\
DEFINE $PENDANTALD   [C&H1]=O
DEFINE $PENDANTSO2   [S&D3](=O)=O
DEFINE $PENDANTJUNK [$PENDANTALD,$PENDANTSO2]
IF 1- *-[$PENDANTJUNK]     { PRINT "/BOLDID/" HALT "MODSMI:AXE" }
\
\  The final step removes LASTBITE, any non-ring atom bonded to only
\  one other heavy atom that is in a ring.
\
DEFINE $LASTBITE  [!R&D1]-R
IF 1- [$LASTBITE] { PRINT "/BOLDID/" HALT "MODSMI:NIBBLE" }
\
\  The following changes heteroatoms into carbons preserving the
\  aromaticity if present.
\
IF 1- N            { PRINT "/BOLDID/" HALT "MODSMI:REPLACE C" }
IF 1- n            { PRINT "/BOLDID/" HALT "MODSMI:REPLACE c" }
IF 1- O            { PRINT "/BOLDID/" HALT "MODSMI:REPLACE C" }
IF 1- o            { PRINT "/BOLDID/' HALT "MODSMI:REPLACE c" }
IF 1- S            { PRINT "/BOLDID/" HALT "MODSMI:REPLACE C" }
                   { PRINT "/BOLDID/" HALT "MODSMI:REPLACE c" }
```

Figure 7. The MODSMI commands that remove the unwanted substituents and change the non-hit heteroatoms in structures 1–3.

point the best strategy for evaluating the hits is for a human to look at them and sort them by hand. MODSMI reduces the numbers of compounds enough that this final human viewing is possible.

One can sometimes reduce the problem of duplicate suggestions for synthesis by carefully preparing the database on which the design is based. For example, if one wished to design only certain types of molecules such as fused-ring compounds of 6-, 7-, or 8-membered rings, one might prepare a list of 2-D structures of a variety of compounds to be tested with ALADDIN for the design. The 3-D structures of these potential templates could then be generated with CONCORD and searched with ALADDIN. The advantage of this strategy over searching a more heterogeneous database is that only one of each ring type will be searched, not potentially many derivatives as are found in typical databases. To prepare the list of 2-D structures one might transform structures from a database by using MODSMI commands to prune them into an essential core. This process would usually convert several different molecules into the same core. On the other hand, one could use MODSMI commands to generate the 2-D structures of potential templates by iterative transformation of the structures in an arbitrary list.

MODSMI commands do not solve all problems. For example, although structures **12** and **13** are geometrically not identical to **7**, they are identical in the 2-D core that holds the pharmacophore atoms, that is, **7**, **12**, and **13** are identical in the two shortest paths between the pharmacophore atoms. In a medicinal chemistry program it might be strategically sensible to prepare **7** and only if it is observed to be active to prepare **12** and **13**. It would be convenient if the computer could organise the data in this way for us. Thus we are interested in a type of routine that will recognise from the 2-D structure of two molecules that they are identical in the geometric properties of the atoms in the shortest paths between the pharmacophoric atoms. This would allow one to recognise that compounds **4–7** are of one class and that **8–11** are of another, but more importantly that compounds such as **12** and **13** are also of the same class as **7**. Such routines would be a logical complement to the MODSMI routines discussed in this report.

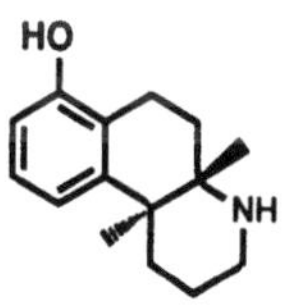

12

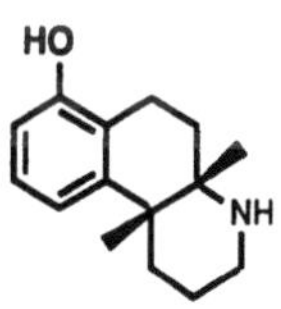

13

Alternatively, one can attempt such classifications by considering the geometric properties themselves. For this purpose ALADDIN produces a table of the values of the geometric properties of the molecules that match the search criteria. This table can then be sorted to identify molecules that are identical. Such a procedure can be useful for dealing with structures that are produced from only one minimiser or structure generator. However, when structures from different programs are in the database it is not always obvious which molecules are 'identical' and which are different. Clustering or other multivariate statistical techniques may be useful, but can require large amounts of computer time for a large data-set. Evaluation of these observations is an area of active research in our laboratories.

REFERENCES

1. Martin, Y.C. 'Computer Design of Potentially Bioactive Molecules by Geometric Searching with ALADDIN'. *Tetrahedron Comput. Methodol.* **1990,** *3,* 15–25.
2. Martin, Y.C.; Bures, M.G.; Willett, P. 'Searching Databases of Three-dimensional Structures'. In *Reviews in Computational Chemistry;* Boyd, D.B.; Lipkowitz, K., Eds; VCH: New York, 1990; Vol. 1, pp.213–263.
3. Van Drie, J.H.; Weininger, D.; Martin, Y.C. 'ALADDIN: An Integrated Tool for Computer-assisted Molecular Design and Pharmacophore Recognition from Geometric, Steric, and Substructure Searching of Three-dimensional Molecular Structures'. *J. Comput.-Aided Mol. Des.* **1989,** *3,* 225–251.
4. Bures, M.G.; Black-Schaefer, C.; Gardner, G. 'The Discovery of Novel Auxin Transport Inhibitors by Molecular Modelling and Three-dimensional Pattern Analysis'. *J. Comput.-Aided Mol. Des.* **1991,** *5,* 323–334.
5. Martin, Y.C.; Danaher, E.A. 'Molecular Modelling of Receptor-Ligand Interactions'. In *Receptor Pharmacology and Function*; Williams, M.; Glennon, R.A.; Timmermans, P.B.M.W.M., Eds; Marcel Dekker: New York, 1989; pp.137–171.
6. Rusinko III, A.; Skell, J.M.; Balducci, R.; McGarity, C.M.; Pearlman, R.S. 'CONCORD, A Program for the Rapid Generation of High Quality Approximate 3-Dimensional Molecular Structures'. Tripos Associates, St. Louis, Missouri, 1988.
7. Anon. Daylight Chemical Information Systems, Irvine, California.
8. Hansch, C.; Leo, A. The 1989 edition of the Pomoma College Medicinal Chemistry Project database of measured octanol-water partition coefficients and pKa's distributed by Daylight Chemical Information Systems, Irvine, California.

CHEMICAL STRUCTURE HANDLING USING THE DISTRIBUTED ARRAY PROCESSOR

Edie M. Rasmussen, Peter Willett and Terence Wilson
Department of Information Studies, University of Sheffield, Western Bank, Sheffield S10 2TN, England

ABSTRACT

The Distributed Array Processor (DAP) is a parallel computer that allows the simultaneous processing of many thousands of data items. The massively parallel nature of the DAP is well suited to the processing of large databases and in this paper we describe its use for the clustering of files of 2-D structures using the Jarvis-Patrick clustering method, for the ranking of output in an experimental substructure searching system for the 3-D macromolecules in the Protein Data Bank, and for the implementation of atom-by-atom searching using Ullmann's subgraph isomorphism algorithm. The experimental results demonstrate that the DAP is considerably faster than a conventional mainframe processor, although the precise degree of speed-up that can be obtained is strongly dependent upon the characteristics of the data that is to be processed.

THE DISTRIBUTED ARRAY PROCESSOR

Parallel processing involves the simultaneous use of some or many processors so as to reduce the elapsed time that is required for the execution of some computational task[1–3]. Parallel computers have traditionally been used for scientific applications that require extensive numeric processing but there is now increasing interest in the use of such machines for non-numeric applications. One such application is that of database searching, where the matching of the query record against the records in a database can be distributed across the available processors.

In this paper, we consider the use of one particular class of parallel computer, an *array processor*, for the processing of files of 2-D and 3-D chemical structures. The specific machine that we have used is the Distributed Array Processor, or DAP[4–7] which was originally developed and marketed by International Computers Limited (ICL) but which is now the responsibility of a spin-off company, Active Memory Technology Limited (AMT). The DAP consists of a single Master Control Unit (MCU) and a large number of very simple processing elements (PEs), each of which has a slow, bit-serial Arithmetic and Logic Unit (ALU) and a small amount of local storage. The MCU broadcasts instructions for execution by the PEs; all of the PEs carry out the instruction except for those that have been *masked* out, i.e., temporarily disabled as a result of previous instructions. The DAP is thus an example of a Single Instruction stream Multiple Data stream (SIMD) computer[8], with a single sequence of instructions being executed in parallel on multiple data items. The data items are stored in the PEs, which are arranged in a 2-D array with each PE being attached to its North, East, South and West neighbours and

W. A. Warr (Ed.)
Chemical Structures 2

with data highways that allow the transmission or selection of data to or from all processors in a row or column (so that the broadcasting of instructions is facilitated by the geometry of the array). The DAP is usually programmed in either a high level Fortran-like language, Fortran Plus (formerly called DAP Fortran), which was used for all of the work reported here, or a low level assembler, APAL (although other languages are becoming available, e.g., Prolog and Ada). Fortran Plus is essentially an extension of Fortran 77 that allows the full exploitation of the parallel capabilities of the DAP and that encompasses many of the features that are expected to become available in Fortran 8X, e.g., vector and matrix handling functions.

The experiments reported here have involved the national DAP facilities that are provided in the UK by the Centre for Parallel Computing at Queen Mary and Westfield College, University of London. This originally consisted of a 64x64 ICL DAP with each PE containing 2 Kbytes of local storage, giving an overall memory size of 8 Mbytes. This machine has recently been replaced by an AMT DAP 610. This again contains a 64x64 array of PEs; however, these have a much faster cycle time and each of them has 8 Kbytes of local storage, giving an overall memory size of 32 Mbytes. The earlier ICL machine was used for the work on clustering described in the next section while the remainder has used the AMT machine.

CLUSTERING OF CHEMICAL STRUCTURES

The Jarvis-Patrick Clustering Method

The last few years have seen rapid developments in the use of structure handling techniques that are based on measures of intermolecular similarity[9–11]. A simple and effective way[12] of measuring intermolecular structural similarity is to compare the lists of fragment substructures characterising each of a pair of molecules to identify those in common; this information can then be used to calculate a similarity coefficient such as the Tanimoto coefficient[13]. One technique that has been suggested is the clustering of chemical structure databases for purposes of compound selection in biological screening programmes[14]. Cluster analysis[15,16] involves the comparison of each molecule with every other molecule in a data-set to identify those that should be clustered together. Hence, for a data-set containing N molecules, at least order $O(N^2)$ similarities must be calculated and thus clustering presents severe computational problems if large data-sets are to be processed, even when highly efficient algorithms are available[16]. Parallel computer hardware offers an attractive way of increasing the efficiency of large-scale clustering since, as Salton and Bergmark[17] note, there is a high degree of parallelism in the calculation of a set of similarity coefficients: the same operation must be repeated many times on different data, and the results of the individual calculations are not interdependent. This type of parallelism is well suited to an array processor since large numbers of similarity coefficients can be evaluated simultaneously by providing each PE in the array with the data required for a single calculation, and then by matching a specified object against all of the PEs in parallel. If N is not greater than the number of processors, then the time complexity of computing the elements of an NxN similarity matrix reduces from order $O(N^2)$ to order $O(N)$ (provided that there are no serial bottlenecks or interprocessor communication costs that interfere significantly with the parallelism).

An extended comparison of different types of clustering method[11] has demonstrated the general utility of the Jarvis-Patrick method[18] for the clustering of files of 2-D chemical structures. This method is based on the concept of shared *nearest neighbours*, where the nearest neighbour for some molecule I is that other molecule that is most similar to it, using some particular measure of similarity. Given a set of N structures, each of which is associated with its K nearest neighbours, two structures, I and J, are assigned to the same cluster if I is one of the K nearest neighbours of J, if J is one of the K nearest neighbours of I and if I and J have at least K(T), a user-specified threshold, of their K nearest neighbours in common. The computation required for the Jarvis-Patrick method is in two separate stages, these being the creation of the lists of K nearest neighbours and the processing of these lists to create the clusters. Both K and K(T) can be varied until clusters acceptable to the user are produced.

The I-th nearest neighbour list ($1 \leq I \leq N$) has the structure I as its own zeroth nearest neighbour, followed by the K nearest neighbours in order. The clustering algorithm is as follows, where LABELTABLE is an N-element integer array that denotes the cluster membership of each of the N structures:

```
FOR I := 1 TO N DO
  LABELTABLE(I) :=I;
FOR I := 1 TO N-1 DO
  FOR J := I+1 TO N DO
  IF I and J should be clustered together, i.e., if they are nearest neighbours of
  each other and if they have K(T) nearest neighbours in common, THEN
    BEGIN
      K := min{LABELTABLE(I),LABELTABLE(J)};
      L := max{LABELTABLE(I),LABELTABLE(J)};
      FOR M := 1 TO N DO
        IF LABELTABLE(M) = L THEN LABELTABLE(L) := K
    END.
```

Thus, when all of the possible pairs of structures, I and J, have been considered, the label table indicates the cluster membership, i.e., the lowest-valued structure in the cluster, for each structure in the data-set.

The Jarvis-Patrick clustering method is computationally elegant since it requires only one pass through the original data-set to identify the nearest neighbours, followed by one pass through the nearest neighbour lists to generate the label table. The first stage is the more demanding, requiring order $O(N^2)$ similarity calculations. Once the nearest neighbour lists have been created, the original data can be discarded to reduce storage requirements. In a clustering run, the I-th nearest neighbour list needs to be compared with K other lists giving a total of only KN comparisons of lists of integers.

Parallel Clustering Algorithm

Previous studies at Sheffield have demonstrated the efficiency of the DAP for nearest neighbour searching in textual databases[7] and we have thus investigated its use for the implementation of the Jarvis-Patrick method[19]. The efficiency of the DAP for non-numeric processing arises in large part from the bit-serial nature of the PE ALUs, which permits very high computational speeds if the processing consists primarily of logical operations on arrays of bits. This characteristic is highly appropriate for the processing of chemical structures represented by fragment bit strings[20]. Assume that the data-set to be clustered contains N molecules, each of which is represented by a bit string containing M bits. The data-set can thus be considered as an NxM-element bit map, BITMAP. The element BITMAP(I,J) is set to True (or False) if the J-th fragment is present in (or absent from) the I-th molecule. Thus, the I-th row and the J-th column of BITMAP are bit strings representing the assignment of fragments to the I-th structure and representing the compounds which contain the J-th fragment, respectively.

Assume that one row of BITMAP is loaded into each PE, so that the same relative location in all of the 4096 PEs can be accessed at the same time in response to instructions broadcast from the MCU. It is thus possible to inspect 4096 structures simultaneously to determine whether a particular bit position is set to True, i.e., to determine whether one particular substructural fragment is present and in common with one of the fragments assigned to the current query molecule, Q. For those PEs where this is the case, increments are made to an array of integers containing the numbers of fragments in common between Q and each of the 4096 structures currently being considered. Hence, if this matching operation is repeated for all of the non-zero elements in the bit string representing Q, the X-th counter ($1 \leq X \leq 4096$) will contain the number of fragments in common between Q and the X-th molecule. Given this information, a similarity coefficient, the Tanimoto coefficient in the experiments reported here, can then be evaluated in parallel for all of the 4096 structures if the number of fragments in the X-th compound, SIZE(X), is stored in the X-th PE. The sequence of operations for a single query molecule, Q, is illustrated below, where QSTRING is an integer array that contains the NQ indices of the non-zero bits in the query bit string and where COMMON(X) is used to cumulate the number of fragments in common between Q and the molecule in the X-th PE:

```
FOR each query molecule, Q, DO
  BEGIN
    COMMON(X) := 0 for all PEs in parallel;
    FOR I := 1 TO NQ DO
      COMMON(X) := COMMON(X)+1 for all PEs in parallel
        where BITMAP(X,QSTRING(I)) is True;
    identify Q's K nearest neighbours by calculating the Tanimoto coefficient,
        COMMON(X)/(NQ+SIZE(X)-COMMON(X)), for all PEs in parallel
  END
```

The nearest neighbour for Q is identified using MAXP, a subroutine in the DAP system library that returns the largest element in a matrix; as K, the number of nearest neighbours required for a Jarvis-Patrick run, is generally greater than one, once the first nearest neighbour has been identified, that PE can be masked out, MAXP called again to identify the second nearest neighbour and so on until K nearest neighbours have been obtained[21].

The use of the non-zero bits in the bit string representing Q has the advantage that only a limited number of locations in each of the PEs needs to be examined. This requires that Q, and hence all of the molecules in the database (since each one of them in turn is used as the query molecule), be stored as a list of fragment numbers, in addition to the bit string, so that the numbers can be used as indices to the locations which need to be checked. If each structure is stored only as a bit string, then the non-zero bits must be located by a scan of the bit string before the compound can be used as the query. This process is inherently serial in character and thus substantially slower than the procedure described here, where the non-zero bits are available as required.

While the creation of the nearest neighbour lists can be accomplished very efficiently, the clustering criteria used in the Jarvis-Patrick method are not overly well suited to parallel processing as only K of the 4096 structures need to be considered for clustering with the current structure (specifically those corresponding to the K nearest neighbours stored for each molecule). The PEs corresponding to these K structures can be isolated by setting a mask, i.e., by deactivating all of the PEs except those K that represent the K nearest neighbours. Each of the K nearest neighbours for the current structure is then broadcast for comparison with each of the K nearest neighbours in the K active PEs. The similarity value is incremented by one in each of the active PEs where a common nearest neighbour is found, and the current structure is then clustered together with those of the K structures where the calculated similarity is greater than the threshold value, K(T). The clustering is effected by updating the label table as described previously. This updating step can be done in parallel, to the extent that all of the PEs storing a value to be updated can be updated together; the detailed implementation of this is described by Rasmussen[21].

Serial Clustering Algorithm

An accurate evaluation of the efficiency of a parallel algorithm requires that it be compared with the best possible serial algorithm. The best such algorithm for nearest neighbour searching, and hence for Jarvis-Patrick clustering, is one that was originally developed for nearest neighbour searching in files of textual documents[22] and that makes use of an inverted file to the database that is to be searched. The algorithm involves the addition of the inverted file postings lists corresponding to the fragments contained in the current query structure, Q; we shall refer to these lists as *query lists*. The addition of the query lists results in a new list, each element of which contains the number of fragments in common between Q and one of the structures in the database. The addition of the query lists may be achieved as follows. When some molecule, X ($1 \leq X \leq N$), is encountered for the first time in a query list, the X-th element of the new list is set to one. This element is incremented by one each time that the X-th structure is encountered in subsequent query lists. Thus, when all of the lists have been processed in this way, the X-th element will contain the number of fragments in common between the query and the X-th molecule. The N numbers can then be used for the calculation of the Tanimoto coefficient and thus for the identification of the nearest neighbour(s).

The algorithm is shown below, where COMMON is the N-element array that is used to cumulate the fragments in common between Q and each of the structures in the database and where NQ is the number of fragments in the query.

```
FOR each query molecule, Q, DO
  BEGIN
    FOR I := 1 TO N DO
      COMMON(I) := 0;
    FOR I := 1 TO NQ DO
      BEGIN
```

```
        retrieve the I-th query list from the inverted file;
        COMMON(X) := COMMON(X)+1 for each identifier, X, in this list
      END;
    FOR I := 1 TO N DO
      update the list of Q's nearest neighbours by calculating the Tanimoto
        coefficient, COMMON(I)/(NQ+SIZE(I)-COMMON(I))
  END
```

An extended discussion of the use of this algorithm for nearest neighbour searching in chemical structure databases is presented by Willett[11].

Experimental Details and Results

The file of structures used in the experiments was a systematic sample of 8192 molecules taken from the *ChemQuest* database. The molecules were represented by 1123-bit strings denoting the presence or absence of the augmented atom, atom sequence and bond sequence substructural fragments used in the CAS ONLINE substructure searching system[23]. The resulting bit map was used as the basis for the DAP processing. In the case of the IBM 3083 mainframe that was used for the serial implementation of the clustering method, the bit map was inverted and the non-zero elements eliminated to create the inverted file. In addition, a 1 in 2 sample of the 8192 structures was used to investigate the effect of database size on the relative efficiencies of serial and parallel processing. This we have evaluated using the *speed-up* measure, S(P)[24]. This is defined by the equation:

$$S(P) = T(S)/T(P)$$

where T(S) is the time required for running the best available serial algorithm on a serial processor and where T(P) is the corresponding time for the best parallel algorithm on a parallel processor. The serial machine that was used was the University of Sheffield IBM 3083 BX mainframe. Tables 1 and 2 show the run times (in CPU seconds) and the speed-ups for generating the nearest neighbour lists and the clusters for 5, 10 and 20 nearest neighbours. It will be seen from Table 1 that the parallel nearest neighbour searching algorithm is from 5.4 to 6.4 times faster than the serial algorithm with the small data-set, and from 6.1 to 7.9 times faster with the large data-set. In both the serial and parallel cases, the incremental time taken to generate 20, rather than 5 or 10, nearest neighbours is

Table 1. Execution times in CPU seconds and speed-ups for the nearest neighbour searching stage of the Jarvis-Patrick method

	4096 Structures			8192 Structures		
K	T(S)	T(P)	S(P)	T(S)	T(P)	S(P)
5	221.1	34.4	6.4	927.8	117.2	7.9
10	222.8	37.3	6.0	935.6	130.5	7.2
20	231.9	43.2	5.4	953.5	157.2	6.1

Table 2. Execution times in CPU seconds and speed-ups for the clustering stage of the Jarvis-Patrick method

		4096 Structures			8192 Structures		
K	K(T)	T(S)	T(P)	S(P)	T(S)	T(P)	S(P)
5	2	28.8	15.6	1.8	120.2	39.2	3.1
10	2	93.3	56.0	1.7	377.0	149.2	2.5
	4	72.2	55.6	1.3	298.5	147.9	2.0
	6	44.6	55.1	0.8	190.0	146.0	1.3
20	6	193.6	210.3	0.9	754.0	625.2	1.2
	8	168.1	209.9	0.8	655.2	623.9	1.1
	10	139.9	209.5	0.7	547.7	622.4	0.9

not large. In the serial case, this represents the housekeeping time needed to maintain a longer list of current nearest neighbours; in the parallel case, this represents the additional time required to call MAXP 20 times, rather than 5 or 10 times. The relative clustering performance of the parallel and serial processors in Table 2 varies over a range from 0.7 to 3.1 as K, K(T) and the data-set size are varied. The time taken for the serial processing is highly dependent on the threshold, K(T), that is chosen, while the parallel processing is affected very little by this factor. This is because a low threshold, which results in large clusters generates considerable updating of the label table; this is less significant on the DAP, where the updating can be done in parallel. The efficiency of the DAP processing is more susceptible to the length of the nearest neighbour lists: every member of the list must be broadcast to every DAP nearest neighbour plane in the matching operation, this requiring a total of K^2 comparisons per structure. In the case of the serial processor, conversely, the lists can be sorted into increasing molecule identifier order, requiring order O(K log K) operations; once this has been done, the comparison of any two lists requires, at most, 2K comparisons.

A further point to note is that the DAP is not able to take full advantage of the fact that only K candidates need to be considered for clustering with the current structure; on the DAP, any number of comparisons up to 4096 is done in equal time, irrespective of K. That said, if all 5, 10 or 20 candidates for clustering can be found in one set of 4096 records, then only that set need be examined; this explains why there is less than a four-fold increase in the clustering times as the data-set size is increased from 4096 to 8192. Thus, in clustering from the nearest neighbour tables, the DAP performs best for a large number of records and short nearest neighbour lists, whereas the serial processor is better for a small number of records and long nearest neighbour lists.

The results for the two stages given above can be consolidated to show the overall time requirements. For 4096 structures, the overall speed-up ranges from 1.5 to 5.0; for 8192 structures, the range is from 1.9 to 6.7. The ratio for the larger data-set is greater than the smaller one for all values of K and K(T) which were tested; this suggests that the use of the DAP would become still more attractive as larger and larger structure databases need to be clustered.

RANKING OF POSSUM OUTPUT

Serial Ranking Algorithm

Elsewhere in these proceedings[25], we have described an algorithm for the ranking of output from our experimental protein retrieval system, POSSUM. The algorithm takes the output from a substructure search of the Protein Data Bank and ranks the motifs that have been retrieved in order of decreasing structural similarity with the query motif, which consists of a pattern of α-helix and β-strand secondary structure elements in 3-D space (the reader is referred to Artymiuk *et al.*[25] for further details of the algorithm, since these will not be repeated here). Although the algorithm can provide useful rankings of the search output, it is extremely time-consuming in execution. The reason for this is evident from an inspection of the description shown below, where the database motif being searched contains NLINES secondary structure elements, the I-th of which contains NPOINTS(I) points, where an integer array FREQ is used to cumulate the distances as they are calculated, and where a comparable array, QFREQ, has previously been calculated to describe the frequency distribution for the inter-line distances in the query motif.

```
FOR each database motif resulting from the POSSUM search DO
  BEGIN
    set to zero all of the elements of an integer array, FREQ;
    FOR I := 1 TO NLINES-1 DO
      FOR J := I+1 TO NLINES DO
        FOR K := 1 TO NPOINTS(I) DO
          FOR L := 1 TO NPOINTS(J) DO
            BEGIN
              calculate the distance between the points LINE(I,K) and LINE
              (J,L);
              increment the element of FREQ corresponding to this distance by one
            END;
        calculate the degree of similarity between FREQ and QFREQ
    END;
display the database motifs in order of decreasing similarity.
```

The computation is dominated by the loop for the calculation of the distances, which has to be executed for each of the motifs in the database. Hence, the computation for each structure has a running time of order $O(NLINES^2 x NPOINTS(I)xNPOINTS(J))$; since NLINES is fixed for a given motif, the main determining factors are NPOINTS(I) and NPOINTS(J), both of which are, in turn, determined by the value of L that is specified by the user prior to executing the search, where L is the separation of the points along the lines representing each of the secondary structure elements. Unless L is set to a large value, i.e., a small number of points per line, huge numbers of inter-point distances need to be calculated and response times can be very lengthy indeed given the MicroVAX hardware on which POSSUM is currently implemented. However, the massively parallel nature of the DAP provides a simple mechanism for substantial improvements in the execution time of this algorithm, as we now demonstrate.

Parallel Ranking Algorithm

The parallel algorithm involves storing one set of co-ordinate points in each of the 4096 PEs. An individual point is then selected and its co-ordinates broadcast from the MCU, thus allowing the parallel calculation of the distances between this distinguished point and all of the other points in the motif. A frequency distribution, FREQ, is stored in each of the PEs and these are updated as the distances are calculated; once all of the distance calculations have been completed, these individual distributions are cumulated to give the overall distribution, TOTFREQ. The algorithm is thus as follows:

```
FOR each database motif resulting from the POSSUM search DO
  BEGIN
    set to zero all of the elements of an integer array, FREQ;
    FOR I := 1 TO NLINES DO
      FOR J := 1 TO NPOINTS(I) DO
        BEGIN
          broadcast the co-ordinates of the point LINE(I,J);
          calculate the distance between LINE(I,J) and the point
            stored in each PE for all PEs in parallel;
          increment by one the element of FREQ that corresponds to each such
          distance for all PEs in parallel
      END
    sum the local FREQ distributions to give TOTFREQ;
    calculate the degree of similarity between TOTFREQ and QFREQ
  END;
display the database motifs in order of decreasing similarity
```

This description assumes that the number of points is not greater than the number of PEs in the DAP. It should be noted that the algorithm results in each distance being calculated twice over. It is possible to mask out PEs that contain points in lines which have been processed already, but this is unnecessary since it takes just as long for the DAP to calculate the distances between the current point and points in some of the lines as in all of the lines (as long as the total number of points is not greater than the number of PEs as mentioned previously). Accordingly, it is simpler just to calculate all of the distances twice over and then to halve each element of FREQ.

An inspection of this DAP algorithm reveals that the computation for each structure has a running time of order O(NLINESxNPOINTS(I)), i.e., a factor of NLINESxNPOINTS(J) faster than that for the serial algorithm presented earlier. It would thus be expected that the advantage of the DAP should grow in line with the number of secondary structure elements in the motif that is being searched for, and with the number of points in each line. This is, indeed, the case as we demonstrate below. There is, however, another factor to be considered in the DAP processing which can be neglected in the case of a serial processor. This is the updating of the FREQ distributions that are carried out once the distances have been calculated for some point. On a serial machine, the updating is effected very simply by treating the calculated distance as an index so that the appropriate element of the array FREQ is incremented by direct addressing. This strategy could be adopted on the DAP, but would make no use of the parallelism in the hardware since it would involve considering each and every PE in turn. An alternative strategy is to take the distance range corresponding to each element of FREQ in turn; the range is broadcast from the MCU and compared with the distance stored in each PE. If a match is obtained in some PE, then the corresponding element of that PE's FREQ is incremented by one. It will be realised that the number of times that this updating operation needs to be carried out on the DAP is determined by the value of R, the distance range for each element of FREQ and QFREQ. Thus, when R is small, there will be very large numbers of elements and very large numbers of updating operations required; conversely, a large value for R implies fewer updates and faster overall DAP processing. On a serial machine, the update time is independent of R and thus the DAP's performance will be affected adversely if FREQ contains large numbers of elements. The individual FREQs are then cumulated to give TOTFREQ once all of the distances have been calculated.

The DAP algorithm is conceptually simple but suffers from a serious defect in that the number of points in the lines comprising a motif is typically much less than the number of PEs in the DAP, i.e., NLINESxNPOINTS$\ll$4096, and the great bulk of the PEs thus remains unused during the calculation of the inter-point distances. This is so even with the smallest values of L used here. There is, however, a simple strategem which overcomes this problem by allowing several motifs to be processed at the same time (so that the total number of points is much closer to 4096 than if only individual motifs are processed in sequence). The strategem is based on the fact that it is only the distances between the points, and not their actual co-ordinates, that are of importance in defining the overall topography of a motif: accordingly the points can be defined in terms of any co-ordinate system whatsoever. Assume, for purposes of simplicity, that just two database motifs are stored in the DAP, these containing a total of A and B points and that each point is denoted by just its X-axis co-ordinate. The arrangement of co-ordinate data in the DAP is hence of the form

$$X(1),X(2)...X(A-1),X(A),X(A+1)...X(A+B-1),X(A+B).$$

We can shift the set of coordinates denoting the second motif into the same coordinate frame as the set denoting the first motif by replacing each value X(I), $(A+1\leq I\leq A+B)$, by X(I)-(X(A+1)-X(1)). All of these B replacement operations are executed in parallel (with the first A PEs masked out so that the X values contained within them remain unchanged). Then, when the point P_1 is broadcast from the MCU to determine its distance from the points $P_2,P_3 \ldots P_A$ (and to update the frequency distribution for the first database motif), it can simultaneously be compared with the coordinates for the points $P_{A+1},P_{A+2} \ldots P_{A+B}$, this corresponding to the calculation of the distances of these points from the point P_{A+1} and thus providing the data for the updating of the frequency distribution for the second database motif. P_2 is then broadcast, then P_3, etc. etc.

The extension of this replacement algorithm to (X,Y,Z) co-ordinate data and to larger numbers of database motifs is trivial (although the detailed implementation requires careful encoding to encompass the differing numbers of points in each of the database motifs[26]).

EXPERIMENTAL DETAILS AND RESULTS

The efficiencies of the parallel and serial versions of the similarity algorithm have been investigated using the three query motifs (calcium binding fold, β-barrel and NAD binding fold) discussed in the paper by Artymiuk *et al.*[25]. The serial algorithm was implemented in Fortran 77 on the IBM 3083 mainframe. The rankings were carried out with L set to values of 0.25, 0.5, 1.0 and 5.0 Å and with R set to 0.5, 1.0, 2.5 and 5.0 Å. The resulting run times for the IBM 3083 and DAP 610 implementations, T(S) and T(P), are listed in Table 3, together with the speed-up as defined previously. An inspection of the results demonstrates that use of the DAP can result in substantial increases in the speed with which the rankings can be generated. The maximum speed-ups listed here are 6.7, 10.1 and 11.3 for the calcium binding fold, β-barrel and NAD binding motifs, respectively. Moreover, it should be noted that these figures are relative to an IBM 3083, which is about ten times faster than the MicroVAX on which POSSUM is currently implemented.

Table 3. Execution times in CPU seconds and speed-ups for the ranking of the outputs from substructure searches for three secondary structure motifs.

Parameters		Calcium binding fold			β-barrel			NAD binding fold		
L	R	T(S)	T(P)	S(P)	T(S)	T(P)	S(P)	T(S)	T(P)	S(P)
5.0	0.5	0.7	0.7	1.0	3.5	3.8	0.9	3.8	1.2	3.1
	1.0	0.7	0.4	1.6	3.5	2.3	1.5	3.8	0.9	4.3
	2.5	0.7	0.3	2.6	3.5	1.4	2.4	3.8	0.7	5.5
	5.0	0.7	0.2	3.4	3.5	1.1	3.1	3.8	0.6	6.1
1.0	0.5	2.1	1.6	1.3	18.0	8.3	2.2	16.9	6.2	2.7
	1.0	2.1	1.1	2.0	18.0	5.7	3.2	16.9	4.5	3.7
	2.5	2.1	0.8	2.8	17.9	4.1	4.3	16.8	3.6	4.7
	5.0	2.1	0.7	3.2	17.9	3.7	4.9	16.8	3.3	5.1
0.5	0.5	6.3	3.0	2.1	60.3	16.9	3.6	54.8	14.4	3.8
	1.0	6.3	2.1	3.0	60.5	11.8	5.1	54.8	10.1	5.4
	2.5	6.2	1.5	4.1	60.5	8.8	6.9	54.7	7.6	7.2
	5.0	6.2	1.3	4.7	60.5	7.7	7.9	54.7	6.9	7.9
0.25	0.5	22.2	7.9	2.8	225.1	50.8	4.4	204.3	43.6	4.7
	1.0	22.4	5.3	4.2	228.5	35.1	6.5	204.2	23.4	8.7
	2.5	22.4	3.8	5.9	228.6	25.7	8.9	203.8	21.1	9.7
	5.0	22.3	3.3	6.7	228.6	22.6	10.1	203.8	18.1	11.3

It will be seen that the execution time on the DAP is crucially dependent on R, the range parameter that specifies the precise form of the frequency distribution that is being calculated. A small value of R corresponds to a frequency distribution with very many elements, and thus very many time-consuming updating operations on the DAP. As an example, consider the DAP run times for the calcium binding fold runs with L=5.0 when R is set to 5.0 and 0.5 Å: the latter run time is no less than 3.6 times as large, whereas the difference in the IBM run times is almost non-existent, owing to the very different way in which updating is implemented on the serial machine. The execution time on the DAP is also crucially dependent on L, the length parameter that specifies the inter-point distances along the linear secondary structure elements. The difference in the expected run times of the serial and parallel run times has been noted previously as being proportional to NLINESxNPOINTS(J); a small value of L corresponds to a very large number of points and thus a greater advantage for the DAP. If we again consider the calcium binding fold runs, but this time with R=5.0 and with L set to 5.0 and 0.25 Å, then the run times are 0.2 and 3.3 for the DAP and 0.7 and 22.3 for the IBM. Thus, in the case of the DAP, the run time has increased by a factor of 16.6 times as L is decreased whereas the corresponding factor for the IBM is 33.3 times.

The relationship between the DAP run times and the values of the parameters L and R follows directly from the analysis of the parallel algorithm that was given in the previous section. This analysis would also suggest that the advantage of the DAP over the IBM should increase with an increase in NLINES, i.e., in the number of secondary structure elements in the motif that is being considered. The calcium binding, β-barrel and NAD binding motifs contain five, seven and eight secondary structure elements, respectively and it certainly is the case that the speed-ups for the the first of these are consistently, and noticeably, less than for the other two motifs. However, the NAD speed-ups are consistently greater than the β-barrel figures, despite the fact that the former contains one less secondary structure element.

The most important conclusion to be drawn from the experimental results is that the DAP provides a very efficient way of ranking the output of POSSUM searches.

SUBSTRUCTURE SEARCHING

Our current work involves the use of the DAP for the implementation of Ullmann's subgraph isomorphism algorithm[27], which has been shown to provide a highly efficient mechanism for substructure searching in files of both 2-D and 3-D chemical structures[28,29].

The Ullmann algorithm operates by means of a backtracking tree search in which database atoms are tentatively assigned to query atoms and the match is extended in a depth-first manner until a complete match is obtained or until a mismatch is detected; in this case, the search then backtracks to the previous assignment and an alternative match is considered. Backtracking search is a common technique for increasing the efficiency of graph matching algorithms: Ullmann's contribution was to identify an heuristic, which he referred to as the *refinement* procedure, that tries to limit the number of levels of the search tree that have to be investigated before a mismatch is identified. Specifically, the algorithm makes use of the fact that if some query atom, Q(X), has another query atom, Q(W), at some specific distance and if some database atom, S(Z), matches Q(W), then there must be some database atom, S(Y), at the appropriate distance from S(Z) which matches Q(X); this is a necessary, but not sufficient, condition for a subgraph isomorphism to be present (except if all of the query atoms, Q(X), have been mapped, in which case the condition is sufficient for subgraph isomorphism). The refinement procedure is called before each possible assignment of a database atom to a query atom; the matched substructure is increased in size by one atom if, and only if, the condition holds for all of the possible values of W, X, Y and Z. Ullmann advocated this particular refinement technique in the context of general graphs, but it seems to be particulary well suited to the processing of chemical graphs[28,29].

The refinement procedure in the Ullmann algorithm involves a large number of logical operations that can be mapped onto the bit-serial nature of the DAP with very great efficiency. Specifically, one bit in each of the 64x64 array of PEs can be used to represent one element in the 64x64 logical matrix that is repeatedly processed during the refinement stage; thus, the DAP allows the simultaneous processing of the entire matrix whereas a conventional processor requires that each element of this matrix is processed in sequence. Accordingly, the refinement stage is executed with quite remarkable speed but this is not, unfortunately, the entire story since it is also necessary to encompass the backtracking component of the atom-by-atom search. This type of operation is inherently sequential in nature and thus much less well suited to a DAP implementation. Even so, our preliminary results suggest that the overall run time of this DAP implementation of the Ullmann algorithm is several times faster than an IBM 3083 implementation.

Our work to date has involved distributing the elements of the adjacency matrix describing a single structure across the array of PEs, so that a query substructure is matched against one database structure at a time. There is, however, an alternative way of implementing the algorithm, in which an entire structure is loaded into each of the PEs. Here, the time to match a single structure against the query will be very much greater than in the previous mapping; however, it does allow up to 4096 structures to be searched in parallel, with the possibility of a substantial increase in the speed-up (although this is at the cost of a markedly more complex algorithm, the efficient execution of which requires extensive synchronisation of the operations that are carried out across the array of PEs). The merits of this approach are currently under active investigation and will be reported shortly.

CONCLUSIONS

In this paper, we have discussed the use of a massively parallel array processor, the DAP, for three chemical structure handling applications. The results that have been obtained suggest that this type of parallel architecture is very well suited to database processing applications, with a substantial degree of speed-up being obtained when compared with efficient serial implementations on a conventional mainframe processor. However, the precise level of performance that can be achieved is critically dependent on the precise characteristics of the data-set that is being processed, e.g., the values of K and K(T) in the Jarvis-Patrick experiments and the values of T and R in the protein searching experiments. This dependency is in accordance with other studies of the use of the DAP[7] and appears to be a general characteristic of the DAP and, by implication, of other types of array processor when they are used for database processing applications.

ACKNOWLEDGMENTS

We thank the staff of the Centre for Parallel Computing, Queen Mary and Westfield College for the provision of computing facilities and Active Memory Technology Ltd., the British Library Research and Development Department and the Science and Engineering Research Council for funding. We thank Dr G.M. Downs and Ms. H.M. Grindley for assistance with the data-sets used in the studies reported here and Dr. S.F. Reddaway for his contributions to our work on the Ullmann algorithm.

REFERENCES

1. *Parallel Computers 2. Architecture, Programming and Algorithms*; Hockney, R.W.; Jesshope, C.R.; Adam Hilger: Bristol, 1988.
2. *Computer Architecture and Parallel Processing*; Hwang, K.; Briggs, F.A.; McGraw-Hill: New York, 1984.
3. *Designing Efficient Algorithms for Parallel Computers*; Quinn, M.J.; McGraw-Hill: New York, 1987.
4. Gostick, R.W. 'Software and Algorithms for the Distributed Array Processor'. *ICL Technical Journal* **1979,** *2,* 116–135.
5. Gostick, R.W. 'Software and Hardware Technology for the ICL Distributed Array Processor'. *Australian Computer Journal* **1981,** *13,* 1–6.
6. Parkinson, D. 'The Distributed Array Processsor (DAP)'. *Comput. Phys. Commun.* **1983,** *28,* 325–336.
7. *Parallel Database Processing. Text Retrieval and Cluster Analysis Using the Distributed Array Processor*; Willett, P.; Rasmussen, E.M.; Pitman: London, 1990.
8. Flynn, M.J. 'Some Computer Organisations and Their Effectiveness'. *IEEE Trans. Comput.* **1972,** *C-21,* 948–960.
9. Carhart, R.E.; Smith, D.H.; Venkataraghavan, R. 'Atom Pairs as Molecular Features in Structure-activity Studies: Definition and Applications'. *J. Chem. Inf. Comput. Sci.* **1985,** *25,* 64–73.
10. Johnson, M.A. 'A Review and Examination of the Mathematical Spaces Underlying Molecular Similarity Analysis'. *J. Math. Chem.* **1989,** *3,* 117–145.
11. *Similarity and Clustering in Chemical Information Systems*; Willett, P.; Research Studies Press: Letchworth, 1987.
12. Adamson, G.W.; Bush, J.A. 'A Method for the Automatic Classification of Chemical Structures'. *Inf. Storage Retr.* **1973,** *9* 561–568.
13. Willett, P.; Winterman, V.; Bawden, D. 'Implementation of Nearest Neighbour Searching in an Online Chemical Structure System'. *J. Chem. Inf. Comput. Sci.* **1986,** *26,* 36–41.
14. Willett, P.; Winterman, V.; Bawden, D. 'Implementation of Non-hierarchic Cluster Analysis Methods in Chemical Information Systems: Selection of Compounds for Biological Testing and Clustering of Substructure Search Output'. *J. Chem. Inf. Comput. Sci.* **1986,** *26,* 109–118.
15. Dubes, R.; Jain, A.K. 'Clustering Methodologies in Exploratory Data Analysis'. *Advances in Computers* **1980,** *19,* 113–227.
16. *Multidimensional Clustering Algorithms*; Murtagh, F.; Physica-Verlag: Vienna, 1985.
17. Salton, G.; Bergmark, D. 'Parallel Computation in Information Retrieval'. *Lecture Notes in Computer Science* **1981,** *111,* 328–342.
18. Jarvis, R.A.; Patrick, E.A. 'Clustering Using a Similarity Measure Based On Shared Nearest Neighbours'. *IEEE Trans. Comput.* **1973,** *C-22,* 1025–1034.
19. Rasmussen, E.M.; Downs, G.M.; Willett, P. 'Automatic Classification of Chemical Structure Databases Using a Highly Parallel Array Processor'. *J. Comput. Chem.* **1988,** *9,* 378–386.

20. *Communication, Storage and Retrieval of Chemical Information*; Ash, J.E.; Chubb, P.A.; Ward, S.E.; Welford, S.M.; Willett, P., Eds; Ellis Horwood: Chicester, 1985.
21. Rasmussen, E.M. *Cluster Analysis on a Highly Parallel Array Processor*. Ph.D. Thesis, University of Sheffield, 1988.
22. Noreault, T.; Koll, M.; McGill, M.J. 'Automatic Ranked Output from Boolean Searches in SIRE'. *J. Am. Soc. Inf. Sci.* **1977,** *28,* 333–339.
23. Dittmar, P.G.; Farmer, N.A.; Fisanick, W.; Haines, R.C.; Mockus, J. 'The CAS ONLINE Search System. Part I. General Design and Selection, Generation and Use of Search Screens'. *J. Chem. Inf. Comput. Sci.* **1983,** *23,* 93–102.
24. Parkinson, D.; Liddell, H.M. 'The Measurement of Performance on a Highly Parallel System'. *IEEE Trans. Comput.* **1983,** *C-32,* 32–37.
25. Artymiuk, P.J.; Grindley, H.M.; Mitchell, E.M.; Rice, D.W.; Ujah, E.C.; Willett, P. 'Representation and Searching of 3-D Protein Structures'. In these Proceedings.
26. Wilson, T. *Implementation of Graph-matching Techniques in Chemical Databases Using a Single Instruction Stream, Multiple Data Stream Array Processor* Ph.D. thesis, University of Sheffield, 1991.
27. Ullmann, J.R. 'An Algorithm for Subgraph Isomorphism'. *J. Assoc. Comput. Mach.* **1976,** *16,* 31–42.
28. Brint, A.T.; Willett, P. 'Pharmacophoric Pattern Matching in Files of 3-D Chemical Structures: Comparison of Geometric Searching Algorithms. *J. Mol. Graphics* **1987,** *5,* 49–56.
29. Downs, G.M.; Lynch, M.F.; Manson, G.A.; Willett, P.; Wilson, G.A. 'Transputer Implementation of Chemical Substructure Searching Algorithms. *Tetrahedron Comput. Methodol.* **1988,** *1,* 208–217.

3-D SEARCHING AND NUMERICAL ANALYSES APPLIED TO FILES OF CRYSTALLOGRAPHIC DATA: METHODOLOGIES, EXAMPLES, AND INTEGRATION WITH 1-D AND 2-D TECHNIQUES.

Frank H. Allen, Olga Kennard, Jean J. Galloy, Owen Johnson and David G. Watson.
Cambridge Crystallographic Data Centre, University Chemical Laboratory, Lensfield Road, Cambridge, CB2 1EW, England

ABSTRACT

Search queries in 3-D chemical information systems are usually formulated in terms of geometrical data items derived from an underlying 3-D co-ordinate set. A very wide range of distances, angles, puckering parameters, etc., are possible. The CSD program GSTAT will locate a fragment, calculate user specified geometry (and linear combinations thereof, if required) and select fragments on the basis of limiting values supplied for any of the derived parameter(s). Even then, statistical analyses of the multivariate data set G(Nf, Np)[Nf = no. of fragments retained, Np = no. of parameters specified] may be required to answer the 3-D query completely. Simple descriptive statistics, cluster analyses, principal component methods, correlation, regression, etc. are essential tools within GSTAT. Integration of 3-D searching with the 1-D and 2-D capabilities of CSD program QUEST are being effected *via* a 1:1 graph matching of chemical and crystallographic connection tables, to be followed by a careful generation of 3-D screening mechanisms. An improved statistics package is also being developed.

INTRODUCTION

The Cambridge Structural Database (CSD)[1] contains experimental results from some 80,000 crystal structure determinations of organo-carbon compounds. Data fields can be divided amongst three broad 'dimensional' categories: 1-D (text and individual numerical items), 2-D (chemical connection tables) and 3-D (atomic co-ordinates). Information is held in a single integrated file (ASER, Figure 1); the program QUEST[2] permits interrogation of the 1-D and 2-D information fields, either singly or in logical combinations, *via* a menu-driven graphics interface. Here search queries are formulated directly, in terms of information which is held *explicitly* in the database. For example: (a) exact location of query text-strings, (b) direct comparison (.gt., lt., .eq., ranges etc.) of a numerical query against a specified database item, (c) location of chemical substructures by matching coded atom and bond properties against the 2-D connection tables in the CSD. Search speeds are enhanced by screening mechanisms[2], which are relatively simple to formulate due to the explicit nature of the 1-D and 2-D data.

In extending search capabilities to the 3-D information in the CSD, we must recognise that queries will be formulated in terms of *implicit* data items, i.e., items

W. A. Warr (Ed.)
Chemical Structures 2

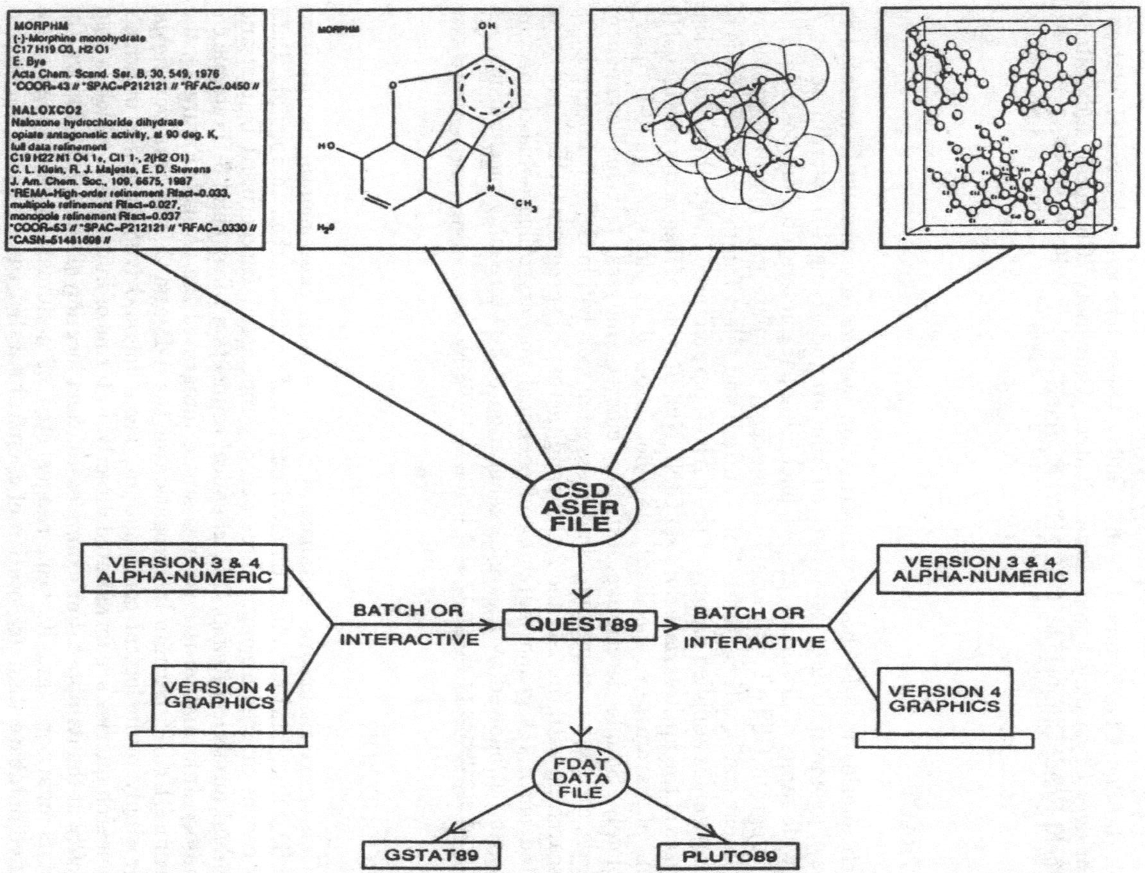

Figure 1. Information summary of an entry in the Cambridge Structural Database and a flowchart of the software system.

which are derived from the 3-D co-ordinate basis. These items are geometric in character and can be related to the chemistry or spatial characteristics of a given crystal structure, molecule or substructure. For the CSD, which has both intramolecular and intermolecular information implicit in its co-ordinate and symmetry records, the range of possible 3-D search terms is very large; typical (and popular) examples are in Table 1. Further, the types of queries posed in 3-D can be far less specific than those in 1-D and 2-D. The query 'find all C-Cl distances in the range 1.68–1.75 Å' rapidly gets refined to 'find all C-Cl distances which exceed the mean value by more than 0.1 Å'. A full range of query types is illustrated in Table 2.

Table 1. Geometrical parameters calculated from 3-D crystallographic co-ordinates by GSTAT

Parameters may relate to a discrete bonded unit (intramolecular) or span two such units (intermolecular) by connectivity extension.

Parameter Type	Specification	
Distance	Atom-Atom, Atom-Centroid, Centroid-Centroid, Atom-Plane Centroid-Plane	
Angles	3 points:	Atoms and/or Centroids
	2 vectors:	Vectors and/or Planes
Torsions	4 points:	Atoms and/or Centroids
Ring Pucker	Cremer Pople (any ring) Altona-Sundaralingam (5 rings)	
Spherical Polar Angles	With respect to sp^2 and sp^3 geometries: directions of approach of one atom to another.	
Modifications and Combinations	25 operators (e.g. SIN, COS, ABS, LOG, +, -, MAX, MIN etc.) can be used to modify or combine existing parameters.	

This rapid generalisation of queries by the user points at once to the scientific attractions of 3-D data, and to the concomitant challenges in providing suitable computing tools. It highlights a need for data analysis and display facilities to be combined effectively with wide-ranging search options – we cannot compose the 'general' C-Cl search until we know the missing comparator, the mean value. This, in turn, highlights our own lack of knowledge of the information content of the 3-D data. No such mystery surrounds the 1-D and 2-D items which are, by and large, a codification of chemical convention. In the CSD these items are an essential key to

Table 2. Typical 3-D search queries in the CSD

Find all C - Cl distances greater than 1.70Å.
Find all C - Cl distances more than 0.1Å longer than the mean.
Find the 10 shortest C - Cl distances
Find the mean C-N-C angle in azacycloheptanes.
Find and rank the 'preferred' conformations of azacycloheptane.
Find the mean O ... O distance in O - H .... O = C hydrogen bonds.
Does this O ... O distance vary with the O - H ... O angle.
Does the O - H vector approach C = O along an O - lone pair direction.

accessing the required experimental 3-D results, results which were the sole reason for the creation of the database in 1965. The development of 3-D search capabilities is therefore a process of search and research, which forms the subject matter of this chapter. The latter feeds back to the former, by indicating improved search terms, better methodologies, and information which can be incorporated in 3-D screening systems.

THREE-DIMENSIONAL SEARCH PROCEDURES

Three-dimensional searches within the CSD System are carried out *via* the program GSTAT. This is an onwardly developing program originating as GEOM[3] (Version 1 of CSD), extensively upgraded by Murray-Rust and Raftery[4,5] as GEOSTAT (Version 2) and further extended as GSTAT for Versions 3 and 4. Figure 1 shows that access to the 3-D data in Version 4 of the CSD is a two stage process. The program QUEST will generate an FDAT interface file comprising 3-D data, crystallographic connection tables, etc., for those entries classified as 'hits' in response to a search of the 1-D/2-D information in the CSD. The screened search capabilities of QUEST[2] therefore act as a primary filter for the 3-D process. At present, no screens specific to the 3-D data are employed.

The GSTAT program has two primary modes of operation. Firstly, an entry-by-entry mode which calculates standard intramolecular and co-ordination-sphere geometry or intermolecular distances for each entry in the FDAT file. The second mode, the fragment geometry mode, permits:

1. Location of a chemical fragment within the crystallographic connection table.
2. Calculation of user-specified geometrical parameters for each occurrence of the fragment.

3. Retention/elimination of the fragment on the basis of geometrical criteria.
4. Generation of a final tabulation of the Np parameters for the Nf retained fragments. This table may then be processed, as a matrix G(Nf,Np), by the statistical analysis routines of GSTAT.

Location of Substructures: the Connectivity Basis

The connection table available to GSTAT describes links between atoms in the 3-D co-ordinate list. Links are established automatically between atoms A, B if d(AB) lies in the range r(A) + r(B)±t, where r is a covalent radius and t is a tolerance value. Suitable r-values have been refined over the years as part of CCDC check and evaluation activities; t is normally 0.4 Å. The crystallographic connectivity representation is, therefore, separate and different from the chemical representation employed in QUEST, as illustrated in Figure 2. In particular it lacks bond-type information, as well as atomic charges, and some or all of the hydrogen atoms may not have been located by the X-ray experiment. This separation of representations is now being rectified, as described later, to provide a direct link between the chemical description and the 3-D co-ordinate set. In current practice, bond type specifications must be approximated by geometric tests e.g., a C=O double bond might be approximated by a distance range of, say, 1.15 to 1.25 Å. Suitable reference data are now available[6,7] to provide suitable distance criteria.

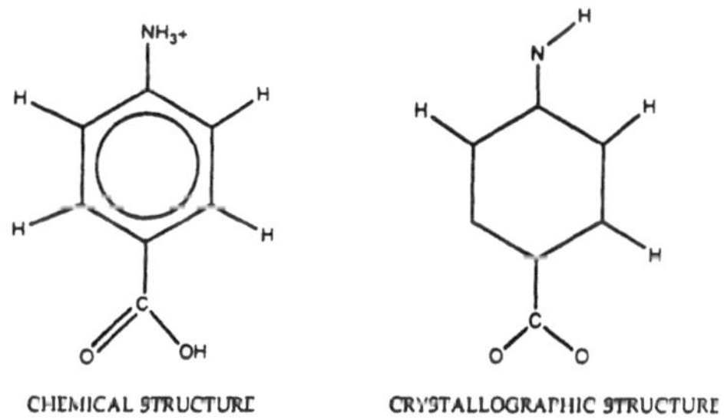

Figure 2. Comparison of 2-D chemical and crystallographic connectivity representations at the molecular level.

The distance-based connectivity can, of course, be extended to include intermolecular interactions using crystallographic symmetry. For a study of H-bonding, for example -O-H...O=C in Figure 3, all O...O intermolecular distances to a specified limit (~3.2 Å) are located. The connectivity representation is then extended to include molecule A′, and the O...O′ 'link'. The H-bonded fragment may now be located *via* normal atom-by-atom, bond-by-bond matching, together with suitable geometric tests (noted above) to approximate required bond types.

Substructure location must also be *exhaustive* for applications in 3-D. All occurrences of the specified fragment are located in each retrieved entry, and the retention of fragments which have atomic overlap is a user decision. This contrasts with the 2-D case (e.g., QUEST) where a single occurrence is normally sufficient to generate a hit.

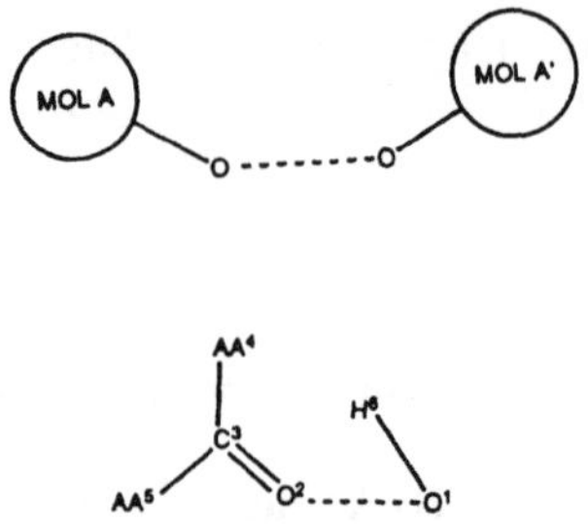

Figure 3. Extension of distance-based crystallographic connectivity to permit location of 'intermolecular' fragments.

Location of Substructures: the Geometrical Basis

The crystallographic connectivities described above are, of course, merely 2-D representations which parallel the chemical description. They differ, however, in having the 3-D co-ordinates as additional atom properties. Hence, each fragment located by an atom-by-atom, bond-by-bond procedure can be tested further to ensure that geometric parameters, derivable from its atomic co-ordinates, fall into required ranges. A given 2-D fragment will normally exhibit a variety of 3-D geometries, each having characteristic patterns of e.g., *interatomic distances, angles* and *torsion angles*. Selection of the required pattern, be it a ring or side-chain conformation, pharmacophore, metal co-ordination mode etc., is the 3-D search problem. User-defined limits on any of the simple parameters above are tested routinely as part of the fragment location procedures in GSTAT. Further selection, based on more complex geometrical parameters, is described below.

In performing the basic distance, angle or torsional tests, we must take account of *topological symmetry* in the 2-D search fragment, and also of structural *enantiomorphic* inversions in 3-D[5,8]. The toposymmetry problem is illustrated in Figure 4. Atom 1 of a cyclohexane search fragment (of 2-D symmetry D_{6h}) may map to any one of the six atoms of a database target, say to number 4. Atom 2 of the fragment may now map to two alternatives in the target, numbers 3 or 5. This gives rise to $6 \times 2 = 12$ possible and equivalent mappings for the fragment. If we regard the atomic enumeration of the fragment as fixed, then each mapping represents one of the 12 possible ways in which we can impose this enumeration onto each database target. In 2-D these enumerations are equivalent and any one of them, chosen at random by QUEST or GSTAT, will satisfy the query at this level. In moving to 3-D, however, each enumeration can give rise to different sequences of the derived geometrical parameters: only one of these may correspond to the sequence specified by the user for test purposes. If we are searching for half-chair forms of cyclohexane, then we might specify the torsional test sequence of Figure 4. The initial (random) fragment mapping in GSTAT might then give rise to the initial torsional sequence of Figure 4, which fails the tests. GSTAT will then continue an exhaustive atom mapping until either (a) an enumeration is found which generates an acceptable torsional sequence (always a permutation of the original sequence) or (b) all 12 possible enumerations have failed the tests.

A further complication (Figure 4) is introduced by structural inversion in 3-D: the co-ordinate set may define one or other of the two equivalent enantiomorphs

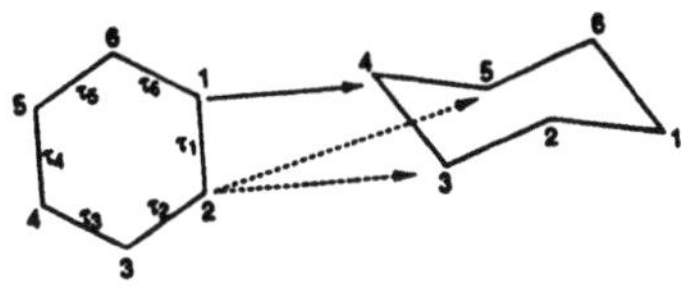

USING RING TORSION ANGLES AS CONSTRAINTS :

	T1	T2	T3	T4	T5	T6
INITIAL MAPPING	61	-43	12	-3	17	-49
TEST SEQUENCE: FROM	-5	-10	40	-55	40	-10
RANGES : TO	5	-20	50	-75	50	-20
FOURTH MAPPING	-3	17	-49	61	-43	12
FOURTH MAPPING INVERSE	3	-17	49	-61	43	-12

Figure 4. Effects of 2-D topological symmetry and enantiometric inversion on fragment location in 3-D (see text).

for achiral compounds. The *sign* sequence of the torsional test values specifically defines only one of these two alternatives. Thus, for each enumeration of each database target, torsional comparators must be generated from both the original co-ordinate set *and* its inverse. This inversion is under user control in GSTAT, since it should not be performed for chiral sets of molecules, where the co-ordinates should reflect the correct absolute stereochemistry. Further, inversion is not necessary for tests involving distances (a 1-D quantity) or inter-vector angles (a 2-D quantity), but only for geometrical parameters, such as torsion angles, which embody true 3-D information in each individual value.

For the example of Figure 4, the inverse of the fourth mapping (representing a torsional permutation 4,5,6,1,2,3 and sign inversion) will satisfy the test sequence.

Calculation of Fragment Geometry

The search procedures described above result in the imposition of a user-defined atomic numbering scheme on each fragment located in the retrieved CSD entries. The user may now specify the required geometrical calculations in terms of his or her original (fixed) fragment enumeration scheme. A wide variety of parameters (Table 1) may be obtained by specifying sequences of atom numbers (2 for a distance, 3 for an angle, etc.). The user also supplies a mnemonic name by which the parameter may be referenced in subsequent instructions. The atomic co-ordinate basis may be augmented by the addition of co-ordinate and vector information which define specified centroids, interatomic vectors and normals to mean planes within the fragment. A variety of 'special' parameters (Table 1) may also be calculated, for example the Cremer-Pople[9] and Altona-Sundaralingam[10] descriptions of ring pucker, spherical polar descriptors of directionality in H-bonded and other intermolecular systems[11], etc. A further feature[4,5] allows for the linear combination of parameters *via* a series of FORTRAN-like instructions. A complete, annotated GSTAT instruction set is shown in Figure 5.

The geometrical results are stored as a matrix G(Nf,Np), i.e., Np user-defined parameters for the Nf occurrences of the fragment. Inclusion, or not, of a particular fragment as a matrix row may be controlled by a SELECT instruction (Figure 5): limiting values may be specified for any or all of the Np individual data items. Suitable numerical limiting values may already be known to the user. Usually, however, limits are assessed by an initial pass through the data-set, followed by inspection of the displayed matrix G (e.g., Figure 6) or of simple statistical indicators derived from its content. We would stress that *any* numerical parameter that can be defined for inclusion in G, including linear combinations of simpler parameters, can be invoked by name in the selection process. This provides additional, generalised and very powerful 3-D search capabilities over and above the simple distance, angle and torsional tests which may be invoked in the fragment location process.

```
FRAG Cyclopropyl-Carbonyls
AT1 C 3
AT2 C 2
AT3 C 2
AT4 C 2
AT5 O 1             Definition
BO 1 2
                    of
BO 2 3
BO 1 3              Fragment
BO 1 4
BO 4 5
TEST DIST 4  5 1.15 1.25
TEST DIST 2  3 1.40 1.56     Geometrical distance constraints
END
SETUP X1  2  3               X1 is midpoint of C2 - C3
DEF #D12  1  2
DEF #D13  1  3               Calculate distances C1-C2, C1-C3
TRA #DM = #D12 + #D13        Generate the mean of the distances
TRA DMEAN = #DM / 2.0        D12, D13;  call the result DMEAN
DEF D23  2  3
DEF D45  4  5
DEF #TAU1  5  4  1  X1       Calculate torsion angle C5-C4-C1-X1
TRA TAUX = ABS #TAU1         Generate absolute value of TAU1
SEL DMEAN 1.40 1.56          Select fragments on basis of DMEAN
HIST TAUX                    Draw a histogram of TAUX
SCAT D23 TAUX                Draw a scattergram of D23 vs. TAUX

Parameter names preceded with a  #  symbol are 'temporary' for internal
use by the program (here in TRAnsform commands).  This instruction  set
will tabulate DMEAN,  D23,  D45,  TAUX  for each fragment located.
```

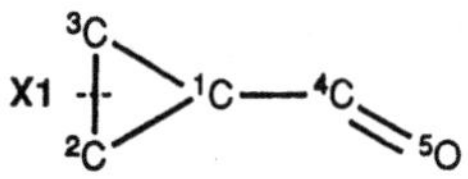

Figure 5. Alphanumeric instruction set for fragment location and geometrical calculations *via* GSTAT.

It may be that the GSTAT facilities summarised so far will be sufficient for some applications, e.g., for molecular modelling. In this case the atomic co-ordinates for the retained fragments may be output in a variety of forms for external use. However, in many cases further analysis of the G matrix is required, particularly to answer some of the more general queries posed in Table 2.

STATISTICAL AND NUMERICAL ANALYSES

The matrix G(Nf,Np) exemplified by Figure 6 provides the basis for a variety of univariate, bivariate and multivariate statistical tests and graphical display facilities which are all contained within the GSTAT program. The most important of these will be exemplified briefly in the following sections.

Nfrag	Refcod	DMEAN	D23	D45	TAUX
7	AIMCTY	1.507	1.512	1.237	146.695
8	ARITOL	1.520	1.500	1.226	150.295
9	BARTUS10	1.526	1.477	1.197	10.474
12	BAXTID	1.537	1.519	1.192	19.962
13	BAXTOJ	1.540	1.528	1.194	15.100
14	BAZLET	1.510	1.492	1.210	120.111
15	BEJVUH	1.509	1.478	1.210	0.422
16	BEJVUH	1.501	1.493	1.242	5.867
17	BEKDAW	1.531	1.524	1.195	5.909
	Mean	1.522	1.494	1.206	72.209
	S.D.Sample	0.014	0.025	0.015	66.984
	S.D.Mean	0.001	0.002	0.001	4.228
	Minimum	1.479	1.416	1.168	0.000
	Maximum	1.554	1.557	1.247	180.000
	Nobs	251	251	251	251

Figure 6. Section of the data matrix G(Nf,Np) generated by the instruction set of Figure 5.

Univariate Statistics and Display

A simple statistical summary of the distribution of each of the Np variables is automatically provided when the G matrix is displayed (Figure 6). Individual variables can be defined, by parameter name(s), and their distribution(s) displayed as simple histograms (e.g., Figure 7). These are vital facilities for (a) assessment of suitable geometric test and selection criteria, (b) assessment of the chemical composition of retrieved data-sets, (c) detection of outlying (often incorrect)

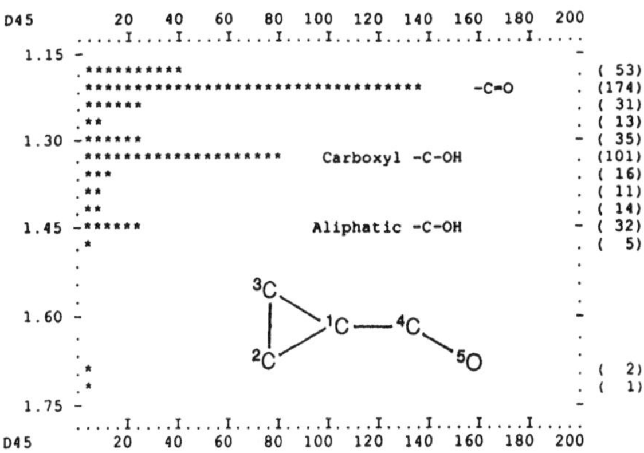

Figure 7. Univariate statistics: histogram of the C-O bond length used to assess suitable geometric restriction embodied in Figure 5.

observations in the experimental data, and (d) derivation of reliable mean values as structural comparators or for use in model building. Items (a), (b), (c) are all exemplified by Figure 7. This histogram was derived from an early pass for the fragment of Figure 5, in which the distance constraint to define the C=O bond was omitted. The histogram is readily analysed in chemical terms and C=O double bond limits (e.g 1.15 to 1.25 Å) can be established by inspection. The histogram also shows three observations (out of 488) which can immediately be classified as outliers. The GSTAT program has provided the computational basis for recent publications[6,7] of mean bond lengths observed in organic and metallo-organic crystal structures. These literature values, or mean values derived by direct use of GSTAT, either answer a given 3-D query directly or are necessary preliminary knowledge for answering more general (Table 2) queries.

Bivariate Statistics and Display

It is often of interest to discover and quantify relationships amongst pairs of the Np parameters. To assist in this process, GSTAT will optionally generate a pairwise correlation/covariance matrix. Large correlations are highlighted and can be further analysed *via* simple linear regression analyses. Bivariate distributions can be displayed graphically *via* 2-D scatterplots. Figure 8 is a typical example for the -N-H...S=C fragment analogous to that in Figure 3. The N-H...S angle is plotted against the H....S distance to show that, even for a small random subset, the hydrogen bonds prefer to be linear and become weaker (longer) as the angle deviates from 180°. Information about directional preferences in intermolecular interactions[11] is vital in the modelling of molecular recognition processes. Generalised queries of this type (see Table 2) can only be answered systematically by careful examination of crystallographic results.

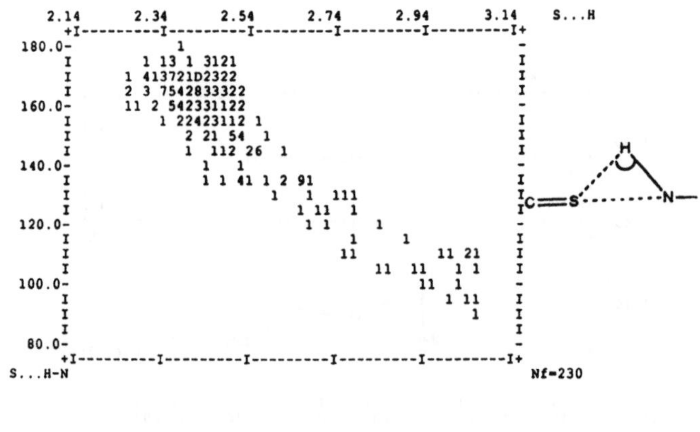

Figure 8. Bivariate statistics: scatterplot of S...H distance *vs.* S...H – N angle in C = S...H –N hydrogen bonds.

Multivariate Analyses

One common query to the CSD (Table 2) involves the classification of fragment conformations – a recognition of the different 3-D shapes that are exhibited by a given 2-D chemical substructure. Here the number of geometrical parameters required to define the 3-D conformation is normally $\gg 2$. For example, 15 distances or six torsion angles provide suitable descriptors for cyclohexane, and we must resort to multivariate statistical methods[12] to analyse the appropriate G-matrix. Two techniques may be selected within GSTAT: principal component analysis and cluster analysis.

Principal component analysis[13–16] performs an eigenanalysis of the correlation matrix derived from G and attempts to describe the Np-dimensional data-set by a new set of m mutually orthogonal axes. These principal component axes are placed along the directions of maximum variance exhibited in matrix G. If m is very much smaller than Np, then we have effected a reduction in dimensionality: scatterplots based on pairs of the new axes can reveal the underlying structure of the original data. For a representative sample of cyclohexanes[17] described by Np = 6 torsion angles, there are m = 3 principal components. A plot of data referred to PC axes 2 and 3 (Figure 9) shows 6 peaks, identified as symmetry-variants of the boat conformation, distributed in a circle around a central large peak, identified as a superposition of chair and half-chair conformers.

The PC method, although numerical in nature, results in a visual description of the data structure. Considerable work would be required to generate, manually or automatically, the co-ordinate data for a 'most representative fragment' for each conformation identified in the plots. Cluster analysis[15,18–20], on the other hand, is a numerical technique which preserves the dimensionality of the original data-set, i.e., it operates on the raw data and performs no mathematical transformations to

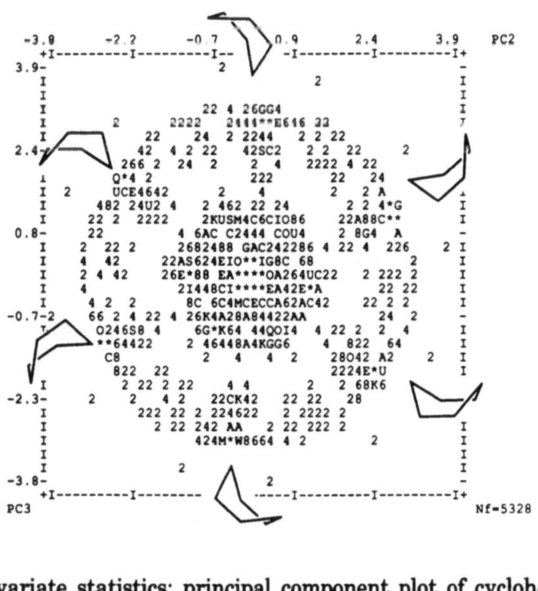

Figure 9. Multivariate statistics: principal component plot of cyclohexane conformation derived from a symmetry-expanded torsional data-set.

other axial systems. For cyclohexane, for example, the technique attempts to form discrete clusters of fragments in which members of each cluster have closely similar co-ordinates in the 6-dimensional space spanned by the torsion angles. Recently a variety of clustering algorithms has been introduced into GSTAT[8,21,22]. These algorithms have been specially modified to deal correctly with torsion angle descriptors in their dissimilarity calculations, and to make use of toposymmetry/ inversion processes during cluster formation. Although some work remains to be done, it is now possible to generate output of the type illustrated in Figure 10 for each discrete cluster, output which should be of direct utility in model building.

RANK= 1 CLUSTER NO.= 6 POPULATION= 51

Mean values	-55.1	58.6	-57.9	53.3	-50.0	51.1
ESD of Means	0.4	0.4	0.4	0.5	0.7	0.7

Most representative fragment (at D= 5.6 deg from mean) is FRAGMENT NO. 20

MRF Data	-55.3	57.5	-56.3	52.8	-48.5	50.5

Coordinate data for most representative fragment:

ACAMYA	**FRAG**	20**CLUS**	4**RANK**	1
C17	7.55081	3.55694	-7.00195	
C18	6.06258	3.45253	-6.62817	
C19	5.36696	2.15941	-7.16700	
C20	5.60514	2.02631	-8.68815	
C21	7.09969	2.02975	-9.01484	
C22	7.74325	3.28042	-8.49906	

Figure 10. Multivariate statistics: section of ouput from conformational clustering algorithms based on torsional descriptors.

INTEGRATED SEARCHING OF THE CSD

Our current research effort is now directed towards an integration of the 3-D search capabilities of GSTAT with the 1-D/2-D facilities of QUEST (see Figure 1). The aim is to develop an extended QUEST program which will be able to generate the G(Nf,Np) matrix, and include all of the geometrical test and selection criteria described above. The statistical functions would then be split off as a separate package. This improved functionality requires upgrades to both the software and to the ASER database itself.

Unified Connectivity Representation

Earlier sections have shown that the underlying reasons for the two stage (Figure 1) QUEST-GSTAT search system are the differences between the chemical and crystallographic connectivity representations (Figure 2). This is not surprising: they emanate from two different sources. Chemical connectivities are input by Cambridge Editorial Staff, crystallographic connectivities are derived from published co-ordinate data. Inevitably these representations have different enumerations of the same molecule. We have now established a graph theoretical matching by which a fixed, canonical enumeration is imposed on both representations. This means that the 3-D co-ordinates of each atom can now be associated with their full chemical description, and the distance-based 'links' of the crystallographic connectivity can be associated with a chemical bond type.

The matching procedures are not straightforward due to the nature of the crystallographic results. Principal problems have been:

1. The presence of more than one bonded unit (residue) in the crystal structure. Residues can be chemically different, in which case a 1:1 mapping to the chemistry is preserved, or they may be chemically identical, in which case all chemical atoms will map to more than one crystallographic atom.
2. The presence of disorder in the crystal structure: a few chemical atoms of an affected residue will map to more than one crystallographic atom.
3. The crystal structure is polymeric; matching to the chemical monomer is complex.
4. The lack of some or all H-atoms in the crystal structure.
5. The presence of local or complete toposymmetry in the hydrogen-depleted crystallographic connectivity but not in the chemistry. Bond-length:bond-type comparisons are needed to resolve this problem.

Continued development of the routines has generated perfect connectivity matching information for some 89% of CSD entries for which 3-D co-ordinates are available. This information is now being included in an extended ASER file.

Screening Mechanisms for 3-D Searches

Current GSTAT use relies heavily on a preliminary retrieval of database entries *via* the 1-D/2-D bit-screened search mechanisms of QUEST. In some cases this can result in very large FDAT subsets for processing by GSTAT. Further bit-screens, relating to information derivable from the 3-D co-ordinates, are therefore an essential adjunct to the 1-D/2-D screens already stored in ASER. Given the extensive list of possible 3-D information which is calculable (Table 1), decisions must be made on which quantities should be covered (at least initially) by a 3-D bit-vector. Experience gained from a variety of research projects is vital in assessing user needs and in providing knowledge that can be embodied in the screens. We are beginning to apply the general philosophy of distance screening of Jakes and Willett[23]; for crystallographic data these concepts will ultimately be extended to include intermolecular descriptors. A logical extension of these concepts to (generalised) inter-vector angles is also in progress[24]. Recent experiences with conformational clustering algorithms[22] indicate that results from these researches can be embodied as shape descriptors for an extendable list of common chemical fragments.

Software Enhancements

The database enhancements noted above, particularly the connectivity unification, mean that full integration of 1-D, 2-D and 3-D search capabilities is now possible. The 2-D substructure search algorithm in QUEST now has access to the 3-D co-ordinates as additional atom properties. Extended features, i.e., exhaustivity of fragment location and considerations of toposymmetry and of enantiomorphs, which underpin 3-D geometrical searches, are now being added to QUEST. Routines for the calculation of geometrical parameters (Table 1) have been transferred from GSTAT, and the QUEST instruction set expanded to cover the required definitions. The menu-driven graphics interface will be upgraded when functionality of the extended QUEST program is finalised.

Statistical Functions

Experience shows that many queries asked of the CSD (see Table 2) can only be answered by statistical and numerical analyses of retrieved data. At present a selection of these techniques is provided within GSTAT. However, some of these methods can be highly demanding of both core storage and computer time. There is a logical need to split this functionality to a separate package so that it can develop more freely. Whilst GSTAT will remain as a separate stand-alone program, there is conflict between the computational requirements of 3-D search functions and data analysis functions. Thus, many more statistical techniques can be added to a specialist package. Some of these are already available as separate subroutines in the CCDC statistical library CAMAL[25].

CONCLUSION

In this paper we have highlighted some of the problems associated with the retrieval of information from databases of 3-D chemical structures, particularly by comparison with their 2-D counterparts. Some of these problems, e.g., allowance for toposymmetry and for achiral or chiral environments at the search stage, and decisions on suitable screening mechanisms, are common to all 3-D systems. Particular problems associated with crystallographic data, especially those originating from differing (but equivalent) 2-D and 3-D chemical bases, are noted and solutions described.

There is no doubt that the word 'search' has a much broader meaning in a 3-D database context than it has for 1-D (text) or 2-D (chemical) databases. This is due to the almost limitless variety of parameters that can be generated from 3-D coordinate sets. The types of query that can be posed are also limitless, ranging from the highly specific to the highly general. Answers to many 3-D 'searches' can only be obtained by careful numerical analyses of retrieval subsets of data. Often the results of such searches represent research results in their own right, each adding to our small but growing knowledge of the systematics of 3-D structure. Research projects suggest new methodologies and provide new information which can, in turn, be used to improve the generalised search process. The available crystallographic databases and their associated software systems[26] must not be viewed solely as search systems answering queries within rigid imposed constraints. They should be viewed more as developing research tools whose effective use is only just beginning.

ACKNOWLEDGMENTS

We would like to acknowledge Dr Sam Motherwell (now of the University Library, Cambridge, UK) and Dr Peter Murray-Rust (now at Glaxo Group Research, Greenford, UK) for their fundamental contributions to the development of GEOM (1978) and GEOSTAT (1982), from which the present GSTAT code continues to develop.

REFERENCES

1. Allen, F.H.; Kennard, O.; Taylor, R. 'Systematic Analysis of Structural Data as a Research Technique in Organic Chemistry'. *Acc. Chem. Res.* **1983,** *16,* 146–153.
2. Allen, F.H.; Davies, J.E. 'File Structures and Search Strategies for the Cambridge Structural Database'. In *Crystallogaphic Computing 4*; Isaacs, N.W.; Taylor M.R., Eds; Oxford University Press: Oxford, 1988; pp.271–289.
3. Allen, F.H.; Bellard, S.; Brice, M.D.; Cartwright, B.A.; Doubleday, A.; Higgs, H.; Hummelink, T.; Hummerlink-Peters, B.G.; Kennard, O.; Motherwell, W.D.S.; Rodgers, J.R.; Watson, D.G. 'The Cambridge Crystallographic Data Centre: Computer-Based Search, Retrieval, Analysis and Display of Information'. *Acta Crystallogr.* **1979,** *B35,* 2331–2339.
4. Murray-Rust, P.; Raftery, J. 'Computer Analysis of Molecular Geometry VI: Classification of Differences in Conformation'. *J. Mol. Graphics,* **1985,** *3,* 50–59.
5. Murray-Rust, P., Raftery, J. 'Computer Analysis of Molecular Geometry VII: The Identification of Chemical Fragments in the Cambridge Structural Data File'. *J. Mol. Graphics* **1985,** *3,* 60–68.
6. Allen, F.H.; Kennard, O.; Watson, D.G.; Brammer, L.; Orpen, A.G.; Taylor, R. 'Tables of Bond Lengths Determined by X-ray and Neutron Diffraction. 1: Bond Lengths in Organic Compounds'. *J. Chem. Soc., Perkin Trans. 2* **1987, S1-S19.**
7. Orpen, A.G.; Brammer, L.; Allen, F.H.; Kennard, O.; Watson, D.G.; Taylor, R. 'Tables of Bond Lengths Determined by X-ray and Neutron Diffraction. 2: Organometallic Compounds and Coordination Complexes of the d-and f- Block Metals'. *J. Chem. Soc., Dalton Trans.* **1989,** S1-S83.
8. Allen, F.H.; Doyle, M.J., Taylor, R. 'Automated Conformational Analysis from Crystallographic Data. 1: A Symmetry-Modified Single-Linkage Clustering Algorithm for 3-D Pattern Recognition'. *Acta Crystallogr.,* **1991,** *B47,* 29–40.
9. Cremer, D.; Pople, J.A. 'A General Definition of Ring Puckering co-ordinates'. *J. Am. Chem. Soc.* **1975,** *97,* 1354–1358.
10. Altona, C.; Sundaralingam, M. 'Conformational Analysis of the Sugar Ring in Nucleosides and Nucleotides: A new Description Using the Concept of Pseudorotation'. *J. Am. Chem. Soc.* **1972,** *94,* 8205–8212.
11. Taylor, R.; Kennard, O. 'Hydrogen-bond Geometry in Organic Crystals'. *Acc. Chem. Res.* **1984,** *17,* 320–326.
12. *Introduction to Multivariate Analysis*; Chatfield, C.; Collins, A.J.; Chapman and Hall: London, 1980.
13. Murray-Rust, P.; Bland, R. 'Computer Retrieval and Analysis of Molecular Geometry. 2: Variance and its Interpretation'. *Acta Crystallogr.* **1978,** *B34,* 2527–2533.
14. Murray-Rust, P.; Motherwell, S. 'Computer Retrieval and Analysis of Molecular Geometry. 3: Geometry of the beta-1'-Aminofuranoside Fragment'. *Acta Crystallogr.* **1978,** *B34,* 2534–2546.
15. Taylor, R. 'The Cambridge Structural Database in Molecular Graphics: Techniques for the Rapid Identification of Conformational Minima'. *J. Mol. Graphics* **1986,** *4,* 123–131.
16. Auf der Heyde, T.P.E.; Buergi, H.-B. 'Molecular Geometry of d^8 Five Co-ordination. 3: Factor Analysis, Static Deformations and Reaction Co-ordinates'. *Inorg. Chem.* **1989,** *28,* 3982–3989.
17. Allen, F.H.; Doyle, M.J.; Auf der Heyde, T.P.E. 'Automated Conformational Analysis from Crystallographic Data. 6: Correlation of Principal Component and Ring Puckering Analyses for Carbocycles of Size 4, 5 and 6'. *Acta Crystallogr.,* **1991,** *B47,* 412–424.
18. *Cluster Analysis*; Everitt, B.; Halstead-Heinemann: London, 1980; 2nd ed.
19. Norskov-Lauritsen, L.; Buergi, H.-B. 'Cluster Analysis of Periodic Distributions: Application to Conformational Analysis'. *J. Comput. Chem.* **1985,** *6,* 216–228.
20. Auf der Heyde, T.P.E., Buergi, H.-B. 'Molecular Geometry of d^8 Five Co-ordination. 2: Cluster Analysis, Archetypal Geometries and Cluster Statistics'. *Inorg. Chem.* **1989,** *25,* 3970–3981.
21. Allen, F.H.; Doyle, M.J.; Taylor, R. 'Automated Conformational Analysis from Crystallographic Data. 2: Symmetry-Modified Jarvis-Patrick and Complete-Linkage Clustering Algorithms for 3-D Pattern Recognition'. *Acta Crystallogr.,* **1991,** *B47,* 41–49.

22. Allen, F.H.; Doyle, M.J.; Taylor, R. 'Automated Conformational Analysis from Crystallographic Data. 3: 3-D Pattern Recognition Within the Cambridge Structural Database System, Implementation and Practical Examples'. *Acta Crystallogr.*, **1991**, *B47,* 50–61.
23. Jakes, S.E.; Willett, P. 'Pharmacophoric Pattern Matching in Files of 3-D Chemical Structures: Selection of Interatomic Distance Screens'. *J. Mol. Graphics* **1986,** *4,* 12–20.
24. Cringean, J.K.; Pepperrell, C.A.; Poirrette, A.R.; Willett, P. 'Selection of screens for 3-D substructure searching'. *Tetrahedron Comput.Methodol.* Submitted for publication.
25. Taylor, R. 'CAMAL: A New Component of the Cambridge Structural Database Software System'. *J. Appl.Crystallogr.* **1986,** *19,* 90–91.
26. *Crystallographic Databases*; Allen, F.H.; Bergerhoff, G.; Sievers, R., Eds; International Union of Crystallography: Chester, 1987.

AN INTEGRATED APPROACH TO 2-D AND 3-D SIMILARITY SEARCHING FOR THE CAMBRIDGE STRUCTURAL DATABASE (CSD)

Eleanor M. Mitchell, Frank H. Allen and Gary F. Mitchell
Cambridge Crystallographic Data Centre, University Chemical Laboratory, Lensfield Road, Cambridge CB2 1EW, England

R. Scott Rowland
Department of Biochemistry, University of Alabama, Alabama 35294, U.S.A.

ABSTRACT

Similarity searching in chemical databases depends crucially upon the chosen molecular attribute sets. The current 2-D implementation in the Cambridge Structural Database System uses substructural bit screens as attributes. These contain chemical information at a restricted connectivity level around each atom or bond; the only larger pattern units represented are rings and ring systems. Gross pattern attributes can, however, be assigned in terms of inter-nodal bond separation frequencies established using a shortest path algorithm. This information can be used alone (or in combination with the chemical attributes) to provide an alternative (or enhanced) approach to the 2-D problem. In 3-D, similarity concepts have meaning at both the substructural and full structural levels. A specific chemical substructure may exist in a variety of 3-D conformations. A modified Minkowski metric based on torsion angle descriptors is used to compare 3-D shapes. This results in a 1-D 'conformational spectrum' graphical representation in which different conformers often appear in well separated groups for rapid identification. At the full molecular level, comparison of complete distance matrices provides the most complete solution. However, due to the vast computational effort this requires, the distance matrix may be reduced to a distance-frequency distribution. Ultimately it is planned that the 2-D (inter-nodal bond separations) and 3-D (Å distances) approaches will be combined to provide suitable descriptors for similarity work.

INTRODUCTION

The majority, if not all, of the currently available two dimensional (2-D) similarity search systems rely on the comparison of bit screen records. Bit screens record the presence/absence of a set of chemical structure characteristics, usually small atom- and bond-centred fragments, together with some larger pattern units, principally rings and ring systems. Studies by Willett and Winterman[1] showed that the best measurement of similarity based on these screen sets is the Tanimoto (or Jaccard) coefficient. Consequently most 2-D similarity systems, including SOCRATES[2,3] and the Cambridge Structural Database (CSD) System[4], employ screen-based similarity searching using the Tanimoto coefficient.

W. A. Warr (Ed.)
Chemical Structures 2

However, if the provision of a 'browsing' type of search, where new structures are happened upon by serendipity, is considered to be one of the main uses of similarity searching[2], it is curious that the user is restricted in performing a 'good' search by the degree of chemical information he/she can attach to the query structure. This problem was brought to our attention by one of our users who wished to carry out a similarity search for a 'skeleton' structure which had little chemical information present – the nodes were allowed to be any one of a number of atomic types and the bond types were also non-specific. Only a very small number of chemical structure-based bit screens could be set for this structure. However enough were set so that a search of the CSD could be carried out and the top 100 most similar structures presented to him for browsing. He professed himself dissatisfied with the hits that the search ranked to the top, and did not consider these hits relevant to his query structure.

It would appear that a search system which is intended to cater for a wide diversity of query structures should provide a means by which all attributes of a structure can be weighted according to their importance. In addition to the chemical attributes that are already used in similarity calculations, searches could be enhanced by adding the full 2-D and 3-D pattern information about a query structure. We have started to explore ways of extending the current 2-D structure-based similarity searching abilities in our system.

CURRENT 2-D SIMILARITY SEARCHING USING CHEMICAL ATTRIBUTES

To illustrate existing limitations, a number of searches were carried out using the current 2-D similarity system. A 3000 structure subset of the January 1990 Version of the CSD was created for these searches to help the analysis of the performance. The subset was generated using the CSD System QUEST program to search for entries from the following chemical classes which represent a wide variety of shapes and sizes:

Class 44	Pyrimidines and purines
Class 45	Carbohydrates
Class 47	Nucleosides and nucleotides
Class 50	Antibiotics
Class 51	Steroids

The test query structures were selected from this subset to display a range of pattern attributes. The chemical diagrams of the structures were entered to the similarity search system using the Version 4 graphics interface, to imitate the normal situation when a user presents a structure to the system for searching. Using structures known to be in the database will allow the effectiveness of the search procedures to be monitored, since they should appear at the top of the ranked hit lists. Due to limited space, the results from only three of the test structures that were chosen are reported here.

Currently, a CSD similarity search will result in the 'Distribution Table for 100 entries' shown in Figures 1–3, together with a brief statistical analysis of the results detailing the best and worst score, the average and the range. The user can then decide whether to look at a table listing the top 100 refcodes with their scores, or to browse immediately through the structure diagrams of these entries.

The problem encountered using the existing similarity search system is

AADRIB (Fully Defined)

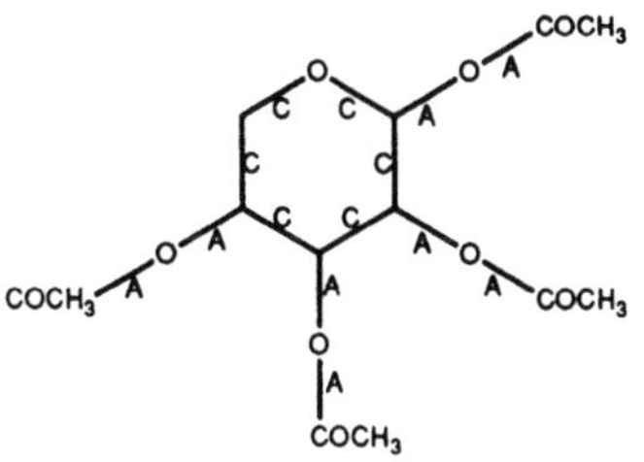

80 screens set

Distribution Table for 100 entries

TANI SCORE	Entries
100 - 88	4
87 - 84	11
83 - 80	7
79 - 76	49
75 - 72	29

Best score	91%
Worst score	73%
average	78%
range	18%

Top 5 hits

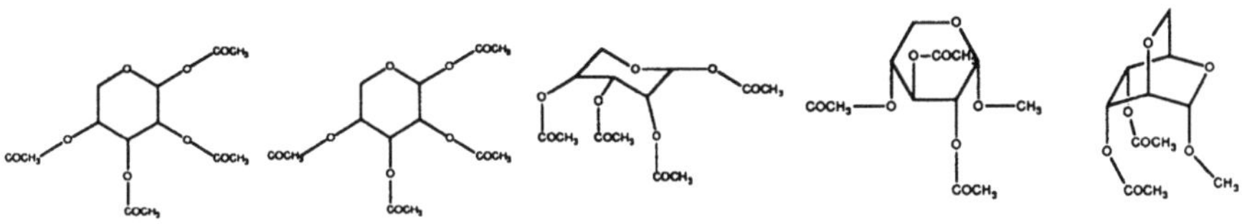

AADRIB	ACARAP	BIBWUE	DOJDUB	ACPYNS
91%	91%	91%	91%	87%

(a)

AADRIB (Partially Defined)

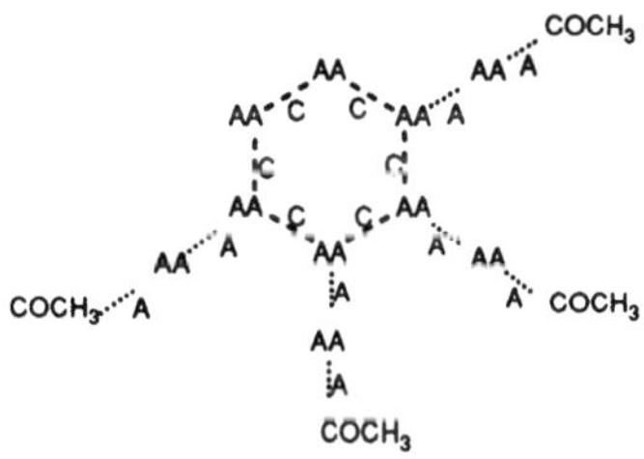

27 screens set

Distribution Table for 100 entries

TANI SCORE	Entries
100 - 35	2
34 - 32	4
31 - 29	30
28 - 26	57
25 - 23	7

Best score	37%
Worst score	25%
average	28%
range	12%

Top 5 hits

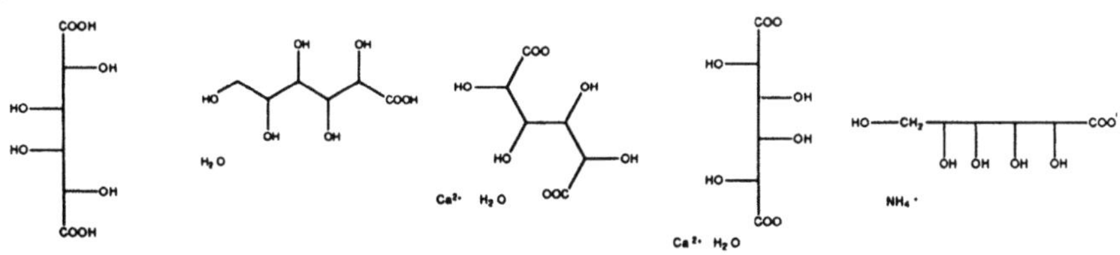

BIVTUV	CAKZAP	CAGLAR	CAGLAR01	BAGDOC
37%	36%	34%	34%	33%

(b)

Figure 1. CSD similarity search for the structure with refcode AADRIB. (a) Results of a similarity search for a fully defined structure. (b) Search for a partially defined structure

AARBOX (Fully Defined)

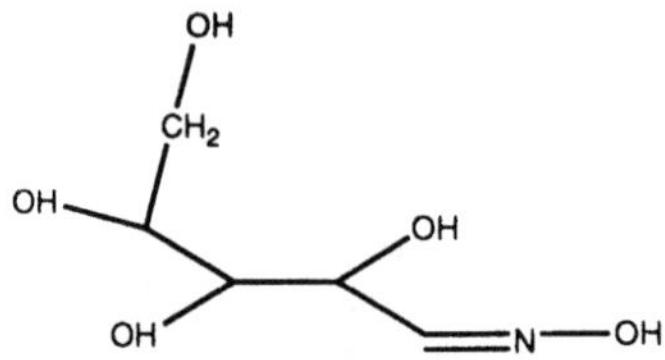

73 screens set

Distribution Table for 100 entries

TANI SCORE	Entries
100 - 87	1
86 - 74	0
73 - 61	13
60 - 48	51
47 - 35	35

Best score	99%
Worst score	46%
average	52%
range	53%

Top 5 hits

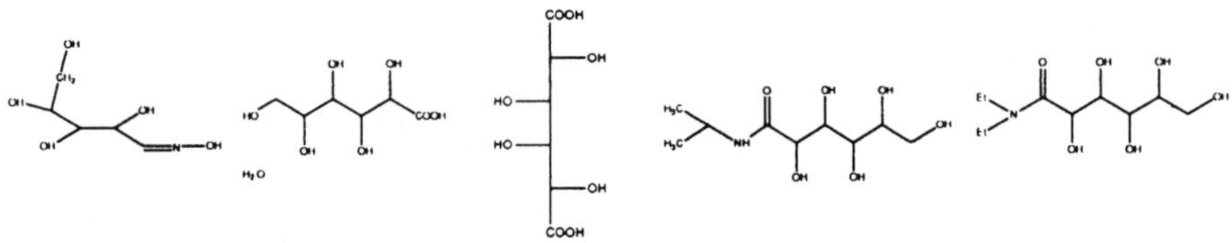

AARBOX 99% | CAKZAP 66% | BIVTUV 65% | DAYVUU 64% | DAYWAB 63%

(a)

AARBOX (Partially Defined)

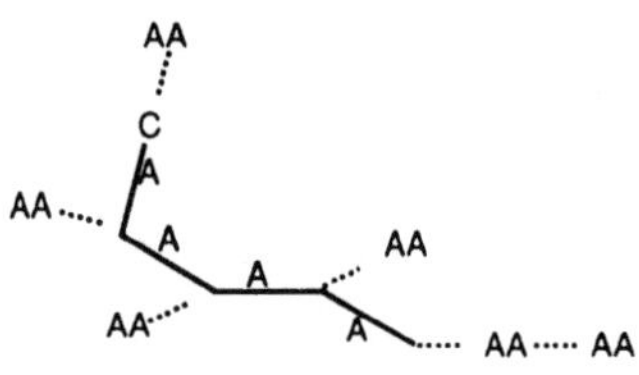

16 screens set

Distribution Table for 100 entries

TANI SCORE	Entries
100 - 29	8
28 - 26	6
25 - 23	4
22 - 20	13
19 - 17	39

Best score	31%
Worst score	16%
average	19%
range	15%

Top 5 hits

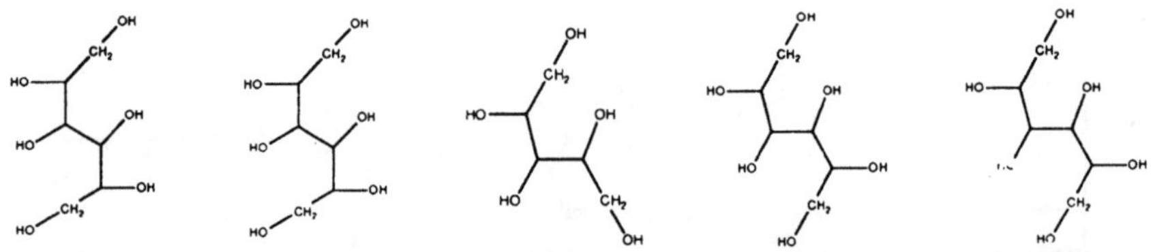

ALITOL 32% | ALITOL01 32% | ARABOL 32% | DLMANT 32% | DLMANT01 32%

(b)

Figure 2. CSD similarity search for the structure with refcode AARBOX. (a) Results of a similarity search for a fully defined structure. (b) Search for a partially defined structure

demonstrated by the examples in Figures 1–3. In these figures, the results of two similarity searches for structures with the same 2-D bonded pattern are shown. In the first search, a structure which has all its atoms and bonds fully defined is entered as the query. In the second search a partially defined structure is used where some of the atoms and bonds are given broad definitions. In the diagrams representing these 'partial' structures, 'AA' represents any atom; bonds which can be single, double or aromatic are represented by '------'; and 'any bond' is represented by '..........'. Fewer structure-based screens can be set for these 'partial' structures, therefore the number of screens representing the fully and partially defined structure are very different.

The similarity searches for the completely defined structures rank the database entry corresponding to that structure to the top position; the top five structures hit are shown in the Figures. It is readily apparent that they are similar to the query. In the case of the partially defined structures however, none of the query structures rank their corresponding database entry in the top five hits and it is difficult to see why some of the structures in the top five were considered similar. What is particularly interesting is that the partially defined search for AADRIB and ACSARM find almost the same top five structures, suggesting that the long tail and extra carbon on the side chain of ACSARM are being ignored in the search for the partially defined structure. To try to take more account of the 'pattern features' of query structures, such as the long tail in ACSARM, we have started to look at similarity searching systems which embody more pattern information in the attribute set.

2-D STRUCTURAL SIMILARITY USING INTER-NODAL BOND SEPARATION (INBS) FREQUENCIES

Calculation of INBS Frequencies

The problems experienced with similarity queries with low chemical specificity identify a need for 2-D attribute sets which consider larger connectivity pattern units of the molecular graph. Indeed, they should describe the 'skeletal' graph (i.e., a graph devoid of chemical embellishment) as a complete pattern unit. In 3-D, the molecular shape is completely defined by the full (Å) distance matrix (e.g., Crippen[5]) involving all atom pairs in the structure, whether they are bonded or not. In 2-D, a similar description can be expressed (e.g., Bersohn[6]) by considering the minimum number of bonds that must be traversed to reach atom *j* from atom *i*. A number of authors (e.g., Randic and Wilkins[7]; Carhart, Smith and Venkataraghavan[8]) have used these inter-nodal bond separations (INBSs) to derive atom codes for structure-activity studies. They would also seem to form a valid basis for 2-D similarity analyses.

The INBS values for the 3000 structure subset were calculated using a shortest-path algorithm similar to that used by Carhart *et al.*[8]. For each molecule, the number of INBSs of length n are scanned to form a frequency distribution for n=1-20. The frequency of INBS values >20 is also recorded. Thus INBS(1) records the number of bonded (1:2) interactions, etc. Unbranched acyclic chains have frequency distributions which decrease linearly with increasing INBS (e.g., 5,4,3,2,1 for n-hexane), whilst any branching or ring formation results in a distribution which is restricted to lower INBS values (e.g., 6,6,3,0,0 for cyclohexane).

ACSARM (Fully Defined)

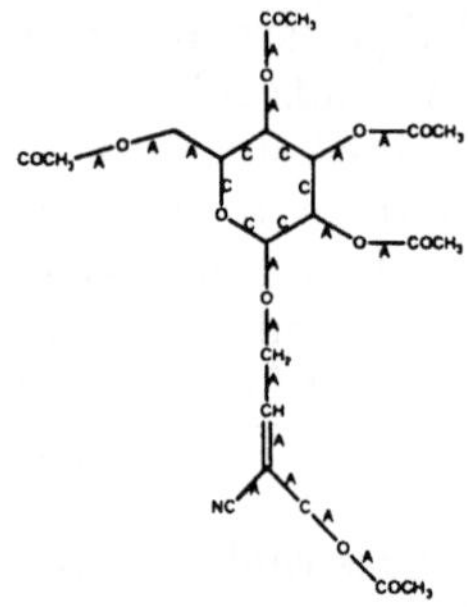

94 screens set

Distribution Table for 100 entries

TANI SCORE	Entries
100 - 88	1
87 - 82	0
81 - 76	19
75 - 70	50
69 - 64	30

Best score	93%
Worst score	68%
average	72%
range	25%

Top 5 hits

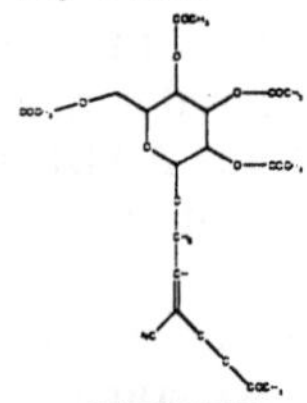

ACSARM 93%

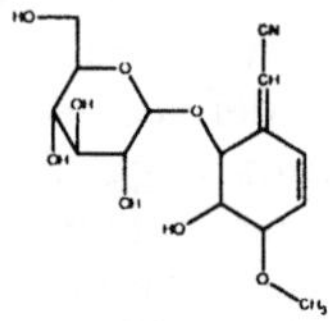

DIZPEH 81%

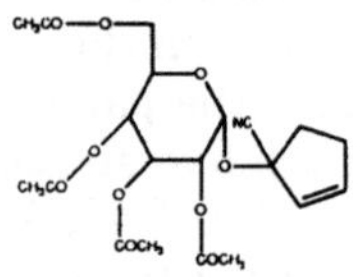

DIDYUK 80%

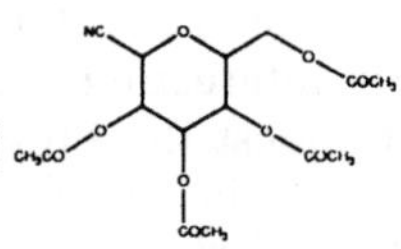

ACYGAP 79%

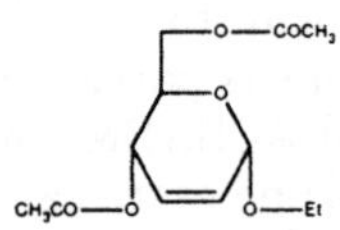

EACXHP02 78%

(a)

ACSARM (Partially Defined)

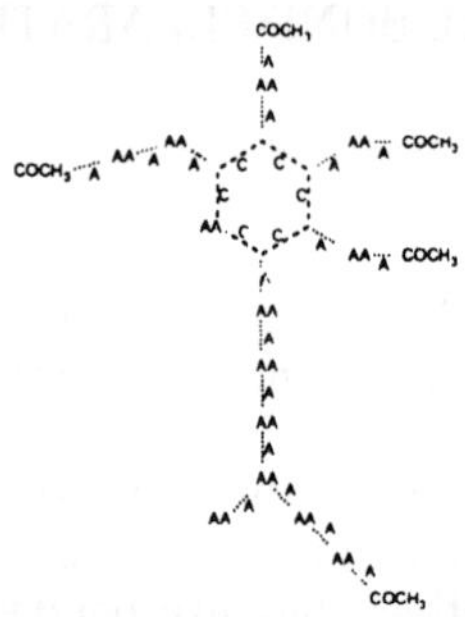

29 screens set

Distribution Table for 100 entries

TANI SCORE	Entries
100 - 42	2
41 - 40	0
39 - 38	6
37 - 36	17
35 - 34	34

Best score	43%
Worst score	32%
average	34%
range	11%

Top 5 hits

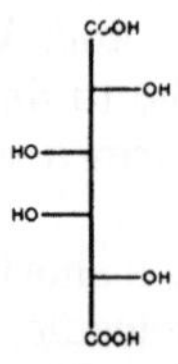

BIVTUV 43%

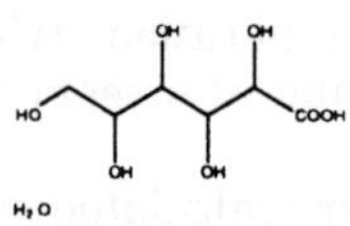

CAKZAP 42%

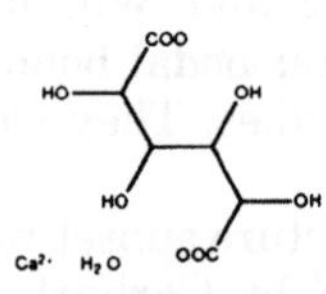

CAGLAR 39%

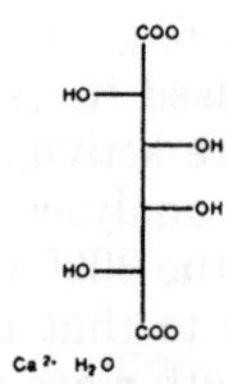

CAGLAR01 39%

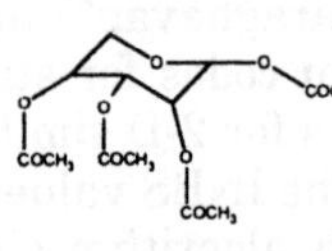

AADRIB 39%

(b)

Figure 3. CSD similarity search for the structure with refcode ACSARM. (a) Results of a similarity search for a fully defined structure. (b) Search for a partially defined structure

Nodal Seps	Occurrences
1:2 (1)	10
1:3 (2)	12
1:4 (3)	13
1:5 (4)	10
1:6 (5)	6
1:7 (6)	3
1:8 (7)	1
1:9 (8)	0
1:10 (9)	0
1:11 (10)	0
1:12 (11)	0
1:13 (12)	0
1:14 (13)	0
1:15 (14)	0
1:16 (15)	0
1:17 (16)	0
1:18 (17)	0
1:19 (18)	0
1:20 (19)	0
1:21 (20)	0

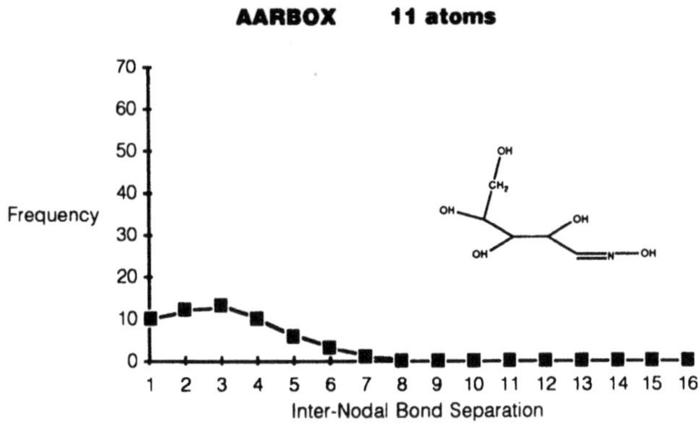

(a)

Nodal Seps	Occurrences
1:2 (1)	22
1:3 (2)	30
1:4 (3)	30
1:5 (4)	36
1:6 (5)	40
1:7 (6)	32
1:8 (7)	25
1:9 (8)	12
1:10 (9)	4
1:11 (10)	0
1:12 (11)	0
1:13 (12)	0
1:14 (13)	0
1:15 (14)	0
1:16 (15)	0
1:17 (16)	0
1:18 (17)	0
1:19 (18)	0
1:20 (19)	0
1:21 (20)	0

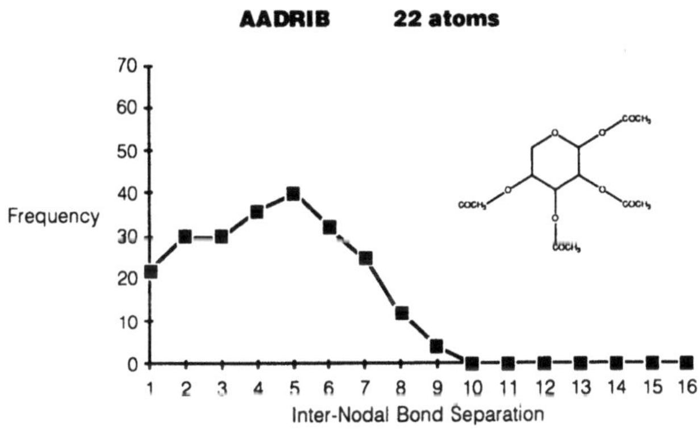

(b)

Nodal Seps	Occurrences
1:2 (1)	34
1:3 (2)	45
1:4 (3)	47
1:5 (4)	55
1:6 (5)	65
1:7 (6)	62
1:8 (7)	57
1:9 (8)	48
1:10 (9)	41
1:11 (10)	36
1:12 (11)	29
1:13 (12)	22
1:14 (13)	12
1:15 (14)	8
1:16 (15)	0
1:17 (16)	0
1:18 (17)	0
1:19 (18)	0
1:20 (19)	0
1:21 (20)	0

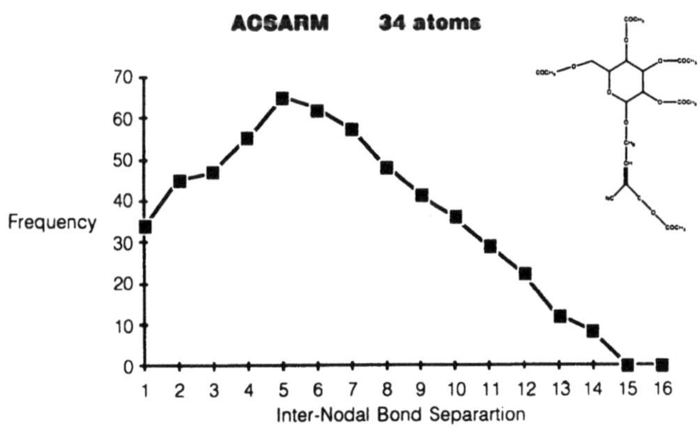

(c)

Figure 4. Inter-nodal bond separation frequencies for three structures of different sizes: (a) AARBOX 11 atoms (b) AADRIB 22 atoms (c) ACSARM 34 atoms

For example, Figure 4 shows the three test query structures used earlier with a listing of their INBS frequencies. It is apparent that the fewer atoms a structure has, the smaller the magnitude of its INBS frequency curve. However, these graphs also demonstrate that structures with different topological shapes display very different curves, irrespective of the magnitude of the curve. The graphs for AADRIB and ACSARM actually display quite similar curves which, if the difference in magnitude is ignored, could be almost exactly superimposed on each other for most of their length. This is because both structures have the same ring system. However, one of the smaller side chains in AADRIB is replaced by a long, acyclic tail in ACSARM and this is reflected in the graphs by the long tail-off on the right hand side of the curve for ACSARM.

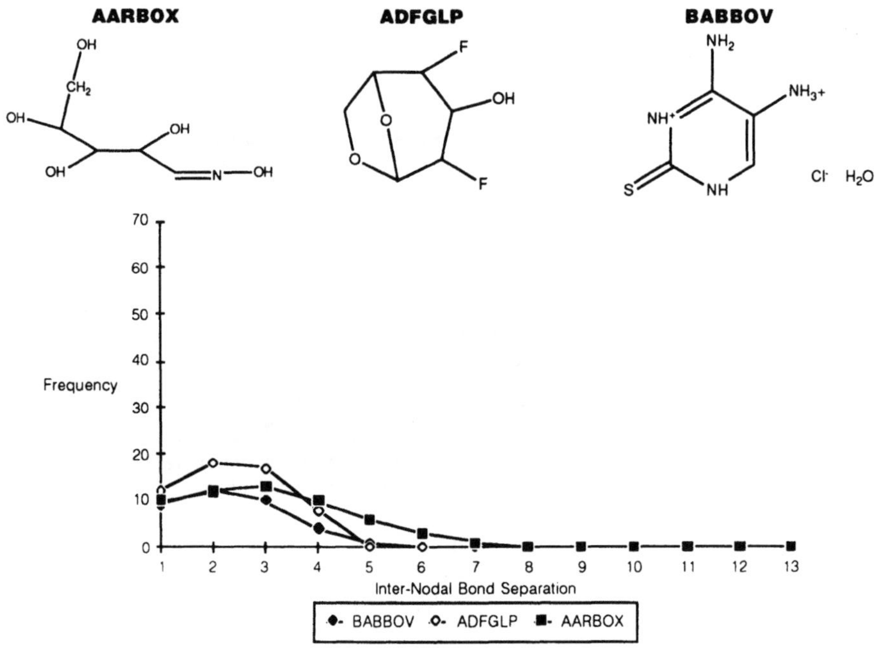

Figure 5. Inter-nodal bond separation frequencies for three structures with 11 atoms

Figures 5 and 6 demonstrate that even if structures have the same number of atoms, their INBS curves look very different. Figure 5 compares the curves of three structures which have 11 atoms; Figure 6 compares the curves from three 22-atom structures. It would appear that the use of INBS frequencies may prove to be of use in a pattern-based similarity search. We report below a series of preliminary experiments where INBS frequencies were used to describe query structures in a similarity search.

Metrics for Comparison of INBS Frequencies

The best similarity coefficient to use for comparison of the INBS frequencies was not clear. Willett's original work used attribute frequencies with an integer

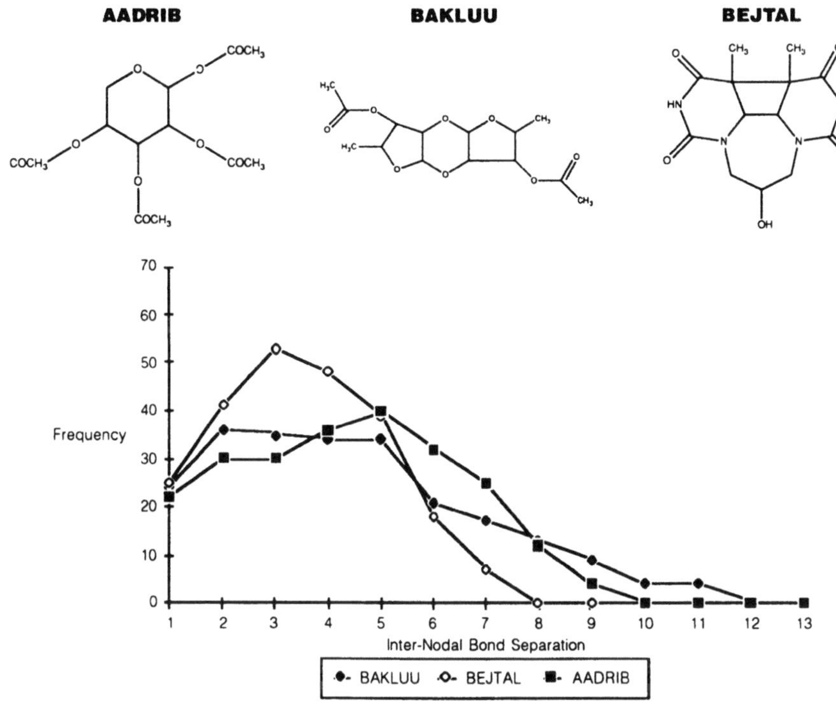

Figure 6. Inter-nodal bond separation frequencies for three structures with 22 atoms

Tanimoto measure[9]. The adoption of bit-encoded information led to the use of association coefficients for comparison of equal-length bit-vectors. In the case of our pattern-based similarity search, it is not the presence or absence of a feature that is being compared, but its frequency of occurrence. Discussions of the different types of similarity measures available for numerical, rather than binary, data can be found in books by Willett[10] and Everitt[11]. From these we chose to test the frequency Tanimoto coefficient and the Minkowski metric:

Frequency Tanimoto

$$S(PQ) = \frac{\upsilon\, D(Q,J) * D(P,J)}{\Sigma\, [D(Q,J)^2 + D(P,J)^2 - D(Q,J) * D(P,J)]} \tag{1}$$

Minkowski Metric

$$D(PQ) = \Sigma\, [|V(Q,J) - V(P,J)|^R]^{1/R} \tag{2}$$

where Q is the query structure, P is the target structure and J ranges from 1 to M, the number of INBSs. It is very unusual for a structure to have INBSs longer than 20 (atom 1 : atom 21), therefore M is currently set to 20. In the Minkowski Metric calculation, R can take different values, the most commonly used are $R = 1$, city block metric, and $R = 2$, Euclidean metric.

As discussed earlier, the frequency curves in Figures 5 and 6, from structures with the same number of atoms but different patterns, are completely different. However, if the frequency curves of structures which have similar patterns but different numbers of atoms are compared it is obvious that, although the curves have different magnitudes, their overall shape is very similar. Figure 7(b) contains a chart showing the curves from nine steroids, all with the same basic pattern (shown in 7(a)) but differing in the number and type of side chains that they have. The shape of each of these curves is, indeed, similar. On this evidence, it was decided to adapt the Minkowski Metric shown above to compare the rate of change (slope increments) of different curves as a means of comparing their similarity. Formula (2) then becomes:

$$D(PQ = \Sigma\ [/\{V(Q+1,J) - V(Q,J)\} - \{V(P+1,J) - V(P,J)\}/^{R}]^{1/R} \qquad (3)$$

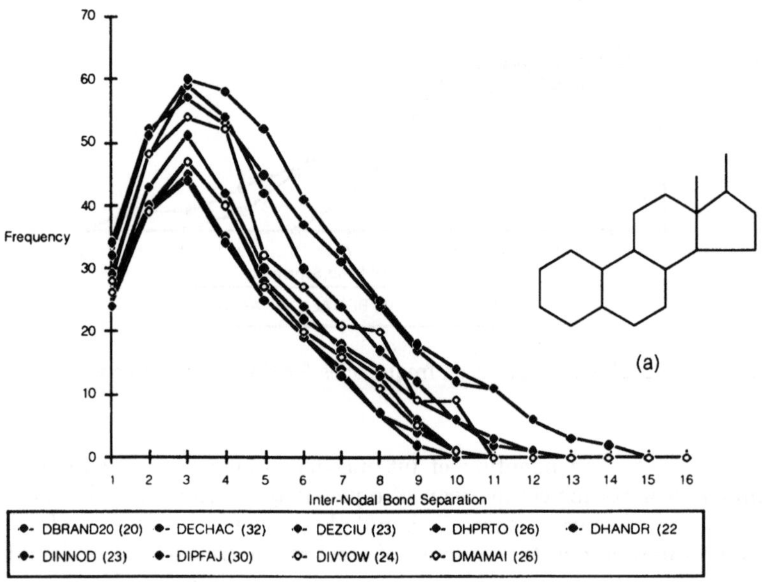

(b)

Figure 7. (a) Basic steroid skeleton for the nine steroids used to draw the graph in (b). Steroids are DBRAND20 (20 atoms); DECHAC (32 atoms); DEZCIU (23 atoms) DHANDR (22 atoms); DHPRTO (26 atoms); DINNOD (23 atoms); DIPFAJ (30 atoms); DIVYOW (24 atoms); DMAMAI (26 atoms); (b)Inter-nodal bond separation frequencies for the steroids conforming to the basic skeleton in (a)

Results from the INBS Method

The pattern-based searches which have been tested so far are:

1. All INBSs with a frequency Tanimoto similarity measure (1);
2. All INBSs with a city block metric similarity measure (3);
3. All INBSs with a Euclidean distance similarity measure (3);

4. INBSs from the most connected atom in the structure with a city block metric measure (3).

The fourth search was carried out using only the paths from the most connected atom in the structure in a city block metric calculation. Preliminary results from the first three searches indicated that the use of *all* INBS frequencies in the calculation resulted in a few 'false drops'. This could be due to the loss of information resulting from the summations of INBSs from every atom in the structure. It was decided to investigate the effect of using only the INBSs from a 'focal atom' in the structure. The central atom in a structure is perhaps best located by the Morgan algorithm[12] All of the structures in the CSD are passed through the Morgan algorithm to produce a canonically enumerated structure, therefore the Morgan 'root atom' was chosen as the 'focal atom'.

Each of these searches was carried out on the 3000 structure subset of the CSD described earlier. Five query structures were chosen to reflect a range of different sizes and structural patterns – AADRIB, AARBOX and ACSARM that were described in detail earlier, together with ACCELL10 and ADENTP (shown in Figure 8). It was of particular interest to see how well the different measures performed on structures of different sizes and shapes because of the theory on which this new similarity measure is based (see Figure 9).

In the discussion of the results, the 'similarity' between the input query structure and the ranked hits was determined based on subjective decisions made by one of the Authors. The decision was based on the topological pattern of the query and that of the ranked structure. However, in a number of cases when a

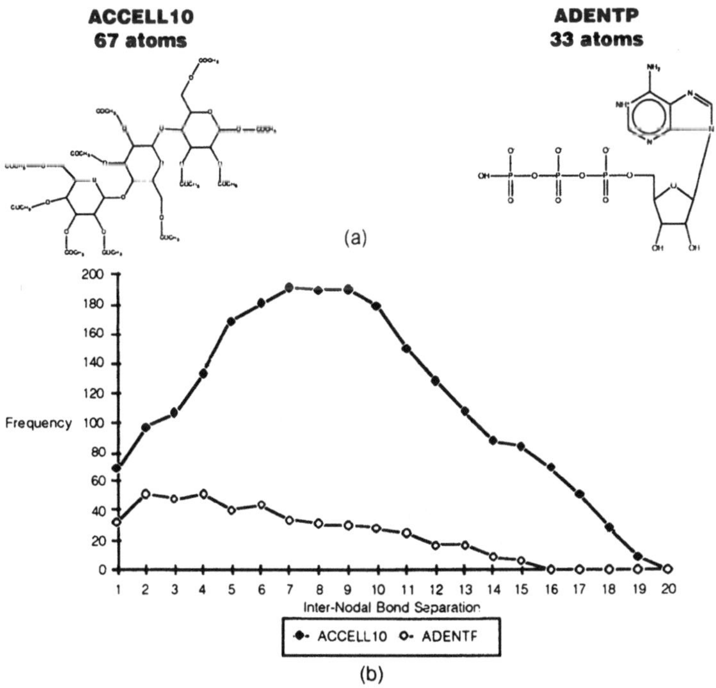

Figure 8. (a) Search structures used in the inter-nodal bond separation frequency pattern similarity searching (b) Graph showing the frequency curves for the two structures in (a)

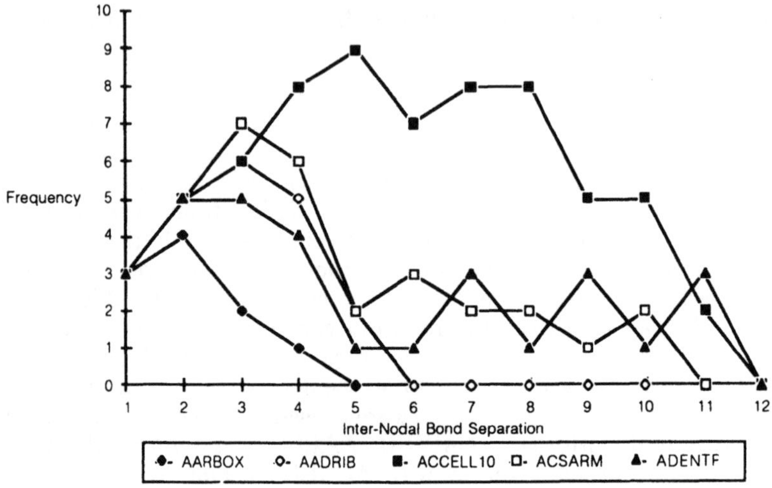

Figure 9. Chart of the inter-nodal bond separations for paths from the most highly connected atom (i.e., Morgan number 1) in the five query structures

structure, which at first appeared to be totally different on a topological level from the query structure, was considered more carefully, it became apparent that although the overall pattern was different from the query, all of the requisite parts of the query pattern were present albeit in a different orientation. In these cases, the ranked structure *was* considered similar.

The first three searches compared the performance of different similarity measures with the full INBS frequency data. When the performance of the frequency Tanimoto measure and the city block metric was compared, it was found that the hits ranked to the top using the Tanimoto measure were not as similar to the query structure as those ranked top by the city block metric. Generally it was found that the city block and Euclidean distance measures ranked the hits equally well, however in two cases the city block ranking was better than the Euclidean ranking, so use of the city block metric may be preferred. The fourth search used only the paths from the most connected atom in the structure in a city block metric calculation. A comparison of the performance of the INBS search for individual query structures using the city block metric for all paths and for the Morgan root atom alone is discussed below.

The all-path search for query structure AARBOX, a small, totally acyclic structure, resulted in a number of small, acyclic structures being ranked to the top (with AARBOX ranked as the most similar molecule), however there was also a small number of cyclic structures with the correct side chains. The canonical atom search performed less well, with 21 structures considered by the program to be identical to the query structure (one of the 21 is AARBOX). However most of these structures have five- or six-membered rings, some with correct side chains and others with side chains that are too long.

The majority of the top ten structures found using an all-path search for AADRIB, a fairly small-sized, cyclic structure with acyclic side chains, were very similar to the query (again, the query structure, and the two other occurrences of the structure under different refcodes in the file, were ranked to the top of the list).

Some of the top structures however, had extra side chains, or had side chains which were longer or shorter than those in the query. Searches based on the canonical root atom did not produce good results. Although the three existing occurrences of the query structure were ranked in the top five structures, the other structures ranked similar to the query were in fact much more complex, with much larger patterns and numbers of atoms.

The all-path search for query structure ADENTP, a medium-sized structure with two fused rings for the main body and a tail composed of a ring and a repeating acyclic unit, produced very good results (with ADENTP ranked to the top of the list). The majority of structures in the top ten were very similar to the query structure, in some cases the patterns were identical, but in a few cases part of the repeating acyclic unit was missing. The canonical atom search was almost as good as the full search, with ADENTP at the top of the list and the other top nine all with very similar patterns to the query, only differing in the length of the tail (sometimes too short, sometimes too long).

The all-path search for ACSARM, a medium-sized structure with a single monocyclic unit and varying lengths of acyclic side chains, one of which produces a long acyclic tail, resulted in a varied ranking (ACSARM was ranked to the top of the list). Most of the structures at the top of the list had the acyclic tail replaced by a cyclic unit, although some corresponded nearly exactly to the query pattern. Some of these cyclic units contained similar side chains to the acyclic tail, while other were too complex. Searches on the canonical root atom produced a higher proportion of structures that were totally different from the query, but there was also a number of structures which were quite similar to the query (ACSARM was ranked to the top).

The results of the search for the structure ACCELL10, a larger structure with a repeating substituted cyclic unit, were very interesting. Both searches ranked the query structure to the top, and then ranked a number of structures that were composed of two out of the three repeating substituted cyclic units (sometimes the third unit was replaced by an unsubstituted ring and other times there was no replacement). However, both searches then found a number of structures that were very different from the query.

Current Status of the INBS Project

It was found that the best method currently tested is that which compares the incremental rates of change in the distribution of the INBSs in a structure using the city block metric similarity measure. The performance of this measure did differ slightly for each of the chosen query structures which were selected to represent a variety of molecular sizes and structural patterns. However, the success of the 'pattern' search appears to be on the same level as the current 2-D similarity search available in the CSD System based on bit-screens. Interestingly, those structures which are considered most similar to the query are not the same in the two searches presumably because they are based on different attribute sets.

Reducing the number of INBSs by reducing the level of connections to those which emanate from the most connected atom resulted in limited success. However, other areas that are currently being explored include the partition of path lengths into cyclic, acyclic and combined cyclic/acyclic paths. Methods of minimising the change in the shape of a graph resulting from the addition of side chains to a structure are also being considered. This should enable a further degree of control over the 'general' pattern features of a query.

Ultimately the aim is to combine the chemical and INBS search strategies together in the CSD System and to allow user control over the relative importance of the chemistry and/or pattern attributes in deriving ranked lists.

SIMILARITY ANALYSIS OF 3-D FRAGMENT CONFORMATIONS

Similarity analysis of 3-D chemical structures is obviously highly desirable at the level of the complete molecule to identify database entries which have topographical shape characteristics similar to those of an input query. The initial approaches to this area are summarised in the next section. However, 3-D similarity analysis is also important in the CSD for discovering the number of different conformations exhibited in the database for a given chemical fragment, isolated by a preliminary 2-D substructure search.

Conformational Descriptors and their Metrical Comparison

Similarity analysis at the 3-D fragment level involves the metrical comparison of an equal number of descriptors for each fragment. In the conformational similarity studies, N_t torsion angles for each of the N_f fragments were used, where N_t is sufficient for a unique definition of fragment shape.

The dissimilarity *D(PQ)* between fragments *P* and *Q* is assessed using the Minkowski metric (equation 2 above), where *V(P,J)* and *V(Q,J)* are the *J*th torsion angles for fragments *P* and *Q*, and summation is from $J = 1 - N_t$. Because torsion angles are circular functions, there is a phase restriction of $-180 < V < 180$, and the torsional dissimilarity measure must be expressed as:

$$D(PQ) = \Sigma\, d(P,Q)_J \qquad (4)$$

where the incremental differences *d(PQ)* are the minimum of:

$$d(P,Q)_J = /V(P,J) - V(Q,J)/ \qquad \text{or} \qquad d(P,Q)_J = 360 - /V(P,J(- V(Q,J)/$$

A further complication is introduced by fragments which exhibit 2-D topological symmetry. This is related to the 3-D search problem fully described in the paper by Allen *et al.* in this publication. Each atomic permutation arising from fragment topological symmetry will give rise to a similar permutation of the torsional descriptors (see Figure 4 of the paper by Allen *et al.* in this publication). In these situations *D(PQ)* must be minimised by keeping fragment *P* static and allowing the torsional sequence for fragment *Q* to adopt all possible symmetry variants and inversions. The minimum value of *D(PQ)* obtained using this method is the value which enters into the similarity analysis.

Preliminary Results

Initial applications of torsional dissimilarity measures concentrated on their use in conformational clustering[13]. Here, all dissimilarity values are calculated for each of the $N_f(N_f - 1)/2$ fragment pairs in a subset of the database comprising the

3-D co-ordinate information for all examples of a query fragment. The unique $D(PQ)$ values are input to either single linkage, complete linkage or Jarvis-Patrick clustering algorithms. This results in the N_f different conformations being classified into N_C clusters, where all members of a given cluster should exhibit closely similar 3-D conformations.

A simple modification of the dissimilarity calculation permits a query conformation (K) to be compared with experimental results (T) derived from the CSD. The torsion angles for conformation K may already be known from the literature. More commonly, however, they will have been obtained by some independent computational procedure (e.g., energy minimisation, *ab initio* calculations, etc.). In this case, it is valuable to be able to compare a computed conformation with the actual experimental knowledge in the CSD.

To test this simple procedure, a subset of 222 six-membered carbocycles were used as targets (T). These were selected from the CSD to include known major conformational variants: (a) chair, (b) half-chair, (c) boat and (d) planar phenyl rings (Figure 10). The query fragment (K) was a chair conformation described by the torsional sequence 60°, −60°, 60°, −60°, 60°, −60°. The 222 $D(KT)$ values were calculated using formula 4. The resultant histogram of these values (Figure 11) shows a 1-D representation of the conformational spectrum of the chosen subset. A large peak at $D(KT) = 0$ (i.e., no dissimilarity between these structures and K) confirms the presence of chair conformations in the subset. When the torsional angle sequences of representative structures in the other peaks in the spectrum are examined, it is found that they can be closely correlated with the other known shape variants ((b) – (d)).

Further trials with more complex fragments have shown the need for some modification of the dissimilarity basis, which will be reported when the work is complete.

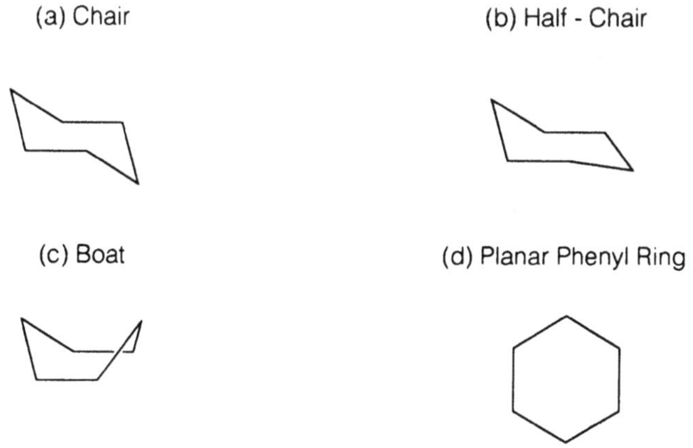

Figure 10. Examples of different conformations for six-membered carbocycles

```
DSIM          5    10   15   20   25   30   35   40   45   50
         .....I....I....I....I....I....I....I....I....I.....
   0.0  -***                                              - (  3)
        .*****************************          CHAIRS    . ( 29)
        .********************                             . ( 20)
        .**********                                       . ( 10)
        .                                                 .
  62.5  -**                                               - (  2)
        .**                                               . (  2)
        .*********************                            . ( 21)
        .************                       HALF-CHAIRS   . ( 12)
        .*****                                            . (  5)
 125.0  -******                                           - (  6)
        .**********                                       . ( 10)
        .****************************       PHENYL        . ( 28)
        .**                                               . (  2)
        .***********                                      . ( 11)
 187.5  -*******************                 BOATS        - ( 19)
        .*************************************            . ( 37)
        .****                                             . (  4)
        .*                                                . (  1)
        .                                                 .
 250.0  -                                                 -
         .....I....I....I....I....I....I....I....I....I.....
DSIM          5    10   15   20   25   30   35   40   45   50
```

Figure 11. Histogram of $D(KT)$ values for 222 six-membered carbocycles in the June 1990 Cambridge Structural Database

APPROACHES TO GENERALISED 3-D SIMILARITY ANALYSIS

For a structure containing n atoms, the 3-D shape is described by the $n(n-1)/2$ interatomic Å distances. A comparison of the distance matrix for a query with that for each target structure is certainly the most exhaustive method of locating matching (or similar) structures. It is also certain to be very demanding on computer resources since (a) matrices of different dimensions, n_q (number of query atoms) and n_t (number of atoms in the target), are being compared, and (b) the atoms may be enumerated differently in each target. As in the 2-D case, some type of data reduction must be effected and the 'false drops' that may be produced due to the resultant loss of information must be accepted. Transformation of the matrices into distance-range frequency distributions is an obvious way of eliminating problems (a) and (b) above. Indeed, this is the basis on which 3-D distance screens would normally be calculated[14].

The effectiveness of such screens, and/or the underlying distance frequency distributions, as attributes for similarity calculations is currently being investigated. Ways in which information from the 2-D (INBS) and 3-D (Å distances) matrices may be combined to form attributes which are more 'information-rich' are also being examined. Figure 12 shows some preliminary results of this work where the 3-D distance frequencies are further partitioned on the basis of INBS values. Figure 12(a) shows the distance distribution for carbon-nitrogen separated chemically by a minimum of four bonds. It shows a very tight range of *ca.* 3.0–6.5 Å. As the INBS value increases however, the range of Å distances shows an increasingly broad distribution. The 1:14 distances of Figure 12(b) span some 13 Å.

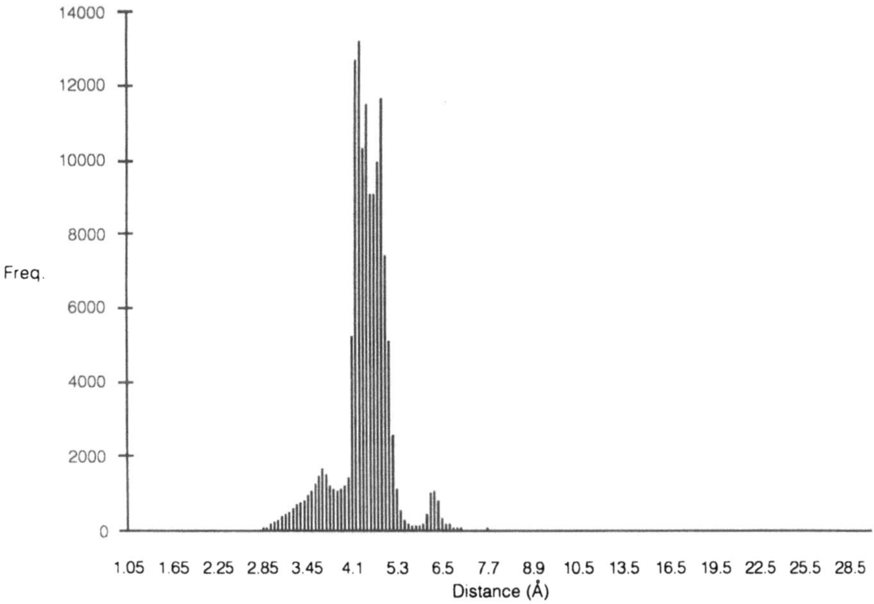

(a)

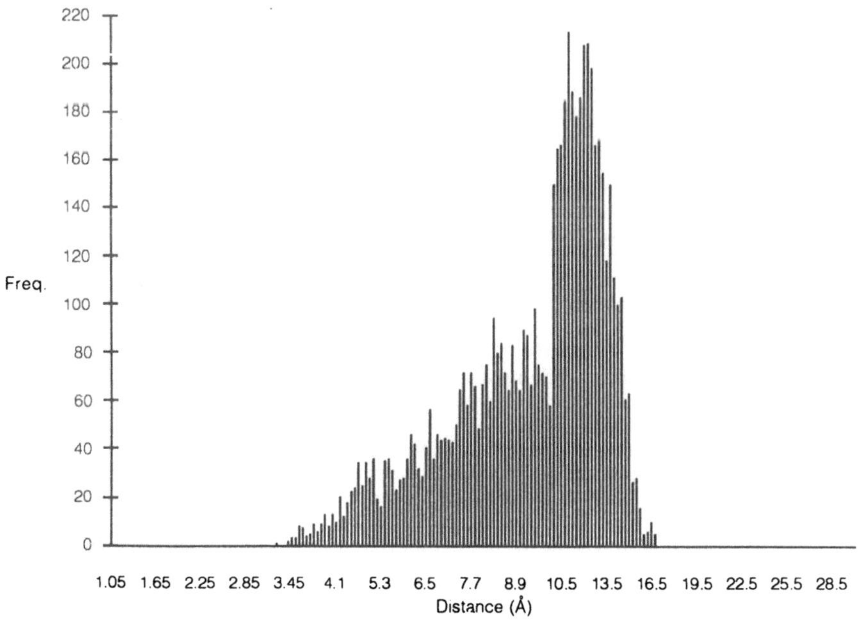

(b)

Figure 12. (a) Distribution of 1:5 C-N distances in the July 1989 CSD (b) Distribution of 1:14 C-N distances in the July 1989 CSD Note that the X-axis is not a linear Å scale

It would seem that the occurrence of very short distances at longer path lengths is an attribute that can be employed meaningfully in 3-D similarity calculations.

CONCLUSIONS

The investigations reported in this paper are only preliminary. They are driven by the obvious importance of similarity analyses in large databases and by a perceived need for a very flexible approach to the problem for both 2-D and 3-D molecular structures. The results obtained so far indicate that similarity searches can be effective at various levels of query definition: chemical, 2-D bonding pattern and 3-D distance distribution, and that these levels can be combined or integrated depending upon the information coded in the query. What is important are (a) the selection of suitable attribute sets for each level of discrimination, (b) the compact storage of attributes for each database entry and (c) the selection of effective metrics for assessment of similarity at the various levels.

REFERENCES

1. Willett P.; Winterman V. 'A Comparison of Some Measures for the Determination of Intermolecular Structural Similarity: Measures of Intermolecular Structural Similarity'. *Quant. Struct. Act. Relat.* **1986,** *5,* 18–25.
2. Bawden D. 'Browsing and Clustering of Chemical Structures'. In *Chemical Structures. The International Language of Chemistry*; Warr W.A., Ed.; Springer-Verlag: Berlin, 1988; pp. 145–150.
3. Willett P.; Winterman V.; Bawden D.'Implementation of Nearest-neighbour Searching in an Online Chemical Structure Search System. *J. Chem. Inf. Comput. Sci.* **1986,** *26,* 36–41.
4. CSD System User's Manual Part I : QUEST89; Cambridge Crystallographic Data Centre: Cambridge, England, 1989.
5. *Distance Geometry and Conformational Calculations*; Crippen, G.M.; Research Studies Press: Letchworth, 1977.
6. Bersohn M.'A Fast Algorithm for Calculation of the Distance Matrix of a Molecule'. *J. Comput. Chem.* **1983,** *4,* 110–113.
7. Randic M.; Wilkins C.L. 'Graph-based Fragment Searches in Polycyclic Structures'. *J. Chem. Inf. Comput. Sci.* **1979,** *19,* 23–37.
8. Carhart R.E.; Smith D.H.; Venkataraghavan R. 'Atom Pairs as Molecular Features in Structure-activity Studies: Definition and Applications. *J. Chem. Inf. Comput. Sci.* **1985,** *25,* 64–73.
9. Willett P. 'Similarity Coefficients and Weighting Functions for Automatic Document Classification: an Empirical Comparison'. *Int. Classif.* **1983,** *10,* 138–142.
10. *Similarity and Clustering in Chemical Information Systems*; Willett, P.; Research Studies Press: Letchworth, 1987.
11. *Cluster Analysis*; Everitt B.; Halstead-Heinemann: London, 1980.
12. Morgan H.L. 'The Generation of a Unique Machine Description for Chemical Structures – a Technique Developed at Chemical Abstracts Service. *J. Chem. Doc.* **1965,** *5,* 107–113.
13. Allen F.H.; Doyle M.J.; Taylor R. 'Automated Conformational Analysis from Crystallographic Data'. *Acta Crystallogr.* **1991,** *B47,* 50–61.
14. Jakes S.E.; Willett P. 'Pharmacophoric Pattern Matching in Files of 3-D Chemical Structures: Selection of Interatomic Distance Screens'. *J. Mol. Graphics* **1986,** *4,* 12–20.

CALCULATION OF THREE-DIMENSIONAL STRUCTURAL SIMILARITY

Catherine A. Pepperrell and Peter Willett
Department of Information Studies, University of Sheffield, Western Bank, Sheffield S10 2TN, England

ABSTRACT

This poster reports work to date on a project to evaluate techniques for measuring the degree of similarity between pairs of 3-D chemical structures represented by interatomic distance matrices. The four techniques that have been tested use the distance information in very different ways and have very different computational requirements. Preliminary experiments with two small data-sets, for which both structural and biological activity data are available, suggest that the most cost-effective technique is based on a mapping procedure that tries to match pairs of atoms, one from each of the molecules that are being compared, that have neighbouring atoms at approximately the same distances.

INTRODUCTION

There is increasing interest in the use of computational techniques to determine the degree of structural similarity between pairs of molecules[1–3]. The extensive work that has been carried out to date has focused on the calculation of similarity when the molecules are represented by 2-D connection tables, such as form the basis for the great majority of computer-based chemical information systems[4,5]. In this case, the degree of similarity between a pair of structures is determined by the substructures that they have in common; these substructures are generally the fragment screens that are used to increase the efficiency of substructure searching[3]. Developments in molecular modelling mean that it is now relatively easy to generate sets of 3-D atomic co-ordinates from a 2-D connection table and several companies have created in-house databases of 3-D structures from their databases of 2-D structures[6,7]. There is accordingly a need to investigate methods for determining the degree of similarity between pairs of 3-D molecules. In this poster, we report the preliminary results of such an investigation using two small sets of structures for which both 3-D co-ordinate data and biological activity data are available.

SIMILARITY MEASURES

In this section, we describe four ways of determining the degree of similarity between a pair of structures for which atomic 3-D co-ordinates are available. Given such data, it is possible to visualise determining molecular similarity using either distance or angular information or both. Our initial studies have focused on the

W. A. Warr (Ed.)
Chemical Structures 2

first of these, since we have considerable previous experience of the use of interatomic distances for 3-D substructure searching[8]; future work will consider the use of angular information for this purpose. Specifically, the degree of similarity between a pair of molecules, A and B, is calculated by a comparison of their distance matrices. Let A contain $N(A)$ non-hydrogen atoms; then the distance matrix for A, DA, is an $N(A)$ times $N(A)$ matrix such that the IJ-th element, $DA(I,J)$, contains the distance between the I-th and J-th atoms in A (and similarly for the matrix DB representing the $N(B)$-atom molecule B). Given the matrices DA and DB, four types of similarity measure have been defined as described in the remainder of this section in order of increasing sophistication (and increasing computational requirements). The first two measures provide a global measure of similarity between A and B, i.e., a measure that denotes the overall degree of similarity without any attempt being made to specify which parts of them are structurally related; the second two measures, conversely, provide a local measure of similarity in which the structurally related features are specified. The advantages of local measures in the determination of similarity have been noted by Klopman and Raychaudhury[9].

As well as summarising the steps that are needed to compute each of these four similarity measures, we also discuss the expected time complexity in each case for the matching of a pair of molecules.

Distance Distribution

The first approach represents an individual molecule by the frequency distribution of the interatomic distances in its distance matrix; the similarity between a pair of molecules can then be calculated by the goodness-of-fit between the two frequency distributions. Given a distance matrix, DA, the method involves sorting the elements $DA(I,J)$ $(1 \leq I \leq N(A)-1; I \leq J \leq N(A))$ into increasing order of distance and representing them as a frequency distribution FA such that the K-th element of FA, $FA(K)$, contains the number of elements of DA that occur in the K-th distance range that has been defined. The reader is referred to the paper elsewhere in these proceedings by Artymiuk *et al*[10]. for a detailed description of the way in which this is carried out.

The use of a single frequency distribution to characterise a molecule means that all of the interatomic distances are assumed to be of the same kind. This will only be the case if the molecule that is being characterised contains only a single elemental type. More useful results may be expected if different frequency distributions are specified for the carbon-carbon, carbon-heteroatom and heteroatom-heteroatom distances. We shall denote these three types of distribution for the molecule A by FA_{CC}, FA_{CX} and FA_{XX}, respectively (and similarly for the molecule B with which A is to be compared). Then, given some goodness-of-fit criterion, $G(FA,FB)$, we can define the overall degree of similarity between A and B by

$$\alpha G(FA_{CC}, FB_{CC}) + \beta G(FA_{CX}, FB_{CX}) + \gamma G(FA_{XX}, FB_{XX})$$

where α, β and γ are user-defined weights reflecting the relative degree of importance attached by the user to these three types of distribution.

This method involves the calculation of all of the distances in both A and B, this requiring $O(N(A)^2 + N(B)^2)$ operations, and a similar number to create the frequency distributions FA and FB. The actual comparison of the two distributions requires M operations, where M is the number of elements in FA or FB. M will be less, or very much less, than the number of distances in either of the molecules so that this comparison stage can be neglected: thus the computation is dominated by the time required for the calculation of the distances, giving an expected time complexity of $O(N(A)^2 + N(B)^2)$.

Individual Distances

Rather than using the distance distributions, which take no account of the actual distances, the second method calculates the degree of similarity between A and B on the basis of the number of components of DA and DB that are identical, i.e., it compares each element of DA in turn with each of the elements of DB to see if any of them match; if so, then the matching element of DB is deleted from further consideration and the next element of DA used to initiate a further scan of DB. A distance match occurs if the distances are the same (to within any allowed tolerance such as plus or minus 0.5 Å) and if the elemental types associated with this interatomic distance are the same; this definition of a distance match also applies to the third and fourth similarity methods discussed further below. The overall degree of similarity between A and B is then calculated from the number of distinct distances in common, $COMMON$. The specific similarity coefficient used here was the Tanimoto Coefficient[3] which is defined by

$$\frac{COMMON}{N(A)(N(A)-1)/2 + N(B)(N(B)-1)/2 - COMMON}$$

where, as previously, $N(A)$ and $N(B)$ are the numbers of non-hydrogen atoms in A and B.

This method requires the comparison of the $O(N(A)^2)$ distances from A with the $O(N(B)^2)$ distances from B; this is done most efficiently if the two sets of distances are first sorted into increasing order, this requiring $O(N(A)^2 logN(A)^2 + N(B)^2 logN(B)^2)$ operations if an appropriate sorting algorithm is used. Once the lists have been sorted in this way, the comparison to identify the matching distances can be accomplished in $O(N(A)^2 + N(B)^2)$ time; thus the computation is dominated by the sorting operation, giving an expected time complexity of $O(N(A)^2 logN(A)^2 + N(B)^2 logN(B)^2)$. [If the lists are not sorted, then the comparison operation dominates, with an expected time complexity of $O(N(A)^2 \times N(B)^2$.]

Atom Mapping

The third method attempts to map the individual atoms from A onto those atoms from B that are most similar to them. We assume, without loss of generality, that $N(A) \leq N(B)$, i.e., that B contains at least as many atoms as does A. Consider the I-th atom in A, $A(I)$: then the I-th row of the distance matrix DA contains the distances from I to all of the other atoms in A. This set of distances is compared with each of the rows, J, from the distance matrix DB to identify the distance matches and hence the the similarity between the atoms $A(I)$ and $B(J)$, where the similarity is calculated using the Tanimoto coefficient as defined in the previous section; however, the similarity in this case is based just upon the distance matches that involve the atoms $A(I)$ and $B(J)$ (rather than all of the atoms in A and B as previously). In this way, we can calculate an $N(A) \times N(B)$ matrix, T, the IJ-th element of which contains the similarity, $T(I,J)$, between the pair of atoms $A(I)$ and $B(J)$. This matrix is referred to subsequently as the *atom match* matrix.

Once the atom match matrix has been created, the interatomic similarities contained within it are used to determine that atom from B which is most similar to each atom from A, i.e., the matrix is used to establish a set of equivalences of the form $A(I) \equiv B(J)$ where $B(J)$ is that atom in B that lies at the centre of the most similar area of 3-D space, as defined by the atoms surrounding it, as does the atom $A(I)$. The basic matching algorithm that is used to establish these equivalences is as follows:

1. Scan the atom match matrix to find the pair of atoms that have the largest calculated value for $T(I,J)$.
2. Store the resulting equivalences as a tuple of the form $\{A(I) \equiv B(J);\ T(I,J)\}$.
3. Remove $A(I)$ and $B(J)$ from further consideration.
4. Return to Step 1 if there are atoms in A that have not yet been mapped to an atom in B.

The overall degree of similarity between A and B is then calculated by the mean similarity when averaged over all of the atoms in A using the information in the tuples that were stored in Step 2 of the scanning algorithm above. Thus, the overall similarity is given by:

$$\frac{1}{N(A)} \sum_{I=1}^{N(A)} T(I,J).$$

This mapping procedure involves the comparison of the entire set of distances for each atom in A with the entire set of distances for each atom in B, requiring $O(N(A)^2 \times N(B)^2)$ comparisons. Once these have been executed and the Tanimoto coefficients evaluated, the method involves repeatedly scanning the elements of the atom match matrix to find the next pair of atoms that should be regarded as being equivalent. The scanning step needs to be carried out $N(A)$ times, once for each atom in A, and the matrix contains $N(A) \times N(B)$ elements; this stage thus requires $O(N(A)^2 \times N(B))$ operations. Accordingly, the computation is dominated by the comparison stage with an expected time complexity of $O(N(A)^2 \times N(B)^2)$.

Maximal Common Substructure

The final measure of similarity that we have used is the maximal common substructure (MCS), i.e., the largest pattern of atoms in 3-D space that are identical in the two molecules that are being compared. If the MCS contains $MCS(A,B)$ atoms, then the similarity between A and B is defined to be

$$\frac{MCS(A,B)}{N(A) + N(B) - MCS(A,B)} .$$

The use of the MCS as a measure of structural similarity has been studied by several previous workers: the particular MCS algorithm used here is one based on the detection of cliques, which is known to be particularly efficient in the context of 3-D chemical graphs[1,11].

MCS detection is equivalent to maximal common subgraph isomorphism: Levi[12] notes that this belongs to the class of NP-complete problems since the identification of all subgraphs containing $MCS(A,B)$ atoms requires

$$O\left(\frac{N(A)! \times N(B)!}{MCS(A,B)!(N(A) - MCS(A,B))!(N(B) - MCS(A,B))!}\right)$$

operations in the worst case. This last approach is thus far more demanding of computational resources than those described previously.

EXPERIMENTAL DETAILS AND PRELIMINARY RESULTS

The similarity methods described above are currently being tested on ten small data-sets from the medicinal chemistry literature that have been used previously by Ormerod *et al.* in studies of fragment weighting schemes for substructural analysis[13]. Each of the data-sets contains between one hundred and two hundred compounds for which both 3-D structural and biological activity data are available.

Each active molecule in a data-set was taken in turn and its similarity calculated with each of the other molecules in that data-set. The molecules were then ranked in decreasing order of similarity and cut-offs applied to retrieve the top 5, top 10 and top 20 compounds from the ranking. These were then checked to determine whether they were active or inactive in the particular biological test system associated with that data-set. The overall utility of a similarity measure was calculated by taking the mean numbers of active molecules at the three cut-off positions when averaged over all of the active compounds.

The experiments are currently in progress. To date, we have fully investigated two of the data-sets, these being a set of 114 structurally heterogenous molecules, of which 66 were mutagenic and 48 non-mutagenic in the Ames reversion test, and a set of 209 9-anilinoacridines, of which 150 showed anti-tumour activity and 59 were inactive. Our preliminary conclusions are that the first two methods give broadly comparable levels of performance and that both are noticeably inferior to the atom mapping method. The best results are given by the MCS method but the time requirements here are quite exhorbitant, and totally unacceptable in the context of an operational similarity searching system where an individual 3-D target molecule might need to be compared with many tens of thousands of structures. Our initial experiments thus suggest that the atom mapping routine is the most cost-effective method that we have tested for the calculation of 3-D structural similarities.

ACKNOWLEDGMENTS

We thank ICI Agrochemicals and the Science and Engineering Research Council for funding and Dr Robin Taylor for providing the 3-D co-ordinate data.

REFERENCES

1. Johnson, M.A. 'A Review and Examination of the Mathematical Spaces Underlying Molecular Similarity Analysis'. *J. Math. Chem.* **1989,** *3,* 117–145.
2. *Concepts and Applications of Molecular Similarity*; Johnson, M.A.; Maggiora, G.M.; John Wiley: New York, 1990.
3. *Similarity and Clustering in Chemical Information Systems*; Willett, P.; Research Studies Press: Letchworth, 1987.
4. Barnard, J.M. 'Recent Developments in Chemical Structure Handling'. *Perspectives in Information Management* **1989,** *1,* 133–168.
5. Lipscombe, K.J.; Lynch, M.F.; Willett, P. 'Chemical Structure Processing.' *Annu. Rev. Inf. Sci. Technol.* **1989,** *24,* 189–238.
6. Martin, Y.C.; Bures, M.G.; Willett, P. 'Searching Databases of Three-dimensional Structures. In *Reviews in Computational Chemistry*; Lipkowitt, K.B.; Boyd. D.B., Eds., VCH: New York, 1990.
7. Rusinko, A.; Sheridan, R.P.; Nilakantan, R.; Haraki, K.S.; Bauman, N.; Venkataraghavan, R. 'Using CONCORD to Construct a Large Database of Three-dimensional Co-ordinates from Connection Tables.' *J. Chem. Inf. Comput. Sci.* **1989,** *29,* 252–255.
8. Jakes, S.E.; Willett, P. 'Pharmacophoric Pattern Matching in Files of 3-D Chemical Structures: Selection of Interatomic Distance Screens'. *J. Mol. Graphics* **1986,** *4,* 12–20.
9. Klopman, G; Raychaudbury, C. 'Vertex Indicies of Molecular Graphs in Structure-activity Relationships: a Study of the Convulsant-anticonvulsant Activity of Barbiturates and the Carcinogens of Unsubstituted Polycyclic Hydrocarbons'. *J. Chem. Inf. Comput. Sci.* **1990,** *30,* 12–19.
10. Artymiuk, P.J.; Grindley, H.M.; Mitchell, E.M.; Rice, D.W.; Ujah, E.C.; Willett, P. 'Representation and Searching of 3-D Protein Structures'. In these Proceedings.
11. Brint, A.T.; Willett, P. 'Upperbound Procedures for the Identification of Similar Three-dimensional Chemical Structures'. *J. Comput.-Aided Mol. Des.* **1988,** *2,* 211–320.
12. Levi, G. 'A Note on the Derivation of Maximal Common Subgraphs of Two Directed or Undirected Graphs'. *Calcolo* **1972,** *9,* 341–352.
13. Ormerod, A.; Willett, P.; Bawden, D. 'Comparison of Fragment Weighting Schemes for Sub-structural Analysis'. *Quant. Struct. Act. Relat.* **1989,** *8,* 115–129.

MOLECULAR DISSIMILARITY IN CHEMICAL INFORMATION SYSTEMS

David Bawden*
Pfizer Central Research, Sandwich, Kent, CT13 9NJ, England

(*present address: Department of Information Science, The City University, Northampton Square, London EC1V 0HB, England)

ABSTRACT

The concept of molecular dissimilarity is introduced, and shown to be a powerful complement to the well-established notion of molecular similarity. It provides a quantitative assessment of structural variation and diversity. Applications within chemical information systems are discussed. These include ranking of search output, selection of representative sets of structures, file screening, data analysis, and creativity stimulation.

INTRODUCTION

The concept of molecular similarity calculation within chemical information systems was pioneered at Sheffield University[1,2] and first introduced into operational systems at Pfizer Central Research (UK)[3,4] and Lederle[5]. It is now well-established within chemical information systems, both commercially available and developed in-house, as described in other papers in this volume. It is a powerful concept, which may be applied to both 2-D and 3-D structural representations, and to chemical reaction information, and used to support similarity searching and browsing, and file clustering and partitioning.

In this presentation, I wish to introduce the related topic of molecular dissimilarity, and to suggest that it may prove to be of equal, or perhaps even greater, value. This concept is not so well-established as its converse. It has been implemented in the Pfizer in-house systems, and has also been investigated at Upjohn. Of the commercial systems, only ORAC/OSAC currently offers a specific dissimilarity option.

THE SIGNIFICANCE OF MOLECULAR DISSIMILARITY

The idea of similarity matching, finding structures with a lot in common with a target, is intuitively attractive, and its practical value easily understood. On the other hand, at first sight it may be difficult to understand why the ability to find very dissimilar, i.e., very different, structures to a target is useful.

I suggest that, on the contrary, molecular dissimilarity may actually be an even more powerful concept than similarity. This is because calculation of dissimilarity gives quantitative access to the twin concepts of structural variation and

W. A. Warr (Ed.)
Chemical Structures 2

structural diversity. These in turn are intimately linked with innovative use of chemical information systems; using them to help generate genuinely novel ideas and concepts. Browsing, *via* molecular similarity, is a powerful aid to innovation; dissimilarity analysis is an ideal complement.

MEASUREMENT OF MOLECULAR DISSIMILARITY

Molecular dissimilarity may be defined in many different ways. For the purposes of this presentation, I shall regard it as simply the converse of molecular similarity. High dissimilarity implies low similarity, and *vice versa*.

As with similarity calculation, it is open to system designers and users to decide upon the sort of structural attributes to be taken into account in calculating dissimilarity. It may be based on particular types of structural feature (rings, functionality, overall size and shape, etc.) or may be an overall measure, may be weighted to features which the user considers important.

Again, molecular dissimilarity may be numerically defined and calculated in a variety of ways. Taking the simple definition above, however, the calculation is precisely the same as for molecular similarity. The same numerical value is used to describe similarity and dissimilarity. In applying the concept of molecular similarity in chemical information systems at Pfizer Central Research, we use the Tanimoto coefficient as the quantitative similarity measure[4]. This measure, first used for the purpose at Sheffield University[1], has been widely adopted as the standard measure of molecular similarity, its value varying between 1.0 (identity) and 0.0 (no similarity). We use it, without adaption, as a measure of dissimilarity also. High values implies low similarity, and *vice versa*.

So, a list of structures ranked in decreasing order of Tanimoto coefficient relative to some target will show the most similar structures first. Conversely, ranked in increasing order of Tanimoto coefficient, i.e., decreasing dissimilarity, the list shows the least similar structures first. Although this sort of dissimilarity ranking has some practical value, as we shall see later, it is of rather limited use, so that a somewhat more sophisticated form of ranking is generally required.

This is carried out so as to give a ranking in which each structure is introduced into the list in that order which makes it most dissimilar from any structure already there. Note that the criterion is not most dissimilar from the target, which would be the strict analogy to similarity ranking, nor most dissimilar from the previous entry in the list. This implies that, in principle, the full matrix of dissimilarity measurements between every pair of structures in a set must be calculated. In practice, various stratagems may be adopted to make the procedure more efficient.

The result is a list, of which the first N structures are the N most diverse which could be chosen for that set. Clearly one structure must be chosen to start the process. This will usually be a random choice, and in practice it makes little difference to the ordering of the list, after the first few entries. This is, of course, quite different from the construction of a similarity list, where the target, or starting, structure crucially determines the ordering of the list.

This procedure applies to a set of structures which is to be ranked, with no *a priori* indication of how. Sometimes, however, there will be some initial requirement. It may be that the requirement is not merely 'structures as different from each other as possible', but 'structures as different from each other, and from an existing set, as possible'. This will most often occur in the selection of representative sets, where the existing set will comprise compounds which have already been tested, or for some other reason are unwanted.

This is easily catered for. The existing structures are treated as though they had already been selected in the ranked list, so that the first structure to be chosen, rather than being random, is that which has least similarity with any in the existing set.

The general procedure for constructing a dissimilarity ranking is shown in outline in Figure 1. For clarity, the 'brute force' procedure is shown, without any modifications to improve efficiency.

1 Construct a list containing the total set of structures from which the representative list is to be selected. This is the CANDIDATE list.

2 Construct a list whose structural types are to be avoided. If there are stuctures to be avoided *a priori*, then they are placed in this list, in random order; otherwise this list is initially empty. This is the EXCLUSION list.

3 Create a list to hold the set of representative structures, which initially will be empty. This is the REPRESENTATIVES list.

4 Choose one compound from the CANDIDATE list (on *a priori* grounds or randomly). Place it in the first position of the REPRESENTATIVES list, and add it to the EXCLUSION list.

5 For each remaining structure in the CANDIDATE list, calculate its dissimilarity with each structure in the EXCLUSION list, and take the largest of these values to be LDIS. Find the structure with the smallest LDIS, and add this to both the REPRESENTATIVES and EXCLUSION lists.

6 Repeat the last step until the required number of representatives has been selected.

Figure 1. Outline scheme for construction of dissimilarity ranking

COMPARISON WITH CLUSTERING OF STRUCTURES

It will be clear that dissimilarity ranking may be used for much the same reasons as file clustering[1,6]. The ranking has two clear advantages, and one disadvantage.

The first advantage is that there need be no prior specification of the number of 'clusters' required, as is necessary with most clustering procedures. Suppose that, in the selection of representatives application (see later), the user needs 500 compounds. Using clustering, 500 clusters would be generated, and an example chosen from each. If subsequently another 200 representative compounds are required, there is in effect no way of obtaining them. The only solution is to generate immediately the maximum number of compounds which could be needed, by forming that number of clusters. But then, if a smaller number is needed, there is no way of choosing, other than randomly, from the set available. In short, clustering requires that the number of representatives required is known *a priori*.

Conversely, with a dissimilarity ranking, the maximum number required, or for that matter the whole file, can be ranked, and the compounds needed taken 'off the top'. If more are subsequently required, then they will simply be the next in the list. The ranking is inherently more flexible than clustering.

The second advantage is that it is much simpler to include the 'different from an existing list' criterion in dissimilarity ranking than in clustering.

The disadvantage of dissimilarity ranking is that each structure in the ranked list appears by itself, without being part of a cluster of similar structures. This can be overcome only by combination of similarity and dissimilarity measures, as discussed below.

APPLICATIONS OF DISSIMILARITY RANKING

I shall now outline some applications for this sort of dissimilarity ranking. Most of these are in regular use within the SOCRATES chemical databank system at Pfizer Central Research, Sandwich.

Ranking Search Output

Perhaps the most obvious application is the ranking of the output from a substructure search. Particularly on a large database, a well constructed substructure search can easily generate hundreds, or even thousands, of valid answers. The user requires some way of analysing and comprehending these, without the need for examining each structure. Clustering can be used for this purpose[1,6], but with the disadvantages noted above.

Looking at, say, the first 20 or 30 structures in a dissimilarity ranking offers a simple and effective way of quickly seeing the structural variation and the major structural classes within a large output. In particular, it quickly shows up the case where the bulk of hits are from a single structural class, with some (perhaps especially interesting) exceptions. We have found this valuable both for 2-D substructure searches, and particularly for 3-D pharmacophoric pattern searches, when dissimilarity calculation is based on the 2-D structure. Dissimilarity ranking is included as a standard user option for processing search results in the Pfizer SOCRATES (2-D) and SOLON (3-D) search systems.

Seeing Structural Attributes in Common

Looking at the top of a dissimilarity ranked list shows the most diverse few compounds in the set. If these all share some structural attributes then it is safe to assume that these are common throughout the file. This can be a useful method of checking out the true 'diversity' or 'randomness' of a set of structures so described.

File Screening

Most chemical databases have a proportion of 'oddities', e.g., inorganics in organic files. These may be errors, or may be genuinely needed 'outlier' entries: in either case, it is useful for the database administrators to know about them. A dissimilarity ranking, starting from an arbitrary structure, gives all such oddities, or an example if several are similar to each other, close to the top.

Selecting Structural Representatives

This is perhaps the most important single application of dissimilarity ranking. It is used to select, from a large file of candidate structures, a subfile of 'representatives' for some purpose, typically for testing for some property. The requirement is to select the representatives so that they are as different from each other as possible, so as to span the structure space as effectively as can be achieved. It may also be desirable to choose the representatives so as to be as different as possible from some existing set: those already tested, known to be inactive, known to be harmful, etc. It is also desirable that the set of representatives is ordered so that the number actually tested in immaterial, from the point of view of spanning structure space. That is to say, the N compounds chosen from the list should be the most diverse N possible, regardless of the actual value of N. Further, when the N have been tested, and it is desirable to select the next M structures from the list, these will give the most diverse overall set of (N+M) structures.

Dissimilarity ranking, as described above, fulfils all these requirements, and is a highly useful method for selecting representative sets of compounds for any purpose. Clustering can also be used for this purpose[1,6], but suffers from the disadvantages noted above.

Distribution of Dissimilarities

A graphical display of the distribution of dissimilarity values between all pairs of structures in a file give a clear picture of the homogeneity, or otherwise, of the file, indicating whether or not clear groupings are present.

Creativity Stimulation

Advocates of the lateral thinking approach for stimulation of creativity and innovation suggest that considering the opposite or converse of an interesting fact or observation can be a useful formal way of sparking new ideas; the 'reversal' technique[7,8]. Clearly, this can be achieved easily when the interesting idea involves a chemical structure, and a dissimilarity calculation routine is to hand.

An example could involve a set of compounds with some property in common, for which a mechanism of action was being hypothesised. Looking at pairs of dissimilar structures within the set could help provoke new ideas, free from the bias introduced by the natural tendency to focus on the commonest, or best known, structural type.

This is one occasion where the simpler form of dissimilarity ranking, i.e., showing those structures with least in common with a target, is appropriate.

COMBINING SIMILARITY AND DISSIMILARITY

Similarity and dissimilarity are opposite sides of the same coin. When they are used explicitly in combination, they can give a uniquely powerful way of seeing into the structural relationships within a data-set.

The most straightforward way of accomplishing this is to construct a matrix, with the first column being the dissimilarity listing of structures, constructed as in Figure 1. Each row is then a similarity listing of the structures from the file most similar to that in column 1. According to the purpose, each structure may occur only once in the matrix, beyond column 1, or may occur as many times as its similarity value dictates.

This sort of matrix can be used for selection of representative compounds. If the 'ideal' compound, in column 1, is unavailable or unsuitable, the similarity listing along the row gives a series of 'next best' compounds.

For data analysis, this matrix can be used to produce graphical representations of structural similarity within the data-set.

CONCLUSIONS

Molecular dissimilarity is a powerful concept with many practical applications. In providing access to information on structural variation and diversity, it gives scope for stimulation of creativity and innovation. Facilities for generating and using dissimilarity information should be included within all chemical information systems.

ACKNOWLEDGMENTS

I am grateful to my colleagues at Pfizer, particularly Trevor Devon and Jeremy Fisher, for support and assistance; to co-workers at Sheffield. University, especially Peter Willett and Vivienne Winterman; and Drs. Johnson and Lajinis of Upjohn for helpful discussions.

REFERENCES

1. *Similarity and Clustering in Chemical Information Retrieval*; Willett, P.; Research Studies Press: Letchworth, 1987.
2. Willett, P.; Winterman, V. 'A Comparison of Some Measures for the Determination of Intermolecular Structural Similarity'. *Quant. Struct. Act. Relat.*, **1986**, *5*, 18–25.
3. Willett, P.; Winterman, V.; Bawden, D. 'Implementation of Nearest Neighbour Searching in an Online Chemical Structure Search System'. *J. Chem. Inf. Comput. Sci.* **1986**, *26*, 36–41.
4. Bawden, D. 'Browsing and Clustering of Chemical Structures'. In *Chemical Structures: The International Language of Chemistry*; Warr, W.A., Ed.; Springer Verlag: Heidelberg, 1988, pp. 145–150.
5. Carhart, R.E.; Smith, D.H.; Venkatararghavan, R. 'Atom Pairs as Molecular Features in Structure Activity Studies – Definition and Application'. *J. Chem. Inf. Comput. Sci.* **1985**, *25*, 64–73.
6. Willett, P.; Winterman, V.; Bawden, D. 'Implementation of Non-heirarchic Cluster Analysis Methods in Chemical Information Systems: Selection of Compounds for Biological Testing and Clustering of Substructure Search Output'. *J. Chem. Inf. Comput. Sci.* **1986**, *26*, 109–118.
7. *The Use of Lateral Thinking*; de Bono, E.; Penguin: London, 1990.
8. Bawden, D. 'Information Systems and the Stimulation of Creativity'. *J. Inf. Sci.* **1986**, *12*, 203–216.

SIMILARITY AND ANALOGY BASED ON DISCRIMINATION NET

Takashi Okada, Information Processing Research Centre, Kwansei Gakuin University, Uegahara, Nishinomiya, 662 Japan

ABSTRACT

A discrimination net mechanism is introduced to search for similar molecules. Similarity between molecular structures reflects the viewpoint defined by an anchor component. Two analogical reasoning methods are formulated to estimate a molecular property. These are implemented as software DNET/MS, which is applied to the acidity estimation and to the qualitative recognition of structure-activity relationships.

INTRODUCTION

Most of the current expert systems employ the production system architecture. However, the inferences executed by this type of system are limited to deductive reasoning. So, many researchers make efforts to put other types of reasoning to practical use. Among them is the analogical reasoning based on the judgement of similarity.

When we wish to develop an intelligent system which knows chemistry well, we cannot avoid the problem of similarity and analogy, too. Since a chemist's way of thinking deeply depends on the perception of structural formula, similarity and analogy in chemistry have to be formulated on the basis of the structural formula: an international language of chemistry. We have developed knowledge base software DNET/MS (Discrimination NET system for Molecular Structure)[1,2]. It searches for molecules of similar structure and executes analogical reasoning. Discrimination net is employed as the basic mechanism of the software.

This paper gives a brief overview of discrimination net and the results of its applications. An architecture of discrimination net is first introduced using as an example organic oxyacids. The molecular structures represented as graphs are ordered in a net employing the acidic OH component as the viewpoint. Similarity search of a query molecule becomes possible on this discrimination net. Then, we review two analogical reasoning methods as well as their applications to the acidity estimation. The inference methods utilise the results of similarity search. The next application of DNET/MS system is the study of structure-activity relationships. Pyrimidine derivatives with hypoglycaemic activity, which were investigated by the ordinary QSAR method, are ordered in a net. Browsing through the net is shown to be a very effective way of recognising the qualitative nature of the structure-activity relationship. Lastly, we discuss various features of discrimination net and DNET/MS.

W. A. Warr (Ed.)
Chemical Structures 2

DISCRIMINATION NET AND SIMILARITY SEARCH

Figure 1 shows an example of a discrimination net consisting of 15 organic oxyacids. The OH components in these molecules are fundamental for the appearance of acidity. There is H_2O at the root-node on the top of the net. Every

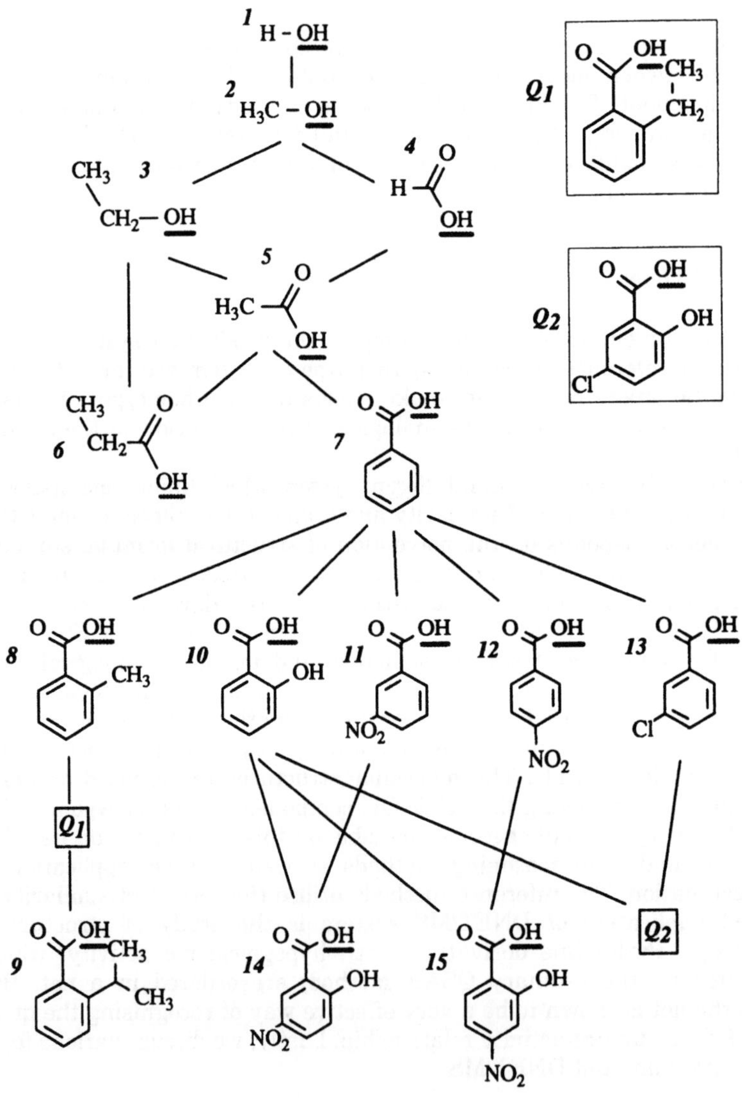

Figure 1. Sample discrimination net of oxyacids[1]

molecule in the net has an underlined OH component. This OH is called an *anchor* which defines the viewpoint of similarity judgement. The two molecules connected by a link, for example, formic acid and acetic acid, are in substructural relationship after their *anchors* are matched to each other. This *anchored* substructure relationship defines a kind of partial order between the molecules. That is, this discrimination net is an expression of the partial orders among molecular structures.

If our interest lies in the acidity of a molecule, similarity search is possible on this net in the following way. Let us employ $\mathbf{Q_1}$ as a query molecule, and place it on the root-node: H_2O. The structures of these two molecules are compared after their anchors are matched to each other. H_2O has no other atoms other than the anchor oxygen, as we neglect hydrogens in structure expressions. So, H_2O is judged to be a substructure of $\mathbf{Q_1}$. Then $\mathbf{Q_1}$ moves down to methanol, which is a substructure of $\mathbf{Q_1}$, too. In the next step, $\mathbf{Q_1}$ goes to both **3** and **4**, then to **5**, and then to **7**. As **6** is not the substructure of $\mathbf{Q_1}$, it does not move onto **6**. The process repeats until $\mathbf{Q_1}$ reaches **8**. Here, $\mathbf{Q_1}$ is a superstructure of **8** and it is a substructure of **9**. Then, the final location of this query is uniquely determined to be between **8** and **9**. In the case of query $\mathbf{Q_2}$, a similar procedure sets its final location under **10** and **13**. Then we can choose the molecules **8** and **9** as the similar molecules to $\mathbf{Q_1}$, while **10** and **13** are selected for $\mathbf{Q_2}$. These selections are reasonable, since we are concerned with the acidity of anchor OH components, and since the selected molecules share the biggest common substructure with the query molecule.

The software DNET/MS has the capability to locate a query in the discrimination net as described above. The query location mechanism is useful not only in the similarity search but also in the construction of the discrimination net itself. That is, we can make up the whole net by starting from the trivial net consisting of the root-node alone, and by repeating the location and the registration process for each molecule. So, DNET/MS can organise molecules in the form of a discrimination net automatically, if the system knows their connection tables as well as their anchor designations. The query location algorithms and the organisation of DNET/MS are fully described in our paper[2].

ANALOGICAL REASONING AND ACIDITY ESTIMATION

The next subject is to introduce methods of analogical reasoning which make use of the results of similarity search. Here, we review two analogical inference methods implemented in DNET/MS[1]. The results of their applications to the acidity estimation are summarised too.

First we introduce the simplest inference method: the direct method. The basic assumption is that the molecules with similar structures show similar properties. So, the property of a query molecule is estimated to be the median of those of the neighboring molecules. For example, the *pKa* of query $\mathbf{Q_1}$ is estimated to be 3.772, that is the median of the two *pKa's* of the adjacent molecules: **8** and **9** as shown in Figure 2.

The second is named the parallelogram method. Let us assume $\mathbf{Q_2}$ as the query molecule in Figure 3. If we select the molecule **13** as the reference structure of $\mathbf{Q_2}$, then the structural change from **13** to $\mathbf{Q_2}$ is the substitution of OH for H at the ortho-position of the anchor. The same type of structural changes can be found at the following three positions: **7** to **10**, **11** to **14**, and **12** to **15**, in the net. Now, we can

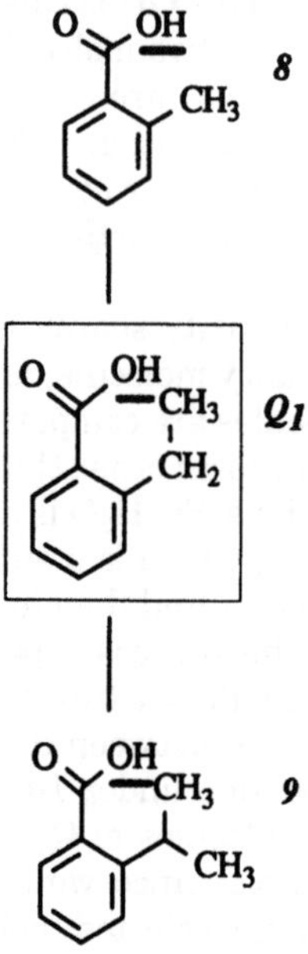

Estimated property value

= median of all the property values on the adjacent nodes

pKa-8=3.908

pKa-9=3.635

pKa-Q_1(estimated)=(3.908 + 3.635)/2

=3.772

pKa-Q_1(observed)=3.793

Figure 2. Direct method

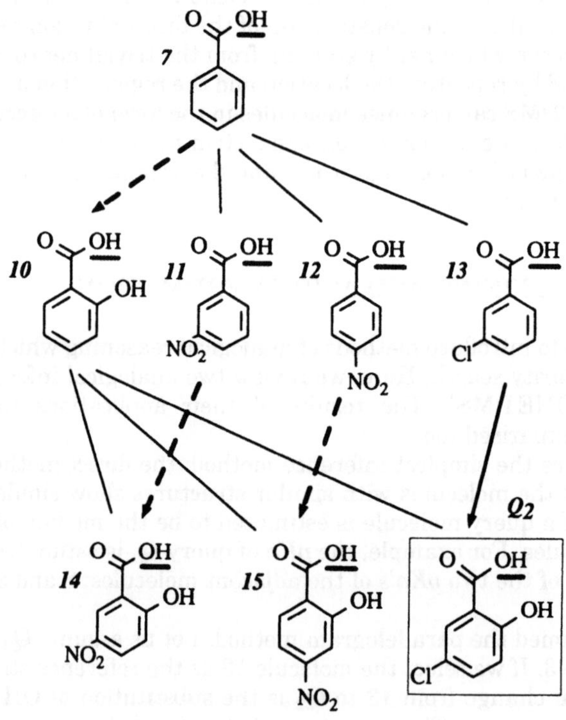

Figure 3. Parallelogram method

visualise three parallelograms in the net. Assuming that the same structural changes cause the same *pKa* differences, we can estimate the *pKa* of Q_2. The estimated *pKa's* from these three parallelograms are shown in Table 1. As fairly good results are obtained here, the usefulness of the parallelogram method is anticipated.

Table 1. *pKa* estimations of Q_2 by the parallelogram method

pKa(molecule) ···▸	*pKa*(molecule)	*pKa*-13	⟶ Estimated *pKa*-Q_2	error*
4.200(7)	2.754(10)	3.821	2.375	-.254
3.493(11)	2.121(14)	3.821	2.449	-.180
3.425(12)	2.231(15)	3.821	2.627	-.002

* The observed *pKa*-Q_2 is 2.629.

The two reasoning methods mentioned above are applied to the acidity estimation, the details of which will appear in a forthcoming paper[3]. Here, we give a short survey of the results. Figure 4 lists the steps of this process. First, we input the *pKa's* and the structural formulae of 160 oxyacids listed in a chemistry handbook[4]. The discrimination net is constructed automatically in the following step. The necessary cpu time is about 2 minutes using an Hitachi M-680D mainframe. The last step is the *pKa* estimations by the two reasoning methods. Leave-one-out estimations are executed for all the molecules in the net. The direct method takes about 6 minutes of cpu time to get 160 estimations. The median of these errors is 0.63. The execution of the parallelogram method for all the molecules takes an hour of cpu time setting a relatively restricted range for parallelogram search. This process results in making 153 *pKa* estimates for 37 molecules. Though we cannot always get an estimate by this method, the median of the resulting *pKa* errors has a small value: 0.12. These results indicate the usefulness of the analogical reasoning methods implemented in DNET/MS software.

1. 160 oxyacids: *pKa* and structural formula
from Basic Handbook of Chemistry, 3rd ed.

2. Automatic Net Construction (cpu time: 2 min.)

3. Leave-One-Out *pKa* Estimation for 160 molecules

A. Direct Method (cpu time: 6 min.)
Median of 160 Estimation Errors = 0.63

B. Parallelogram Method (cpu time: 1 hour)
Median of 153 Estimation Errors for 37 Molecules = 0.12

Figure 4. Acidity estimation

RECOGNITION OF QUALITATIVE STRUCTURE-ACTIVITY RELATIONSHIP

A group of compounds with a biological activity often share a common component responsible for the activity appearance. Then we can construct a discrimination net of the molecules, employing this common component as the anchor. The resulting net might be useful for recognising the qualitative relationships between structures and biological activities.

Sekiya *et al.* investigated the structure-activity relationship of 36 hypoglycaemic pyrimidine derivatives by the adaptive least-squares method[5]. The descriptors shown in Figure 5 are employed for ALS analysis in the first stage. The best recognition of the activity was obtained in the discriminant function *L* in the figure. Their conclusion is that the 2-(1-piperazinyl)pyrimidine moiety is the key structure for the activity, and the basicity of the piperazinyl group is the important factor. However, we think that the existence of many indicator variables in the above functional form gives some difficulty in its chemical interpretation.

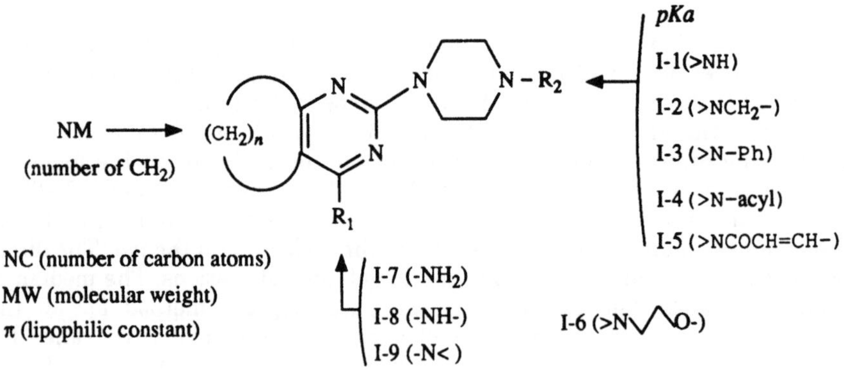

$$L = -0.00286\,MW + 0.241\,pK_a + 0.741(I\text{-}4) + 2.099(I\text{-}5) + 0.358(I\text{-}6) - 0.602$$

Figure 5. Quantitative structure-activity relationship of hypoglycaemic pyrimidine derivatives[5]

We make up a discrimination net of the 36 compounds employing the pyrimidine skeleton as the anchor, expecting that the net gives us some insight into the effects of structural modifications on hypoglycaemic activity. Figure 6 shows most of the net which includes 32 compounds with NM=4. Each compound on a node is represented by molecule-ID and a pair of substituents [R_1,R_2]. Also denoted under a node are the properties: activity-class: pKa, indicator-variable (if any) of the compound.

We first search characteristic node pairs where a relatively large activity gap exists. We can find 7 links along which the activity decreases from 3 to 1. The associated structural modifications are the introductions of electron-withdrawing or conjugating substituents at R_2 (SO_2:32→7, CS:23→5, CO:**28**→1, N:**32**→**12** and **10**, pyrrolidinyl:**23**→**9**, phenyl:**23**→**8**). The basicity of the piperazinyl group must be an important factor, since all these structural changes have the effect of

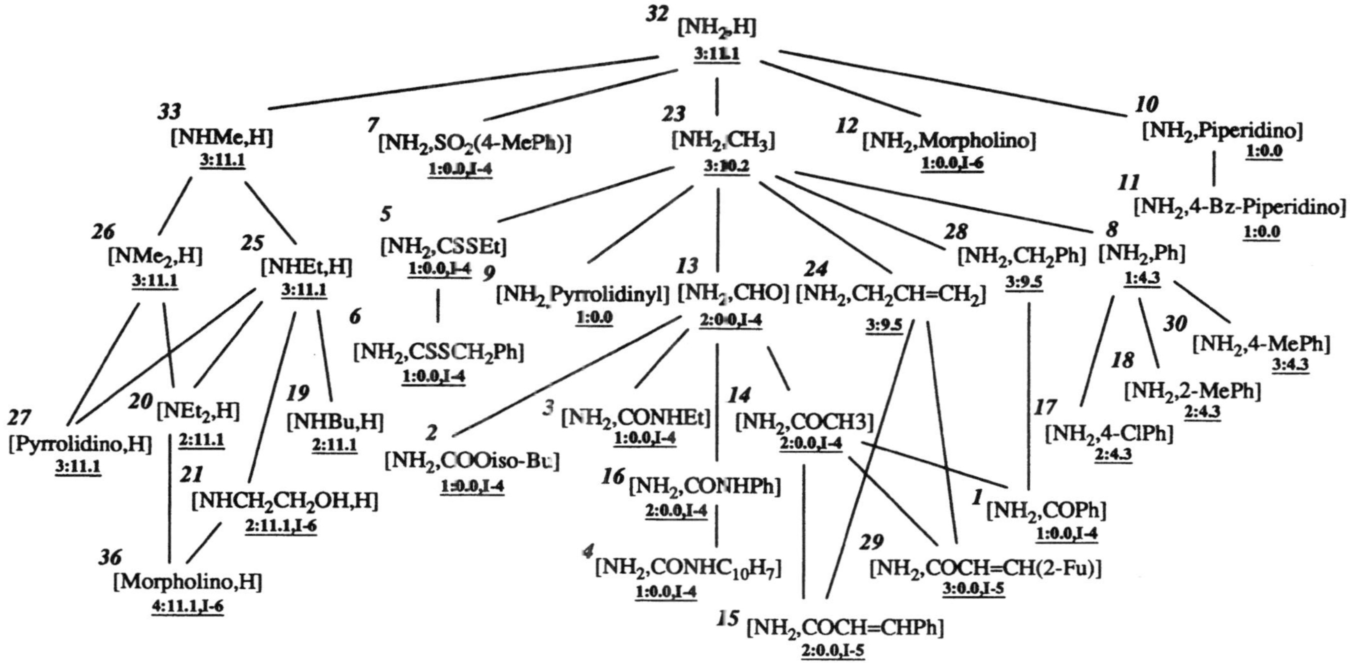

Figure 6. Discrimination net of hypoglycaemic pyrimidine derivatives

decreasing it. The same type of structural change at **23**→**13** also reduces the activity level from 3 to 2. Here, let us focus our attention on the *pKa* values expressing the basicity of the piperazinyl group. We notice that the compounds within a major branch of the net share an identical *pKa* value. Further, rough correlation exists between the *pKa* values and the activities of the compounds under the node **32**. The same correlation is found among the molecules under the node **23**, too. Further, the improvement of the basicity expression is expected to give better interpretations of some activities. For example, if electronic substituent effects are adequately considered, the basicity seems to explain the activity differences among the molecules **2**, **3**, and **16**, though their *pKa's* cited are equal to each other. Thus, we can conclude that the basicity of the piperazinyl group is the most important factor.

There still remain a few compounds, for which activities need additional reasoning. They are (1) the improvement of activity from **20** and **21** to **36**; (2) the activity differences among **15**, **29**, and **1**; (3) those among **8**, **17**, **18**, and **30**. Though we cannot give proper interpretations to these problems currently, some steric model seems to be necessary to explain the phenomena, as the substituents concerned are fairly large in these cases.

The descriptors appearing in the discriminant function L are examined so that we give them chemical interpretations using the structural modifications observed in the discrimination net. First, the existence of the descriptor *pKa* agrees well with the speculation on the net. Next, we see that the activity always decreases, when the descriptor I-4 changes from 0 to 1 along a link in the net. The direction of this activity change, however, is opposite to that expected from the sign of I-4 coefficient in L. Careful examination of these links shows that great decreases in *pKa* always take place. The same can be said for the descriptor I-5. Therefore, these two variables seem to give corrections which should be expressed as higher order terms in *pKa*. We can find no common factor associated with the change in the value of I-6. After all, three indicator variables have no reasonable chemical interpretations on the net, though they directly represent structural characteristics. Molecular weight effect is hard to interpret from the net of structures.

The recognition of structure-activity relationships is so important that there may exist various methods to explore it. The discrimination net can be a new tool to handle the problem. Browsing a net in detail and considering the activity changes caused by structural modifications, we can make working hypotheses for molecular design. It will be also effective to explore the chemically meaningful descriptors to be used in usual QSAR studies. Sometimes one may find that the numerical approach is more effective, when applied to a set of molecules on the top of major branches.

DISCUSSION

Similarity and analogy of chemical structures are successfully treated by the discrimination net mechanism implemented in DNET/MS. The analogical reasoning methods used in the acidity inference can be used, as it is, to estimate various molecular properties such as C=O bond length or ^{13}C chemical shift. Inference by DNET/MS does not require expert knowledge of chemistry but it requires a database of molecular structures and the associated properties. Anchor designations are also necessary to indicate the local component responsible for the property appearance. This type of analogical estimation will be useful in many cases, as it can avoid the difficulty of the knowledge acquisition problem.

There exist many other properties which current DNET/MS cannot treat. Such properties as log *P* and heat of formation, however, are roughly represented as a sum of local component properties. So, the refinement of the current inference method is promising for these properties. Chemical reactions can be seen as a kind of molecular property, and they can be ordered in a discrimination net. The anchor position is set at the atoms where the bond interchange occurs. DNET/CR system is currently under development in our laboratory to estimate the feasibility and the mechanism of a given reaction.

Discrimination net architecture has also shown its usefulness in helping chemists' qualitative recognition of structure-activity relationships. Our experience shows that arranging molecules in a tree or in a net based on the substructure relationship is a natural way to perceive the characteristics of a group of molecules. Once we get a net of molecules using a computer, it will be an easy and interesting task for chemists to focus characteristic variations of property and to give reasons from the structural viewpoint.

The difference between the methodologies of conceptual clustering employed in the CLUSMOL[6] and of the discrimination net in DNET/MS is stated here. Both of them receive a set of specific molecules and arrange them in a tree or in a net following the substructure relationship. CLUSMOL does not use the anchor component information in building up a tree, but it is the principal objective for the method to discover an anchor substructure from the resulting tree. On the other hand, DNET/MS deals with a group of molecules of which the key component is already known. It is used to find factors affecting the property of the anchor. We should select a proper methodology to study the structure-activity relationship from (1) conceptual clustering, (2) discrimination net, and (3) numerical QSAR. The necessary amount of preliminary knowledge increases in this order, while the computation time decreases.

The EMIL project[7] by Fujita *et al.* at Kyoto University collects examples of structural transformations and the associated activity changes, which are used as a knowledge base to propose meaningful structural modifications in lead compound evolution. However, efforts to collect the knowledge are enormous, and some kind of assisting system seems to be desirable. Since the net given by DNET/MS can visualise the structural transformations among the given drugs, this system might be also helpful in the accumulation of this kind of knowledge.

Software DNET/MS is written in common LISP, and it runs by loading the entire net in LISP working space. So, the necessary memory becomes very large if we have to treat a lot of molecules. An improvement of the system planned is to hold the net on a disk.

Similarity judgement and analogical reasoning are extremely general capabilities necessary for a wide range of problem solving in chemistry. The discrimination net mechanism covers a very limited portion. For example, it cannot judge the similarities among halogen elements. Benzene and pyridine are completely different entries too. Further research efforts are expected to solve these problems.

ACKNOWLEDGMENTS

This work was partly supported by Special Co-ordination Funds of the Science and Technology Agency of the Japanese Government. The author wishes to thank Dr

Tetsuo Sekiya of Mitsubishi Chemical Co. Ltd. for suggesting the application of DNET/MS to the structure-activity studies of pyrimidine derivatives. Thanks are also due to Prof. Yukio Yoneda of Tokai University for his continuous encouragements and discussions, and to Mr. Toshikazu Kawai and Miss Etsuko Kinoshita of Kwansei Gakuin University for their help in this study.

REFERENCES

1. Okada, T.; Kawai, T. 'Analogical Reasoning in Chemistry. 1. Introduction and General Strategy'. *Tetrahedron Comput. Methodol.* **1989,** *2,* 327–336.
2. Okada, T.; Kawai, T. 'Analogical Reasoning in Chemistry. 2. DNET/MS System'. *Tetrahedron Comput. Methodol.***1989,** *2,* 337–347.
3. Okada, T.; Kinoshita E. 'Analogical Reasoning in Chemistry. 3. Estimation of Acidity'. To be published.
4. *Basic Handbook of Chemistry, 3rd ed.*; Chemical Society of Japan, Ed.; Maruzen: Tokyo, 1984; Vol. 2, p. 339.
5. Sekiya, T.; Hata, S.; Yamada, S. 'Pyrimidine Derivatives. VII. Structure-Activity Relationship of Hypoglycaemic 4-Amino-2-(1-piperazinyl)pyrimidines Investigated by the Adaptive Least-Squares Method'. *Chem. Pharm. Bull.* **1983,** *31,* 2432–2437.
6. Okada, T.; Wipke, W. T. 'CLUSMOL: A System for the Conceptual Clustering of Molecules'. *Tetrahedron Comput. Methodol.***1989,** *2,* 249–264.
7. Fujita, T. private communication.

SIMILARITY SEARCHING IN THE DEVELOPMENT OF NEW BIOACTIVE COMPOUNDS. AN APPLICATION.

Guenter Grethe and W. Douglas Hounshell
Molecular Design Limited, 2132 Farallon Drive, San Leandro, CA 94577, U.S.A.

ABSTRACT

Similarity searching in the reaction and molecule domain is applied to the design and development of a hypothetical bioactive compound. Potential synthetic pathways are generated utilising reaction and molecule similarity features in REACCS. Lead optimisation of the target compound is made possible by similarity searches over a database of compounds of known activity and by subsequent manipulation of the results. It is shown that similarity searching can be used as a powerful, complementary tool in the design of new chemical compounds.

INTRODUCTION

Similarity searching has been widely used in the molecule sector[1–5], particularly in search for similar and more effective compounds for drug research. Less has been reported on similarity searching in the reaction domain[6,7]. Both searching capabilities have been implemented recently in Molecular Design Limited's mainframe programs MACCS-II and REACCS[7]. In general terms, similarity in these programs is defined as follows.

Molecular similarity

The similarity of two objects can be described as the number of features they have in common divided by the number of features present in either. We have implemented a molecule similarity function where a similarity value is calculated based on the presence or absence of a set of 933 structural features:

$$\mathrm{SIM(mol)} = \frac{Q \sqcap C}{Q \sqcup C}$$

where Q = set of weighted features in query
C = set of weighted features in candidate

The limiting cases are:

(a) Q and C have no features in common → SIM = 0
(b) Q and C have all features in common → SIM = 100

For a similarity search, we include as hits all candidate molecules whose

W. A. Warr (Ed.)
Chemical Structures 2

similarity value, measured against the query molecule, is greater than some user-specified threshold:

$$\text{MOL—SIM—HIT if SIM(mol)} \geqslant \text{Threshold(mol)}$$

By definition, any molecule compared to itself has a similarity value of 100. In this sense, a similarity search can be viewed as a 'fuzzy' exact match search.

Reaction similarity

We have also defined and implemented a reaction similarity search[7] by making use of the presence or absence of the 933 molecule features in the participants of the reactions to obtain a 'molecule' component of the similarity value, and an additional set of of 230 features associated with the reacting centres and their immediate environment to obtain a 'reaction-centre' component of the similarity value.

The user specifies threshold similarity values for each of the two components of reaction similarity. For the candidate reaction to be considered a hit, both molecule and reaction-centre similarity values must exceed their respective user-specified thresholds:

$$\text{RXN—SIM—HIT if SIM(mol)} \geqslant \text{Threshold(mol)} \textbf{ AND } \text{SIM(r-c)} \geqslant \text{Threshold(r-c)}$$

Sub- and Super-Similarity

A sub-similarity can be defined by only considering those features present in the query:

$$\text{SUBSIM} = \frac{Q \sqcap C}{Q}$$

By definition, any query which is a substructure of the candidate, will give a similarity value of 100. In this sense, a sub-similarity search can be viewed as a 'fuzzy' substructure search.

A super-similarity can be defined by only considering those features present in the candidate:

$$\text{SUPERSIM} = \frac{Q \sqcap C}{C}$$

By definition, any query which is a superstructure of the candidate, will give a similarity value of 100. Therefore, a super-similarity search can be viewed as a 'fuzzy' superstructure search.

The application of these different similarity searches in the design and development of new chemical compounds is illustrated by an example from the pharmaceutical industry.

APPLICATION

Synthesis

The design of the lead compound, a potential phosphordiesterase inhibiting cardiotonic agent, is based on information from the published literature (Figure 1). Synthetic pathways towards this target are developed in retrosynthetic fashion employing reaction similarity searching over large databases. We begin with a similarity/conversion search with the target as query structure and the search command shown in Figure 2. To obtain the maximum number of reasonable suggestions a global search is conducted over 10 REACCS databases with a combined total of 225,000 reactions. [We used the following public databases: MDL's THEIL, ORGSYN, JSM-REACCS, CLF, CHIRAS, METALYSIS, and the CCR databases (1986-1989) from the Institute of Scientific Information (ISI), Philadelphia.] Reactions with high relevancy to our problem are chosen and represent the first level of the synthesis tree in Figure 2. The starting materials of these reactions in turn then become the new targets for the same type of similarity search.

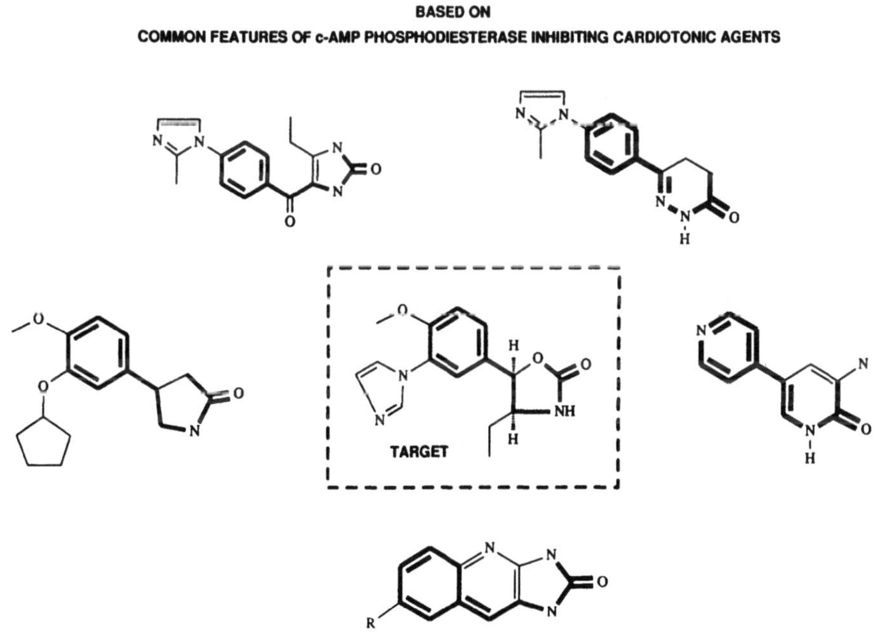

Figure 1. Target construction

By using exact molecules as queries in these searches, users do not have to learn a complex query language and, more importantly, can avoid any bias based on preconceptions which is inherent in substructure searching. Substructure searching, however, is subsequently used to fine-tune the results and to reduce the initially large number of hits from the similarity search to manageable size. The result is a synthesis tree consisting of many reasonable pathways, one of which is shown in

Figure 3. Each step in this route is solidly supported by data from the reaction databases.

Derivatisation

The target compound, after successful synthesis and positive test results, becomes the new lead compound. This requires the preparation of a series of derivatives for patent purposes and for the study of structure-activity relationships. In this first step of lead optimisation, structural changes are mostly confined to the periphery of the original molecule. It is important that these derivatives can be prepared along the lines of the original synthetic route and that the required starting materials are readily available, either in-house or commercially.

To find a wide variety of potential candidates we use similarity and super-similarity searching techniques to browse through the Fine Chemical Directory, a database of 65,000 commercially available chemicals. With the first method we use one of the starting materials from the original synthetic sequence as the query and set the threshold to 50 after some experimentation (Figure 4). The implementation of adjustable parameters, or thresholds, reflects on the very subjective nature of similarity searching and serves the need of end-users to create their own environment suited to their task. Super-similarity searching generally provides simpler molecules than that used in the query. Therefore, with the initial target as the query (Figure 5), it is an alternative to finding new starting materials which can be used in the original synthesis or which might generate new ideas for the design of related routes. Interestingly, a large number of the suggestions are very similar to those obtained earlier in the similarity searches indicating the complementary nature of the two searches.

Lead Optimisation

During the derivatisation phase the basic skeleton of the original target is seldom affected and stays the same. But in order to find new compounds with a better activity profile which possibly can serve as new leads, structural changes are frequently required. Since time-consuming and costly syntheses of these potential candidates are to be avoided, databases of existing compounds should be searched first. Fortunately, most companies maintain a large inventory of proprietary compounds which have only undergone limited testing and very often have not been tested for the activity under investigation. A large number of compounds is even available in quantities sufficient for today's effective testing methods.

While substructure searches which consider important structural features of molecules from previous studies can provide new insights, similarity searching over large proprietary databases has the potential of retrieving a wide selection of new leads with markedly different structural features. We prove this point by using in place of an in-house database the MACCS-II Drug Data Report, a database of compounds with known biological activity. The results are shown in Figure 6. Comparison of substructure and similarity searching results clearly show the advantages of the latter for generating substantially new leads and ideas.

To increase the probability of finding compounds with new structural features

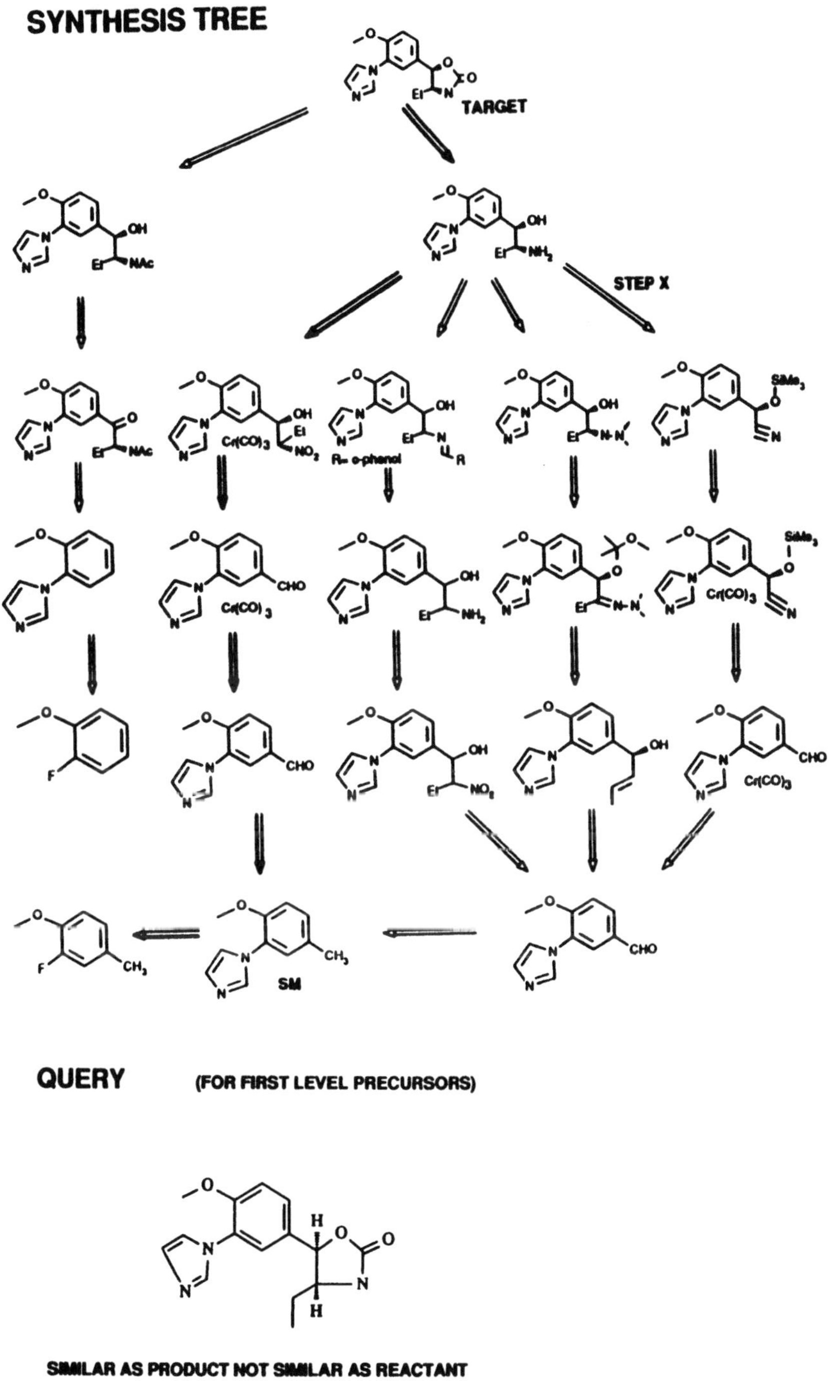

Figure 2. Synthesis tree

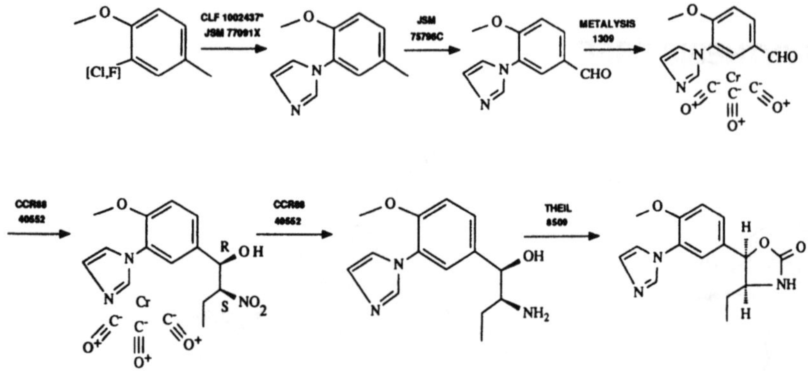

* reactions are represented by external registry numbers

Figure 3. Preferred synthetic sequence

and improved activity, several similarity searches with known drugs as queries are carried out. Intersection of the hit lists provides a small number of structurally different, active compounds (Figure 7). Additional important information is gained by a sub-similarity search over the same database (Figure 8). Though searches of this type generally generate a large number of hits, the potential of this method should not be neglected, Rapid mass-screening tests are available today to handle a large number of compounds. Alternatively, list manipulation techniques similar

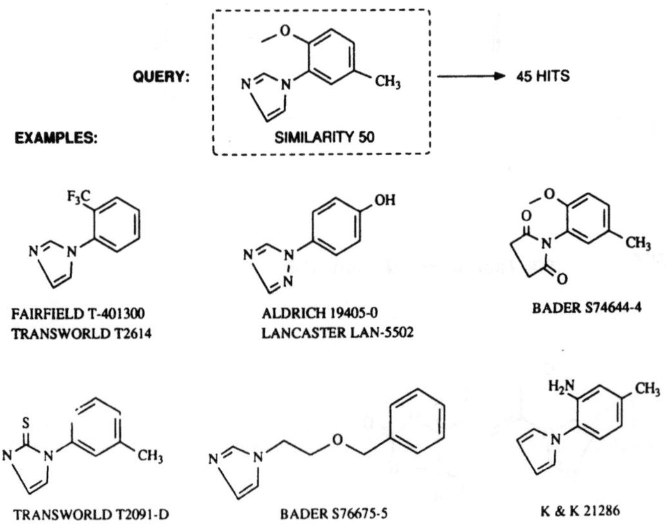

Figure 4. Searching for starting materials in the Fine Chemicals Directory

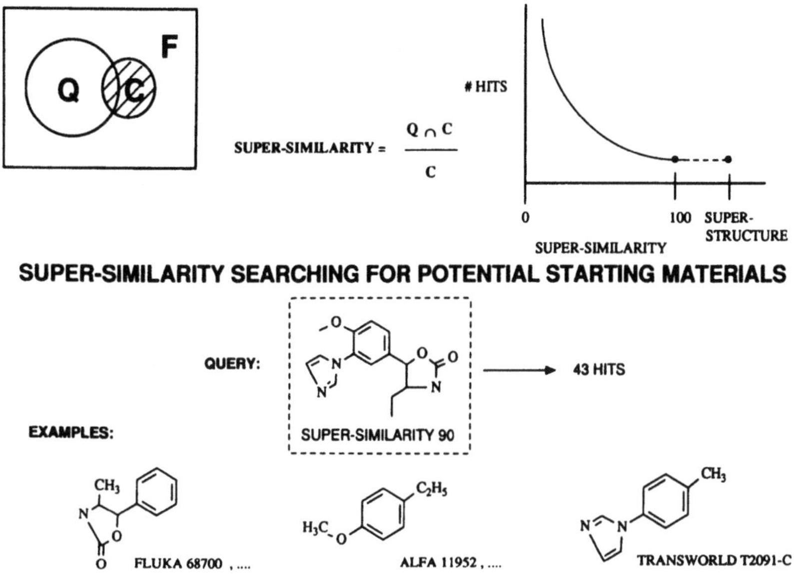

Figure 5. Super-similarity and super-similarity searching for potential starting materials

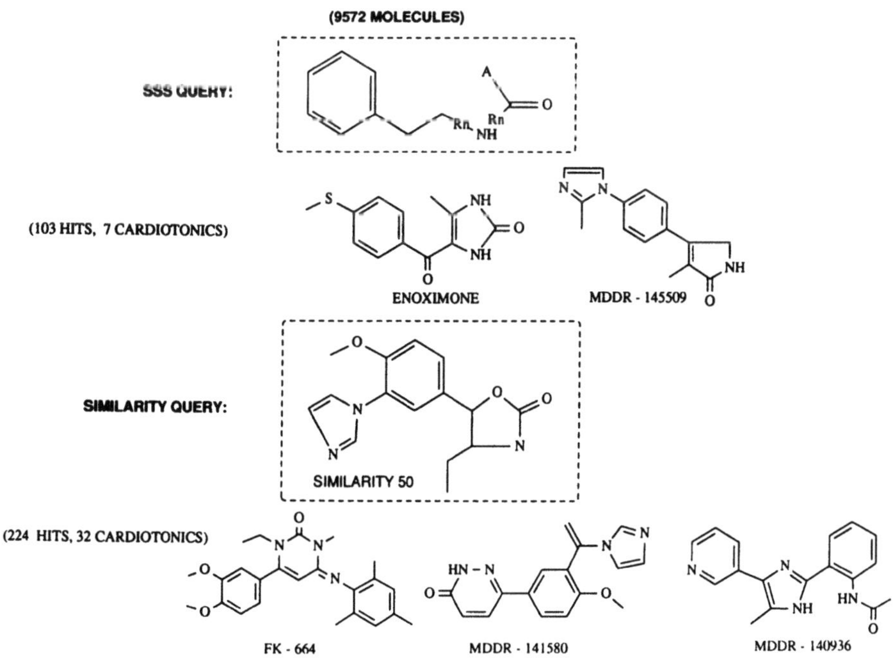

Figure 6. Substructure *versus* similarity search over MACCS-II Drug Data Report

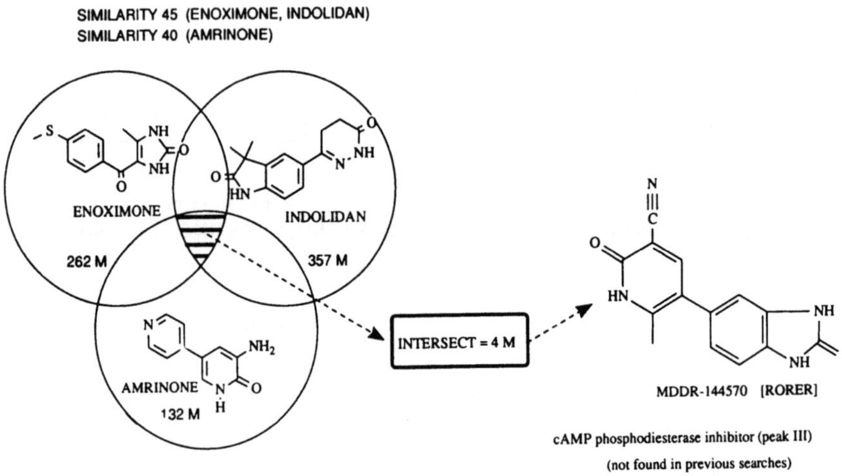

Figure 7. Similarity searching and hitlist intersection: development of a new lead compound

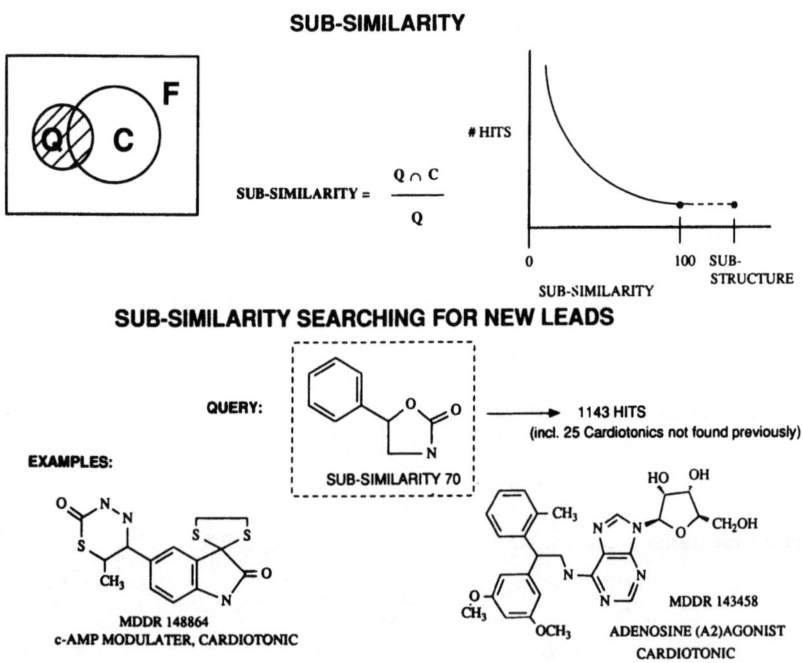

Figure 8. Sub-similarity and sub-similarity searching for new leads

to those described before can be used to narrow down the number of candidates. The fact that our search results in the identification of 25 active compounds not found previously with other searching techniques suggests sub-similarity searching as a useful complementary method to select new leads from existing databases.

CONCLUSION

We demonstrate the usefulness of similarity searching over both the reaction and the molecule domain as an important tool in the design and development of new chemical compounds. Reaction similarity searching is instrumental in designing viable synthetic routes to a proposed target molecule. Molecule similarity searching and its sub- and super-similarity variations are effectively used for derivatisation and lead optimisation studies.

REFERENCES

1. *Similarity and Clustering in Chemical Information Systems*: Willett, P.; Research Studies Press: Letchworth, 1987, and references therein.
2. Carhart, R.W.; Smith, D.H.; Venkatararghavan, R. 'Atom Pairs as Molecular Features in Structure Activity Studies – Definition and Application'. *J. Chem. Inf. Comput. Sci.* **1985,** *25,* 64–73.
3. Nilakantan, R.; Bauman, N.; Dixon, J.; Venkataragharan, R. 'Topological Torsion: A New Molecular Descriptor for SAS Applications. Comparison with other Descriptors'. *J. Chem. Inf. Comput. Sci.* **1987,** *27,* 82–85.
4. Sheridan, R.P.; Venkatarahvaran, R. 'New Methods in Computer-aided Drug Design'. *Acc. Chem. Res.* **1987,** *20,* 322–329.
5. Willett, P.; Winterman, V.; Bawden, D. 'Implementation of Non-hierarchic Cluster Analysis Methods in Chemical Information Systems: Selection of Compounds for Biological Testing and Clustering of Substructure Search Output'. *J. Chem. Inf. Comput. Sci.* **1986,** *26,* 109–118.
6. Vladutz, G. 'Do We Still Need a Classification of Reactions ?'. In *Modern Approaches to Chemical Reaction Searching*; Willett, P., Ed.; Gower: Aldershot, 1986, pp. 202–220.
7. Moock, T.E.; Grier, D.L.; Hounshell, W.D.; Grethe, G.; Cronin, K.; Nourse, J.G.; Theodosiou, J. 'Similarity Searching in the Organic Reaction Domain'. *Tetrahedron Comput. Methodol.* **1988,** *1,* 117–128.

EVALUATION OF SIMILARITY SEARCHING METHODS USING ACTIVITY AND TOXICITY DATA

Geoffrey M. Downs, Andrew R. Poirrette, and Peter Willett
Department of Information Studies, University of Sheffield, Western Bank, Sheffield S10 2TN, England

Peter T. Walsh
Research and Laboratory Services Division, Health and Safety Executive, Broad Lane, Sheffield S3 7HQ, England

ABSTRACT

In this paper, we discuss the use of similarity searching methods with files of 2-D chemical structures characterised by fragment bit strings. The first part of the paper considers the use of a weighting scheme in which the importance of a match on a particular fragment type is a logarithmic function of the reciprocal of that fragment's frequency of occurrence in the database that is being searched. Empirical studies with three large data-sets provide only equivocal support for the utility of this weighting scheme, despite the fact that recent substructural analysis studies provide a semi-quantitative rationale for its use. The second part of the paper considers the utility of similarity searching methods for the prediction of carcinogenicity, mutagenicity, skin irritation and toxicity for compounds in the RTECS database. Both discrete and continuous data are used to demonstrate that structurally similar molecules also exhibit comparable toxicities.

INTRODUCTION

Computer techniques for searching files of chemical structures have been under active development since the early Sixties and there is now a well-established body of theory and practice underlying the design and implementation of chemical structure retrieval systems[1–4]. Such systems have traditionally provided two main types of searching facility, these being structure search, which involves the search of a database for the presence or absence of a specified query compound, and substructure search, which involves the retrieval of all molecules in a file containing some specified query substructure, irrespective of the environment in which the query substructure occurs. More recently, there have been several reports of systems for similarity searching, in which those molecules are retrieved that are most similar to some input query molecule. There are very many ways in which the similarity between a pair of molecules might be defined[5,6]: current similarity searching systems use an approach first suggested by Adamson and Bush[7], in which the similarity between a pair of substances is based on the fragment substructures that they have in common (these fragments typically being those used for the screening stage of a conventional substructure searching system). The rapid development of techniques for fragment-based similarity

W. A. Warr (Ed.)
Chemical Structures 2

searching that has taken place over the last few years has led to such facilities becoming available in many chemical information systems as a complement to conventional substructure searching[8–11].

A similarity search involves the user inputting some target molecule of interest. The similarity is calculated between the target and each compound in the database, by comparing the corresponding sets of fragment screens to identify those in common, and the structures sorted into order of decreasing similarity with the target. Interesting compounds from this ranking can then be used as the basis for subsequent searches. Such a browsing-like approach to chemical retrieval removes the need for a carefully delineated substructural query, since the chemist can input just a structure of known interest, and allows complete control over the volume of output produced by the search, since the chemist can inspect as many or as few compounds as necessary from the ranking. Similarity searching can also be used to cluster files of structures[12,13].

In addition to providing novel types of retrieval facility, similarity methods may also be used for structure-property correlation studies. Given a structure with known property or activity characteristics, a similarity searching system can be used to identify structurally related molecules that may be expected to exhibit similar property or activity characteristics. Previous work has shown the general validity of this expectation, but the experiments that have been reported have involved only small data-sets, containing a few tens of compounds for which property or activity data are available, or only qualitative results[7,9,14–16]. This paper reports a quantitative evaluation of the effectiveness of similarity methods using several data-sets where both structural and activity data are available for some hundreds or thousands of molecules. The paper is in two main parts. The first of these reports an evaluation of a fragment weighting scheme that has been suggested for the calculation of molecular similarity, using data-sets derived from high-throughput primary biological screens that are in use at Pfizer Central Research (U.K.). The second part then reports the use of similarity methods on a range of data-sets extracted from the Registry of Toxic Effects of Chemical Substances (RTECS) database.

COMPARISON OF SIMILARITY MEASURES

Inverse Database Frequency Weighting

Willett and Winterman[16] have reported an extended comparison of similarity coefficients and weighting schemes for similarity searching, using simulated property prediction to evaluate the effectiveness of the various measures that were tested. The experiments showed that association coefficients gave the best results of the various similarity coefficients that were tested and that taking account of the number of occurrences of a fragment type in a molecule gave better results than the other weighting schemes that were tested. However, this latter finding is of less practical importance in the context of similarity searching, since the systems that are available utilise fragment bit strings that indicate merely the presence or absence of a fragment type in a structure.

One of the weighting schemes considered by Willett and Winterman was an *inverse database frequency* (IDF) weighting scheme. The idea here is that if two structures have some relatively infrequently occurring fragment in common, they should be regarded as being more similar to each other than if they have a very frequently occurring fragment, e.g., a benzene ring. This idea seems intuitively reasonable but it was not found to give particularly good results in practice, although this may have been due, at least in part, to the very small numbers of compounds in the data-sets used in this study[16]. Willett and Winterman's findings are in stark contrast to recent work by Moock *et al.* on similarity searching in large databases of chemical reactions[17], where it was found that an IDF weighting scheme gave much better results than if unweighted fragment occurrence data were used. Support for the use of an IDF weight comes from a recent comparison by Ormerod *et al.*[18] of fragment weighting schemes for use in drug discovery programmes. This supporting evidence is presented in detail in the Appendix to the present paper and suggests the use of a weight

$$\log\left(\frac{N}{FREQ(I)}\right)$$

where $FREQ(I)$ is the number of structures that have been assigned the I-th screen in a data-set of N structures: this IDF weight is similar to that suggested on purely empirical grounds by Moock *et al*[17]. In the remainder of this section, we report experiments to test whether this weight, for which the Appendix provides a semi-quantitative rationale, is in fact better than weights that do not have such a rationale.

Data-sets

The experiments used three large data-sets, each containing *ca.* 10K structures, that had been tested for activity in a particular biological screen at Pfizer Central Research and that had been assigned sets of both 2-D and 3-D fragment screens. The data-set structures were originally in 2-D connection table form and were as follows:

1. PFIZER-1: 10,057 compounds of which 22 were active.
2. PFIZER-2: 10,481 compounds of which 824 were active.
3. PFIZER-3: 10,688 compounds of which 1,622 were active.

These sets of structures were input to the CONCORD program to obtain 3-D co-ordinates. It was not possible to obtain co-ordinate data for some of the structures and the 3-D data-sets were thus as follows:

1. PFIZER-1: 9,042 compounds of which 18 were active.
2. PFIZER-2: 10,211 compounds of which 793 were active.
3. PFIZER-3: 10,403 compounds of which 1563 were active.

The molecules in these data-sets were then characterised by the presence or absence of the screens that are used for substructure searching in SOCRATES, the Pfizer Central Research chemical and biological databank system[19]. In the case of the 2-D structures, there were 1315 atom-centred, bond-centred and atom sequence fragment screens; in the case of the 3-D structures, there were 1471 inter-atomic distance screens[20].

Experimental Results and Discussion

Assume that a query structure, Q, is represented by an M-element vector

$$(q(1), q(2) \ldots q(I) \ldots q(M))$$

where $q(I)$ $(1 \leq I \leq M)$ denotes the weight assigned to the I-th screen. A similar representation is used for a database structure, D, with the exception that $d(I)$ is restricted to binary weights, i.e., either one or zero depending upon whether the I-th screen has or has not been assigned to D. Two query weighting schemes were used, these being simple binary weighting, i.e., $q(I) = 1$ or 0, or the IDF weight defined previously. Two similarity coefficients were used, these being the vector dot product

$$\sum_{I=1}^{M} q(I) \times d(I)$$

and the Tanimoto coefficient

$$\frac{\sum_{I=1}^{M} q(I) \times d(I)}{\sum_{I=1}^{M} q(I) + \sum_{I=1}^{M} d(I) - \sum_{I=1}^{M} q(I) \times d(I)}$$

The two weighting schemes and the two similarity coefficients give a total of four different similarity measures.

Given a target compound that is known to be active, a good similarity measure will be one that leads to a ranking in which active compounds cluster towards the top of the ranking above the inactive compounds: in a perfect analysis, all of the active compounds in the database will precede all of the inactive compounds. Accordingly, each active molecule in a data-set was considered in turn as the query structure for a similarity search. The fast inverted file algorithm described by Willett[6] and, in modified form, by Moock *et al*[17]. was used to rank the data-set in order of decreasing similarity with this query structure, using each of the four similarity measures. The resulting ordered list was then inspected to identify the number of active molecules in:

1. the top 10, the top 20 and the top 40 molecules in the ranking;
2. the top decile, the top two deciles and the top three deciles in the ranking.

The mean values for these six numbers were then obtained by averaging over the complete set of active structures for each data-set; the results are listed in Tables 1–4.

The effectiveness of the similarity measures may be compared by seeing how well they process the data-sets, using each of the evaluation criteria described above to rank the four measures in order of decreasing effectiveness. There is a total of 18 such rankings for the 2-D screen records and a further 18 for the 3-D records, i.e., three data-sets, each of which has rankings of the similarity measures based on the top 10, 20 and 40 structures in the output from the similarity search and on the top one, two and three deciles of the output from the similarity search. The degree of association between these eighteen rankings may be determined by means of the Kendall Coefficient of Concordance, W[21]. If a statistically significant value of W is obtained, i.e., if a significant association is found to exist between the rankings, it may be concluded that there are statistically significant differences

Table 1. Mean numbers of actives in the first 10, 20 and 40 positions of the output from the similarity searches obtained with the 2-D data-sets

Dataset	Binary Weight						IDF Weight					
	Dot Product			Tanimoto			Dot Product			Tanimoto		
PFIZER-1	0.5	0.7	0.7	1.0	1.1	1.4	0.6	0.8	1.1	0.6	0.8	1.0
PFIZER-2	2.9	5.3	9.3	3.0	5.3	9.0	3.0	5.5	9.8	2.9	5.3	8.7
PFIZER-3	4.9	9.5	17.7	4.4	8.3	15.2	5.2	9.8	18.2	5.0	9.3	16.8

Table 2. Mean numbers of actives in the first 10, 20 and 40 positions of the output from the similarity searches obtained with the 3-D data-sets

Dataset	Binary Weight						IDF Weight					
	Dot Product			Tanimoto			Dot Product			Tanimoto		
PFIZER-1	0.7	0.7	0.8	1.2	1.2	1.6	0.7	0.7	0.8	1.1	1.3	1.5
PFIZER-2	3.3	5.9	9.9	3.8	6.8	12.6	3.2	5.6	10.1	3.5	6.6	12.5
PFIZER-3	4.5	8.6	16.1	4.6	8.4	15.7	4.4	8.3	16.3	4.8	9.0	16.6

Table 3. Mean numbers of actives in the first, the first two and the first three deciles of the output from the similarity searches obtained with the 2-D data-sets

Dataset	Binary Weight						IDF Weight					
	Dot Product			Tanimoto			Dot Product			Tanimoto		
PFIZER-1	4.3	6.1	7.3	6.1	8.4	9.7	5.8	7.1	9.3	6.6	7.7	9.1
PFIZER-2	114.5	214.8	309.4	111.3	205.6	294.9	114.1	212.0	303.4	109.7	202.6	290.9
PFIZER-3	240.2	407.9	570.6	204.9	362.7	514.9	252.7	443.1	605.6	233.0	421.8	584.5

between the performances of the various fragment weights. Furthermore, if a set of rankings does show a significant association, then the best ranking of the objects is given by the mean of the ranks for each object when averaged across all of the rankings.

The results of carrying out this procedure are shown in Table 5, where it will be seen that statistically significant associations are obtained with both the 2-D and the 3-D screens. The best average rankings of the four measures are:

2-D: IDF/Dot Product > Binary/Tanimoto > Binary/Dot Product > IDF/Tanimoto
3-D: IDF/Tanimoto > Binary/Tanimoto > IDF/Dot Product > Binary/Dot Product

Table 4. Mean numbers of actives in the first, the first two and the first three deciles of the output from the similarity searches obtained with the 3-D data-sets

Dataset	Binary Weight						IDF Weight					
	Dot Product			Tanimoto			Dot Product			Tanimoto		
PFIZER-1	3.8	5.4	7.1	6.2	7.6	8.6	4.3	6.2	8.4	6.1	7.9	9.2
PFIZER-2	133.2	234.1	324.5	117.3	207.7	295.3	137.5	233.5	325.0	125.6	213.8	301.9
PFIZER-3	217.6	396.7	570.0	224.8	398.3	562.3	235.8	418.7	589.1	241.1	422.8	589.7

where > denotes 'performs better than'. Thus, for both data-sets, a similarity measure that uses the IDF weight gives the best average level of performance; unfortunately, the similarity coefficient is not the same in the two cases. Indeed, the best measure for the the 3-D data-sets is actually the worst for the 2-D data-sets, a result that is particularly surprising: previous work has shown the general utility of the Tanimoto coefficient when compared with other similarity coefficients and the analysis here has suggested that the IDF weight should be better than the binary weight. One might hence have assumed that the best overall measure would be the IDF/Tanimoto combination (as indeed it is in the case of the 3-D data-sets).

The analysis in the Appendix provides at least some rationale for why the IDF weighting scheme should give better results than a simple binary weighting of the query screens in a similarity searching system based on fragment co-occurrence data. However, the experimental results that we have obtained do not provide unequivocal support for this finding. The most widely used similarity measure is the Binary/Tanimoto combination: this performed well in both sets of analyses and we have accordingly chosen to use this for the more extended studies reported in the remainder of this paper.

Table 5. Degree of association between the rankings of the four similarity measures using Kendall's Coefficient of Concordance, *W*. The right-hand part of the table lists the average ranks of the four measures

Dataset	*W*	Significance	Binary/ Dot Product	Binary/ Tanimoto	IDF/ Dot Product	IDF/ Tanimoto
2-D	0.17	$p < 0.05$	2.72	2.61	1.67	2.78
3-D	0.16	$p < 0.05$	3.00	2.39	2.72	1.72

ANALYSIS OF TOXICITY DATA

Introduction

The Registry of Toxic Effects of Chemical Substances (RTECS) is produced by the National Institute for Occupational Safety and Health (NIOSH), a US government-sponsored organisation. RTECS is a compilation of toxicity information from the open literature, providing brief descriptions of substances for which toxic effects have been reported. The data in RTECS have not, in general, been subjected to peer review, although the Editorial Boards assesses a limited number of citations to resolve ambiguities[22]. The database is of interest for investigating the use of similarity methods in activity prediction since it provides a large body of toxicological data in association with machine-readable structure representations, specifically Wiswesser Line Notation (WLN) records.

Data-sets

The experiments used several subsets of the RTECS database as follows:

1. RTECS-1 (animal carcinogens): these were 536 compounds for which carcinogenicity data were available from International Agency for Research on Cancer (IARC) monograph reviews. These reviews categorise the evidence for carcinogenicity as sufficient, limited, inadequate or no evidence. In this work, the first two categories and the last two categories were taken to represent carcinogens and non-carcinogens, respectively. In all there were 352 carcinogens and 184 non-carcinogens.
2. RTECS-2 (mutagens): these were 611 compounds that had been tested in the Ames reversion test as part of the Environmental Protection Agency's GENETOX programme. In all, there were 491 structures classified by the programme as mutagens and 120 classified as non-mutagens.
3. RTECS-3 (skin irritants): these were 968 compounds that had been tested on rabbits in the open version of the Draize test for skin irritation. Only skin irritants are listed but these are divided into severe, moderate and mild irritants; for the work reported here, the data-set contained 491 severes and moderates and 477 milds.
4. RTECS-4 (toxic compounds): these were 269 compounds for which time-weighted average and control limit Threshold Limit Values (TLVs) in ppm are available. The data are continuous and were converted into a measure of toxicity, T, given by

$$T = log_{10}\frac{1000}{TLV}$$

   where T was in the range 0.0 to 5.3.
5. RTECS-5 (toxic compounds): these were 6631 compounds for which rat oral LD50 values in mg/kg are available. The data are again continuous and were converted into a measure of toxicity, T, given by

$$T = log_{10}\frac{1000 \times MW}{LD_{50}}$$

   where MW is the molecular weight and where T was in the range 0.5 to 2.5.

These data-sets were extracted from the CD-ROM version of RTECS by specifying the retrieval of those records that had one of the particular types of data listed above. Records having WLNs were then extracted from these downloaded files; the WLNs were converted to connection tables and these tables were then used to generate fragment screen records for each of the structures. Specifically, each of the molecules was characterised by the presence or absence of 157 ring screens[23] and of 1123 augmented atom, atom sequence and bond sequence fragments that are used for substructure searching in CAS ONLINE, the Chemical Abstracts Service substructure searching system. These screen records were used to create an inverted bit map, which was then used as the basis for the similarity searching experiments; full details of these, and other, experiments are presented by Walsh[24].

Experimental Results and Discussion

In view of the results presented above, the experiments here were carried out using the conventional similarity measure based on binary weighted fragments and the Tanimoto coefficient. Each structure in a data-set was taken in turn and used as the target structure in a nearest neighbour search, i.e., its similarity was calculated with every other molecule in the data-set to identify those that were most similar to it. Once the nearest neighbour searches had been executed, the resulting rankings were examined to determine the extent to which structurally similar molecules exhibited comparable activity characteristics.

In the case of RTECS-1, RTECS-2 and RTECS-3, the effectiveness of the rankings was carried out by means of simulated nearest neighbour classification. The activity class of the target structure was assumed to be unknown and its class was then assigned by the majority vote of its k nearest neighbours, where k was a small odd number. The overall effectiveness of the similarity searches was evaluated by recording the number of correctly and incorrectly assigned members of the activity classes. We shall illustrate this procedure with respect to a classification of the 536-member carcinogen data-set, which contains 352 carcinogens and 184 non-carcinogens. If the nearest neighbour procedure results in correct assignments of 242 of the carcinogens and of 105 of the non-carcinogens, then the number of false positives, i.e., non-carcinogens incorrectly assigned as carcinogens, is 79 and the number of false negatives, i.e., carcinogens incorrectly assigned as non-carcinogens, is 110. This situation is illustrated in Table 6. The fraction, f, of correct carcinogen assignments is then 242/(242 + 79), as against a fraction of 352/536 carcinogens in the entire data-set. The probability that such a

Table 6. Analysis of nearest neighbour assignment experiments using RTECS-1 data as an example

		Actual		Total
		Carcinogen	Non-Carcinogen	
Assigned	Carcinogen	242	79	321
	Non-Carcinogen	110	105	215
Total		352	184	536

fraction of correct assignments could have arisen by chance can be tested by using the Normal approximation to the Binomial Test; alternatively the distribution of carcinogens and non-carcinogens can be tested by means of the χ^2 test. In the case of RTECS-4 and RTECS-5, where continuous data were available, the predicted toxicity value, T, was calculated for each compound in turn by taking the arithmetic mean of the T values for its k nearest neighbours. The sets of observed and predicted values were then compared using the product moment correlation coefficient.

The results of the experiments are detailed in Tables 7–10. All of the values listed in all of the tables are significant at the 0.05 level of statistical significance. Thus, despite the acknowledged limitations of much of the data in the RTECS database and the very simple structural representation that we have used, all of the data-sets gave statistical significant relationships between the toxicity data and molecular structure. It is important to emphasise that the similarity-based methods studied here do not perform as well as more sophisticated approaches to the prediction of molecular toxicity, e.g., the work of Klopman and co-workers with the CASE system[25,26] and of Enslein and co-workers with the TOPKAT system[27,28]. However, the similarity approach is extremely simple to implement and does not require parameterisation. It should also be noted that the molecular descriptors used here are just fragments developed to optimise screenout in 2-D substructure searching systems; one would expect that better results might well be obtained if a more sophisticated type of structural representation were to be used, e.g., one including fragments describing the 3-D structure of a molecule or fragments that had been developed to maximise the separation of toxic and non-toxic structures.

Table 7. RTECS-1 assignment results. Carcinogens = 352 and non-carcinogens = 184

k	Carcinogen		Non-Carcinogen		Total	f	χ^2
	Agree	%Agree	Agree	%Agree	%Agree		
1	242	68.8	105	57.1	64.7	0.75	33.9
3	277	78.7	97	52.7	69.8	0.76	54.8
5	272	77.2	88	47.8	67.2	0.74	33.4
9	284	80.7	73	39.7	66.6	0.72	25.0

Table 8. RTECS-2 assignment results. Mutagens = 491 and non-mutagens = 120

k	Mutagen		Non-Mutagen		Total	f	χ^2
	Agree	%Agree	Agree	%Agree	%Agree		
1	426	86.8	66	55.0	80.5	0.89	99.9
3	478	97.4	55	45.8	87.2	0.88	181.8
5	475	96.7	48	40.0	85.6	0.87	138.8
9	476	96.9	40	33.3	84.5	0.86	107.9

Table 9. RTECS-3 assignment results. Non-mild, i.e., moderate or severe, skin irritants = 491 and mild skin irritants = 477

k	Non-Mild Irritant		Mild Irritant		Total	f	χ^2
	Agree	%Agree	Agree	%Agree	%Agree		
1	280	57.0	297	62.3	59.6	0.61	36.1
3	301	61.3	283	59.3	60.3	0.61	41.2
5	306	62.3	274	57.4	59.9	0.60	37.9
9	321	65.4	278	58.3	61.9	0.62	54.5

Table 10. RTECS-4 and RTECS-5 product moment correlation coefficients between observed and predicted TLV and LD_{50} toxicities.

k	RTECS-4	RTECS-5
1	0.48	0.61
3	0.49	0.66
5	0.52	0.66
9	0.45	0.64

CONCLUSIONS

In this paper, we have described two quantitative studies of the use of intermolecular similarity for the prediction of activity in large data-sets of chemical structures characterised by substructure searching screens. Our results may be summarised as follows:

1. Despite theoretical arguments supporting the use of an inverse frequency weighting scheme for fragment occurrences, we have been unable to demonstrate that this scheme gives more effective rankings than binary weights.
2. Similarity searching provides a simple, but relatively effective, means of identifying molecules in a database that have similar activity or toxicity characteristics to a chosen target molecule.

ACKNOWLEDGMENTS

We thank the Department of Education and Science and the Health and Safety Executive for funding, Dr. David Bawden and Mr. Jeremy D. Fisher (Pfizer Central Research, U.K.) for the provision of the Pfizer screening data, Fraser Williams (Scientific Systems) Ltd. for connection table software, and Dr. John K. W. Davies (Health and Safety Executive) and Dr. Mark A. Johnson (The Upjohn Company) for statistical advice.

APPENDIX

Ormerod *et al.* have recently reported a comparison of fragment weighting schemes for use in *substructural analysis*[18]. This is an empirical technique that seeks to use the results of previous biological screening tests to rank potential lead compounds in a drug discovery programme in order of decreasing probability of activity. If an effective ranking can be obtained, then synthetic and testing activities may be restricted to those molecules that occur at the top of the ranked list. Substructural analysis involves the calculation of fragment activity weights that denote the contribution of an individual fragment to the activity of a molecule containing it. The overall score for a molecule is obtained from the weights of its constituent fragments and this score then determines where the structure occurs in the final ranking. The technique is totally empirical in character but has the advantage over more conventional structure-activity correlation techniques that it can be used with very large, structurally heterogeneous data-sets. Ormerod *et al.* compared 14 different fragment weighting schemes, using ten small data-sets from the QSAR literature[18] and identified the general effectiveness of a class of fragment weights derived from studies of relevance feedback in document retrieval systems[29,30]. Of these *relevance weights*, the most generally useful was found to be the following:

$$\log\left(\frac{ACTS(I)/NACT}{INACTS(I)/NINACT}\right)$$

where $ACTS(I)$ and $INACTS(I)$ are the numbers of active and inactive compounds, respectively, in the data-set that contain the I-th fragment and where $NACT$ and $NINACT$ are the total numbers of active and inactive compounds, respectively, in the data-set. The weight thus represents the ratio of the proportion of active molecules in which the I-th fragment occurs to the proportion of inactive molecules in which it occurs.

Ormerod *et al.* suggest that this is the most appropriate weight for the ranking of molecules in substructural analysis studies[18]. The rankings in such studies are, of course, based on activity data that have been obtained previously in a drug discovery programme; in a similarity search, conversely, one typically has just a single active compound, i.e., the one that is used as the target compound in the search. In this case, it is known that $ACTS(I)$ is at least one; it may be more but this information is not available. For similar reasons, $INACTS(I)$ is not more than $FREQ(I)$-1, where $FREQ(I)$ is the frequency of the I-th fragment in the complete data-set of N compounds, where $N = NACT + NINACT$. Accordingly, we may rewrite the weight as:

$$\log\left(\frac{1/NACT}{(FREQ(1)-1)/NINACT}\right)$$

i.e.,

$$\log\left(\frac{NINACT}{NACT(FREQ(I)-1)}\right).$$

In a research environment, *NACT* will be very much less than *NINACT* and thus $NINACT \approx N$. In addition, the screens that are used in chemical information systems for substructure and similarity searching are typically of medium frequency in the database that is being searched, so that $FREQ(I) \gg 1$. The weight is hence:

$$\log\left(\frac{N}{NACT \times FREQ(I)}\right).$$

NACT is unknown in a research environment, but is a constant for a given data-set. Thus, in the absence of extensive activity data, the weight is a simple inverse logarithmic function of the number of structures in the data-set that have been assigned the *I*-th screen.

REFERENCES

1. *Communication, Storage and Retrieval of Chemical Information*; Ash, J.E.; Chubb, P.A.; Ward, S.E.; Welford, S.M.; Willett, P., Eds; Ellis Horwood: Chichester, 1985.
2. Barnard, J.M. 'Recent Developments in Chemical Structure Handling'. *Perspectives in Information Management* **1989,** *1,* 133–168.
3. Lipscomb, K.J.; Lynch, M.F.; Willett, P. 'Chemical Structure Processing'. *Annu. Rev. Inf. Sci. Technol.* **1989,** *24,* 189–238.
4. *Chemical Structures: The International Language of Chemistry*; Warr, W.A., Ed.; Springer-Verlag: Heidelberg, 1988.
5. Johnson, M.A. 'A Review and Examination of the Mathematical Spaees Underlying Molecular Similarity Analysis'. *J. Math. Chem.* **1989,** *3,* 117–145.
6. *Similarity and Clustering Methods in Chemical Information Systems*; Willett, P.; Research Studies Press: Letchworth, 1987.
7. Adamson, G.W.; Bush, J.A. 'A Method for the Automatic Classification of Chemical Structures'. *Inf. Storage Retr.* **1973,** *9,* 561–568.
8. Bawden, D. 'Applications of Two-dimensional Chemical Similarity Measures to Database Analysis and Querying'. In *Concepts and Applications of Molecular Similarity*; Johnson, M.A.; Maggiora, G.M., Eds; John Wiley: New York, 1990.
9. Carhart, R.E.; Smith, D.H.; Venkataraghavan, R. 'Atom Pairs as Molecular Features in Structure-activity Studies: Definition and Applications'. *J. Chem. Inf. Comput. Sci.* **1985,** *25,* 64–73.
10. *Concepts and Applications of Molecular Similarity*; Johnson, M.A.; Maggiora, G.M., Eds; John Wiley: New York, 1990.
11. Willett, P.; Winterman, V.; Bawden, D. 'Implementation of Nearest Neighbour Searching in an Online Structure Search System'. *J. Chem. Inf. Comput. Sci.* **1986,** *26,* 36–41.
12. Hodes, L. 'Clustering a Large Number of Compounds. 1. Establishing the Method on an Initial Sample'. *J. Chem. Inf. Comput. Sci.* **1989,** *29,* 66–71.
13. Willett, P.; Winterman, V.; Bawden, D. 'Implementation of Nonhierarchic Clustering Methods in Chemical Information Systems: Selection of Compounds for Biological Testing and the Clustering of Substructure and Search Output'. *J. Chem. Inf. Comput. Sci.* **1986,** *26,* 109–118.
14. Johnson, M.A.; Lajiness, M.; Maggiora, G. 'Molecular Similarity. A Basis for Designing Drug Screening Programmes'. *Prog. Clin. Biol. Res.* **1989,** *291,* 167–171.
15. Lajiness, M.S.; Johnson, M.A.; Maggiora, G.M. 'Implementing Drug Screening Programmes Using Molecular Similarity Methods'. *Prog. Clin. Biol. Res.* **1989,** *291,* 173–176.
16. Willett, P.; Winterman, V. 'A Comparison of Some Measure for the Determination of Intermolecular Structural Similarity'. *Quant. Struct. Act. Relat.* **1986,** *5,* 18–25.

17. Moock, T.E.; Grier, D.L.; Hounshell, W.D.; Grethe, G.; Cronin, K.; Nourse, J.G.; Theodosiou, J. 'Similarity Searching in the Organic Reaction Domain'. *Tetrahedron Comput. Methodol.* **1988,** *1,* 117–128.
18. Ormerod, A.; Willett, P.; Bawden, D. 'Comparison of Fragment Weighting Schemes for Substructural Analysis'. *Quant. Struct. Act. Relat.* **1989,** *8,* 115–129.
19. Bawden, D.; Devon, T.K.; Faulkner, D.A.; Fisher, J.D.; Leach, J.M.; Reeves, R.J.; Woodward, F.E. 'Development of the Pfizer Integrated Research Data System SOCRATES'. In *Chemical Structures: The International Language of Chemistry*; Warr, W.A., Ed.; Springer-Verlag: Heidelberg, 1988, pp. 63–75.
20. Jakes, S.E.; Watts, N.; Willett, P.; Bawden, D.; Fisher, J.D. 'Pharmacophoric Pattern Matching in Files of 3-D Chemical Structures: Evaluation of Search Performance'. *J. Mol. Graphics* **1987,** *5,* 41–48.
21. *Nonparametric Statistics*; Siegel, S.; McGraw Hill: Tokyo, 1956.
22. Kissman, H.M.; Wexler, P. 'Toxicology Information Systems: A Historical Perspective'. *J. Chem. Inf. Comput. Sci.* **1985,** *25,* 212–217.
23. *Computer Storage and Retrieval of Generic Structures in Patents: Ring Perception and Screening to Extend the Search Capabilities;* Downs, G.M.; PhD Thesis, University of Sheffield, 1988.
24. *Similarity and Clustering of Chemical Structure Information: Application to the RTECS Database to Explore Structure-activity Relationships for Chemical Toxicity Prediction*; Walsh, P.T.; MSc Thesis, University of Sheffield, 1989.
25. Frierson, M.R.; Klopman, G.; Rosenkranz, H.S. 'Structure-activity Relationships (SARs) Among Mutagens and Carcinogens: a Review'. *Environmental Mutagenesis* **1986,** *8,* 2830327.
26. Rosenkranz, H.S.; Klopman, G. 'CASE, the Computer-automated Structure Evaluation System, as an Alternative to Extensive Animal Testing'. *Toxicol. Ind. Health* **1988,** *4,* 533–540.
27. Enslein, K. 'An Overview of Structure-activity Relationships as an Alternative to Testing in Animals for Carcinogenicity, Mutagenicity, Dermal and Eye Irritation, and Acute Oral Toxicity'. *Toxicol. Ind. Health* **1988,** *4,* 479–498.
28. Enslein, K.; Borgstedt, H.H.; Blake, B.W.; Hart, J.B. 'Prediction of Rabbit Skin Irritation Severity by Structure-activity Relationships'. *Toxicol. In Vitro* **1987,** *1,* 129–147.
29. Robertson, S.E.; Sparck Jones, K. 'Relevance Weighting of Search Terms'. *J. Am. Soc. Inf. Sci.* **1976,** *27,* 129–146.
30. *Document Retrieval Systems*; Willett, P., Ed., Taylor Graham: London, 1988.

SIMILARITY CRITERIA FOR CHEMICAL STRUCTURES AND REACTIONS

Johann Gasteiger, Wolf D. Ihlenfeldt
Institute of Organic Chemistry, Technical University Munich, W-8046 Garching, Germany

ABSTRACT

New definitions of similarity of chemical structures are presented that are based on finding building blocks for synthesis and on general types of reactions. The merits of these similarity criteria in analysing a database of structures and in designing organic syntheses are illustrated. Reaction similarities are based on values for electronic and energy effects. They allow novel search strategies for reaction databases and inferences on reaction conditions.

THE IMPORTANCE OF SIMILARITY

The concept of similarity has recently received a lot of interest as a method for finding new lead structures with a desired biological activity, for clustering large data sets, and as a method of browsing. The planning of chemical reactions and the planning of organic syntheses are other important applications of similarity measures.

Several contributions in these Proceedings address the topic and the prospects of structural similarity. Two main approaches can presently be distinguished:

1. Those based on comparing the occurrence of constitutional fragments.
2. Those based on analysing the frequency distribution of interatomic distances.

Our interest in synthesis design in general, and in the further development of the EROS (Elaboration of Reactions for Organic Synthesis) system in particular, has led us to explore various definitions of structural similarity. The recognition of similarity between a target structure and an available starting material can be an important guideline in synthesis design and can drastically reduce the number of synthesis intermediates that have to be evaluated. A stepwise retrosynthetic analysis is the most commonly used procedure in synthesis design to find a pathway from the target structure to available starting materials. As indicated in Figure 1, this approach asks for the generation and evaluation of many reaction steps and intermediates. However, if structural similarities between the target molecule and some starting materials A and B can be recognised directly, the search for the reaction steps leading from A and B to the target compound can be initiated from both ends allowing a guided search without much branching.

This paper also treats the topic of similarity of chemical reactions. Databases of chemical reactions are becoming of increasing interest and use in the planning process. Incomplete as these databases still are – and always will be with regard to new reactions – problems arise when a query does not find an entry in the reaction

W. A. Warr (Ed.)
Chemical Structures 2

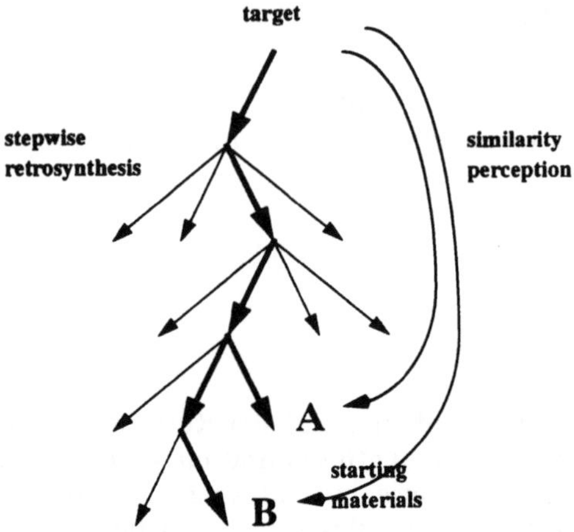

Figure 1. The importance of similarity in reducing the search space in synthesis design

database. In such a situation, it would be quite helpful to find a similar reaction and thus be able to draw some information from the reaction database despite its incompleteness.

Novel criteria that help to plan organic syntheses have been defined for the *similarity of structures*. Likewise, criteria that can be used to search reaction databases have been conceived for the *similarity of reactions*.

SIMILARITY OF STRUCTURES – DEFINITIONS AND IMPLEMENTATION

From the beginning, we believed that there is no simple measure of similarity of structures that is equally powerful over the whole range of organic structures. We therefore defined a series of similarity measures and explored them for different classes of compounds.

The substructures of a compound are an obvious starting point for defining structural similarity. Indeed, we have taken measures of similarity that are based on substructural features. However, we did not dissect a molecule into all conceivable small substructural units. We, rather, tried to keep the *building blocks* for a synthesis intact and make the cuts at places in a molecule where reactions usually set in. These definitions of similarity extract the carbon skeleton, the ring system, a combination of both, etc.

A second class of similarity measures puts more emphasis on the *synthesis reactions*. Basic reaction types are grouped together and thus comprise structural transformations defining similarity. Substitutions, reductions and oxidations are among these similarity measures.

All definitions of similarity can be viewed as transformations that are imposed on a structure and convert it into a new, reduced structure. Table 1 lists the transformations that are presently used to define structural similarity.

If two structures have to be analysed for similarity, both have to be changed through transformations that correspond to the respective similarity criterion. For example, when two structures are similar on the basis of the element exchange

Table 1. Structural transformations used to define similarity. BO : bond order is taken into account (+) or not (−); DB : double bonds are treated as substituents (+) or not (−); AR : heteroatoms in aromatic rings are substituents (+) or not (−); A1 : heteroatoms bonded to aromatic rings are substituents (+) or not (−);

1)	Full structure
2)	Element exchange
3)	Element and ZH exchange
4)	Ring skeleton
5)	Pure ring system
6)	Carbon skeleton
7)	Without terminal alkyls
8)	Without enclosed CH_2
9)	Ring system substitution
10)	Biggest molecule
11)	Symmetry cracker
12)	Maximum oxidation
13)	Maximum reduction/no aromates
14)	Maximum reduction
15)	Maximum reduction/no c-aromates
16)	Aromatic Frame
17)	Full chiral structure
18)	Carbon chain + α–heteroatom
19)	Carbon chain + α-heteroatom + aromates
20)	Substitution (-BO+DB-AR-A1)
21)	Substitution (+BO-DB-AR-A1)
22)	Substitution (+BO+DB-AR-A1)
23)	Substitution (-BO+DB-AR+A1)
24)	Substitution (+BO-DB-AR+A1)
25)	Substitution (+BO+DB-AR+A1)
26)	Chain + α-hetero + oxidation
27)	Chain + α-hetero + reduction
28)	Elimination
29)	Elimination + substitution (-BO+DB-AR-A1)
30)	Chain + α-hetero + element exchange
31)	Chain + α-hetero + aromates + elimination
32)	Chain + α-hetero + aromates + reduction + elimin

criterion, all halogen atoms are exchanged against chlorine in both structures. If the two transformed structures thus obtained are identical, the two initial structures are considered similar by this element exchange criterion. Or, when the maximum oxidation similarity criterion is to be applied, each carbon atom in the two structures to be compared that bears an oxygen atom (either singly or doubly bonded) is put into its maximum oxidation state. Primary alcohols and aldehydes are converted into carboxylic acids, secondary alcohols into keto-groups. Complex transformations, like oxidations, are extremely difficult to code as substructure queries, especially if the formation of quinones and other complex changes in structure are involved. Again, the resulting structures are compared and if they are identical,the initial two structures are similar.

Identity matching is based on a comparison of hashcodes. We have developed a hashcode algorithm[1] that converts the connection table of a structure into a signed 32 bit integer number. Care was taken to avoid collisions, i.e., as of now we have not found an example where two different structures lead to the same hashcode.

In summary, the search for similarity between two structures involves two steps:

1. Transformation of the two structures according to the rules underlying a given similarity criterion.
2. Checking for identity between the two transformed structures.

This two-step procedure is illustrated in Figure 2 for two structures that are checked for similarity based on the maximum oxidation criterion.

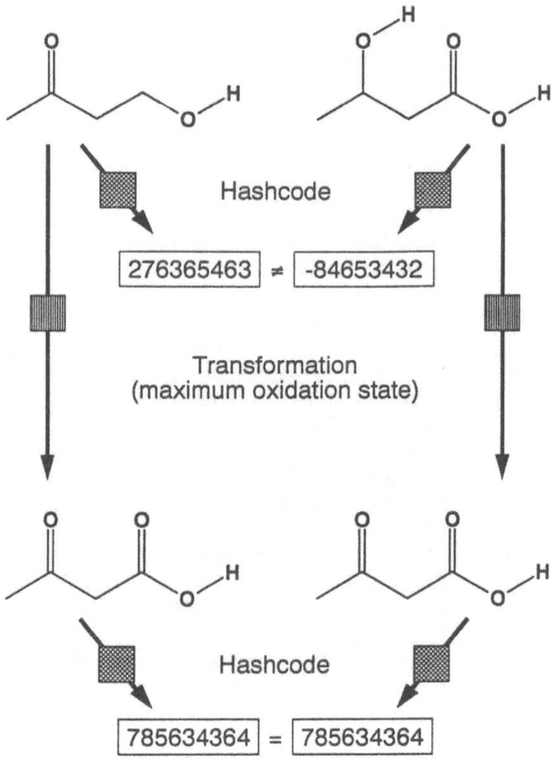

Figure 2. The two-step process for similarity checking based on the criterion of maximum oxidation state

Initial comparison of the hashcode of the two structures shows that they are different. Then, all carbon atoms bearing oxygen atoms are put into the maximum oxidation state. The hashcode of the resulting two structures is found to be identical. Thus, we can conclude that the two initial structures are similar.

As already mentioned, our main objective in defining similarity criteria was to study their utility in developing strategies for designing organic syntheses. In particular, we wanted to use these similarities for finding suitable starting materials for a given target structure. As a basis for comparison a catalogue of available starting materials obtained from Janssen Chimica, comprising approximately 9,000 organic compounds, was used. All structures in this catalogue were submitted to the transformations corresponding to all of the similarity criteria in Table 1. So after hashcoding we obtained 32 signed 32-bit hashcodes for each structure. These hashcodes can be stored on a file. However, the concise nature of these hashcodes even allows them to be kept in core memory (Figure 3).

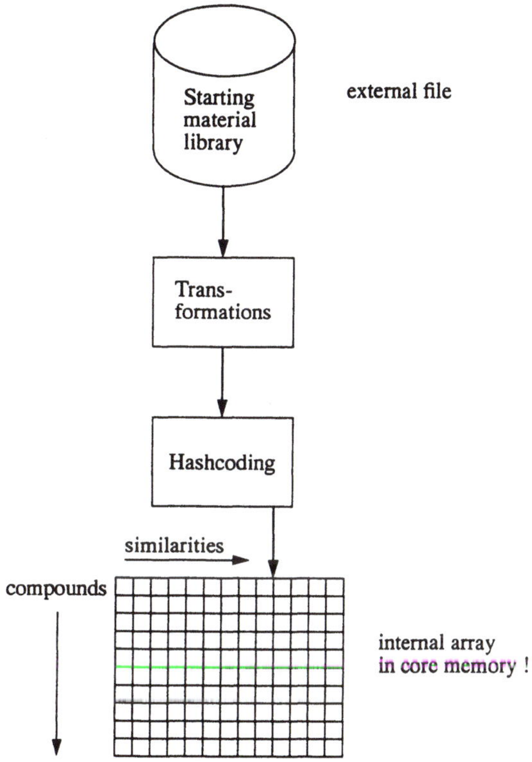

Figure 3. Storing the hashcodes for similarity comparison in the memory

In fact, in our implementation, up to 15 different hashcodes can be stored simultaneously in the core memory. This allows a very rapid determination (1-2 secs for each transformation of the whole Janssen Chimica catalogue)of up to 15 different similarities. The two-step process of transformation and hashcoding is made once for the entire starting material catalogue. The starting materials and their transformed structures are stored as an array of hashcodes in the memory. Then, for similarity matching of a query structure only this single compound has to be processed by transformation and hashcoding. All similarities are rapidly found by comparing the hashcodes of the transformed query structure with the array of hashcodes in the memory (Figure 4).

The entire set-up has been integrated into the WODCA system[2] (Workbench for the Organisation of Data for Chemical Applications). WODCA has been implemented on SUN workstations with peripheral graphical input and output programmes that make heavy use of the window techniques of the Sun View

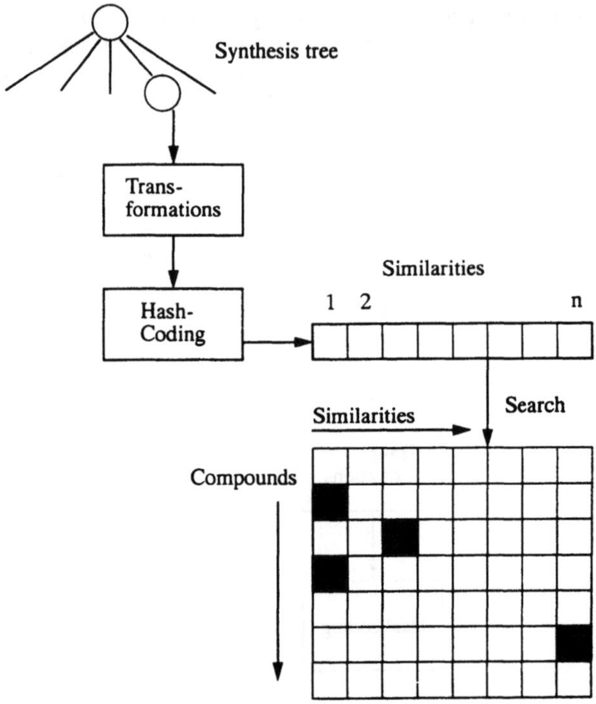

Figure 4. Searching for structures in the starting material library that are similar to a query structure

Graphics library. The WODCA main programme is written in FORTRAN, the peripherals in C. Furthermore, a FORTH subsystem is used for the definition of flexible system commands. Figure 5 gives a typical picture of the screen. The WODCA shell has icons for the various subsystems and functionalities. The molecule editor MED was used to specify the query structure, o-chloroaniline. The CTX File Browser gives the results of a similarity search in the Janssen Chimica catalogue that considers molecules as similar that have the same aromatic system and the same α-atoms on the aromatic rings. Additional information on these compounds is given in the window of the WODCA shell.

EXAMPLES OF STRUCTURAL SIMILARITIES

In the following, some of the similarity criteria will be illustrated with examples from the Janssen Chimica catalogue. The query structure of these studies is highlighted against a black background.

Figure 6 shows all compounds of the catalogue that can be interconverted through halogen exchange.

If exchange of XH groups is also allowed, additional compounds as shown in Figure 7 are obtained.

Clearly, the underlying principle of these similarity criteria is the fact that halogens and XH groups can easily be interconverted through nucleophilic aliphatic substitution reactions. Any organic chemist will use this concept more or less intuitively when browsing through a catalogue of chemicals.

Figure 8 shows all the chemicals in the catalogue that have a benzene ring and a chain of four carbon atoms attached to this ring.

The structures in this list show various substituents at different sites of the benzene ring as well as different substituents and substitution patterns at location 1, 2, 3, 4, 1 and 2, 1 and 3, etc. of the carbon chain. Thus one search command immediately gives the entire variety of structures containing the phenylbutane skeleton. All fourteen positions of this skeleton would have to be specified as open to obtain the same result by a conventional substructure search. Such a search would be rather time-consuming.

Another transformation expels all CH_2 groups from a carbon chain but not those of a ring. In this way, all homologues are perceived as being similar. Figure 9 illustrates this with cyclohexylalkanoic acids.

Elementary transformations can be combined. A series of those combinations that we have explored can be seen with the criteria 18–27 and 29–32 of Table 1. As an example, the procedure for extracting the carbon chain and all α-heteroatoms

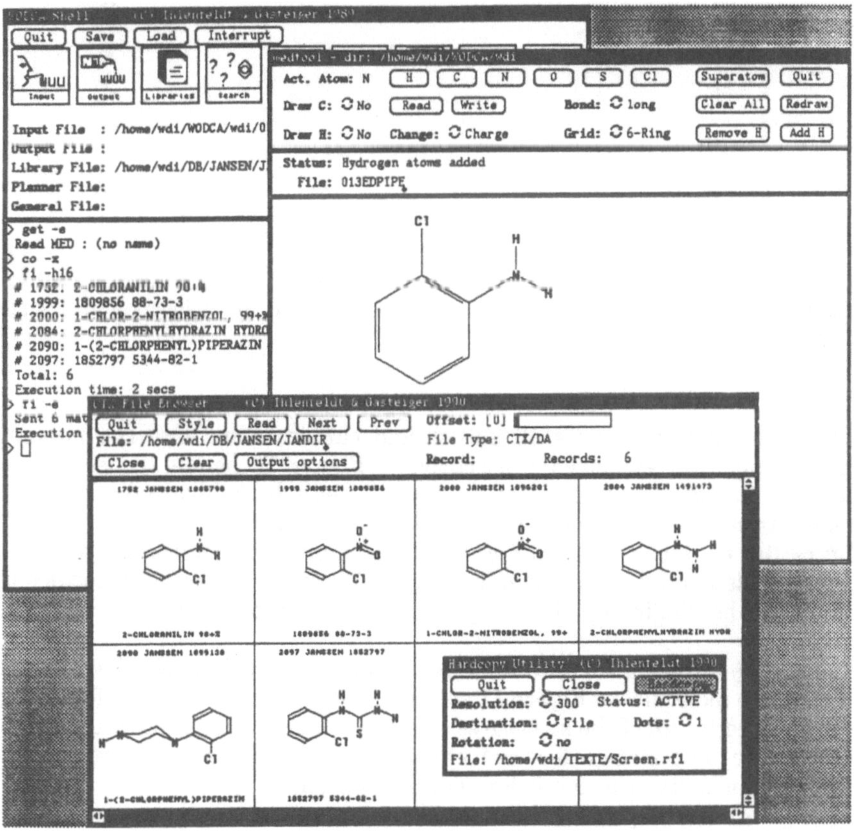

Figure 5. Typical screen image of a WODCA study

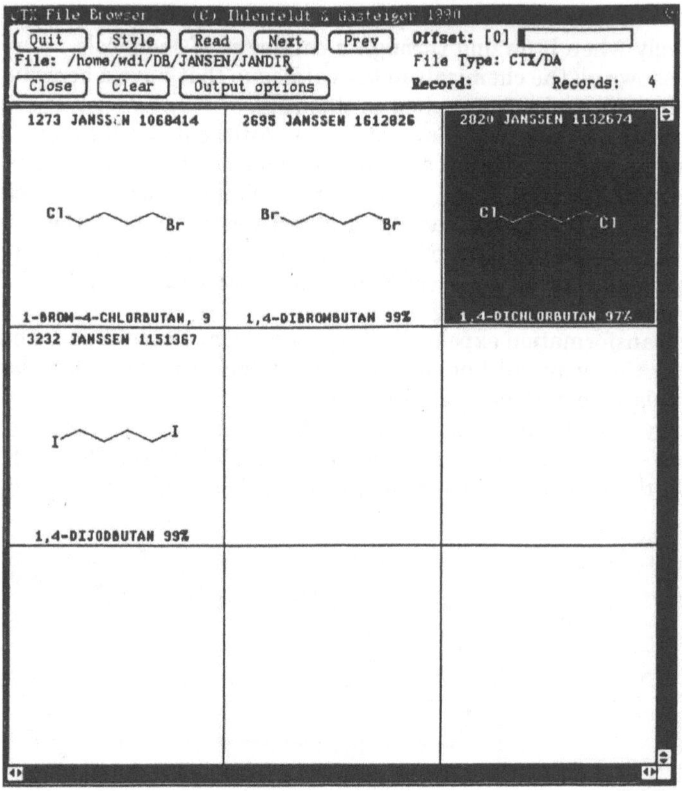

Figure 6. Structures considered similar when exchanging all halogen atoms against chlorine

and the procedure that puts all substituted carbon atoms into the maximum oxidation state are employed consecutively. Figure 10 shows the sequence of structural changes that are made with 2-methyl-4,5-dihydrofuran by this composed transformation and eventually lead to laevulinic acid.

Figure 10 is a clear indication how the similarity criteria chosen by us incorporate elementary reaction steps. Figure 11 shows the set of molecules that are found to be similar by the composed transformation explained in detail by Figure 10.

It should be stressed that the same set of structures will be obtained by using any one of the molecules in Figure 11 as a query structure. This is true for all of the similarity criteria. All of the structures of Figure 11 can be interconverted through rather simple and straightforward chemical reactions. This underscores how useful this similarity criterion is for synthesis design.

Figure 12 should emphasise the manifold nature of structural changes embodied in the transformations used to define similarity. It shows a variety of structures that are obtained from 2-acetamino-4-methyl-thiazolo- 5-sulphonic acid chloride by some of the similarity criteria of Table 1.

It is left to the reader to contemplate the usefulness of the various similarity criteria for finding precursors for a synthesis of 2-acetamino-4-methyl-thiazolo-5-sulfonic acid chloride. Some of the structures of Figure 12 are attractive starting materials for the synthesis of this target compound.

To increase the potency of the above similarity criteria for synthesis design, they have been integrated into a system that allows a user to specify strategic bonds. Bonds chosen by the chemist as strategic for a synthesis are broken and the fragments (synthons) are converted into neutral molecules. These building blocks for a synthesis are then analysed by the similarity measures.

The use of these methods is illustrated in Figure 13. The target structure chosen was 2-methyl-2-hepten-6-one. The bond between carbon atoms 4 and 5 was selected as strategic. The file of starting materials was then searched for those compounds that are similar to the two fragments obtained by breaking this bond. The transformation applied here is the isolation of the carbon backbone plus α-heteroatoms followed by an elimination sequence (number 31).

The seven structures obtained from the Janssen Chimica catalogue are shown in the window at the lower right hand side. Additional information on these compounds is in the window to the left of the structures. The arrow in the target

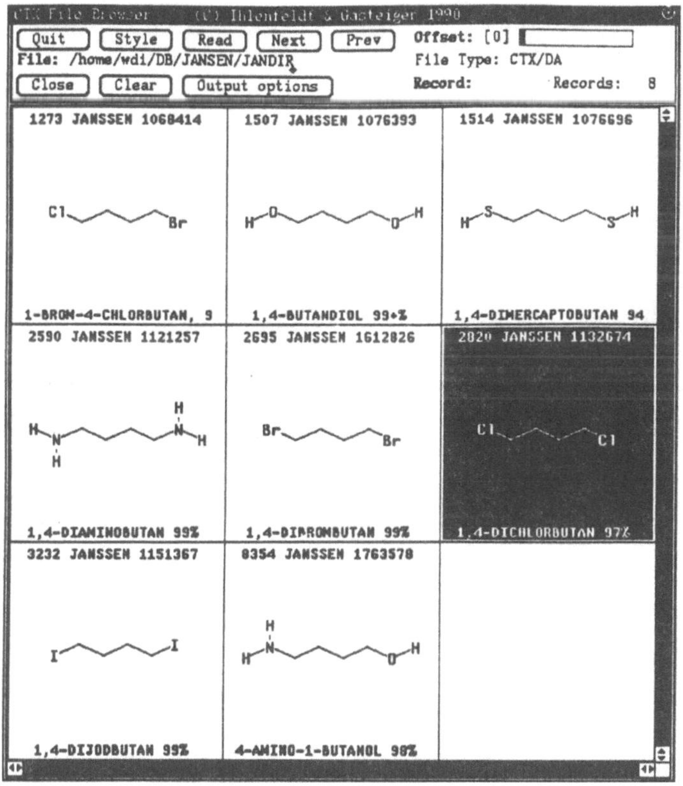

Figure 7. Structures that turn out to be similar when halogen atoms and XH groups are exchanged

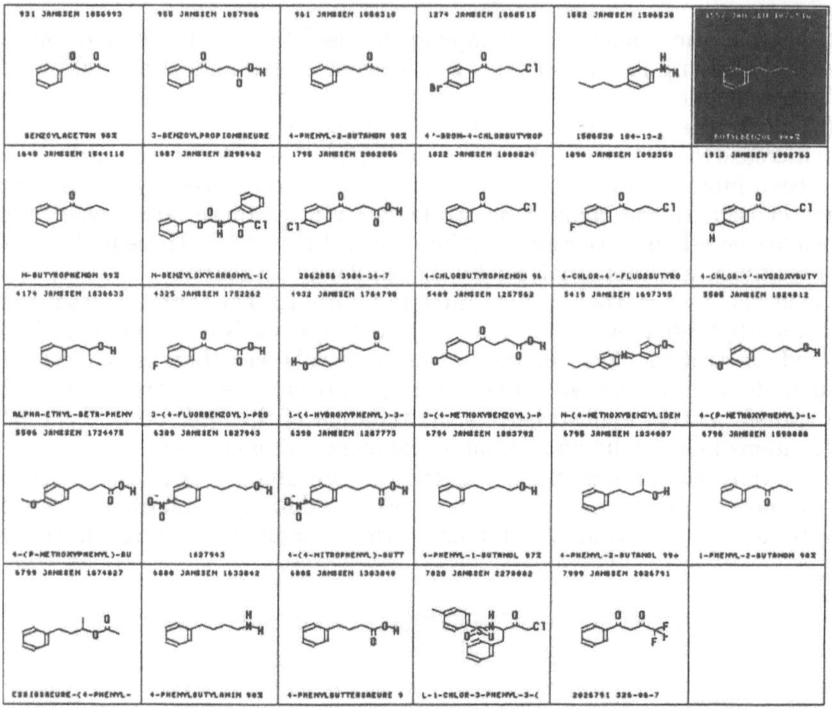

Figure 8. Structures that are similar as they have the same ring system and carbon skeleton

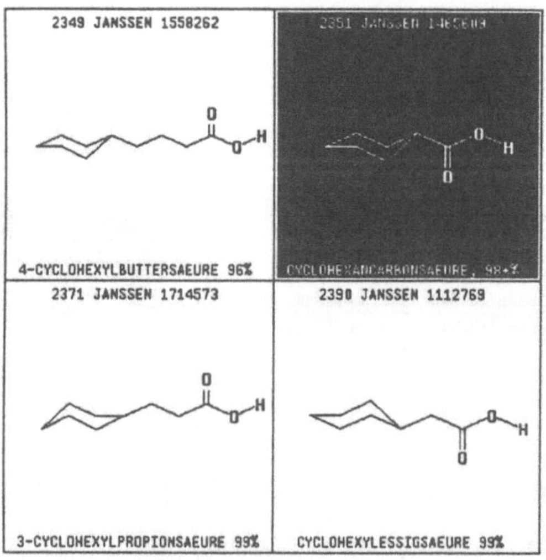

Figure 9. Structures being perceived as similar by cutting out all CH_2 groups from a carbon chain

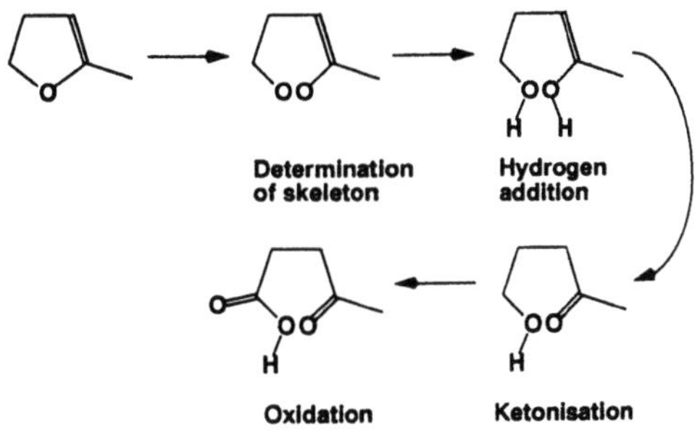

Figure 10. Structural changes made with 2-methyl-4,5-dihydrofuran by the composed transformation: chain + α-atom + oxidation

structure points to the left-hand structure. Therefore those compounds similar to that part of the target structure are highlighted. In the example isoprene is recognised as a starting material for the left side of the target and various acetone derivates are candidates for the right side. Note that only the location of the strategic bond was specified, not the direction this bond should be broken heterolytically. This was determined automatically by evaluating resonance stabilisation effects. No halogen-substituted acetone derivatives were found. The time needed for this search was less than two seconds.

Strategic bonds can be selected at the user's discretion as was done in the example above. In addition, procedures are being developed that automatically determine strategic bonds. Both approaches are built into SESAM (System for the Excogitation of Syntheses for Aliphatic Molecules), that is currently under development.

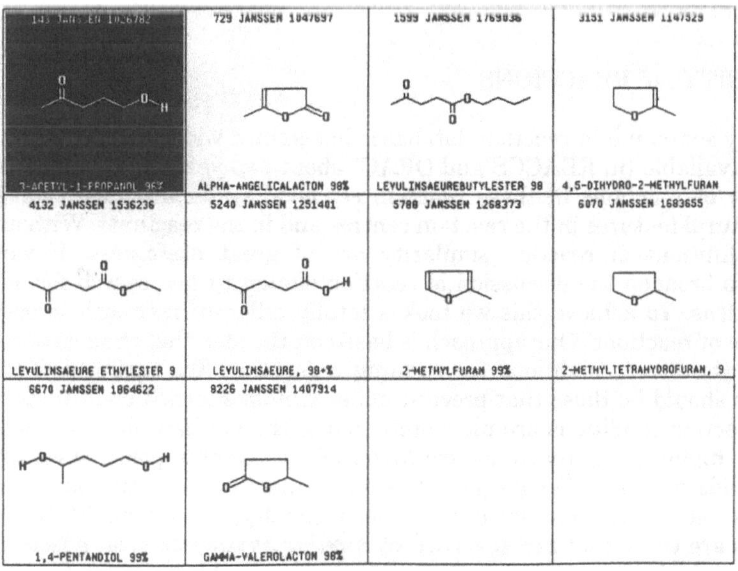

Figure 11. Structures transformed into the same basic structure by the similarity criterion no. 26 (chain + α-atom + oxidation; c.f. Fig. 10)

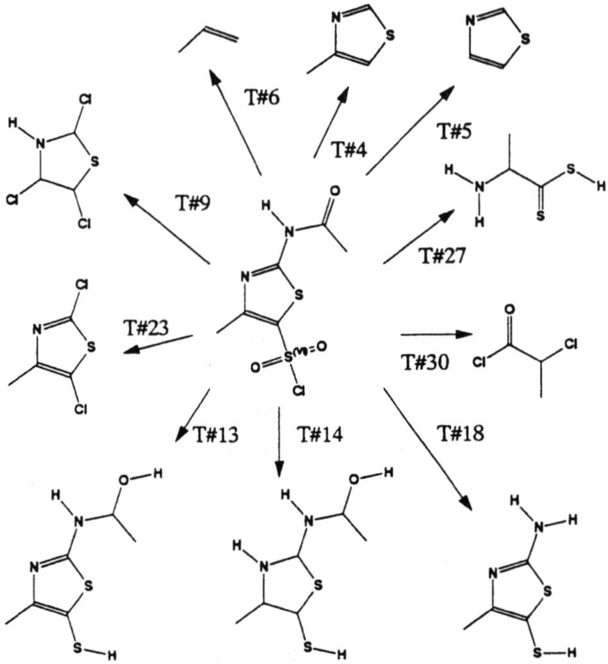

Figure 12. Structures obtained from 2-acetamino-4-methyl-thiazolo-5-sulphonic acid chloride by the transformations used to define similarity. (The numbers refer to the criteria listed in Table 1.)

SIMILARITY OF REACTIONS

Similarity searching in reaction databases has gained widespread interest since it became available (in REACCS and ORAC) about two years ago. The definition of similarity of reactions in these reaction retrieval systems is based on common substructural features in the reaction centres and in the reactants. Without doubt, these definitions of reaction similarity are of great usefulness. However, we wanted to broaden the discussion of reaction similarity and search for additional helpful ideas. To achieve this we took a totally different approach to specify the similarity of reactions. Our approach is based on the idea that chemists often want to know the exact conditions for running a reaction. We assumed that similar reactions should be those that proceed under *similar reaction conditions*.

The reaction conditions are ultimately tied to the mechanism of a reaction. For example, highly polar reactions are favoured by strongly polar solvents and by strong acids or bases. For its part, the mechanism of a reaction is governed by electronic, steric and energy effects. Consequently, we concluded that similar reactions are those that are governed by *similar physicochemical effects*.

Let us look at some examples. We are in the favourable position of being able to calculate important electronic and energy effects by empirical methods that have been developed in our group[3,4]. These allow rapid processing of sizeable data sets of fairly large molecules and can thus determine values for inductive, resonance and polarisability effect, as well as for charge distribution and bond dissociation energies. In addition, functions have been developed, that combine the electronic effects to quantitative estimates for the ease of heterolysis of a bond (polar reactivity).

Table 2 gives values for the reactivity of breaking a methyl alkyl or methyl aryl ethers into methoxide ion and an alkyl or aryl carbocation. These values show the increase in reactivity going from a phenyl group to a methyl and further to a *tert*-butyl or benzyl group. According to these data a methyl phenyl ether will not cleave at the oxygen-phenyl bond but rather at the oxygen-methyl bond. Furthermore, the fact that a *tert*-butyl ether and a benzyl ether obtain nearly the same value for the ease of heterolysis indicates that their reactivities and reaction mechanisms are closely related: they are similar reactions. The reactions being similar, we can conclude that they require similar, in this case acidic, conditions.

To accentuate the potential of the values on physicochemical effects for searching reaction databases, a reaction query is constructed. Let us assume that the reaction database contains the cleavage of methyl benzyl ether but not the one of methyl *tert*-butyl ether. With no entry to be found when searching for the cleavage of a tert-butyl ether, the physicochemical parameters and polar reactivity for the query will be calculated. Searching the database now with this value gives

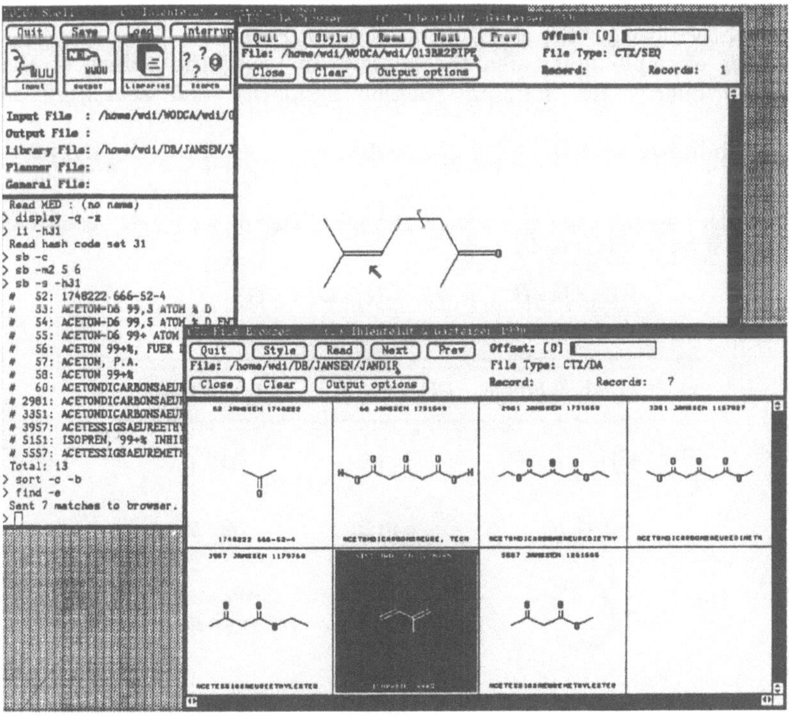

Figure 13. Searching for starting materials that bear similarity to the two fragments obtained through cleavage of a target molecule at a selected bond

Table 2. Estimates for the reactivity in polar cleavage of methyl alkyl or aryl ethers

$$CH_3-O-R \longrightarrow CH_3-\overline{\underline{O}}|^{\ominus} + R^{\oplus}$$

	Polar reactivity [%]
R = $-CH_3$	70.4
$-C(CH_3)_3$	84.1
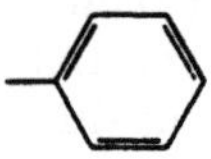	36.9
$-CH_2-$ 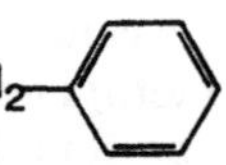	83.7

a close hit: the cleavage of the benzyl ether. The conclusion to be drawn from this is that the *tert*-butyl ether should be cleavable under reaction conditions that resemble those used for the cleavage of the benzyl ether.

The next example illustrates the benefits of using the values of the bond dissociation energy for defining reaction similarity and searching reaction databases.

Table 3 indicates that the bond dissociation energies of the O-alkyl (aryl) bond

Table 3. Values for the bond dissociation energies (BDE) in the homolysis of the oxygen alkyl or aryl bond in esters and ethers

$$CH_3-O-R \longrightarrow CH_3-O\bullet + R\bullet$$

BDE [kJ/mol]	CH_3-O-R	CH_3-COOR
R = $-CH_3$	344.5	342.1
$-C(CH_3)_3$	336.2	327.7
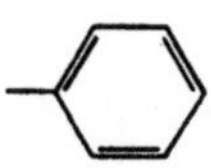	391.3	422.8
$-CH_2-$ 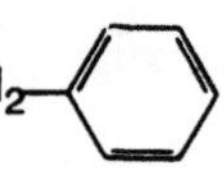	292.0	289.0

heavily depend, both for esters and for ethers, on the nature of the group to be split off. However, for a given group, the values of BDE in esters and ethers are very similar.

Again, a search scenario is constructed to highlight the potential application of the values of the BDE in searching a reaction database. Let us assume that our reaction database contains entries for the cleavage of methyl esters and ethers, as well as benzyl ethers but none for a benzyl ester. Let us further assume that we are interested in cleaving an ester without using acidic or basic conditions as other groups in the molecule would not survive such conditions. Searching the database for ester cleavage we only find the example of a methyl ester. Now we calculate the electronic and energy effects for this query of a methyl ester and use the value of the bond dissociation energy (344.5 kJ/mol) to search for similar reactions, as homolytic reactions do not need acidic or basic conditions. The closest entry that is found is the cleavage of a methyl ether (342.1 kJ/mol). However, the reaction conditions for the cleavage of a methyl ether are not satisfactory for our query because they would need strongly acidic conditions. So we search the database for other cases of ether cleavages and find an example of cleaving a benzyl ether through palladium-catalysed hydrogenolysis. For the methyl case, we have already noted the similarity between an ester and an ether cleavage. We can now apply this similarity to the benzyl case: we decide to use a benzyl ester and try to cleave it in a palladium-catalysed hydrogenolysis. Benzyl esters can actually be cleaved in this way.

From the above, it is obvious that the efficient use of the physicochemical parameters in searching reaction databases requires some chemical insight on the side of the user. However, these parameters can become powerful tools when used intelligently and allow us to draw inferences on the reaction conditions for a desired transformation.

CONCLUSIONS

The new definitions of the *similarity of structures* that are based on building blocks for synthesis and on general types of reactions have a large potential in synthesis design. They allow a rapid access to starting materials that are useful for a synthesis problem.

In addition, these similarity criteria show promise to facilitate ordering and browsing large data sets of structures. Furthermore, they could be used for encoding and searching Markush-type data-sets in patent documentation.

The definitions of the *similarities of reactions*, that are based on physicochemical parameters, give new tools for searching reaction databases. They allow us to draw inferences on the reaction conditions for a desired transformation.

ACKNOWLEDGMENTS

We are grateful for the financial support W. D. Ihlenfeldt received by a scholarship from the 'Studienstiftung des Deutschen Volkes'. We would also like to thank

Janssen Chimica for providing us with their catalogue of commercially available compounds in computer-readable form.

REFERENCES

1. Ihlenfeldt, W.D.; Gasteiger, J. 'Der Einsatz von Hashcodes zur Erkennung der Strukturellen Ähnlichkeit'. In *Software-Entwicklung in der Chemie 2*; Gasteiger, J., Ed.; Springer-Verlag: Heidelberg, 1988, pp. 13–33.
2. Gasteiger, J.; Ihlenfeldt, W.D. 'The WODCA System – an Integrating Environment for the Chemist'. In *Software Development in Chemistry 4*; Gasteiger, J. Ed.; Springer-Verlag: Heidelberg, 1990.
3. Gasteiger, J.; Hutchings, M.G.; Christoph, B.; Gann, L.; Hiller, C.; Löw, P.; Marsili, M.; Saller, H.; Yuki, K. 'A New Treatment of Chemical Reactivity: Development of EROS, an Expert System for Reaction Prediction and Synthesis Design'. *Top. Curr. Chem.* **1987**, *137*, 19–73.
4. Gasteiger, J. 'Empirical Methods for the Calculation of Physicochemical Data of Organic Compounds'. In *Physical Property Prediction in Organic Chemistry*; Jochum, C.; Hicks, M.G.; Sunkel, J., Eds; Springer-Verlag: Heidelberg, 1988, pp. 119–138.
5. Gasteiger, J.; Weiske, C. 'ChemInform – an Integrated Information System on Chemical Reactions'. In *Proceedings of the 13th International Online Information Meeting*; Learned Information: Oxford, 1989, pp. 147–154.

THE *ANSA* MATRIX, A USEFUL TOOL FOR CODING CYCLIZATION REACTIONS

Josef Brandt
Institut für Organische Chemie, Technische Universität München, Lichtenbergstraße 4 D-8046 Garching (DE)

ABSTRACT

The *ansa matrix* is proposed as a new concept for coding, storing and retrieving ring changes (ring opening, formation of rings) in the CASTOR-system of reaction documentation. The *ansa* matrix contains entries of reaction-invariant paths between members of the *reaction core* that exist in the molecular graph outside the reaction core. It does not add or alter any information that is contained in the atom vector, $^{12}\mathbf{a}$, and the BE-matrices, $^{1}\mathbf{B}$ and $^{2}\mathbf{B}$, of the *ensemble of molecules* of reactant(s), **1**, and product(s), **2**, but – like *e.g.* the reaction matrix, $^{12}\mathbf{R}$, and the quantities relating to the reaction core, *i.e.* the irreducible R-matrix, $^{12}\mathbf{R}^{I}$, the *intact BE-matrix*, $^{12}\mathbf{B}^{I}$, and its associated *intact atom vector*, $^{12}\mathbf{a}^{I}$, – it presents relevant pieces of that information in a form more suitable for mathematical treatment, and for storage and retrieval operations in practical applications. Adaptations of the details of its definition under the aspect of practical implementation are being investigated.

THE CASTOR-SYSTEM

The CASTOR-system[1,2] is a system for the documentation of chemical reactions. It relies on a procedure of non-prejudiced coding, indexing, classification, storage and retrieval of reaction and structure information. Its conceptual and mathematical basis is the algebraic model of chemical structure of J. Dugundji and I. Ugi, the *DU-model* (see next section).

An earlier implementation of the CASTOR system on a CDC-Cyber computer system[3], using the CDC-IMF-facility (formerly EDMS) had proven the practical applicability of the underlying theory as well as the usability of the ancillary systems, *e.g.*, data capture and conversion, semi-automatic addition of by-products, graphical output[2]. The database model of EDMS was CODASYL-based. Our implementation, however, used the hierarchical components, *e.g.*, the member-owner construct, for the sole purpose of data storage optimisation. The main parts used a relational model. The relational model became dominant in an ADABAS implementation, where member-owner sentences were replaced by dynamic field constructs. The present goal of development is a purely relational implementation under a portable relational DBMS.

W. A. Warr (Ed.)
Chemical Structures 2

THE DUGUNDJI-UGI-MODEL, A MATHEMATICAL MODEL OF CHEMICAL STRUCTURE

The Fundamental Equation

The formalism of Ugi and Dugundji, the *DU-model*[4–7], provides an unambiguous code for any transition from one chemical structure **1** (the [ensemble of] reactants) into another structure **2** (the [ensemble of] products):

$$\mathbf{1} \Longrightarrow \mathbf{2} \tag{1}$$

Structure **1** is represented by a BE-matrix $^{1}\mathbf{B}$ (the *bond-electron matrix* of structure **1**, *i.e.*, of the [ensemble of] reactants). Structure **2**, *i.e.* the [ensemble of] products, is represented by a BE-matrix $^{2}\mathbf{B}$. Such a transition is commonly called a 'reaction'. The fundamental equation of the Dugundji-Ugi-model is:

$$^{1}\mathbf{B} + {}^{12}\mathbf{R} = {}^{2}\mathbf{B} \tag{2}$$

Together with the atom vector $^{12}\mathbf{a}$, it represents the full information about the structural changes that are effected by the reaction (1).

We note that (1), or its inverse:

$$\mathbf{2} \Longrightarrow \mathbf{1} \tag{2}$$

are genuine *reaction equations*, *i.e.*, they are stoichiometrically balanced. In chemistry, we frequently abbreviate a reaction equation, *e.g.*:

$$\mathbf{A} + \mathbf{B} \Longrightarrow \mathbf{C} + \mathbf{D} \tag{4}$$

into something like:

$$\mathbf{A} \Longrightarrow \mathbf{C} \tag{5}$$

because we consider **A** the 'substrate' or 'main reactant', and **C** the 'main product'[8–15]. Obviously, formula (5) is not a reaction *equation*[†]. *The term reaction* is thus strictly reserved to the 'complete' (*i.e.*, stoichiometrically balanced) equation (4) where **A** + **B** = **1** is the reactant-EM, and **C** + **D** = **2** is the product-EM of the reaction (1). This reaction is represented by an operation on $^{1}\mathbf{B}$, leading to $^{2}\mathbf{B}$. In the framework of the valence-bond model this operation is equivalent to the addition of a matrix $^{12}\mathbf{R}$, the reaction matrix (R-matrix) leading from **1** to **2**.

The reverse reaction (3) is represented by:

$$^{2}\mathbf{B} + {}^{21}\mathbf{R} = {}^{1}\mathbf{B} \tag{6}$$

with the corollary:

$$^{12}\mathbf{R} = -{}^{21}\mathbf{R} \tag{7}$$

* The IUPAC Division for Organic Chemistry, Commission on Physical Organic Chemistry, has clarified this point by calling the abbreviated form (5) a *transformation*:

> 'A *transformation* is distinct from a *reaction* in that it describes only those changes that are involved in converting the structure of a substrate into that of a product, regardless of the reagent or the precise nature of the substrate, or (with some exceptions) the mechanism.'[16]

The quantities $^{1}\mathbf{B}$, $^{2}\mathbf{B}$, $^{12}\mathbf{R}$ and $^{21}\mathbf{R}$ are square matrices of rank n, where n is the number of atoms in **1** and **2**. The atoms are collected in an *atom vector* **a**. We note that the atom vector is reaction-invariant:

$$^{1}\mathbf{a} = {}^{12}\mathbf{a} = {}^{2}\mathbf{a} \tag{8}$$

An atom vector **a** is simply an enumeration of the atoms of the molecular formulae of ensembles **1** and **2**.

Coding of the Reaction Core in the CASTOR System

For the purpose of classification, three auxiliary quantities have found to be useful[17–19].

- $^{12}\mathbf{R}^{I}$, the *irreducible R-matrix*
- $^{12}\mathbf{B}^{I}$, the *intact BE-matrix*
- $^{12}\mathbf{a}^{I}$, its associated atom vector

These quantities define the **reaction core.** They represent some generalisation, *i.e.*, a representation of the equivalence classes of reactions under the aspect of bond order* and of changes of bond order. In order to make the BE- and R-matrix formalism amenable to a practical implementation of a reaction documentation or indexing system, there has to be a convention concerning an unambiguous ('canonical') numbering of the entries of $^{1}\mathbf{a}$ and $^{2}\mathbf{a}$ and hence of $^{1}\mathbf{B}$ and $^{2}\mathbf{B}$, and a convention concerning a canonical numbering of the entries of $^{12}\mathbf{R}$ and $^{21}\mathbf{R}$, and hence of of $^{12}\mathbf{a}$. Such a convention is commonly called a *canonicalisation*†

For BE-matrices, various proposals for canonicalisation procedures have been made[20–23]. These are not, or not meaningfully, applicable to R-matrices[18]. For R-matrices we have therefore developed a procedure[2,3,17–19,24], which relies on a maximalisation of (appropriate mappings of) (side- and main-) diagonals of $^{12}\mathbf{R}^{I}$. This procedure leads not only to a unique, but in most common cases also to a *chemically* meaningful numbering scheme‡ for the entries of $^{1}\mathbf{a}^{I}$ and $^{2}\mathbf{a}^{I}$.

THE *ANSA* MATRIX: DEFINITIONS

Rings are Secondary Information in the DU-Model

Rings are a particularly interesting structural feature of a chemical structure, and thus ring openings and closures are frequently considered a particular class of reactions. In the formalism of the DU-model (see above), rings do not, *a priori*, receive a particular treatment, *i.e.*, the notion of a 'cycle' is not a primary or fundamental one in the DU-model.

The BE-matrix does, of course, contain all information that is necessary to determine the presence of cycles. Detection of cycles, then, is a standard operation

* including bond order 0, which means bond formation or rupture.

† There seems to have been a discussion at a recent CSA-meeting, whether the term 'canonisation' should also be allowed.

‡ An adaptation of *e.g.*, Randić's numbering algorithm[23], which relies on a maximilisation of the row strings of $^{12}\mathbf{R}$, or $^{21}\mathbf{R}$, respectively, might also be suitable, but it does not have this semantically important property, and furthermore requires a greater computational effort in this particular application.

borrowed from graph theory. The R-matrix contains all the information that is needed, in conjunction with the BE-matrix, to determine the openings and closures of cycles. We could say, then, that the notion of 'cycles' is a secondary, or derived, notion in the framework of the DU-model.

The entries of the fundamental equation (2) of the Dugundji-Ugi-model (together with the atom vector $^{1}\mathbf{a}$) represent the full information about the structural changes effected by a reaction. The ring information is contained in the structure code, and can be extracted if necessary. The same applies to the encoding of the reaction.

For practical applications in chemistry, however, the presence, formation or disappearance of cycles is of primary importance, and it would be tedious, to say the least, if one had to go through the whole process of cycle analysis so often. In a practical implementation this would be completely unacceptable. It is therefore desirable to have the ring information immediately available to a computer program, and, if possible, also to present it in an easily perceivable form to a human reader. It is therefore justifiable to devote a separate item in the structure code, and also in the reaction code, to ring information*

We will thus try to find a method of coding the presence of rings and of their changes in a meaningful and usable way, both for the purpose of machine processing and of classical (printed) documentation.

Ring-Changing Reactions

In the present paper we only will treat the aspect of coding *reactions*, thus of ring *changes*. We will do this in conjunction with the string notation for reactions that is based on the formalism of irreducible R- and intact BE-matrices[17,18]. We will define a new entity which we call an '*ansa* matrix'†, and we will propose several representations of this entity in varying degrees of generalisation: the *ansa*-existence matrix, the *ansa-length matrix* and the *ansa-member matrix*. The *ansa*-existence matrix or the *ansa*-length matrix may be expanded into a string, or linear notation, analogous to the expansion of the irreducible R-matrix $^{12}\mathbf{R}^{I}$ or the intact BE-matrix $^{12}\mathbf{B}^{I}$ as described in[2,17,18]. In addition to economising on computer storage space, the linear notation allows a quick inspection for ring changes in many common cases.

Obviously, an elaboration about ring-changing reactions‡ should, in principle, be double-tracked, one track discussing bond formation, *i.e.*, cyclisation, the other discussing bond opening, *i.e.*, ring cleavage. As one track is simply a mirror image of the other, we shall henceforth only discuss bond formation (cyclisation), trusting that the reader can easily construct the other track, following the arguments in an appropriate fashion.

Vocabulary

Unless ring changes occur strictly *within* the reaction core, they involve structural features that are relatively remote from the reaction core. Therefore, they become

* Note that this has been done from a very early period in the development of rational reaction codes, see *e.g.*[25,26]

† in order to avoidconfusion with the term '*path* matrix', as it is used in graph theory.

‡ By *ring change* we shall denote the creation and destruction of cycles in a molecular graph.

apparent at relatively high levels of specificity in the scheme of reaction categories, RB-classes, RA-classes, R1-classes *etc.*, thus at relatively high levels of n in the hierarchy[17,18] of Rn-classes.

'Internal' and 'External' Ring Changes For the purpose of the present discussion we shall (temporarily) introduce the terms of 'internal' and 'external' ring changes. By 'internal' ring changes we shall denote all changes of such rings that consist exclusively of atoms that are members of the reaction core. These changes can be determined from the information contained in $^{12}\mathbf{R}^I$ *and* $^{12}\mathbf{B}^I$ alone, and will not be treated here*

All other ring changes are 'external'. They are changes of rings that contain some atoms that are *not* members of the reaction core.

'External' Paths, or Ansæ For 'external' ring changes to occur, the presence of a path between two members of the reaction core which is *external* to the reaction core, is a necessary (but not a sufficient) condition. We shall call such an external path an *'ansa'*. We wish to have information about *ansæ* between members of the reaction core readily available. It is precisely that kind of information that we deposit in the '*ansa* matrix' (AM).

For the discussion in later sections let us adopt the following convention:

By i,j,k,l,m,n we shall denote members of the reaction core, thus:

$$i,j,k,l,m,n \in \mathcal{I}^I \tag{9}$$

where $\mathcal{I}^I$ is the index set of the reaction core. By t,u,v,w,x,y,z we shall denote atoms that are *not* members of the reaction core, thus:

$$t,u,v,w,x,y,z \notin \mathcal{I}^I \tag{10}$$

'Internal' Paths. A necessary condition for *any* ring change is the presence of a *reaction-variant* path between a pair of members of the reaction core which is 'internal' to the reaction core *i.e.*, that consists only of atoms belonging to the reaction core. 'Reaction-variant' means that such a path exists either *before* the reaction occurs, or else afterwards†, the former case denoting a ring opening, the latter a ring closure.

Existence of such 'internal paths' may be stored and coded in a matrix, the 'internal path matrix' (IPM), following procedures similar to those for *ansa* matrices. In usual practice, this will not be done, since reaction cores usually consist of few, typically 4 . . . 6, atoms, and the computation of the IPM will be faster than storage and retrieval operations, thus the dichotomy of 'compute or store' will be in favour of 'compute' in this case.

Both necessary conditions, taken together, are sufficient for the occurrence of ring changes.

* We note that all (reaction-invariant) links between members of the reaction core are completely coded in the *intact BE-matrix*, $^{12}\mathbf{B}^I$, and that changes of links between them are completely coded in the *irreducible R-matrix*, $^{12}\mathbf{R}^I$.

† *but not both before and afterwards!*

The Representations of the *Ansa* matrix

The Ansa-existence matrix. The *ansa*-existence matrix, ${}^{12}\mathbf{L}^I$, which is associated with the chemical reaction (1), or its mathematical representation (2), is a matrix of rank n, where $=$ card(I^I), *i.e.*, n is the rank of ${}^{12}\mathbf{a}^I$ (*i.e.*, the number of atoms in the reaction core).

The entries ${}^{12}l^I_{ij}$*O*/) *of the ansa*-existence matrix ${}^{12}\mathbf{L}^I$, are Boolean values:

$$\forall i,j \ {}^{12}l^I_{ij} \in \{\mathbf{false}, \mathbf{true}\} \tag{11}$$

An entry l_{ij} is **false** if there is no 'external' path (*ansa*) between atoms i and j, *i.e.* if between atoms i and j there is no path that contains only atoms $t, u, \ldots$ If there exist(s) any *ansa* between atoms i and j, then the value of I_{ij} is **true**.

In practical applications, $\mathbf{L}^I$ will not be stored as such, but will be computed 'on the fly' from ${}^{12}\mathbf{P}^I$ (*cf* section 3.4.2) with the trivial operations:

$$l_{ij} = \mathbf{false} \text{ for } p_{ij} = 0 \tag{12}$$
$$l_{ij} = \mathbf{true} \text{ for } p_{ij} > 0 \tag{13}$$

The *ansa*-existence matrix is, however, useful for the subsequent discussions of other quantities, *e.g.*, the *ansa*-length matrix, or the *ansa*-member matrix, (see sections below), and also inside computer code where it can be stored as an array of bit vectors (or sets).

The Ansa-Length Matrix. The *ansa*-length matrix, ${}^{12}\mathbf{P}^I$, which is associated with the chemical reaction (1), or with its mathematical representation (2), is a matrix of rank n, where n is defined as above.

The entries ${}^{12}p^I_{ij}$ *of the ansa*-length matrix ${}^{12}\mathbf{P}^I$, are positive integers or zero:

$$\forall_{i,j} \ {}^{12}p^I_{ij} \in \boldsymbol{N}_0 \tag{14}$$

An entry p_{ij} is zero if $l_{ij} =$ **false**, *i.e.*, if there is no path containing only $t, u, \ldots$ between atoms i and j.

If $l_{ij} =$ **true**, then the value of p_{ij} is the minimum of the number of atoms $t, u, \ldots$ forming such a path, *i.e.*, the smallest number of atoms outside the reaction core that form an *ansa* between i and j. The matrix thus codes the change of the smallest ring(s) in a network of rings. This agrees with the common usage in chemistry where changes of smaller rings, usually, are of higher interest than others.

The Ansa-member Matrix. The *ansa*-member matrix, ${}^{12}\mathbf{Q}^I$, which is associated with the chemical reaction (1), or with its mathematical representation (2), is a matrix of rank n, where n is defined as above. Since, in general, the entries of the *ansa* matrix are vectors, the expression '*matrix*' is not strictly applicable here, but will be retained for the sake of convenience.

The members of the path form an alternating sequence of vertices ('nodes', 'atoms') and edges ('bonds'). For the following discussion, we omit the bonds, dealing with the nodes only.

An entry ${}^{12}q_{ij}$ of the *ansa*-member matrix is zero if ${}^{12}p_{ij} = 0$.

Otherwise the entry ${}^{12}q_{ij}$ of a the *ansa*-member matrix is a vector of length ${}^{12}p_{ij}$, consisting of descriptors of the nodes (*e.g.* atoms, groups, *etc.*) $t, u, \ldots$ which constitute the *ansa* corresponding to ${}^{12}p_{ij}$.

These vectors thus are lists of descriptors of nodes that constitute the path from i to j ordered* by their (graph-theoretical) distance from i.

The particular code for the entries of ${}^{12}q_{ij}$ may be adapted to practical requirements. Thus, the numbers $t, u, \ldots$ may be used, or they may be replaced by more detailed node descriptors at the expense of storage space in favour of execution speed†

USE OF THE *ANSA* MATRIX

We give a short example of the appearance and use of an *ansa* matrix, using a Diels-Alder reaction, namely the reaction of two molecules cyclopentadiene-1,3, (**3** and **4**), yielding [5.2.1.0^{2,6}]-tricyclodecadiene-3,8 (**5**):

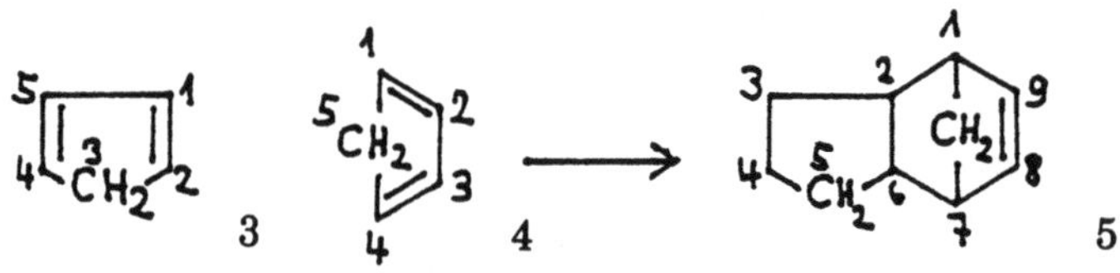

(15)

$$\mathbf{3} + \mathbf{4} \Longrightarrow \mathbf{5} \tag{16}$$

Printed Code

We shall denote an atom a of a molecule **A** by $\mathbf{A}_a$, and the atoms of the reaction core by 1, 2, . . . The graphical representation of the reaction core of reaction (16), using the symbols proposed by Vladutz[25–27] V, is:

(17)

Atoms 1, 2, 3, 4, 5, 6 of the reaction core correspond to $\mathbf{4}_1$, $\mathbf{4}_2$, $\mathbf{4}_3$, $\mathbf{4}_4$, $\mathbf{3}_1$, $\mathbf{3}_2$ on the reactant side, and to $\mathbf{5}_1$, $\mathbf{5}_9$, $\mathbf{5}_8$, $\mathbf{5}_7$, $\mathbf{5}_6$, $\mathbf{5}_2$ on the product side.

The linear code for reaction core (17) is:

$$(111111)11101, \text{CCCCCC} \tag{18}$$

There are two paths in (16) that are *outside* the reaction core: one from 1 to 4 with *one* atom not being a member of the reaction core, thus $\mathbf{P}_{14} = 1$; the other path leads from 5 to 6 and comprises 3 atoms, thus $p_{56} = 3$. If we append the *ansa*-length matrix to the linear code (18), we obtain:

$$(111111)11101, \text{CCCCCC}[00003;;1] \tag{19}$$

This may be further abbreviated to:

$$(1_6)1_301, \text{C}_6[0_43;;1] \tag{20}$$

As an exercise, one might re-convert code (19) into the reaction equation (16). We note that the code (19) does not contain any information about the *names* of

* As a corollary, the entries of ${}^{12}q_{ij}$ are the same as those of ${}^{12}q_{ji}$, but in reverse order. The precedence of i over j is defined by the canonicalization rules for the CASTOR string notation[7,21–22,25]

† In a DARC environment, they could be DELs, or a more detailed quantity in the DARC formalism.

atoms outside the reaction core, nor about any other atoms at all that are not mentioned in **P**. It is advisable to expand the components of linear notation into the full string form, and to write each component properly aligned in one row.

The code thus expanded immediately reveals:

- There is a cyclic shift of bonds which, in combination with the reaction-invariant bonds, indicates a ring formation *within* the reaction core.
- In spite of the reaction-variant bond 5–6, the external path 5–6 does not correspond to a ring change, but only to a change of bond order within a reaction-invariant ring because of the reaction-invariant bond 5–6.

Computer Code

In the CASTOR-system the storage format resembles closely the format of (19), but here we store the *ansa*-member matrix in addition to the *ansa*-length matrix. The descriptors of the nodes are stored as pointers into the reactant structure(s) and into the product structure(s)*

$$\begin{array}{l} (111111)11101, \\ \mathbf{4}_1\ \mathbf{4}_2\ \mathbf{4}_3\ \mathbf{4}_4\ \mathbf{3}_1\ \mathbf{3}_2 || \mathbf{5}_1\ \mathbf{5}_9\ \mathbf{5}_8\ \mathbf{7}_7\ \mathbf{5}_6\ \mathbf{5}_2 \\ [00003(\mathbf{3}_5\ \mathbf{3}_4\ \mathbf{3}_4 || \mathbf{5}_6\ \mathbf{5}_7\ \mathbf{5}_8);;1(\mathbf{3}_5 || \mathbf{5}_{10})] \end{array} \quad (21)$$

Furthermore, there is a piece of information which establishes the correspondence between the internal numbers of reactants and products and a detailed description of these structures: in the example given here, the numbers **3**, **4**, **5** are associated with external identifiers (*e.g.*, registry numbers) that point to entries in a file of structures. In this particular example (16), two of the molecules, **3** and **4**, happen to refer to the same structure, *i.e.*, their *external* identifiers *e.g.*, their registry numbers, would be the same, but in the reaction they play a different role. Thus the external identifier would appear twice, but in a different context.

CONCLUSION

The various operations on *ansa* matrices, particularly the systematic computation of ring changes from the (external) *ansa* matrix and the internal-path matrix, are being investigated in test implementations. These have so far proven the validity of the method.

Finally, the corresponding program modules are being incorporated into a new version of the CASTOR system that is purely based on the relational database model.

REFERENCES

1. Brandt, J.; von Scholley, A.; Wochner, M.; Stadler, K. 'A Documentation System for Chemical Reactions'. Paper given in a symposium on Chemical Reaction Databases in the Division of Chemical Information at the 188th National Meeting of the American Chemical Society, Philadelphia, August 26–31, 1984.
2. Ugi, I. K.; Brandt, J.; von Scholley, A.; Minker S.; Wochner, M.; Schönmann, H.; Straupe, B. 'Hierarchisch strukturierte Speicherung und Ermittlung von chemischen

* The separating marks (||) between the strings that point into the reactant-EM and those that point into the product-EM are added here only in order to facilitate reading.

Reaktionen'. BMFT-FB-ID 85–005, Fachinformationszentrum Energie, Physik, Mathematik: Karlsruhe, 1985.

3. *Hierarchische Klassifizierung und Dokumentation von chemischen Reaktionen*; von Scholley, A; Dissertation TU München, 1981.
4. Dugundji, J.; Ugi, I. 'Algebraic Model of Constitutional Chemistry as a Basis for Chemical Computer Programs'. *Top. Curr. Chem.* **1973,** *39,* 19–64.
5. Ugi, I.;Bauer, J.; Brandt, J.; Friedrich, J.; Gasteiger, J.; Jochum, C.; Schubert, W. 'Neue Anwendungsgebiete für Computer in der Chemie'. *Angew. Chem.* **1979,** *91,* 99–111.
6. Ugi, I.;Bauer, J.; Brandt, J.; Friedrich, J.; Gasteiger, J.; Jochum, C.; Schubert, W. 'New Applications for Computers in Chemistry'. *Angew. Chem. Int. Ed. Engl.* **1979,** *18,* 111–123.
7. Ugi, I. 'A Qualitative Global Mathematical View of Chemistry – James Dugundji's Contribution to Computer Assistance in Chemistry'. In *Computer Applications in Chemical Research and Education*; Brandt, J.; Ugi, I., Eds; Dr. Alfred Hüthig Verlag: Heidelberg, 1989; pp. 345–366.
8. Blower, P. E.; Dana, R. C. 'Creation of a Chemical Reaction Database from the Primary Literature'. In *Modern Approaches to Chemical Reaction Searching*; Willett, P., Ed.; Gower: Aldershot, 1986; pp. 146–164.
9. *CASREACT User Guide*; Chemical Abstracts Service: Columbus, OH, 1988.
10. French, S. E. 'Our Reaction Access System'. *CHEMTECH* 1987 (Feb), 106–111.
11. Fujita, S. 'Structure-reaction Type Paradigm in the Conventional Methods of Describing Organic Reactions and the Concept of Imaginary Transition Structures Overcoming this Paradigm'. *J. Chem. Inf. Comput. Sci.* **1987,** *27,* 120–126.
12. *Name Index of Organic Reactions*; Gowan, J. E.; Wheeler, T. S.; Longman: London, 1960.
13. *Organic Name Reactions*; Krauch, H.; Kunz, W.; Wiley: London, 1964.
14. *Reaktionen der organischen Chemie. Ein Beitrag zur Terminologie der organischen Chemie*; Krauch, H.; Kunz, W.; Dr. Alfred Hüthig Verlag: Heidelberg, 1976.
15. Lynch, M. F.; Willett, P. 'The Automatic Detection of Chemical Reaction Sites'. *J. Chem. Inf. Comput. Sci.* **1978,** *18,* 154–159.
16. *Nomenclature for Organic Chemical Transformations*; Jones, F. A. Y., Ed.; IUPAC, Organic Chemistry Division, Commission on Physical Organic Chemistry, December 1986 (draft).
17. Brandt, J.; Bauer, J.; Frank, R.M.; von Scholley, A. 'Classification of reactions by Electron Shift Patterns'. *Chem. Scr.* **1981,** *18,* 53–60.
18. *Ein mathematisch begründetes hierarchisches Ordnungssystem chemischer Reaktionen und dessen theoretische und praktische Anwendung*; Brandt, J.; Habilitationsschrift, TU München, 1981.
19. Brandt, J.; von Scholley, A. 'A Systematic Classification of Reactions by Electron Shift Patterns'. Paper given at the Chemical Notation Association Conference in Exeter, September 1982. See also: *Communication, Storage and Retrieval of Chemical Information*; Ash, J. E.; Chubb, P. A.; Ward, S. E.; Welford, S. M.; Willett, P., Eds; Ellis Horwood: Chichester, 1985.
20. Morgan, H. L. 'The Generation of a Unique Machine Description for Chemical Structures – a Technique Developed at Chemical Abstracts'. *J. Chem. Doc.* **1965,** *5,* 107–113.
21. Jochum, C.; Gasteiger, J. 'Canonical Numbering and Constitutional Symmetry'. *J. Chem. Inf. Comput. Sci.* **1977,** *17,* 113–117.
22. Schubert, W.; Ugi, I. 'Constitutional Symmetry and Unique Descriptors of Molecules'. *J. Am. Chem. Soc.* **1978,** *100,* 37–41.
23. Randic, M. 'On Unique Numbering of Atoms and Unique Codes for Molecular Graphs'. *J. Chem. Inf. Comput. Sci.* **1975,** *15,* 105–108.
24. Brandt, J.; von Scholley, A. 'An Efficient Algorithm for the Computation of the Canonical Numbering of Reaction Matrices'. *Comput. Chem.* **1983,** *7,* 51–59.
25. Vleduts, G.E. 'Concerning One System of Classification and Codification of Organic Reactions'. *Inf. Storage Retr.* **1963,** *1,* 117–146.
26. *Abtomatizirovannye Informatsionnye Sistemy dlya Khimii*; Vleduts, G. E.; Geivandov, E. A.; Nauka, Moskva, 1974. [CA 81, 12584d].
27. Vladutz, G. 'Do We Still Need a Classification of Reactions?'. In *Modern Approaches to Chemical Reaction Searching*; Willett, P., Ed.; Gower: Aldershot, 1986; pp. 202–220.

CHEMICAL REACTION RETRIEVAL USING CITATION-BASED RELATIONSHIPS

Daniel E. Meyer
Advanced Research Technologies Inc., P.O. Box 556, Wayne, PA 19087, USA

Nadia F. Abdul-Malik and George E. Vladutz
Institute for Scientific Information Philadelphia, PA 19104, USA

ABSTRACT

It has been reported that a novel method for retrieving reactions similar to a given query reaction is to use relationships created by literature citations. Later publications citing a new reaction are likely to contain pertinent information about related reactions. In this study of all citing papers to 50 new synthetic methods indexed in *Current Chemical Reactions*, approximately half of the papers found by cited reference searching would not have been retrieved with reaction or substructure search approaches.

CHEMICAL REACTION RETRIEVAL

In May, 1989, Wipke and Vladutz presented a paper studying the relationship and relevance of citations to a select group of reactions[1]. They stated that all currently available reaction databases are selective in coverage and thus do not comprehensively include all related reactions. They also determined that 90–95% of the citing papers were relevant to the original reactions. This study is an extension of that work and with the objective scope being to determine the proportion of related reactions which might be found using a cited reference search approach.

A random sample of 50 reactions was selected from the June/July 1986 issues of *Current Chemical Reactions* (CCR). These reactions were published in the primary literature during the first quarter of 1986. This group of 50 reactions served as the core group for identifying the citing papers.

ISI's *Science Citation Index* (SCI) was searched from 1986 – 1989 for all papers which cited any of the core group of reactions. For the 4-year period, 21 (42%) of the original papers were never cited. This percentage is in agreement with studies which indicate that 40% – 50% of all papers published are never cited[2,3]. A total of 169 citing papers were identified for the remaining 29 core papers.

These 169 citing papers consisted of 31 reviews and 138 articles. The distribution of the citing papers over the 4-year period is shown in Figure 1. As seen in Figure 1, the citations peaked between 24–36 months after the original publication which matches the standard citation distribution curve[4].

Figure 2 shows the number of times the original papers were cited. The average of approximately five citations per article over the 4-year period is again close to the average which has been reported by ISI[5]. Two papers were highly cited: one by Stille and Grubbs[6] was cited 15 times and one by Tabuchi *et al.*[7] was cited 21 times.

W. A. Warr (Ed.)
Chemical Structures 2

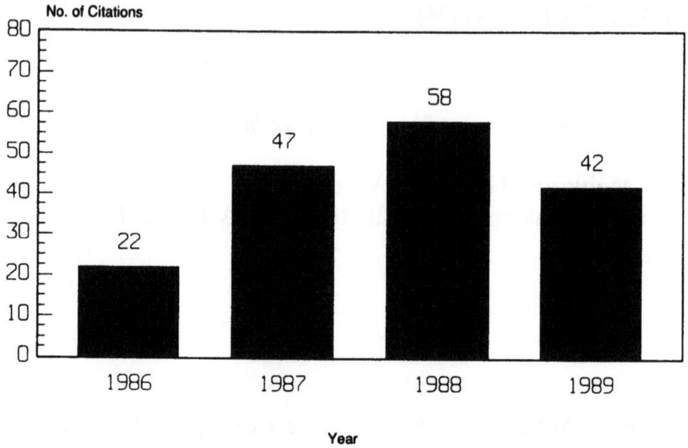

Figure 1. Publication year of citations

Reviews were separated from the citing articles because of their omission in structure and reaction databases. In most cases reviews are indexed and included as bibliographic records but are not included in structure format. Accordingly, these articles would not be retrieved using a substructure search approach.

This study focused on the remaining set of 138 citing papers. In using the 1986–1989 CCR, we found that only 81 (59%) of these articles were included in the CCR database and would be the maximum per cent retrieved in our search. This reflects Wipke and Vladutz's comment about the selective nature of existing reaction databases. Because of this coverage gap, we elected to use a substructure search approach in a comprehensive structure file to determine if the citing articles could be retrieved by an alternative method. This approach was described recently in a paper by R. Synge who found that cited reference searching did identify papers

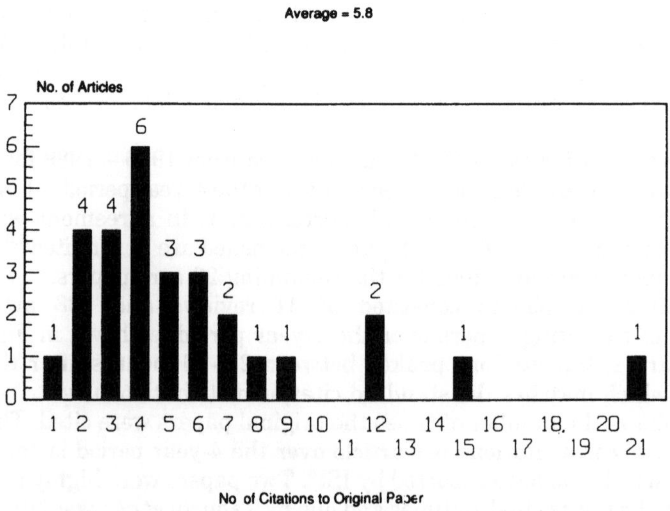

Figure 2. Number of times paper cited

which were not retrieved by substructure or keyword searching[8].

One method of keyword searching would be to search on authors' names to track all follow-up papers by the same researchers. Of the 138 citing papers, only 29% were produced by the one or more of the same authors from the core papers (self-cited).

In using a substructure approach, we determined that only 58% of the 138 citing papers would have been retrieved. Thus we observe that no single approach, reaction, substructure, cited reference, or keyword, would allow for complete retrieval of related papers and therefore a comprehensive search strategy must include multiple approaches.

We divided the 138 citing papers into four groups: Not Self-Cited/ Found by Substructure; Self-Cited/Found by Substructure; Not Self-Cited/Not Found by Substructure; and Self-Cited/Not Found by Substructure. We examined each citing paper for its relationship to the original article and grouped the papers into the following categories.

Not Self-Cited/Found by Substructure

This group included 56 papers with the following six categories, ranked by decreasing order of frequency:

1. Synthesis of the same class of compounds via different routes or different catalysts/conditions.
2. Starting material is a product of the reaction in original paper leading to different classes of products.
3. Studying the mechanism of the original reaction.
4. Synthesis of starting material of the reaction in original paper.
5. Synthesis of synthon.
6. Detailed article following preliminary communication.

Self-cited/Found by Substructure

This group included 24 papers with the same six categories but ranked in a different order:

1. Synthesis of the same class of compounds *via* different routes or different catalysts/conditions.
2. Starting material is product of reaction in original paper leading to different classes of products.
3. Detailed article following preliminary communication.
4. Synthesis of starting material of the reaction in original paper.
5. Synthesis of synthon.
6. Studying the mechanism of the original reaction.

Not Self-Cited/Not Found by Substructure

This group consisted of 42 papers. Of these 42 papers, 16 papers were not included in the CCR reaction database. Thus these 16 (12% of the 138) would only have been

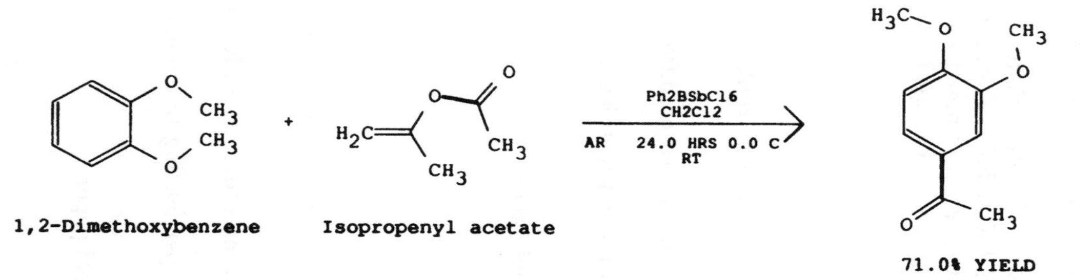

Core Paper THE DIPHENYLBORYL HEXACHLOROANTIMONATE PROMOTED FRIEDEL-CRAFTS ACYLATION REACTION.

MUKAIYAMA T, NAGAOKA H, OHSHIMA M, MURAKAMI M, CHEM LETT (2),165-168(1986).

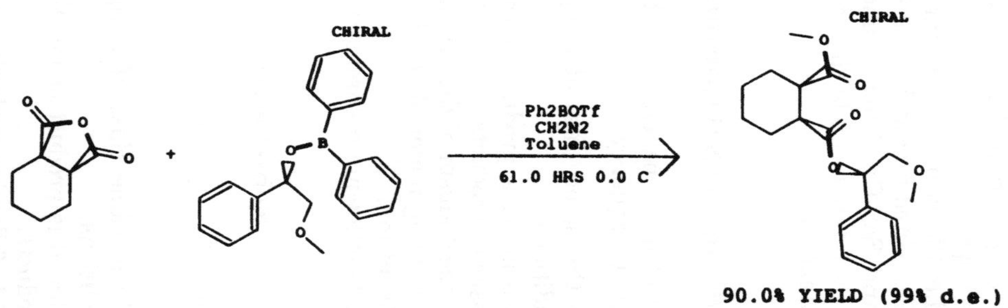

Citing Paper HIGHLY EFFICIENT ASYMMETRIC ESTERIFICATION OF CYCLIC MESO-DICARBOXYLIC ANHYDRIDES CATALYZED BY DIPHENYLBORYL TRIFLATE.

OHSHIMA M, MUKAIYAMA T, CHEM LETT (2),377-380(1987).

Figure 3. Synthesis of different classes of compounds using similar types of catalysts

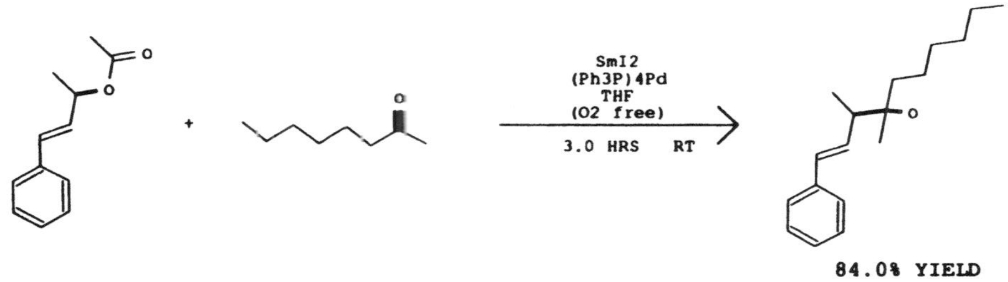

Core Paper — REDUCTIVE COUPLING OF ALLYLIC ACETATES WITH CARBONYL COMPOUNDS BY USING PD(O)-SMI2 SYSTEM.

TABUCHI T, INANAGA J, YAMAGUCHI M, TETRAHEDRON LETT 27(10),1195-1196(1986).

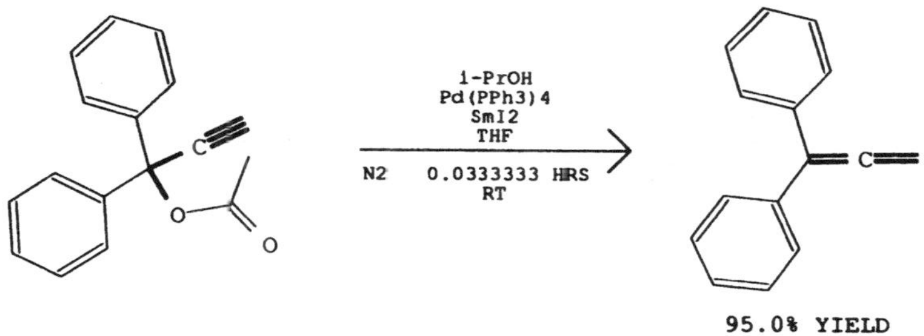

Citing Paper — PALLADIUM-CATALYZED REDUCTION OF PROPARGYLIC ACETATES WITH SMI2. A MILD AND CONVENIENT METHOD FOR THE PREPARATION OF ALLENES.

TABUCHI T, INANAGA J, YAMAGUCHI M, TETRAHEDRON LETT 27(43),5237-5240(1986)

Figure 4. Synthesis of different classes of compounds using similar catalyst system

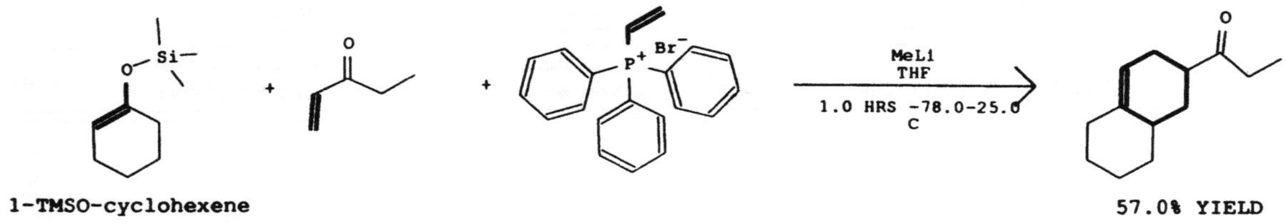

Core Paper SEQUENTIAL MICHAEL-MICHAEL-RING CLOSURE REACTIONS FOR 3-DIFFERENT-COMPONENT, ONE-POT, 2+2+2 CONSTRUCTION OF ACYLCYCLOHEXENES AND AN ACYLCYCLOHEXANOL.

POSNER G H, LU S B, ASIRVATHAM E, TETRAHEDRON LETT 27(6),659-662(1986).

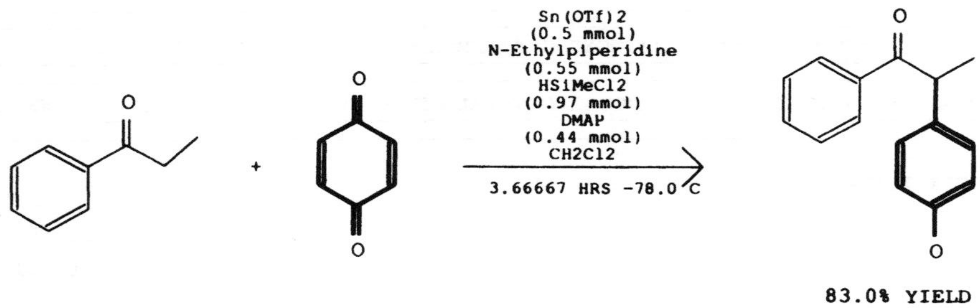

Citing Paper THE TIN(II) ENOLATE ADDITION-REACTIONS TO ALPHA,BETA-UNSATURATED KETONES AND QUINONES.

MUKAIYAMA T, IWASAWA N, YURA T, CLARK R S J, TETRAHEDRON 43(21),5003-5017(1987).

Figure 5. Unexpected products from known types of reactions

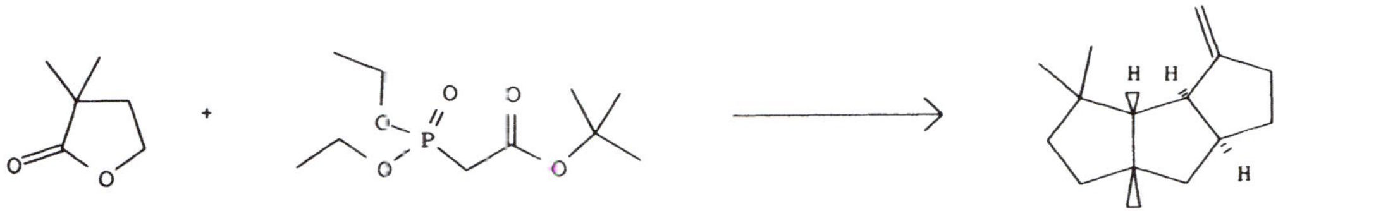

Core Paper	SYNTHESIS OF (+/-)-DELTA(9,12)-CAPNELLENE USING TITANIUM REAGENTS.
	STILLE J R, GRUBBS R H, J AM CHEM SOC 108(4),855-856(1986).

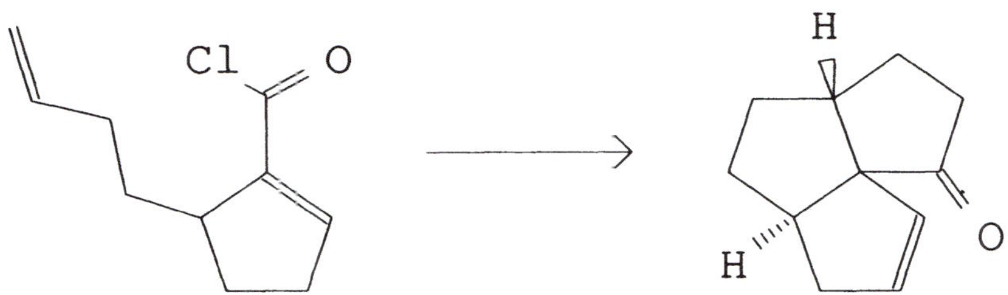

Citing Paper	INTRAMOLECULAR (2+2) CYCLO-ADDITIONS OF KETENES AND VINYLKETENES TO OLEFINS .3. THE SYNTHESIS OF ANGULAR ANNULATED TRIQUINANE DERIVATIVES.
	VEENSTRA S J, DEMESMAEKER A, ERNST B, TETRAHEDRON LETT 29(19),2303-2306(1988).

Figure 6. Synthesis of structural isomers

retrieved by the cited reference search approach. This group was ranked into the following six categories:

1. Synthesis of different classes of compounds using similar types of catalysts.
2. Synthesis of different classes of compounds using similar types of reactions.
3. Unexpected products from known types of reactions.
4. Different methods of creating chiral centres.
5. Application of different reagents for similar types of reactions.
6. Synthesis of structural isomers.

Self-Cited/Not Found by Substructure

This, the smallest group consisted of 16 papers and could be grouped into the following two categories:

1. Synthesis of different classes of compounds using similar types of catalysts.
2. Synthesis of different classes of compounds using similar types of reactions.

Not Found by Substructure

Four examples of reactions which were included in the group are presented in Figures 3–6. The source paper is shown on top with the citing work below.

CONCLUSIONS

Since reaction databases tend to be selective in coverage, cited reference searching can help to identify related reactions which would be missed *via* substructure or keyword searching alone. A summary of the study is shown in Figure 7. Based on the finding of Wipke and Vladutz that 90–95% of the citing papers to reactions are relevant to the source reaction, this study also emphasises that citing reference searching may be an important strategy in finding key related papers missed by other approaches.

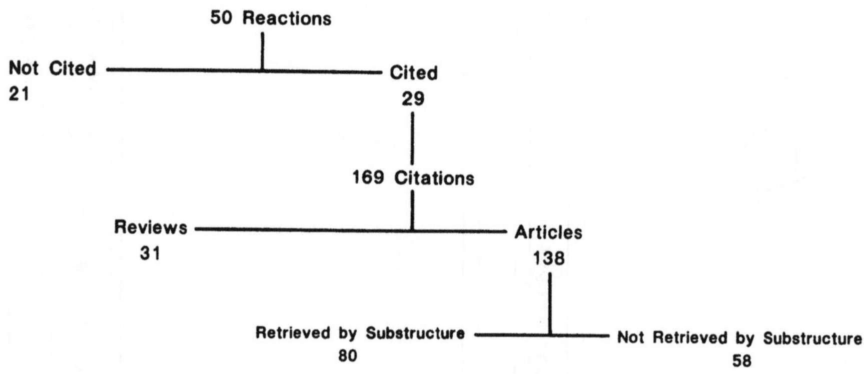

Figure 7. Summary of the data for reaction citations

REFERENCES

1. Wipke, W.T.; Vladutz, G.E. *Tetrahedron Comput. Methodol.* **1990**, *3*(2), 83–107.
2. Garfield. E. *Current Comments/Current Contents*, Oct 28th, 1985.
3. *Science Citation Index*, 1988, ISI Publishing.
4. Small, H., personal communication.
5. *Science Citation Index*, 1989, ISI Publishing.
6. Stille, J.R.; Grubbe, R.H. 'Synthesis of (±)-$\Delta^{(9,12)}$-Capnellene Using Titanium Reagents'. *J. Am. Chem. Soc.* **1986**, *108*, 855–856.
7. Tabuchi,T.; Inanaga, J.; Yamaguchi, M. 'Palladium-catalysed Reduction of Propargylic Acetates with SmI_2. A Mild and Convenient Method for the Preparation of Allenes'. *Tetrahedron Lett.* **1986**, *27*, 5237–5240.
8. Synge,R.L.M. '25 Years of Science Citation Index – Some Experiences'. *J. Chem. Inf. Comput. Sci.* **1990**, *30*, 33–35.

CHEMICAL REACTION SEQUENCE SEARCHING

Glen A. Hopkinson, Anthony P. Cook, Ian P. Buchan and Alison E. Reynolds
ORAC Ltd., 18 Blenheim Terrace, Woodhouse Lane, Leeds, LS2 9HD, England.

ABSTRACT

In this paper we describe how the ORAC reaction database system has been integrated with the LHASA synthesis planning program. This interface allows literature precedents for synthetic steps proposed by LHASA to be retrieved from ORAC and displayed to the user. The problem of sequence searching in reaction database systems is also explored and approaches to this problem are described. Finally, the potential use of synthesis planning systems to identify and classify links between reactions in ORAC for use during sequence searching is considered.

INTRODUCTION

The synthesis of a complex organic molecule can be a challenging task for even the most experienced organic chemist. Success or failure may depend on a particular transformation being effected in good yield. Synthesis planning programs[1], such as LHASA[2], can often suggest interesting approaches to a synthetic problem, but elaboration of the route still requires reference to the literature. The chemist needs to decide what reaction conditions are appropriate; if there is a good literature precedent; if there is need for interfering functionality protection and deprotection steps; the availability, cost and supplier of reactants and reagents. Reaction database systems, such as ORAC[3], have become an indispensable aid to practical organic chemists and information specialists in considering such details. We have developed an interface which allows the ORAC reaction database system to be accessed directly from a synthesis planning program to assist the user in sequence elaboration.

The LHASA/ORAC Interface facilitates the transfer of information between LHASA and ORAC by allowing the synthesis planning program to construct automatically a reaction query to be searched by ORAC. The search results are then ranked in order of similarity of the product structures to the target molecule before being displayed to the user. Occasionally, if a transformation requires more than one synthetic step the reaction database system may fail to locate relevant answers unless the composite reaction or the complete reaction scheme has been registered in the reaction database. We are currently extending the ORAC program to address this problem.

Reaction schemes for making specific compounds from readily available starting materials or solving particular synthetic problems are commonly reported in the literature. It is therefore desirable that a reaction database system allow input and indexing of complete reaction schemes. A reaction database system should be able to search reaction schemes for sequences of reactions that satisfy a query. In addition, the search program should be able to link together individual reactions in a database to find sequences implicitly represented in a reaction database by identifying if the product of one reaction is the reactant of another.

W. A. Warr (Ed.)
Chemical Structures 2

One problem with the above approach is that even in large reaction databases, links between reactions based on an exact structure match may be relatively rare for the majority of reactions. To overcome this problem, a further extension of sequence searching in reaction databases would be to allow the search system to locate sequences which involve 'analogous' links between reaction steps. Such links could be based on the presence of the reacting environment if the reactant in one reaction is the product of another. By classifying reactions, possibly using a synthesis planning system, this kind of searching may be realised in a reaction database system.

Figure 1 summarises the synergic relationship between synthesis planning and reaction database systems that is explored in this paper.

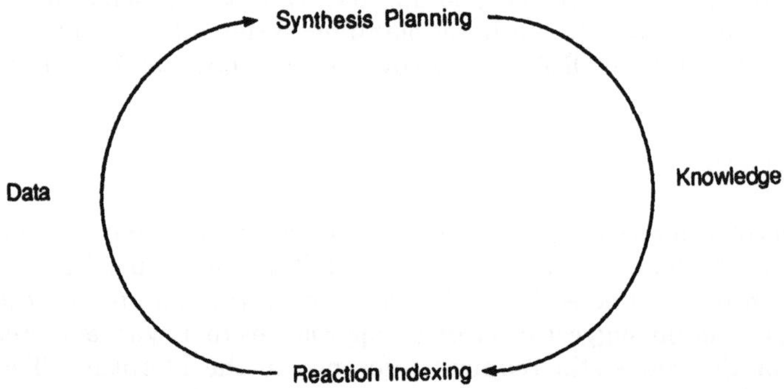

Figure 1. Synergic relationship between synthesis planning and reaction database systems

THE LHASA/ORAC INTERFACE

The primary purpose of a reaction database system is to provide quick and easy access to the reaction literature. Users of synthesis planning programs have a need for such access to the literature in elaborating routes suggested by an analysis[4]. An interface between LHASA and ORAC has recently been developed to provide such a link. The interface allows literature precedents for synthetic steps proposed by LHASA to be retrieved automatically from ORAC's database of over 120,000 reactions and to be displayed to the user on an ORAC Display Form. Company databases can also be accessed to search and display in-house preparative methods and processes in conjunction with databases supplied by ORAC Ltd.

The LHASA/ORAC Interface has been developed using a set of 'toolkit' functions which can be called from the synthesis planning program to perform specific tasks (Figure 2). These functions are fully compatible with the ORAC Host Language Interface[5] (a general purpose programmer's interface to develop custom applications), allowing additional ORAC facilities to be provided as required. The Synthesis Design/ORAC Integration Module provides functions to:

1. Write SMD query files with atom-to-atom mapping for reactions.
2. Read SMD query files and construct a composite query with a specified search logic.
3. Search the ORAC reaction databases.

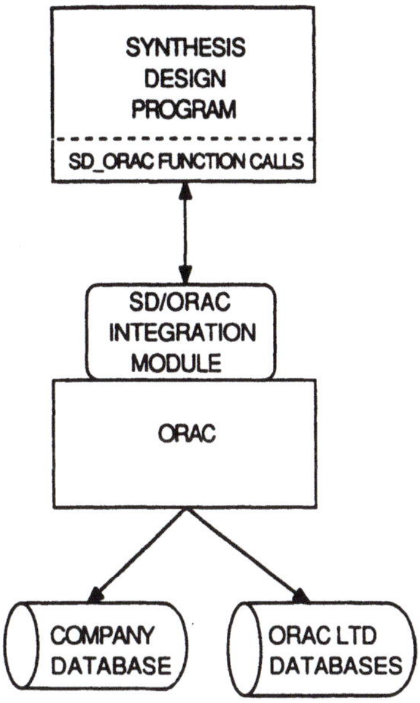

Figure 2. Use of 'toolkit' functions

4. Access the ORAC Query Menu to refine the results of a search.
5. Display answer sets using an ORAC Display Form.

When the ORAC button on the LHASA precursor display is selected by the user, the LHASA/ORAC Interface application is initiated (Figure 3). LHASA constructs an ORAC reaction search query[6] based on the transform details and the target molecule. The search query is transferred to ORAC *via* an SMD file. The query and search progress are presented to the user on the LHASA/ORAC Interface Display (Figure 4). The Interface also provides access to the ORAC Query Menu to allow the search results to be refined. The user can edit the query submitted by LHASA or perform additional searches, generate reports, etc. using the ORAC Query Menu (Figure 5) before returning to LHASA to continue processing.

The reaction query submitted to ORAC includes information relating to interfering functionality, participating groups and the structural change occurring in the transformation. The literature precedents located in the ORAC reaction databases are then ranked in order of similarity of the product structures to the target molecule, before being displayed to the user on an ORAC Display Form (Figure 6). This ranking is particularly useful when an ORAC search generates a large number of hits; the user is able to browse through the hits viewing first the precedents that are structurally similar to the target molecule being analysed by LHASA.

We have found that for most LHASA transforms, good literature precedents can be found by searching ORAC. However, occasionally no hits are found in ORAC for certain transformations that comprise a number of reaction steps. Unless the composite reactions have been explicitly registered in ORAC, the search will fail to locate such sequences. For example, a particular transform in LHASA consists of a Beckmann rearrangement of a ketone to an amide followed by hydrolysis. In

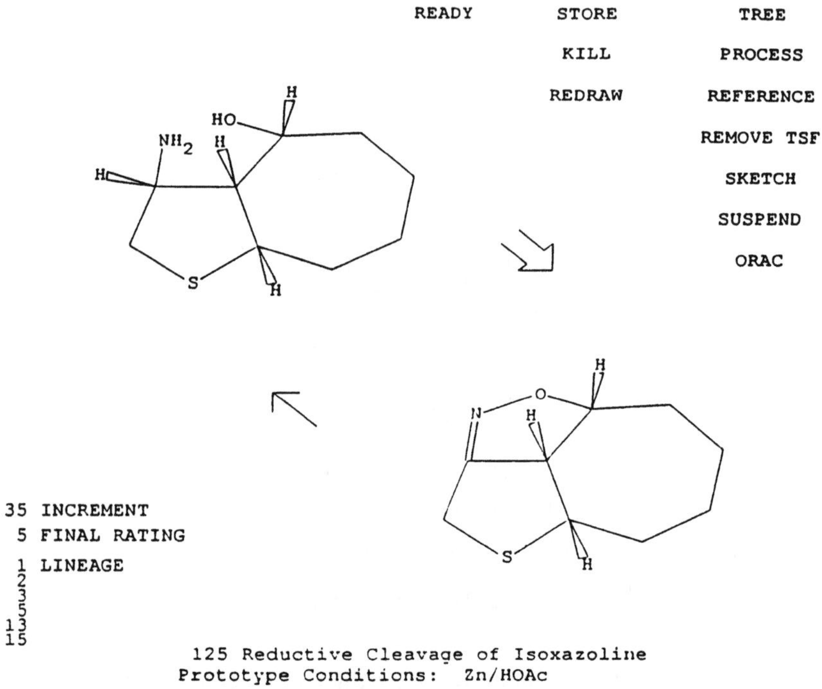

Figure 3. The LHASA precursor menu

ORAC, there are literature precedents for conversion of a ketone to an oxime, Beckmann rearrangement of oximes and for hydrolysis of amides, but no such composite reactions. This simple example highlights a general problem of searching reaction databases: the need to be able to find sequences of reactions that satisfy a user's query, when necessary.

MULTI-STEP REACTION SCHEMES

A multi-step reaction scheme consists of a number of single step reactions linked together. For example, the hypothetical reaction scheme shown in Figure 7 is made up of four separate reactions; the dashed arrows represent the links between each single-step reaction. A reaction scheme may contain linear, convergent and divergent paths.

An 'explicit' reaction scheme is one that has been reported in the literature. Explicit schemes may or may not be indexed in a special way in the reaction database system. Approximately 15% of the reactions in Theilheimer's *Synthetic Methods* are reported as reaction schemes. In the ORAC Theilheimer database, each individual reaction step and the overall transformation is stored as a separate entry. For example, a sequence A→B→C→D is indexed as the three reactions A→B, B→C and C→D, plus the overall transformation A→D. A drawback of this approach is that transformations such as B→D cannot be located in answering a user's query. In CASREACT[7], this problem has been overcome by indexing all possible transformations in a reported reaction scheme as individual reactions. The search program is then able to locate any composite reaction derivable from the scheme using its standard search mechanisms.

REACTION SEQUENCE SEARCHING

Reaction sequences are spanning trees or linear pathways linking two molecules in a reaction database. A sequence starts from a set of reactants and finishes with the generation of one or more products. In the reaction scheme shown in Figure 7, the reactions

$$\begin{aligned} A + B + C &\rightarrow D + E \\ F + D &\rightarrow G + H \end{aligned}$$

form a 2-step reaction sequence from A to G if at least one atom in A is mapped to an atom in G. 'Implicit' links occur between individual reaction or schemes stored in a reaction database when a product of one reaction is a reactant of another reaction, but they are not part of the same explicit scheme. An implicit link may be required to find a reaction sequence if an explicit sequence satisfying the query

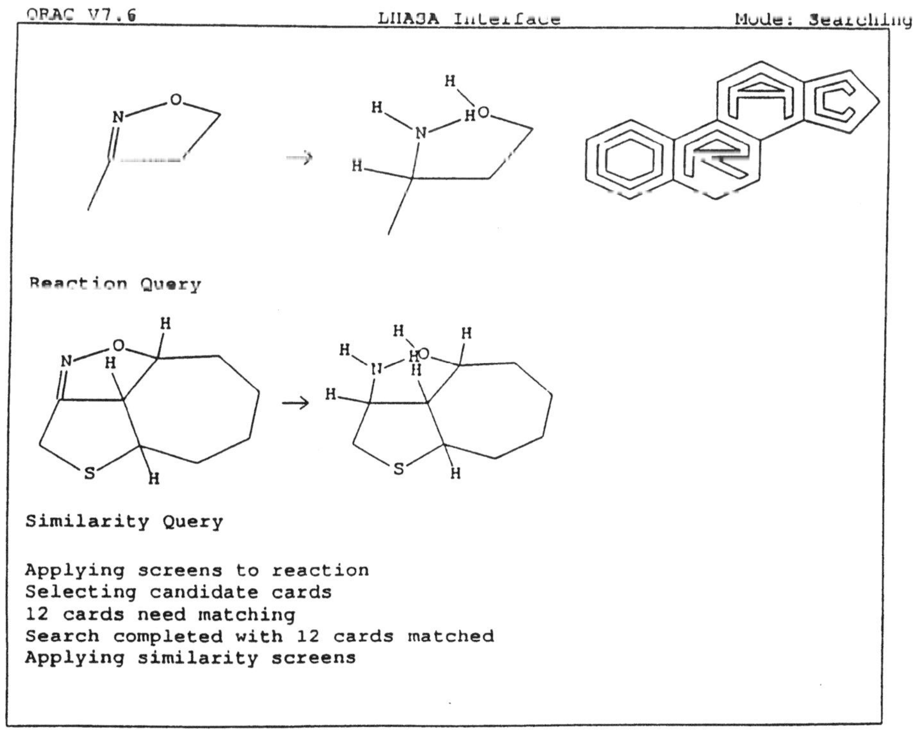

Figure 4. The LHASA/ORAC interface display

ORAC: V7.6 Page: Data Card Mode: Searching

Card:204 CONFALONE P N, LOLLAR E D and PIZZOLATO G
Box: CHC
Skip: 1 J. Am. Chem. Soc., 1978, 100, 6291.
No:1 of:12

Reductive opening of isoxazoles

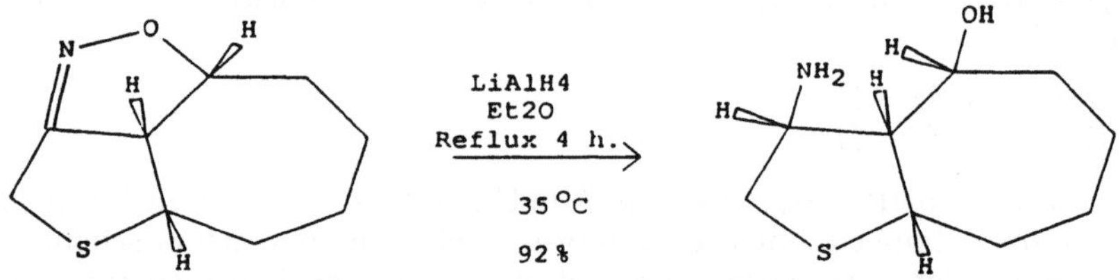

Lithium aluminium hydride not only cleaves the N-O bond in the tricyclic adduct but also reduces the imino functionality. Hydride delivery occurs from the less-hindered convex side of the cup-shaped structure and leads to the amino-alcohol.
The reaction is part of a synthesis of biotin.
Via Comp. Het. Chem., vol. 1, page 459.
Compiled by: CRAGG J E Experimental: Given

NEXT BACK SKIP KEEP DROP STOP KEYS PLOT SITE VIEW HELP

Figure 5. Display of literature precedents

does not exist. However, implicit links do not ensure synthetic relevance or yield of a reaction sequence involving them. This may mean that some hits found using implicit links between reactions may not prove to be useful synthetically.

An analysis of the ORAC CORE reaction database of over 50,000 reactions shows that approximately 15% of the reactions are linked to other reactions *via* implicit links. We estimate that there are about 25,000 implicit links between reactions in the ORAC CORE database. However, implicit links between reactions in different selective databases are much less common due to the variation in the type of chemistry and examples used. The distribution of links between reactions is very dependent on the reactant and product structures. Common reactants, such as benzaldehyde, may account for a relatively high percentage of implicit links in a reaction database.

A reaction sequence search system needs to be able to find sequences in reaction schemes and also sequences involving implicit links between individual reaction steps either in schemes or individually stored reaction entries. The user may wish to indicate when defining a search the maximum number of reaction steps in a sequence and whether implicit links can be used in the search. We are currently extending the Reaction Search facility in ORAC to provide such facilities. One major problem that must be overcome is the ability to transfer mapping information relating the reactant and product structures for each reaction in a

ORAC: V7.6 Page: Query Menu Mode: Searching

QUERY OPTIONS	SETS	HIT No.	KEEP No.	COMMANDS
AUTHORS	SET 1	4		SEARCH
JOURNAL NAME				DISPLAY
YEAR OF PUBLICATION				PLOT
ORAC CARD NUMBER				SAVE SET
EXACT COMPOUND				RESTORE SET
SIMILAR COMPOUND				DELETE SET
SUBSTRUCTURE				CLEAR SETS
REACTION				DEFINE QUERY
SIMILAR REACTION				RESTART QUERY
REACTION SITE				HELP
REACTION KEY				--->
REAGENT KEY				LHASA
SOLVENTS				LOGIC
YIELD				
TEMPERATURE				AND
SAVED SET				OR
SAVED QUERY				AND NOT

Figure 6. The ORAC query menu

sequence. This is required to ensure that the correct reaction change is performed by the overall sequence of reactions.

ANALOGOUS REACTION LINKS

In most reactions, only a part of each reactant molecule is changed. If the environment of the reacting centre is the same in the product of another reaction

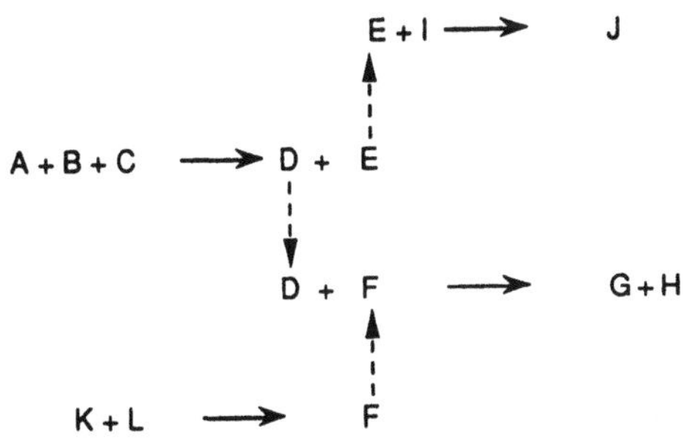

Figure 7. An example reaction scheme

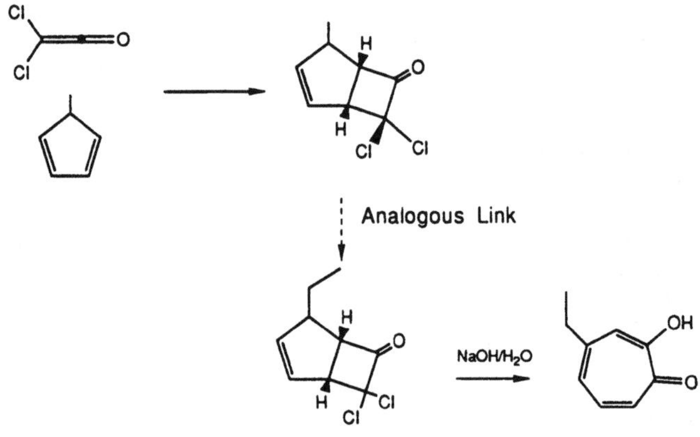

Figure 8. An analogous link

an analogous link can be made between the two reactions (Figure 8). Note that the two structures may be structurally very dissimilar when the whole molecule is considered; what is important is the reaction environment of the reaction being linked to. This type of analogous link is often useful to a chemist when designing a synthesis. However, account must be taken of interfering functionality and additional thought must be given to protection and deprotection steps. Identifying such analogous links could be done by reaction centre matching, i.e., the reacting part of one reaction is present as a substructure in a product of another reaction. Further constraints could be imposed, such as homology of substituents around the reacting centre.

Analogous links provide the reaction indexing analogue of transforms in a synthesis planning program such as LHASA. In a typical reaction database one would expect to find many more analogous links than exact structure links. However, evaluating the correspondence of reaction environments is a very difficult task. In LHASA, each transform defines the scope and limitations of a reaction and attempts to predict whether a transformation is plausible when applied to a particular target molecule. A similar approach would be to classify reactions in ORAC and analyse databases to identify products of reactions that can be indexed as members of the reaction classes[8] (Figure 9). This would provide the necessary information to identify analogous links when searching for sequences of reactions that satisfy a user's query.

CONCLUSION

The synergic relationship between synthesis planning and reaction database systems has been highlighted in this paper. Reaction database systems can be useful in providing literature precedents for transformations suggested by a synthesis planning or reaction prediction program. We are evaluating other opportunities to use ORAC in elaborating synthetic sequences. For example, ORAC could be useful for suggesting functional group protection and deprotection steps. Access to vendor information relating to reagents and starting materials

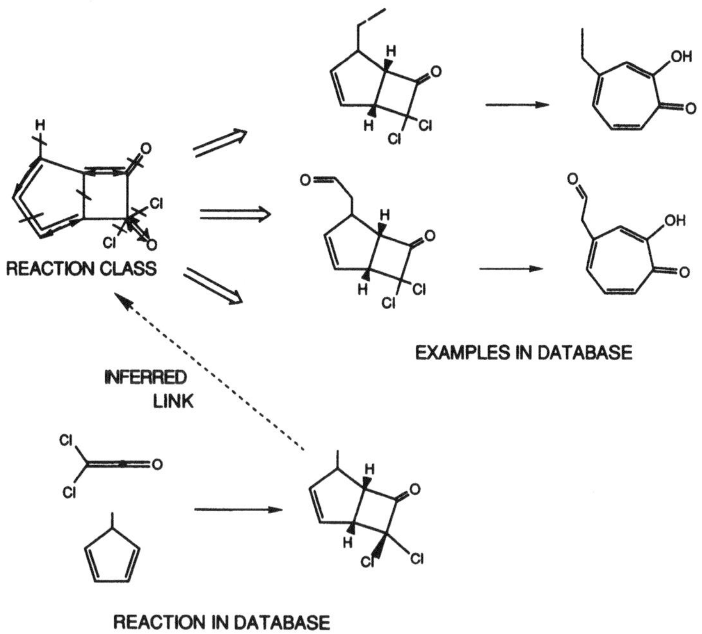

Figure 9. Use of classification of reactions in ORAC

could also be provided through a functional interface to a structure management system, such as OSAC.

Reaction database systems need to be able to store reaction schemes and search for reaction sequences involving both explicit schemes and implicit links between individual reactions in a database. A third class of links between reactions has been defined which involves identifying the presence of the reacting environment of one reaction in the product of another; synthesis design programs could prove useful in classifying and indexing such links.

REFERENCES

1. Zefirov, N.S.; Gordeeva, E.V. 'Computer-assisted Synthesis'. *Russ. Chem. Rev.* **1987,** *56,* 1002–1014 and references within.
2. Corey, E.J. 'Retrosynthetic Thinking – Essentials and Examples'. *Chem. Soc. Rev.,* **1988,** *17,* 111–133.
3. Johnson, A.P. 'Computer Aids to Synthesis Planning'. *Chem. Br.* **1985,** *21,* 59–69.
4. Sieber, W. 'Reaction Retrieval and Synthesis Planning'. In *Chemical Structures: The International Language of Chemistry*; Warr, W.A., Ed.; Springer-Verlag: Heidelberg, 1988; pp. 361–366.
5. Johnson, A.P.; Burt, K.; Cook, A.P.; Higgins, K.M.; Hopkinson, G.A.; Singh, G. 'Integration and Standards: Use of a Host Language Interface'. In *Chemical Structure Information Systems: Interfaces, Communications and Standards*; Warr, W.A., Ed.; ACS Symposium Series No. 400, American Chemical Society: Washington, 1989; pp. 50–58.
6. Hopkinson, G.A. 'Recent Developments in Reaction Searching'. In *Proceedings of the 13th International Online Meeting*; Learned Information: Oxford, 1989; pp. 139–145.

7. Blower Jr., P.E.; Chapman, S.W.; Dana, R.C.; Erisman, H.J.; Hartzler, D.E. 'Machine Generation of Multi-Step Reactions in a Document from Single Step Input Reactions'. In *Chemical Structures: The International Language of Chemistry*; Warr, W.A., Ed., Springer-Verlag: Heidelberg, 1988; pp. 399–407.
8. *Automatic Chemical Synthesis Planning*; Blurock, E.S.; Strelow, T.; Dechema Monograph, 1989; Vol. 116, 1989, pp. 531.

MULTISTEP REACTION SCHEMES IN THE REACTION ACCESS SYSTEM (REACCS)

Brad Christie and Tom Moock
Molecular Design Limited, 2132 Farallon Drive, San Leandro, California 94577, U.S.A.

ABSTRACT

Although in-house reaction database software packages with structural searching capabilities have been available for several years, they are primarily based on single-step reactions. Conventions have been adopted for many of these for the representation of multi-step reactions. However, at the time of this writing there exists no system which uses a general approach for searching multi-step reaction sequences, both within a single document, and for sequences which span multiple documents. A comparison of methods for how this can be achieved within the architecture of the Reaction Access System (REACCS) is described.

INTRODUCTION

Computerised files of organic reactions, and the software to access them, have been commercially available for nearly a decade. Both the software and the database have undergone constant development during that time: the programs available today are far more advanced than the earliest versions, and the creation of new and larger reaction databases continues at a rapid pace. In contrast, the process of laboratory organic synthesis is only somewhat more efficient that a decade ago. The pressure to reduce the costs of organic synthesis has fuelled an increasing interest in computer tools which can improve the productivity of industrial research laboratories.

The Reaction Access System (REACCS)[1,2,3] is a database program for storing and searching chemical reactions, usually relating to organic synthetic methodology. In REACCS' view, a reaction is composed of a collection of molecules, representing reactants, products and reagents. Many of REACCS' capabilities function consistently within the molecular and reaction domains: for example, REACCS supports exact match and substructure searches on both molecules (SSS) and reactions (RSS).

This hierarchical relationship between molecules and reactions was originally meant to be extended further, to include collections of reactions. These ordered collections are called *reaction schemes*, and are intended to represent transformations which cannot be executed in a single step. Since the overall transformation is often more important than any of the individual steps, it is important to treat the reactions as a group, rather than as disjointed single-step reactions.

Such multi-step reaction schemes are common in the literature of organic chemistry. The *Current Chemical Reactions* journal (CCR), published by the Institute for Scientific Information (ISI) is composed largely of extended schemes,

W. A. Warr (Ed.)
Chemical Structures 2

with relatively few single-step reactions. Organic chemists are accustomed to viewing synthetic chemistry this way.

Multi-step reactions as part of searchable databases can come in several forms. A reaction scheme recorded within one document (such as a total synthesis, or a multi-step procedure for performing a single transformation) is called an *explicit* scheme; this type is the most common, and perhaps the most important. A scheme consisting of reactions assembled from different documents is called an *implicit* scheme, and can be located by a computer program, rather than being a record of experimental fact. Implicit schemes can be used to discover new routes to synthetic targets, and can in principle extend the utility of databases of reactions.

We can further subdivide implicit scheme searches into two categories. If the *exact* product of one reaction is used as the reactant in another reaction, then the connection between the two reactions is by exact-structure, and is referred to here as a first-order implicit sequence. If the connection between the two reactions is by *inexact* structure match (such as two highly similar molecules), then we refer to it as a *second-order implicit sequence*. This latter type of search has much in common with synthesis planning programs; it has the most to offer, but poses the greatest technical challenge.

At present REACCS, as well as most other in-house reaction systems, treat reaction schemes by manually identifying and registering overall steps, after each of the component steps have been registered. This method has several disadvantages:

1. Manual creation and registration of overall steps is time consuming and may require expert chemical knowledge.
2. Usually only the broadest overall steps are registered; other summary steps are left out.
3. This method does not address the need to view all components of the transformation at one time, complete with intermediates and data relating to each step.

The first commercial computer system to store, search and retrieve reaction schemes was ChemBase (see Figure 1). ChemBase could represent and display linear schemes, but was limited in its ability to perform scheme-oriented structure searches (see below). The Chemical Abstracts Service (CAS) has announced the availability of their reaction file (CASREACT),[4,5] organised with reaction schemes in mind. Although at present the scheme searching capabilities available at the user interface are limited, CASREACT appears to represent schemes in a more rigorous manner internally, allowing further user interface refinements in the future.

This paper describes a feasibility and prototyping study of multi-step reaction sequence representation and searching, using REACCS as a starting point. The results of this study will show the benefits and costs of a few alternatives, and the utility of such a system in the hands of practising synthetic chemists. Detailed studies were carried out on probable architectures of systems which support Explicit- and First-Order Implicit schemes; options for the support of Second-order Implicit schemes are described in theoretical terms only.

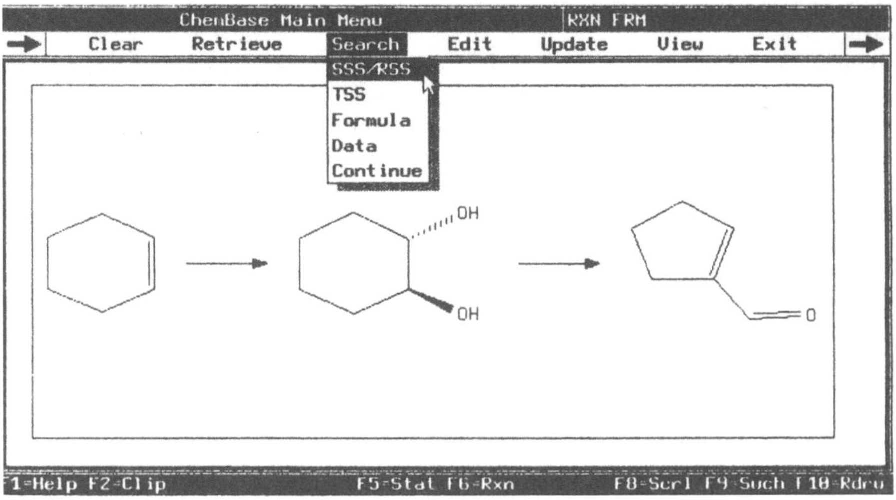

Figure 1. ChemBase can store and search short linear sequences in a database on MS-DOS personal computers

THE REACTION ACCESS SYSTEM (REACCS)

The Reaction Access System has been described in detail elsewhere; only a brief outline will be included here. Figure 2 illustrates the database structure used to represent reactions and data. REACCS maintains two structural registries in a single database: a molecule registry and a reaction registry. The molecule database contains structural and other data relating to single molecules, including connection tables, substructure keys (for substructural and similarity searches), and cross-references to reactions in which they take part. Likewise, the reaction database contains references to the molecules which participate, organised by their rôles as reactants (educts), products, solvents and catalysts (reagents).

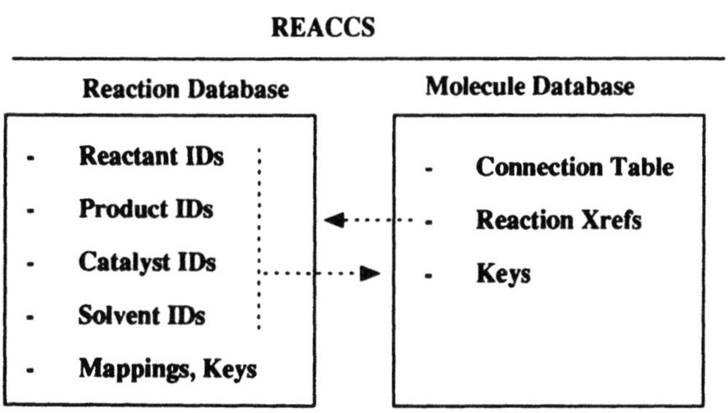

Figure 2. The structure of a REACCS reaction file. It consists of two linked databases: one for reactions, and one for molecules

Additional structural information includes atom-to-atom mappings and reaction-centre keys.

A molecule which is involved in several reactions is not represented redundantly; each reaction points to the same connection table in the molecule database. The storage requirements for a reaction are thus limited to *identifiers to the molecule database*; increasing the number of reactions, while holding the number of molecules constant, will therefore result in only a minor increase in the overall size of the database. This fact will become important when weighing alternative methods of representing and searching reaction schemes.

Although many of the features and considerations mentioned here will apply to many reaction indexing systems, any proposed solution must take into account the technology of the existing single-step program. A developer using a database system organised quite differently from REACCS might therefore come to different conclusions as to which approach is optimal.

EXPLICIT SCHEMES

Figure 3 describes several types of schemes important to a user of a reaction indexing program. The type of schemes most often illustrated in the literature are explicit schemes. Such a scheme is usually thought of as being stored *explicitly* into the database by its builder. Often the overall transformation is more important than its individual steps. Since the reactions are not assembled randomly, strategic knowledge is embodied in the reaction scheme: the author had some plan, or goal in mind, when the sequence of reactions was carried out. The number of answers resulting from a search over explicit schemes will always be smaller than if the reactions were assembled at random.

Reaction schemes in general can be thought of as graphs, where the nodes represent molecules, and the edges represent reactions. Schemes can therefore take on a variety of shapes, including linear, convergent, divergent, cyclic, and any combination thereof. A general representation of schemes must include all of these.

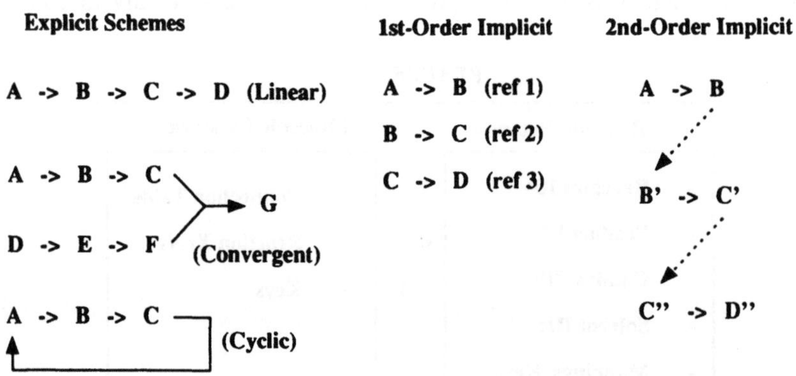

Figure 3. Illustration of difference types of reaction schemes. Explicit schemes are registered into the database; implicit schemes are found by computer programs

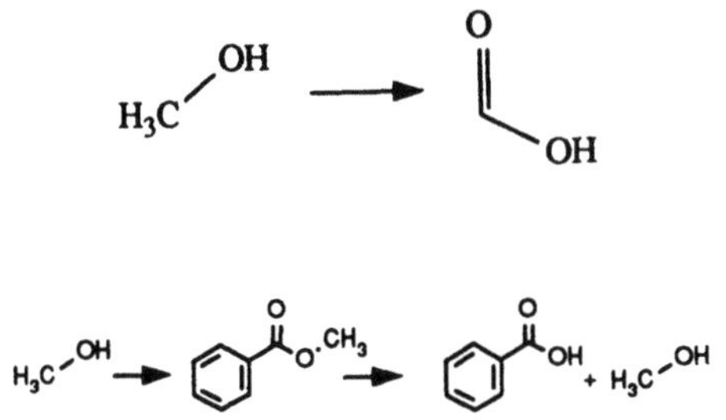

Query

Candidate Scheme (False Hit)

Figure 4. Atom-atom mappings in muti-step reaction queries. The establishment of the fate of the atoms in a query is necessary to avoid false hits like the reaction shown here

As illustrated in Figure 4, propagation of atom-atom mappings across multi-step reactions is important to eliminate false hits. It is especially easy for atoms to be lost or diverted in unimportant directions over the course of many reactions.

FIRST-ORDER IMPLICIT SCHEMES

If one removes the restriction that steps in a scheme be contained within a single document, a computer program can be used to find schemes which are contained *implicitly* in a database. If the product of a preceding step is used directly as a reactant in the following step, we refer to it as the first-order type. The reactions may never have been carried out as a sequence in the laboratory; nonetheless, these sequences are always viable, although not necessarily valuable or interesting.

Any program that can be used to find implicit schemes will also find explicit schemes. The added value of a search over implicit schemes will be based on the number of additional answers an implicit search provides, which in general is determined by the number of independent reactions sharing common molecules. The diversity of the chemistry, the 'unusualness' of the molecules, and the ratio of reactions to molecules in the database will all have an effect on the volume of results of implicit sequence searches.

SECOND-ORDER IMPLICIT SCHEMES

Often precedence for a proposed reaction scheme cannot be found using a single molecular system. Valuable answers to questions can be missed if part of a proposed scheme operates on a molecular system only slightly different than other parts; often the difference between the two molecular systems is inconsequential to the success of the proposed scheme. Schemes whose connections between reactions are based on non-identical structures can be referred to as Second-Order Implicit sequences. The connections between reaction participants can be based on a variety of relationships:

1. Common substructure. This can be the presence of one or more functional groups shared by the two molecules, or a common carbon skeleton or ring system.

2. Structural similarity[3,6]
3. Small chemical difference.
4. Generalised molecules (abstraction).
5. Generalised reactions (transforms).

Second-order implicit scheme searching can be quite valuable to the discovery process, but can also generate the greatest number of 'noise schemes'. In its purest form such a searching program becomes very similar to an unconstrained synthesis planning program. Users of retrosynthesis programs, such as CASP, LHASA or SECS, will recognise this as an important problem; directing and reducing the output is critical to their use in practical settings.

CURRENT METHODS OF REPRESENTING SCHEMES

At the time of writing only ChemBase and CASREACT store explicit schemes in databases as a coherent group of reactions. Nonetheless, all reaction indexing programs, including those which represent only single-step reactions, must use scheme-oriented literature as the source of their reaction data. The most common approach has been to describe manually and register summary reactions in addition to each single step. Although some data must be stored redundantly, it has proven itself a viable method in database building in the past several years, especially when the database covers literature containing few or short schemes, such as *Theilheimer* and the *Journal of Synthetic Methods*. In many respects, users of such systems (as opposed to database builders) may find this method perfectly adequate for explicit scheme searching.

ChemBase includes a relatively efficient and easy-to-use interface for creating short, linear schemes containing several intermediates. ChemBase does not allow the user to assign atom-atom mappings, and its earlier releases did not allow the assignment of reaction centres. Reaction searches in ChemBase can also retrieve false hits where the direction of the arrow is reversed, relative to the search query, since its internal data structure does not differentiate the sequential rôles of intermediates.

The developers of CASREACT[4,5] have published methods of assigning and searching both reaction centres and explicit reaction schemes. At the time of writing its user interface does not allow the assignment of atom mappings or reaction centres. CASREACT also can retrieve false hits where the direction of the arrow is reversed, similar to the limitation in ChemBase. Improvements in its user interface may resolve some of these limitations.

OPTIONS FOR REPRESENTING EXPLICIT SCHEMES

Representing explicit schemes in REACCS will necessarily require changes to its hierarchical organisation. Figure 5 describes a new database architecture for storing explicit schemes. In this new organisation, the REACTION hierarchy has been sublimated under a new SCHEME hierarchy. A single-step reaction thus becomes a single-step scheme. Data is divided between scheme-oriented data, such as literature references, and step-oriented data, such as reaction conditions.

We examined two methodologies for searching explicit sequences: prior

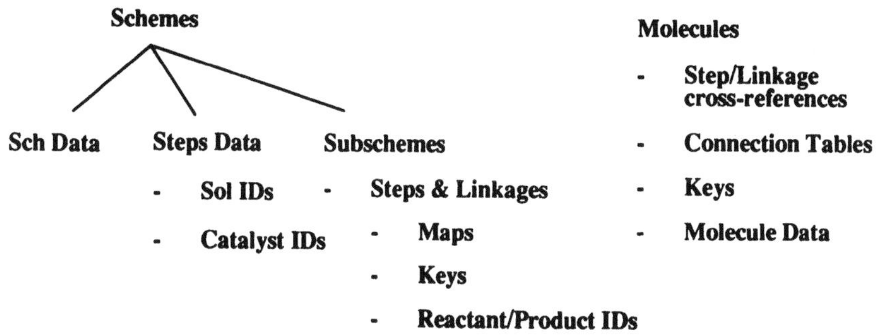

Figure 5. Proposed new database structure for the representation of explicit reaction sequences. Much of the definition of reactions is now under the general heading of reaction schemes. The molecule database remains separate

registration, and search-time discovery. The relative merits of these two methods was discussed in theoretical terms by Fugmann[7] *et al.*

The developers of CASREACT[5] chose the former method. For a given sequence of reactions, the challenge is to register each single step, then enumerate and register all possible transformations between each pair of components. Figure 6 illustrates this process and how it was accomplished in this study.

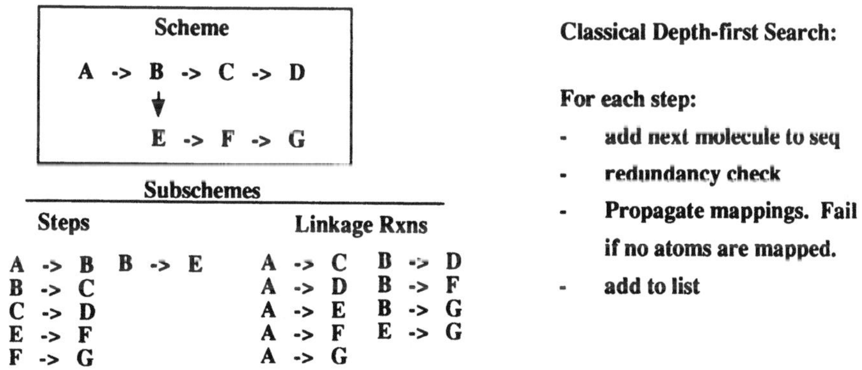

Figure 6. Multi-step reaction schemes are broken down into their component subschemes: single-step and linkage reactions. A depth-first search through the component molecules is used to traverse all pathways and check for loops and redundancies

As mentioned previously, propagation of atom-atom mappings across multiple steps is important in eliminating irrelevant schemes. Under normal circumstances, this is straightforward: given a mapping of atoms from A to B and from B to C, one can establish a mapping of all atoms surviving from A to C. However, if an intermediate contains symmetry properties not shared in the reactants and final products, ambiguities in mappings may arise which can cause the program to fail to hit valid schemes in the database. Figure 7 illustrates this problem. The mapping routines must be able to resolve these conflicts.

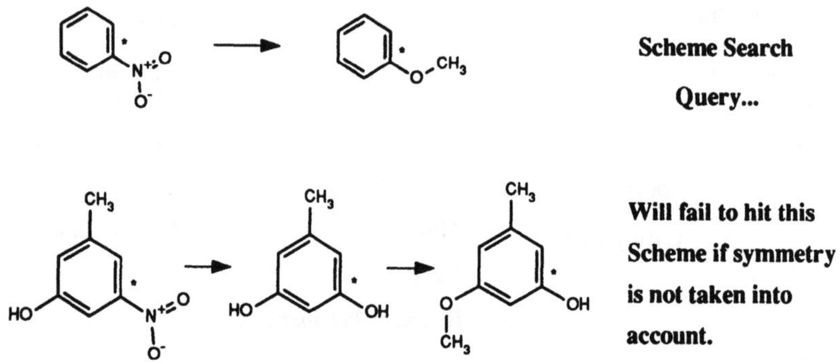

Figure 7. Symmetry in propagation of mappings. If an intermediate contains symmetry properties not contained in the reactants and products, valid hits can be lost if not handled carefully

For a linear scheme of n steps, one can determine the total number of subschemes (both single-steps and linkage reactions) using the formula

$$\text{Subschemes} = \frac{n(n-1)}{2}$$

For an existing database containing few and/or short schemes, such as *Theilheimer* or *The Journal of Synthetic Methods* (JSM), this would represent an increase in the total number of reactions of about 20% over that for single-step reactions alone, and a negligible increase over manual registration of summary reactions (the increase is due to the inclusion of some summary reactions that were not registered manually). The time required to execute a search will increase at the same rate, since REACCS' RSS search time is generally linear with the number of reactions.

The increase in the size of the database is also negligible, since the added reactions do not require redundant registration of their component molecules; for each linkage reaction, only molecule pointers, maps and keys need be registered.

The other most likely method, run-time discovery of subschemes during the search, will be described in greater detail in the following section on implicit schemes. The most important trade-offs of the two methods are increased database size (for prior registration of subschemes) and searching time (for search-time discovery). For most ordinary databases the increase in database size using prior registration is small, but could become large for databases consisting of total syntheses or other long sequences. Prior registration also has the advantage of being able to assign reaction keys to both the query and the subschemes. As we will describe later, this is impossible using the search-time discovery method.

Surprisingly, Fugmann expected the prior-registration approach to be infeasible, in spite of announcements and demonstrations by Chemical Abstracts of a system using that method. A likely explanation is that Fugmann assumed redundant storage of connection tables, not necessary in either REACCS or CASREACT.

SEARCHING OF FIRST-ORDER IMPLICIT SCHEMES

As described in the preceding section on explicit schemes, we considered two approaches to searching for implicit schemes: prior registration of all possible subschemes, and search-time discovery. It was considered desirable by the authors to create a single algorithm to handle both cases, so we first examined prior registration as a method of searching implicit schemes. A good way to view the problem is to treat an entire database as a single, highly complex scheme, and enumerate all subschemes using the method described in the previous section. The feasibility of this approach would be determined by the number of subschemes generated; if it were too large, the database would grow to unmanageable size.

In a preliminary study, we found that some molecules in all synthetic methodology databases were rather popular as both starting materials and products. Figure 8 lists some of the more popular compounds used in the Theilheimer database, including the number of reactions in which they participate. Benzaldehyde might be considered the holy grail of organic chemistry, participating in 359 reactions as a reactant, and 62 reactions as a product. The product of those two numbers (not correcting for redundant or cyclic schemes) represents 22,258 two-step schemes for benzaldehyde alone; there were 20 molecules in Theilheimer that would generate over a thousand two-steps schemes. When one also considers the number of longer sequences through these molecules, it was apparent to the authors that registration of all possible subschemes would create a database of unmanageable size. Unless a chemically reasonable mechanism is used to prune the generation of subschemes, prior registration is not a feasible method of searching implicit schemes.

	Number of Reactions		
Molecule	**As Reactant**	**As Product**	**Total 2-step Schemes**
Benzaldehyde	**359**	**62**	**22258**
Aniline	**361**	**26**	**9386**
Phenol	**110**	**37**	**4070**
Acetophenone	**105**	**34**	**3570**
Benzoic Acid	**52**	**55**	**2860**
Benzonitrile	**66**	**35**	**2310**
Benzene	**179**	**12**	**2148**
Benzophenone	**88**	**21**	**1848**
Toluene	**52**	**23**	**1196**

Figure 8. Table of the most popular molecules in *Theilheimer*. The large number of reaction schemes through these molecules would create a very large database if all were registered

SEARCH-TIME DISCOVERY OF IMPLICIT SCHEMES

The other method we investigated was to discover the linkages between reactions during the search itself, rather than at registration time. The approach can be viewed as a graph exploration problem, as illustrated in Figure 9. The first step in the process is to perform two substructure searches over the molecule database, to

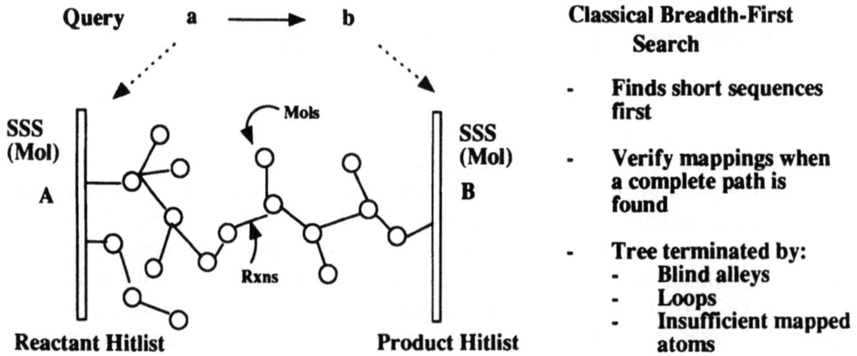

Figure 9. Illustration of search-time discovery of reaction sequences. A breadth-first search is used to find a pathway of molecules linking the reactant hit list to the product hit list

create hit lists for both the reactant and product components of the query. The program then attempts to find linear sequences of molecules which can connect the entries in the reactant hit list with those in the product hit list. The molecule network is searched using a traditional breadth-first search approach, so that shorter schemes are found first. A pathway is terminated if there are no more connections (blind alley), if a loop to a previously explored molecule is found, or if all atoms mapped onto a reactant from the query are lost through propagation. When a complete pathway is found, a summary reaction, complete with propagated mappings and reaction centres, is created and matched against the query.

This method of search-time discovery can be used as well for the searching of explicit schemes. In that case one could consider the molecule network as containing coloured edges, with each colour representing a single literature document. A successful pathway must consist entirely of edges of the same colour. One should expect far fewer hits from a search restricted in this manner, and a shorter search time as well.

COMPARISON OF SEARCHING METHODS

It was our desire that a single method be used for searching both explicit and implicit schemes, but only if the two methods had roughly equivalent search times. We were also interested in the quantity and quality of search results obtained from searches of implicit schemes, since at the time there did not exist such a system published in the literature. We expected that such searches would take longer than reaction substructure searches over single-step reactions, and that this would be acceptable if a concomitant increase in valuable results were obtained as well. Rather than simply proving a methodology, we had the increased burden of showing that the system would represent sufficient added value for industrial chemists to pay a licence fee.

Our goals in this comparison were:

1. Determine the most practical method of searching explicit schemes, either prior registration or search-time discovery;

2. Assess the viability of implicit scheme searching, considering the increased number of answers and the added time required.

Figure 10 shows the results of searches with several reaction substructure queries applied to the Theilheimer database, on a DEC Microvax 2000. For each query, three types of searches were performed:

1. A standard RSS search over the currently released database, consisting of single-step reactions and manually-assigned summary reactions. The performance data of this test is close to that obtained by the method of prior registration of subschemes, since the same searching program is used over a similar number of reactions.
2. A Scheme-RSS search over explicit schemes only, using the method of search-time discovery. The search was carried out for one-, two and three-step schemes.

RSS/Scheme Query	RSS	Search-time disc, Explicit	First-order Implicit
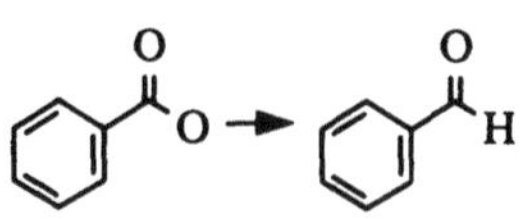	Hits 1-Step **4** Sum. **6** Time: **215**	Hits/Step: **4-2-1** Time: **259** Time/RSS: **1.2**	Hits/Step: **4-14-11** Time: **635** Time/RSS: **2.9**
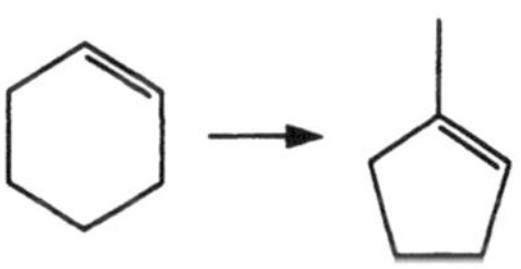	Hits 1-Step **2** Sum. **2** Time: **34**	Hits/Step: **2-2** Time: **414** Time/RSS: **12.2**	Hits/Step: **2-4-4** Time: **1822** Time/RSS: **53.5**
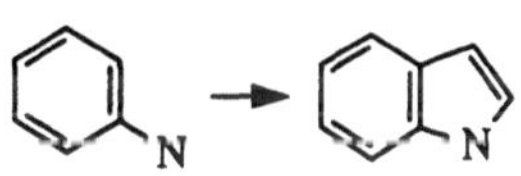	Hits 1-Step **44** Sum. **10** Time: **50**	Hits/Step: **45-3** Time: **181** Time/RSS: **3.6**	Hits/Step: **45-4-3** Time: **569** Time/RSS: **11.3**
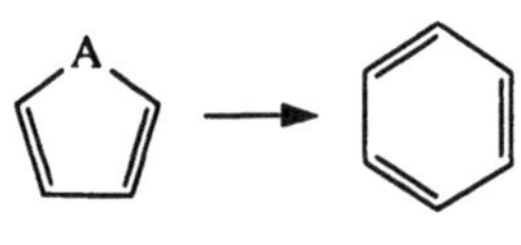	Hits 1-Step **11** Sum. **3** Time: **37**	Hits/Step: **9-4** Time: **143** Time/RSS: **3.8**	Hits/Step: **9-7-1** Time: **1738** Time/RSS: **46.9**
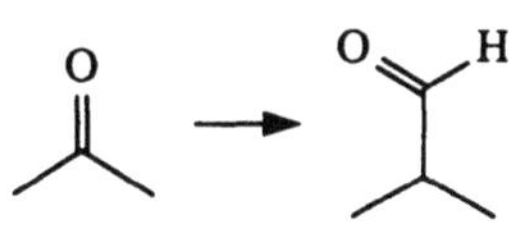	Hits 1-Step **0** Sum. **6** Time: **57**	Hits/Step: **0-8-1** Time: **302** Time/RSS: **5.2**	Hits/Step: **0-9-1** Time: **1429** Time/RSS: **25.0**

Figure 10. Comparison of search times and number of hits between standard REACCS RSS (equivalent to automatic prior registration), search-time discovery of explicit schemes, and first-order implicit schemes

3. A Scheme-RSS search over both explicit and first-order implicit schemes, using the method of search-time discovery. The search was carried out for one-, two-, and three-step schemes.

The first example, the conversion of an aromatic carboxylic acid to an aldehyde, is an excellent demonstration of the value of scheme searching. Four single-step reactions were found, as well as six additional manually-assigned summary reactions. The search time of 215 seconds would be representative for both our current database, and one built using prior registration. Searching for explicit schemes using search-time discovery was quite efficient, requiring only 20% more time than that required for prior registration, covering sequences up to 3 steps. The results for the implicit search (using search-time discovery) were quite impressive, finding 14 two-step schemes and 11 three-step schemes, most of which were not found by explicit scheme searches. The added performance cost, a factor of 2.9, was considered by the authors to be small in comparison to the added answers we obtained. Overall we considered the results of this experiment to be very encouraging, suggesting that search-time discovery was a viable alternative to prior registration.

Unfortunately, further study did not in general duplicate these results. The second example, a six-to-five ring contraction, returned two one-step hits and two summary reactions using non-scheme REACCS. Search-time discovery of explicit schemes returned the same results, but performance suffered by a factor of 12. Searching of implicit schemes returned a few more answers, but the cost in search time was a factor of over 53.

Another example, synthesis of indoles from anilines, fell somewhat in the middle. Search-time discovery of explicit schemes took 3.6 times longer than standard REACCS or prior registration. Searching for implicit schemes took over 11 times as long to complete, and returned only a few extra answers. The low number of additional implicit schemes could be explained by the high degree of historical interest in this reaction, and therefore the relatively large number of explicit schemes in the literature.

Overall, it can be seen that searches of explicit schemes by search-time discovery requires about three to 10 times as much computer resource as a comparable search using the method of prior-registration. Although a modest decrease in performance is acceptable in order to accommodate a single methodology for all types of scheme searches, we are not convinced that chemists using the system would accept the performance penalty.

Although Fugmann and others hinted that the searching of first-order implicit schemes would return a wealth of reaction information, we found the results in many cases to be rather sparse, especially with consideration of the two- to 50-fold increase in computer resources necessary to find them. On the other hand, the added time necessary to complete a search over implicit schemes could be well worth the investment if the number of answers could be increased. Methods for accomplishing this are described in the next section.

METHODS OF SEARCHING SECOND-ORDER IMPLICIT SCHEMES

The paucity of valuable results obtained from searches of implicit schemes can be partially explained by the characteristics of the data in the reaction database, rather than the searching method *per se.* As described earlier, the number of connections between reactions, and therefore the number of answers to a reaction scheme query, is related to the ratio of reactions to molecules in the database:

$$\text{Connections} \sim \frac{\text{Reactions}}{\text{Molecules}}$$

One can think of approaches to second-order implicit schemes in terms of how this ratio can be increased in a database. One approach is to reduce artificially the number of molecules in the database; another is to increase the number of reactions, while holding the number of molecules constant. There is an important difference between approaches to second-order implicit sequences, and the types of schemes described previously. Explicit and first-order implicit schemes consist entirely of *recorded* fact, and are guaranteed to work in the laboratory. In order to rebalance the equation to return more answers to scheme queries, some abstract or empirical chemical knowledge must be used the process; reliability is therefore not guaranteed, and the quality of results obtained will be based largely on the viability and generality of the rules by which the balance of the equation is modified.

Reducing the Number of Molecules.

In order to reduce the number of molecules without eliminating important chemistry, a computer program must be able to assign equivalence to groups of molecules; this group is then replaced by a single molecule, which can undergo all the reactions each member of the group participates in. This can be done by replacing molecules by *abstract structures*, where chemically unimportant parts are removed. In this case the 'unimportant' parts are those substructures which do not participate in or facilitate any chemical reactions found in the database, are chemically non-labile, and contain few functional groups, if any. The more abstract the molecules become, the fewer their number and therefore the higher the number of connections between reactions.

Another approach is to identify equivalent molecules which are quite similar, or are separated by a minimal chemical distance. In this way all members of a group of molecules are replaced by a single member.

An atom correspondence must be established between the representative member and all other members of the group, so that the reaction mappings can be assigned correctly. This is straightforward for the case of structural abstraction, but could be more difficult for molecular similarity if the similarity rules are fuzzy or ambiguous.

Increasing the Number of Reactions.

Since the number of experimental facts in the database must remain fixed, any new reactions to the database must be created artificially. The reliability of the search results obtained from this approach is therefore determined by the quality of the artificially-produced reactions. The reactions can be produced solely by the rules of graph theory, or they can be based on empirical data.

One method of creating new reactions is by the use of chemical transforms. These can be simple transforms, based solely on graph theory or general and well-understood chemistry. A more ambitious approach would be to use one of the available retrosynthesis programs mentioned previously; these systems have the advantage of far more sophisticated transform descriptions, in many cases representing many millions of dollars in development. In either case, the reactions are created by the following steps:

1. Select a molecule in the database.
2. Generate all precursors.
3. For each precursor.
4. If precursor exists in the database, and the new reaction is not already registered, add new reaction.

A more abstract method of creating new reactions is analogous to the similar-molecules method described above. If two molecules are very similar and/or have a low chemical distance, and the difference is of the correct type, assume that they are chemically interconvertible. In this case 'correct type' can mean that the difference is at or near a functional group.

CONCLUSIONS

The results of the study indicate that the best method of searching explicit schemes in REACCS is by exhaustive enumeration and prior registration of all possible subschemes, analogous to the method adopted by the developers of CASREACT. The same conclusion may not be made for all reaction indexing systems, however; both CASREACT and REACCS maintain a separate molecule registry, which limits the growth of the database when prior registration is used.

We found that the number of answers to reaction substructure queries over first-order implicit schemes was disappointing in general, compared to the amount of computer resources required to execute the search. Although many chemists will be willing to wait for those results, the authors are not conviced that a majority of users would consider the added results to be worth the decrease in performance.

It is the opinion of the authors that in order to make implicit schemes a more viable method of searching reaction data, it is necessary to take the next step to include searches of second-order implicit schemes. A program which can accomplish this next step will blur the distinction between current reaction indexing software and traditional synthesis planning systems. Such a program can in principle dramatically increase the utility of reaction database, and could change the way chemists use reaction data software.

REFERENCES

1. Wipke, W.T.; Dill, J.; Hounshell, D.; Moock, T.E.; Grier, D. 'Exploring Reactions With REACCS'. In *Modern Approaches to Chemical Reaction Searching*; Willett, P., Ed.; Gower: Aldershot, 1985; pp. 92–117.
2. Moock, T.E.; Nourse, J.G.; Grier, D.; Hounshell, W.D. 'The Implementation of Atom-Atom Mapping and Related Features in the Reaction Access System (REACCS)'. In *Chemical Structures: The International Language of Chemistry*; Warr, W. A., Ed.; Springer-Verlag: Heidelberg, 1988; pp. 303–314.

3. Moock, T.E.; Grier, D.L.; Hounshell, W.D.; Grethe, G.; Cronin, K.; Nourse, J.G.; Theodosiou, J. 'Similarity Searching in the Organic Reaction Domain'. *Tetrahedron Comput. Methodol.* **1988,** *1,* 117–128.
4. Blower, P.E.; Dana, R.C. 'Creation of a Chemical Reaction Database from the Primary Literature'. In Modern Approaches to Chemical Reaction Searching; Willett, P., Ed.; Gower: Aldershot, 1985; pp. 146–164.
5. Blower, P.E.; Chapman, S.W.; Dana, R.C.; Erisman, H.J.; Hartzler, D.E. 'Machine Generation of Multi-step Reactions in a Document from Single-step Input Reactions'. In *Chemical Structures: The International Language of Chemistry*; Warr, W.A., Ed.; Springer-Verlag: Heidelberg, 1988; pp. 399–408.
6. *Similarity and Clustering in Chemical Information Systems*; Willett, P.; Research Studies Press: Letchworth, 1987.
7. Fugmann, R.; Ploss, G.; Winter, J.H. 'Supply of Information on Chemical Reactions. An Advanced, Topology-Based Method'. *J. Chem. Inf. Comput. Sci.,* **1988,** *28,* 47–53.

INDEX

Springer-Verlag and the Environment

We at Springer-Verlag firmly believe that an international science publisher has a special obligation to the environment, and our corporate policies consistently reflect this conviction.

We also expect our business partners – paper mills, printers, packaging manufacturers, etc. – to commit themselves to using environmentally friendly materials and production processes.

The paper in this book is made from low- or no-chlorine pulp and is acid free, in conformance with international standards for paper permanency.

MIX
Papier aus verantwortungsvollen Quellen
Paper from responsible sources
FSC® C105338

If you have any concerns about our products,
you can contact us on
ProductSafety@springernature.com

In case Publisher is established outside the EU,
the EU authorized representative is:
Springer Nature Customer Service Center GmbH
Europaplatz 3, 69115 Heidelberg, Germany

Printed by Libri Plureos GmbH
in Hamburg, Germany